3D Bioprinting from Lab to Industry

3D Bioprinting from Lab to Industry

Edited by

Prosenjit Saha
JIS University
Kolkata, India

Sabu Thomas
Mahatma Gandhi University
Kottayam, India

Jinku Kim
Hongik University
Sejong, Republic of Korea

Manojit Ghosh
Indian Institute of Engineering Science and Technology (IIEST)
Howrah, India

WILEY

Published by John Wiley & Sons, Inc., Hoboken, New Jersey.
Published simultaneously in Canada.

Library of Congress Cataloging-in-Publication Data applied for:

Hardback ISBN: 9781119894377

Cover Design: Wiley
Cover Image: Courtesy of Dr. Prosenjit Saha

Set in 9.5/12.5pt STIXTwoText by Straive, Pondicherry, India

Contents

List of Contributors

Jaideep Adhikari
School of Advanced Materials
Green Energy and Sensor Systems
Indian Institute of Engineering
Science and Technology
Howrah, West Bengal
India

Noura Al Hashimi
The Vijay Lab, Division of Engineering
New York University Abu Dhabi
Abu Dhabi
UAE

Ananya Barui
Centre for Healthcare Science and
Technology
Indian Institute of Engineering
Science and Technology
Shibpur
India

Hanan H. Beherei
Refractories, Ceramics and Building
Materials Department, Advanced
Materials
Technology and Mineral Resources
Research Institute, National
Research Centre
Cairo
Egypt

Soumyadeep Bera
Centre for Interdisciplinary Sciences
JIS Institute of Advance Studies and
Research
Kolkata, West Bengal
India

Asmita Biswas
School of Medical Science and
Technology
Indian Institute of Technology
Kharagpur
Kharagpur, West Bengal
India

Hema Bora
School of Medical Science and
Technology
Indian Institute of Technology
Kharagpur
Kharagpur, West Bengal
India

Shalini Dasgupta
Centre for Healthcare Science and
Technology
Indian Institute of Engineering
Science and Technology
Shibpur
India

Anish Deb
Centre for Interdisciplinary Sciences
JIS Institute of Advance Studies and Research
Kolkata, West Bengal
India

Ankita Deb
Centre for Interdisciplinary Sciences
JIS Institute of Advance Studies and Research
Kolkata, West Bengal
India

Poonam Debnath
Centre for Interdisciplinary Sciences
JIS Institute of Advance Studies and Research
Kolkata, West Bengal
India

Santanu Dhara
School of Medical Science and Technology
Indian Institute of Technology Kharagpur
Kharagpur, West Bengal
India

Krishna Dixit
School of Medical Science and Technology
Indian Institute of Technology Kharagpur
Kharagpur, West Bengal
India

Mohamed G. Farahat
Botany and Microbiology Department, Faculty of Science
Cairo University
Giza
Egypt

Manojit Ghosh
Department of Metallurgy and Materials Engineering
Indian Institute of Engineering Science and Technology
Shibpur, West Bengal
India

Laura K. Gorwill
Department of Biology
University of Toronto Mississauga
Mississauga, Ontario
Canada

D. A. Gouripriya
Centre for Interdisciplinary Sciences
JIS Institute of Advance Studies and Research
Kolkata, West Bengal
India

Yves Grohens
Laboratoire d'Ingénierie des Matériaux de Bretagne
Université Bretagne Sud
Lorient
France

Laila Hussein
Refractories, Ceramics and Building Materials Department,
Advanced Materials, Technology and Mineral Resources Research Institute,
National Research Centre
Cairo
Egypt

Varnit Jain
Department of Metallurgy and Materials Engineering
Indian Institute of Engineering Science and Technology
Shibpur, West Bengal
India

Maxine Joly-Chevrier
Faculty of Medicine
University of Montreal
Montreal, Quebec
Canada

Josmin P. Jose
Mar Thoma College
Tiruvalla, Kerala
India

Nandakumar Kalarikkal
International and Inter University Centre for Nanoscience and Nanotechnology
Mahatma Gandhi University
Kottayam, Kerala
India

School of Pure and Applied Physics
Mahatma Gandhi University
Kottayam, Kerala
India

Ananda Kalevar
Department of Surgery, Division of Ophthalmology
University of Sherbrooke
Sherbrooke, Quebec
Canada

Michèle Kanhonou
Léonard de Vinci Pôle Universitaire Research Center
Paris La Défense
France

Sofiane Khelladi
Arts et Métiers Institute of Technology, CNAM, LIFSE
HESAM University
Paris
France

Jinku Kim
Department of Biological and Chemical Engineering
Hongik University
Sejong
Republic of Korea

Vidyapati Kumar
Department of Mechanical Engineering
Indian Institute of Technology
Kharagpur West Bengal
India

Mostafa Mabrouk
Refractories, Ceramics and Building Materials Department
Advanced Materials, Technology and Mineral Resources Research Institute, National Research Centre
Cairo
Egypt

Michael Marchand
Department of Surgery, Division of Ophthalmology
University of Sherbrooke
Sherbrooke, Quebec
Canada

Mina Mina
Department of Mechanical and Manufacturing Engineering
University of Calgary
Calgary, Alberta
Canada

Ankita Mistri
Department of Mechanical Engineering
Indian Institute of Technology
Dhanbad, Jharkhand
India

Mona Moaness
Refractories, Ceramics and Building Materials Department
Advanced Materials, Technology and Mineral Resources Research Institute
National Research Centre
Cairo
Egypt

Prajisha Prabhakar
International and Inter University Centre for Nanoscience and Nanotechnology
Mahatma Gandhi University Kottayam
Kottayam, Kerala
India

Baisakhee Saha
School of Medical Science and Technology
Indian Institute of Technology Kharagpur
Kharagpur, West Bengal
India

Prosenjit Saha
Centre for Interdisciplinary Sciences
JIS Institute of Advance Studies and Research
Kolkata, West Bengal
India

Samanta Sam
School of Energy Materials
Mahatma Gandhi University
Kottayam, Kerala
India

Debashis Sarkar
ME Department
Asansol Engineering College
Asansol, West Bengal
India

Aiswarya Sathian
International and Inter University Centre for Nanoscience and Nanotechnology
Mahatma Gandhi University Kottayam
Kottayam, Kerala
India

College of Science and Engineering
James Cook University
Townsville, Queensland
Australia

Vriti Sharma
Centre for Healthcare Science and Technology
Indian Institute of Engineering Science and Technology
Shibpur
India

Mridula Sreedharan
International and Inter University Centre for Nanoscience and Nanotechnology
Mahatma Gandhi University
Kottayam, Kerala
India

M.S. Sreekala
School of Chemical Sciences
Mahatma Gandhi University
Kottayam, Kerala
India

Abbas Tcharkhtchi
Arts et Métiers Institute of Technology
CNRS, CNAM, PIMM
HESAM University
Paris
France

Sabu Thomas
School of Energy Materials
School of Nanoscience and Nanotechnology
School of Polymer Science and Technology
School of Chemical Science and International and Inter University Centre for Nanoscience and Technology (IIUCNN)
Mahatma Gandhi University
Kottayam, Kerala
India

Department of Chemical Sciences
University of Johannesburg
Doornfontein, Johannesburg
South Africa

TrEST Research Park, Sreekariyam
Trivandrum, Kerala
India

Thara Tom
Mar Thoma College
Tiruvalla, Kerala
India

School of Chemical Sciences
Mahatma Gandhi University
Kottayam, Kerala
India

Simon D. Tran
Faculty of Dental Medicine and Oral Health Sciences
McGill University
Montreal, Quebec
Canada

Hamid Reza Vanaei
Léonard de Vinci Pôle Universitaire Research Center
Paris La Défense
France

Arts et Métiers Institute of Technology
CNAM, LIFSE
HESAM University
Paris
France

Saeedeh Vanaei
Department of Mechanical Industrial and Manufacturing Engineering
University of Toledo
Toledo, OH
USA

Shohreh Vanaei
Department of Bioengineering
Northeastern University
Boston, MA
USA

Pravin Vasudeo Vaidya
School of Medical Science and Technology
Indian Institute of Technology Kharagpur
Kharagpur, West Bengal
India

Advanced Technology Development Centre
Indian Institute of Technology Kharagpur
Kharagpur, West Bengal
India

Sanjairaj Vijayavenkataraman
The Vijay Lab, Division of Engineering
New York University Abu Dhabi
Abu Dhabi
UAE

Department of Mechanical and
Aerospace Engineering, Tandon
School of Engineering
New York University
Brooklyn, NY
USA

Kevin Y. Wu
Department of Surgery, Division of
Ophthalmology
University of Sherbrooke
Sherbrooke, Quebec
Canada

Foreword

Ajoy Kumar Ray

JIS Institute of Advanced Studies and Research (JISIASR) Kolkata, India
Department of Electronics and Electrical Communication Engineering, Indian Institute of Technology, Kharagpur, India

Tissue damage due to disease or injury has been a major concern for the entire global population, and its remediation poses serious challenges to the healthcare providers, biomedical researchers, and professionals across the globe. The concern becomes manifold since some organs and tissues have very limited self-renewal or regenerative capacity. Thus for restoring such injured or degraded tissues, we need to regenerate functional living tissue or even the whole organ artificially. This has resulted in an extremely important and emerging field of tissue engineering and regenerative medicine, which offers the potential to provide solutions for bio-fabrication of functional tissues.

Bioprinting, the process of creating complex, living tissues using 3D printing technology has received paramount importance in recent times. The interdisciplinary technology, lying at the intersection of engineering and biology, is the backbone of regenerative medicine and tissue engineering and points to the future of modern medicine.

Using this technology, it is important to precisely position multiple cell types layer by layer using computer-aided additive manufacturing techniques. Bioprinting is essentially a computer-aided transfer process for assembling biologically relevant materials including biomolecules, cells, tissues, and biodegradable biomaterials, resulting in the formation of an engineered bio-functional construct. Indeed it is a revolutionary concept in the regeneration or repair of damaged tissues by automating the layer-by-layer hierarchical fabrication of cell-laden structures both in vitro and in vivo.

Some of the early models of 3D bioprinting technology used computer-controlled ink-jet printer or graphics plotter and were used for precisely positioning the cells on a 2D substrate. This work provided the fundamental platform for

3D bioprinting. From those early efforts in mid-1980s, the interdisciplinary endeavor in 3D bioprinting and regenerative medicine technology has progressed substantially over the past three and a half decades, making personalized medicine a reality.

Since the original native tissues of a patient have their own distinct complex architecture as well as 3D organization and distribution of cells and extracellular matrix, it remains a significant challenge to precisely understand the complexity of native tissues from a structural and functional perspective. It holds endless promise for revolutionizing healthcare and various industries. From regenerative medicine to pharmaceutical testing and beyond, the possibilities offered by bioprinting are vast and continually expanding. However, the growth in this field also holds the numbers of existing challenges that need to be addressed by the scientific community.

In this context, the book *3D Bioprinting from Lab to Industry*, edited by Prosenjit Saha, Sabu Thomas, Jinku Kim, and Manojit Ghosh, has opened up the exciting world of bioprinting and its modern industrial applications.

The book, written by some of the esteemed scientists in this field, will embark on a voyage through the intricate landscapes of bioprinting. The contributors of each chapter have shared their experiences to present the fundamental challenges along with the solutions that lie in each area. From the first chapter, where the authors explain the fundamental principles of 3D bioprinting, the readers will enjoy a journey through the diverse facets of bioprinting – controlling this translational technology to the design, fabrication, and applications of biomaterials and bioinks at industrial scale, the book covers all the important aspects of 3D bioprinting.

The broad overlap of additive manufacturing with tissue engineering has transcended the boundaries of conventional medicine, presenting enormous potential for a future where organs can be printed on demand at laboratory, customized to individual needs with precision. It holds endless promise for revolutionizing healthcare for various industries. From regenerative medicine to pharmaceutical testing and beyond, the possibilities offered by bioprinting are vast and continually expanding. However, the growth in this field also holds the numbers of existing challenges that need to be addressed by the scientific community.

Because of its affordability, commercial viability, and capability to fabricate complex and hollow constructs, the extrusion-based bioprinting, one of the most common techniques in bioprinting field, has been employed to print living cells, tissue constructs, organ modules, and even organ on-a-chip devices. In this book, the extrusion-based 3D bioprinting has been covered in adequate details for the readers. Bioinks are formulation of materials suitable for processing by an automated biofabrication technology. With the advancement of technology – especially the introduction of extrusion-based printing techniques that use a small-diameter nozzle to deposit bioinks for the fabrication of complex constructs – individual cells and cell aggregates lack the ability to withstand the shear stress.

Starting with an introduction of the fundamental principles of bioink, the authors have presented here the translational technology for the design, fabrication, and applications of biomaterials and bioinks at industrial scale.

Furthermore, the book presents the industrial applications of bioprinting, showcasing how this technology can be used in personalized medicine, from planning to implementation, through case studies and laboratory-based research outcomes. The profound impact of AI and machine learning on the design of 3D bioprinting has also been included here. The processing parameters on the design of scaffolds, organoids, and cell/tissue models during printing process have also been discussed in detail by the authors.

With a lucid introduction to the realm of bio printing, followed by cellular requirements and preparation for bio printing, the authors have discussed various materials used for bioprinting. Bioprinting in regenerated organs, 4D bioprinted multiresponsive structure, the controlling factors in bioprinting, in situ bioprinting, machine learning, and in particular deep learning techniques in 3D bioprinting, nanomaterial and design of scaffolds, organoids, and cell/tissue models during printing process have been discussed in detail by the authors.

How to plan a bioprinting project and how it moves from laboratory to industry are two very important issues, which have been discussed here. The industrial applications of bioprinting, showcasing how this technology is utilized in personalized medicine from planning to implementation, have been discussed here through case studies and lab-based research outcomes. Finally, toxicity issues hold paramount importance in 3D bioprinting, and along with this comes the most pertinent ethical issues involved in this area. These have also been adequately discussed in this book.

For researchers, industry professionals, students, or anyone curious about the present and the future of 3D bioprinting and tissue engineering, this book will provide valuable insights into this rapidly evolving field. I hope that the book inspires and informs readers about the remarkable possibilities presented by 3D printing in shaping the future of modern medicine and healthcare industry.

Happy reading!

1

Introduction of 3D Printing and Different Bioprinting Methods

Asmita Biswas[1], Baisakhee Saha[1], Hema Bora[1], Pravin Vasudeo Vaidya[1,2], Krishna Dixit[1], and Santanu Dhara[1]

[1] *School of Medical Science and Technology, Indian Institute of Technology Kharagpur, Kharagpur, West Bengal, India*
[2] *Advanced Technology Development Centre, Indian Institute of Technology Kharagpur, Kharagpur, West Bengal, India*

1.1 Introduction of 3D Printing: Principles and Utility

3D printing (3DP), also known as additive manufacturing (AM), solid-freeform (SFF), and rapid prototyping (RP), is a fabrication technique using model data, where 3D structures are fabricated using controlled layer-by-layer deposition [1]. It was first described by Charles Hull in 1986, followed by production and commercialization by S. Scott Crump and his company Stratasys [2]. The basic subcategories of 3DP are stereolithography, fused deposition modeling, selective laser melting, electronic beam melting, and laminated object manufacturing [3]. 3DP involves scaffold construction by material addition, with high geometric precision reducing material waste. The primary procedure comprises data acquisition and synthesis of meshed 3D computer models in computer-aided design (CAD), followed by surface tessellation language (STL) file creation. This is followed by the slicing of mesh data into multiple 2D layer files and transferring them to a 3DP machine for fabrication [4]. Manufacture of complex designs, low cost, ease of access, and rapid and environment-friendly procedures are some of the advantages of 3DP in industrial, research, healthcare, and biomedical sectors.

3D Bioprinting from Lab to Industry, First Edition. Edited by Prosenjit Saha, Sabu Thomas, Jinku Kim, and Manojit Ghosh.

1.2 Ink Preparation and Printability

The choice of the base material as well as the recipe of its preparation to cater to the need for 3DP are of utmost importance. Bioink is the material used to produce either engineered or artificial living tissue using 3DP. It is the cell-trapping milieu composed of a multicomponent aqueous mixture that usually forms gels. This sol-gel transition of bioink is offered either by ionic bonds, covalent bonds, hydrogen bonds, or van der Waals interactions. Bioinks may be hydrogels, decellularized extracellular matrix, cell pellets, or tissue spheroids, of which hydrogels are the most common [5] due to their cell adhesion, growth, and proliferation capability since they absorb and retain large amounts of water. Ink for 3D bioprinting can be subdivided into two categories: cell-laden inks called bioink and cell-free inks called biomaterial ink. The bioinks usually consist of hydrogel precursors and are directly printed into Petri dishes filled with media and antibodies, whereas the biomaterial inks are usually utilized to print 3D scaffolds wherein the cells can be seeded on the scaffold under controlled conditions [6].

An ideal bioink should provide mechanical stability, stiffness, viscosity, surface tension, structural integrity, and biological ability – biocompatibility and biodegradability [7]. Many natural polymers like alginate, agarose, gelatin, chitosan, collagen, fibrin, and hyaluronic acid and synthetic polymers like polylactic acid (PLA), poly-D,L-Lactic acid (PDLLA), polylactic-co-glycolic acid (PLGA), polyvinyl alcohol (PVA), acrylonitrile butadiene styrene (ABS), polyethylene glycol (PEG), polyether ketone (PEEK), polycaprolactone (PCL), polybutylene terephthalate, and polyurethane (PU) [8] are used as bioinks for 3DP in the form of single or multicomponent.

Printability: The term "printability" is the ability of a bioink to form a 3D structure with accurate fidelity and integrity as per the design and the geometry. However, the terminology modulates itself according to the printing approach. For extrusion printing, the bioink must be able to form continuous filament; for the inkjet technique, it should form well-defined droplets while for laser printing, a prominent jet is required. The different printability indices [5] are the following:

1) *Extrudability*: The minimum extrusion pressure essential for printing at the desired flow rate.
2) *Strand printability*: Comparison of the diameter of printed strands with the CAD-generated parameters.
3) *Integrity factor*: Comparison of the thickness of printed scaffolds with designed geometries.
4) *Pore printability*: Comparison of the printed pores with designed internal geometry.
5) *Irregularity*: Comparison of outer geometry of scaffolds with designed parameters in *X*, *Y*, and *Z* directions.

Rheological properties and gelation kinetics determine the printability of a bioink, which is again dictated by the type of bioprinting. Low-viscosity bioinks are preferred in inkjet bioprinting; rapidly crosslinkable, shear-thinning bioinks are desirable for extrusion bioprinting and photo-crosslinkable bioinks are favorable for stereolithographic printing [9]. Rheological properties like viscosity, viscoelasticity, yield stress, shear thinning, elastic recovery, and viscoelastic shear moduli affect the printability of bioinks. The rheological properties of the crosslinked bioink must facilitate scaffold remodeling to mimic the ECM environment. This process provides physicochemical cues to the cells, promoting their spreading and proper distribution. For instance, substrates that mimic the mechanical properties and Young's modulus (~12 kPa) of native skeletal muscles offer better myogenic differentiation [10]. The key rheological parameters for a "good" bioink are described below:

Viscosity: Viscosity is the ratio of shear stress to shear rate and is governed by the molecular weight and concentration of the polymer. High-viscosity inks are preferable for high-fidelity printing but may limit cell growth within the substrate due to higher shear stress. This shear stress can be overcome by either using hydrogel inks having shear-thinning properties or using pre-gel solutions with lower viscosities. For e.g., alginate-based bioinks are directly extruded into calcium solution leading to ionic crosslinking. Due to higher surface tension, viscous bioinks prevent droplet formation without any merger of the columns with one another. Hence, crosslinking agents come into the picture with the caution of appropriate concentration so as to avoid phase separation and phase change [11]. Temperature-dependent hydrogen bonding or hydrophobic interactions may be exploited, as in case of gelatin, Pluronic, etc. Colloidal-like suspensions of densely packed microgels or jammed gels also prevent exposure of cells to high shear stress [12].

Viscoelasticity and yield stress: Viscoelasticity is the property of retaining elastic shape while allowing viscous flow. It is guided by three parameters – storage modulus (G′), viscous modulus (G″), and yield stress. Tan (δ), the ratio between G′ and G″, gives information about the rheological characteristics of the bioink. Yield stress is the stress limit beyond which deformation occurs. The parameters of G′ and yield stress are governed by the number of crosslinks within the bioink. These crosslinks offer resistance to shape change within the yield stress. Paradoxically, though yield stress of the bioink renders shape and stiffness to the substrate, it can also deter cell encapsulation and further growth. Hence, additives like carrageenan, gellan gum, and hyaluronan are added to the bioinks to improve yield stress [13, 14]. However, in stereolithography and light-assisted bioprinting, low-viscosity bioink is required for easy flow and for each layer to be crosslinked with each other.

Shear thinning: Shear thinning is the phenomenon where increase in shear rate results in the decrease in viscosity. Partially crosslinked hydrogels, colloidal suspensions, polymer melts, or polymer solutions above certain critical concentrations show shear-thinning properties with shape preservation. Shear thinning leads to decrease in viscosity in the extrusion phase, but rise in viscosity after extrusion results in shape preservation. For e.g., shape retention in printed calcium phosphate cement is due to high zero-shear viscosity [15]. PCL and PLA melts, used in polymer-based fused deposition modeling, possess intrinsic shear thinning properties due to shear-induced disentangling and alignment of long polymeric chains [16]. High resting viscoelasticity of pastes, solid suspensions, and colloidal dispersion bioinks arises due to the restoration of interaction between the suspended particles, which had been disrupted due to the shear-thinning process [17]. Hydrogels demonstrate non-Newtonian fluid behavior with shear-thinning features. So, the random polymer chains align themselves in one direction under shear force and become suitable for extrusion process.

Surface tension: Due to surface tension, there is an attraction between the liquid molecules, which ensures a contact angle between each printed strand. When the substrate has a higher surface energy than the surface tension of the bioink, the ink spreads; conversely, lower surface energy results in less spread [18]. For e.g., shape fidelity in printed constructs of ceramic slurries is reduced by both surface tension and gravity. It has been observed that a reduction in surface energy leads to droplet formation instead of a cuboidal structure [19].

Elastic recovery: This property explains how the bioink recovers its original solid-like property without any distortion after undergoing deformation or transition from liquid to solid state [20]. Due to this feature, multi-layered structures can be built up. Elastic recovery is the combination of both viscous flow and elastic recovery where the viscous modulus, G'', explains the fluid-like behavior of bioinks, and elastic modulus, G', defines the solid-like behavior of bioinks imparting elastic shape recovery. While the former allows mixing of cells and extrusion, the latter allows suspension of cells. Often, these moduli vary under different conditions of temperature, stress, and shear rate.

The recovery of solid-like behavior after extrusion through a needle must be fast to ensure good shape fidelity. The rheological evaluation of a bioink is done on the basis of the kinetics of yield stress and elastic recovery. The first step is to evaluate the effect of increasing shear stress and filament-forming capability, followed by the measurement of viscosity as a function of shear rate to evaluate the shear thinning property. Then, recovery tests are done to investigate whether the materials can restore their elastic properties on exposure to alternating low and high sheer stress [17].

1.3 Methods of Bioprinting in Fabrication and Tissue Engineering

The essential types of bioprinting are laser-assisted bioprinting, droplet printing, extrusion printing, and stereolithography. Laser-assisted bioprinting comprises laser-based ejection of material onto the substrate. Droplet printing consists of droplet ejection by different mechanisms leading to deposition on substrate and scaffold fabrication. Low-viscosity ink and cell-laden media-based scaffold fabrication are primary candidates for droplet printing based on enzymatic, chemical, or mechanical-based gelation.

Extrusion bioprinting consists of material extrusion and layer-by-layer deposition on substrate-synthesizing scaffold. By far, it is the most conventional production technique in bioprinting. Stereolithography involves UV or light-assisted photopolymerization-based ink gelation leading to scaffold fabrication.

1.3.1 Laser-Based Printing

Laser-based printing utilizes a digital printing mechanism where a pulse laser is used to create a driving force for depositing biomaterial on substrate. It consists of three parts, that is, laser, ribbon, and collector (Figure 1.1). Laser used for printing has the excitation near UV range of 300–400 nm, or at UV range of 193–248 nm (Table 1.1) [26, 27].

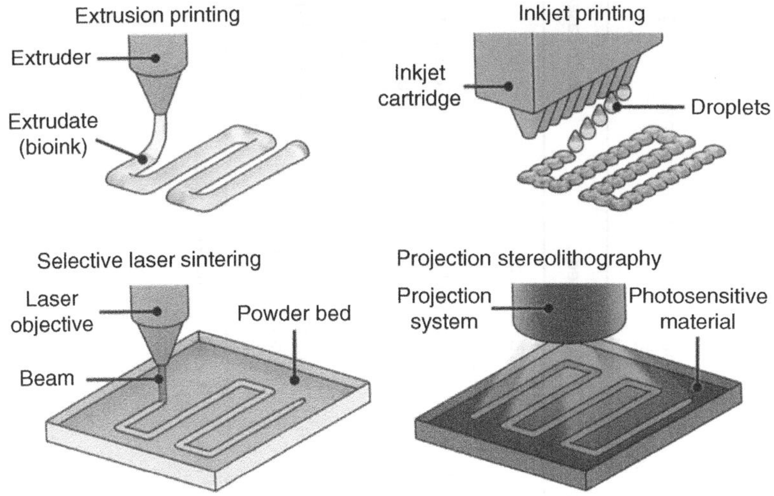

Figure 1.1 Schematic representation of types of 3D bioprinting [21] – extrusion printing, inkjet printing, selective laser printing, and stereolithography. *Source:* Adam et al. [21] / Springer Nature.

Table 1.1 Comparison between the different methodologies of 3DP – laser-based, inkjet-based, extrusion-based, and stereolithography [22, 23].

Parameters	Laser-based	Inkjet-based	Extrusion-based	Stereolithography
Resolution	~20–100 μm	~30 μm	~100 μm	~5–100 μm
Speed	Slow	Fast	Medium	Fast
Surface finish	Very good	Excellent	Fair	Good
Cell viability	—	Yes	Yes	Yes
Materials	Biomaterials	Hydrogels, live cell printing	Ceramics, polymers, metals, hydrogels	Photocurable polymers
Gelation methods	Ionic thermal photopolymerization enzymatic	Ionic pH mediated thermal enzymatic [24]	Photopolymerization, ionic pH mediated thermal enzymatic [25]	Photopolymerization

Source: 22,23 / Placone Jk et al. / John Wiley & Sons.

Ribbon comprises of transparent glass, gold, or titanium as an absorbing material, and bioink consists of cells, polymer mixture, and growth factors. Laser beam pulses at a specific time period cause vaporization of the metal over the bioink, further creating pressure to form a bubble-containing hydrogel, resulting in deposition on substrate. Laser-assisted printing offers resolution ranges from picometer to micrometer in size controlled by viscosity of bioink, thickness of bioink coated over the ribbon, collecting substrate wettability, and air gap between the ribbon and substrate [27].

1.3.1.1 Types of Laser Printing

Laser Direct Writing

Light direct writing (LDW) was reportedly first utilized to manufacture microlens arrays by Gale et al. in 1983. 2D and 3D scaffolds are fabricated by laser beam directed on substrate in a desired pattern. Direct writing prototyping is mediated by target translation, laser translation, or target rotation. Translation and rotation on *X*, *Y*, and *Z* axes result in almost six degrees of freedom [29]. The basic components of LDW consist of an optical system, a mechanical system, and a guidance system [30]. The different types of LDW include selective laser sintering/melting, micro laser sintering, and laser machining (Figure 1.2). Selective laser sintering utilizes continuous wave (Ar^{2+}, CO_2) and long pulse laser with 50 μm resolution using ceramics, polymers, and metals in tissue engineering and prosthetics applications [31]. LDW is utilized to print biological scaffolds and cell-laden ink in customizable 2D and 3D complex geometry with high spatial resolution. Also, it aids deposition on various substrates including hydrogels, microfluidic devices, nanofibers, and living tissue. LGDW is the first laser-based printing technique to print live cells. Interaction between light and cells occurs due to comparable refractive index of cells and media, higher the differences, stronger the interactions. Weak laser beam focused on media-containing cells generates a gradient force resulting in hauling of cells towards laser and deposition on substrate [32]. By this technique, limited cells are deposited on substrate (2.5 cells/min). Another limitation of LDW is that cells can be transferred for limited distance (300 μm) [33]. To overcome this situation, optical fiber is used, which will transmit light as well as transfer the cells to fiber core.

Matrix-assisted Pulsed Laser Evaporation Direct Write (MAPLE DW)

MAPLE was first developed by the US Naval Research Laboratory [34]. In MAPLE, biomaterials are blended with volatile solvents like alcohol and are frozen at −196 °C, followed by laser exposure [35]. Photon energy is converted to thermal energy and absorbed by the biomaterial present in volatile solvent. These vaporized molecules get desorbed over the desired pattern surface, whereas volatile components (possessing limited adhesion coefficient) evaporate during deposition. MAPLE DW can print a wide variety of biomaterials ranging from lower to higher

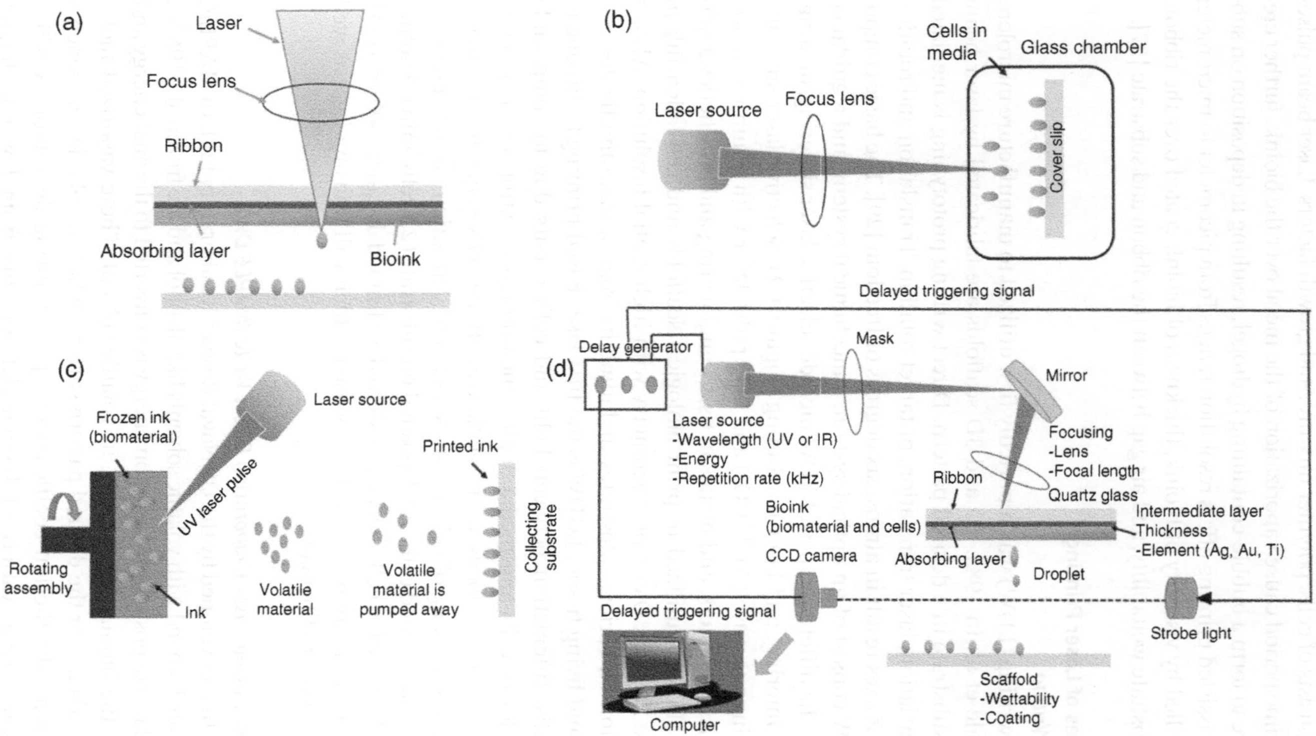

Figure 1.2 (a) Basic schematic of laser-assisted printing (b) laser-guided direct writing– laser force optically guided the cells and deposited over the substrate (c) matrix-assisted pulsed laser evaporation: – laser melted frozen ink converting photon energy to thermal energy depositing over the substrate, (d) laser-induced forward transfer (LIFT) on laser exposure converts photon energy to thermal energy creating a bubble and releases bioink to get deposited over the substrate. *Source:* [27]. Li, 2016; [28]. Ozbolat et al. 2017.

molecular weights with minimum damage from condensed to vaporized state [36]. MAPLE transfers particles without disrupting their chemical and biological functionality. However, MAPLE-based technique cannot be used for transferring live cells as they cannot be entrapped in the volatile solvent due to toxicity [37].

MAPLE DW is a modification of MAPLE in which a thin coating of hydrogel is created over the quartz crystal termed as printed ribbon, absorbs the majority of laser wavelength, thus reducing the chances of genetic alteration caused by UV radiation [38]. Cells encapsulated in ink and coated over the metal ribbon were printed by MAPLE DW. Laser exposure melts down the quartz crystal, thus creating a high vapor pressure and forming a plasma bubble. This bubble bursts and allows the coating material to get absorbed into the collecting substrate. Plasma bubble formation further can be tuned by the viscosity of coated material, thickness of the coating, and laser fluence.

Laser-Induced Forward Transfer (LIFT)

LIFT comprises CAD software, laser, cartridge, and collecting substrate. The cartridge consists of three layers – transparent quartz glass, a thin gold layer, and bottom of gold layer, which is coated with bioink. With the help of spin coating, uniform layer of bioink is created; thickness of bioink plays a great role in the formation of bubble. To maintain the consistency of the process, the rheological parameter should be controlled carefully [39]. Introducing absorbent material has successfully mitigated the detrimental effect on the live cells while printing. During the printing process, all the cells due to gravity, settle at the bottom of the coating; upon laser exposure, the quartz glass absorbs most of the radiation and then interacts with the intermediate layer here, due to which thermal energy vapor pressure is created, which forms the bubble. Then, vapor bubble accumulates energy and expands to form a jet and drive the bioink to collect substrate [40]. Bioink parameters and laser energy play a crucial role in the formation of proper jet [41].

Major limitation of laser-based printing is the cost of fabricating scaffolds. Additionally, the metal film used for adsorption layer can lead to contamination. Choi et al. described that multiple bioink printing is difficult in laser-based printing [42]. Printing consecutive material is time-consuming and resolution gets compromised. The material accumulates at the bottom due to gravity, leading to contamination [43].

1.3.2 Extrusion-Based Printing

In extrusion-based bioprinting, hydrogel, viscous solution, or cell spheroids are extruded from nozzle on a substrate, in a layer-by-layer technique, using a fluid dispensing system and an automated three-axis robotic system [44].

Basic components of extrusion printer consist of a reservoir containing material, printhead extruding material, and substrate where material is deposited. A pneumatic actuator or screw device feeds material through the cartridge into the nozzle for deposition on the substrate (Figure 1.3). Actuators regulate the positioning of nozzle on *X*, *Y*, and *Z* dimensions, controlling ink deposition. Scaffolds with complex geometries require sacrificial support for successful and accurate fabrication. Recent developments include co-axial printing for core shell scaffold fabrication and multimaterial extrusion bioprinting technology proficient in extruding multiple bioinks with concise and fast switching between different reservoirs to fabricate complex multiple-material

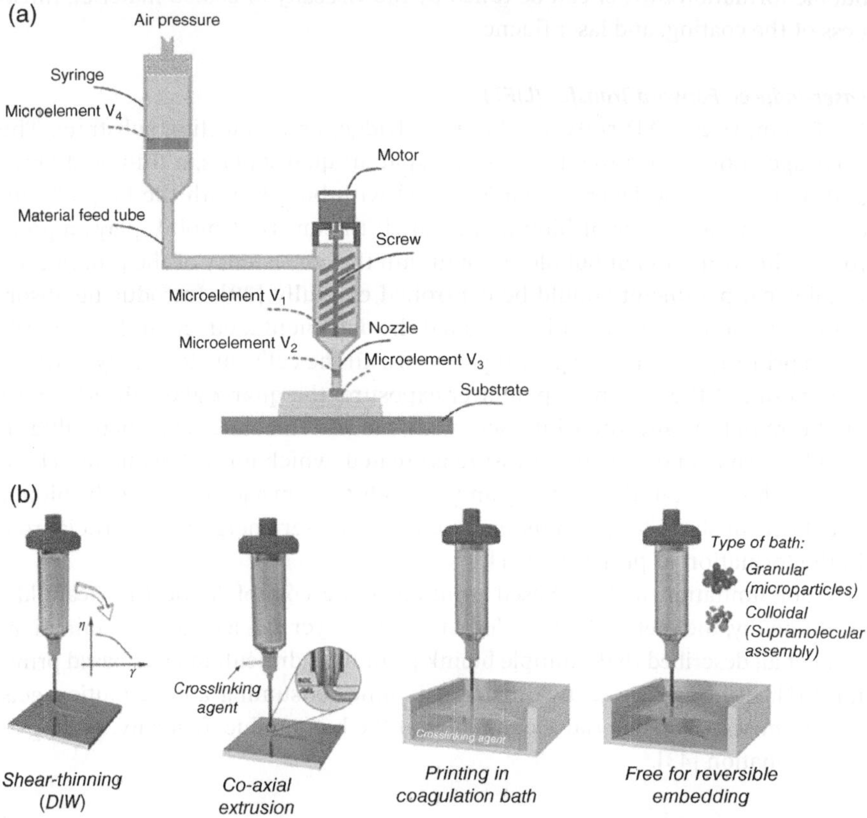

Figure 1.3 Extrusion printing principles. (a) Basic schematic of extrusion printing – pneumatic and screw based, (b) conventional types of extrusion bioprinting – shear thinning (DIW), co-axial extrusion, coagulation bath-based extrusion printing, and freeform reversible embedding. *Source:* [49]. Abdolmaleki H et al. 2021 / with permission of John Wiley & Sons.

scaffolds. The substrate requires high surface roughness and wettability whereas shear thinning, low surface tension, and shape-retentive bioink are essential [47]. Structural integrity of the constructs is relatively superior to extrusion-based bioprinting due to the continuous deposition of cylindrical filaments making it the most appropriate type of printing for complex 3D scaffold fabrication [48].

Extrusion printing is cost-effective and rapid fabrication technique and relatively straightforward mechanism compared to the other printing methods. In addition, geometrical and fabrication parameters can be altered effortlessly according to necessity depending on the required scaffold. However, extrusion printing has restricted resolution hence the minimum scaffold sizes fabricated are between 200 and 1000 μm. Also, live cell printing is relatively complicated using extrusion printing compared to inkjet printing due to extensive shear stress impacting cell viability.

1.3.3 Droplet Printing

Droplet-based printing presents ink deposition and direct patterning of bioink aided by computer-based non-contact pattern reproduction techniques. The droplet-based printing technology has the following subtypes:

1) Aerosol jet
2) Inkjet
3) Electrohydrodynamic jet

Depending on ink viscosity, aerosol jet printing involves aerosol generation in the nebulization step by pneumatic or ultrasonic means (Figure 1.4). The droplets are transported through a mist tube to the print head by inert gas and are

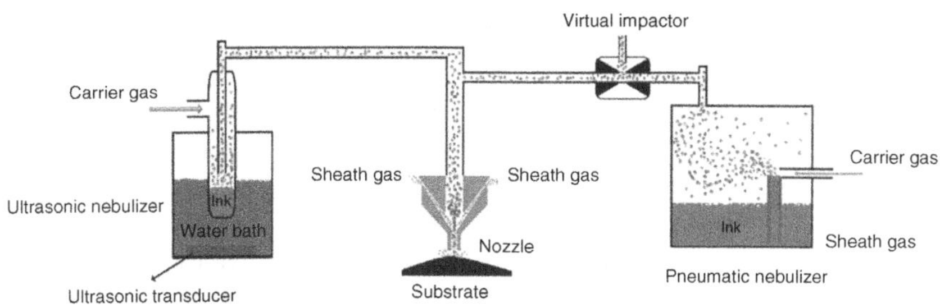

Figure 1.4 Schematic representation of aerosol jet printer with pneumatic and ultrasonic nebulizer. *Source:* Abdolmaleki et al. [49] / with permission of John Wiley & Sons.

encased with a sheath gas flow at the bottom of the nozzle. Aerosols are deposited on moving substrates due to the aerodynamic interaction between carrier gas and sheath gas [49].

Electrohydrodynamic jet printing utilizes an electric field to generate fluid flow from nozzle to the substrate. Ink is forced through the nozzle from ink chamber toward the tip, utilizing air backpressure to initiate printing. The voltage applied between the substrate and the nozzle causes charge accumulation due to electric field, resulting in the pendant droplet coming out of the nozzle tip. A Taylor cone forms due to meniscus deformation as a consequence of shear stress from charge accumulation (Figure 1.5). The surface tension of the ink is overcome by electrostatic forces by exceeding the electric potential of the meniscus above optimum value, leading to discharge of jet from Taylor cone's surface to the substrate [52].

Constraints of droplet printing involve ink flow when printed on a nonplanar surface, hence both inkjet and electrohydrodynamic printing are possible only on planar surfaces. Although aerosol jet printing resolves the issue by aerosol ejection instead of droplet ejection conformal printing remains an unresolved matter in aerosol jet technique on polyhedron geometry substrate. Aerosol and inkjet printing reportedly exhibit low resolution, limiting their applications. However, the E-jet printing technique provides a resolution of around 100 nm, providing rapid and flexible fabrication (Table 1.2).

Table 1.2 Comparison of different modes of droplet-based printing techniques [49].

Characteristic	Inkjet	Aerosol Jet	E-jet
Deposition mechanism	Electrostatic [24]	Aerodynamic [53]	Electrohydrodynamic
Viscosity [cP]	1–20	0.7–2500	<300
Working distance [mm]	1	1–5	<1
Print speed [mm s^{-1}]	Up to 2000	Up to 200	Up to 100
Minimum line width [μm]	30	10	<0.1
Geometry of substrate	planar	Planar and nonplanar	Planar
Nozzle diameter [μm]	10–50	100–300	0.3–30
Particle size of ink [nm]	<100	<700	Variable based on nozzle diameter

Source: Abdolmaleki et al. [49] / with permission of John Wiley & Sons.

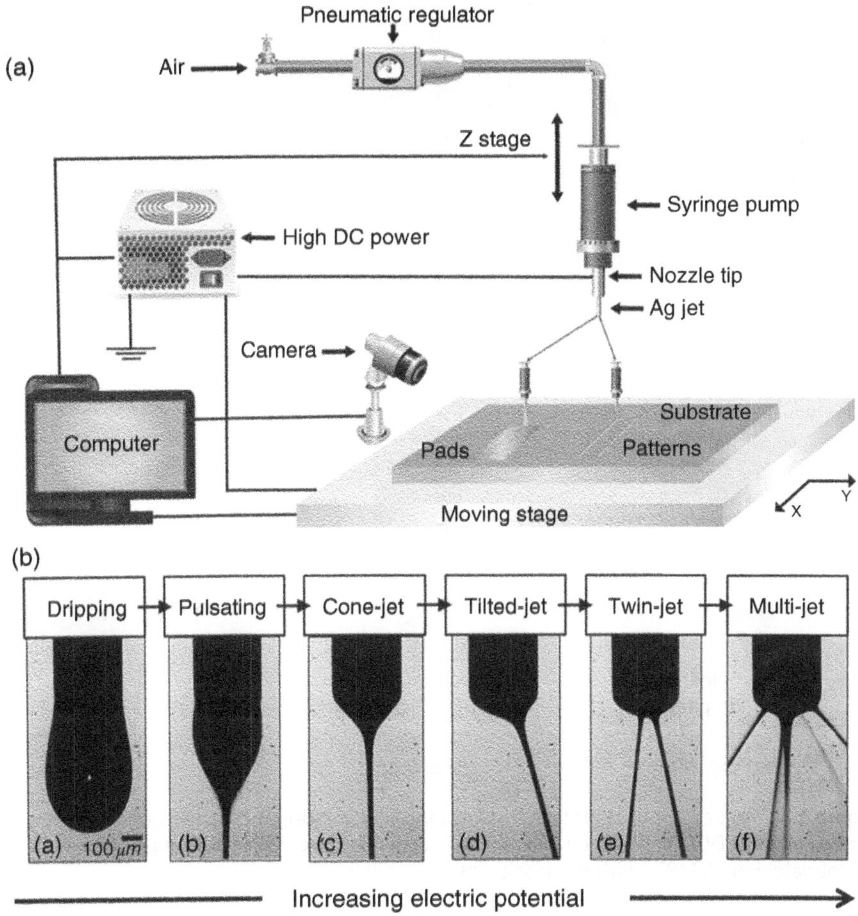

Figure 1.5 (a) Schematic representation of electrohydrodynamic jet printing [50], (b) EHD jetting modes with respect to electric potential. Increasing electric potential exhibits different jetting modes namely (a) dripping, (b) pulsating, (c) cone-jet, (d) tilted jet, (e) twin jet, and (f) multi jet. *Source:* Lee et al. [51] / Adapted with permission of American Chemical Society.

1.3.4 Inkjet-Based Printing

Inkjet-based bioprinting is a subtype of droplet-based printing. It utilizes pneumatic, thermal, or sonic actuation to maneuver cell-laden microdroplets from the nozzle to a substrate [54]. The prerequisites for substrate characteristics include induction of viscous forces, wettability, and high friction coefficient,

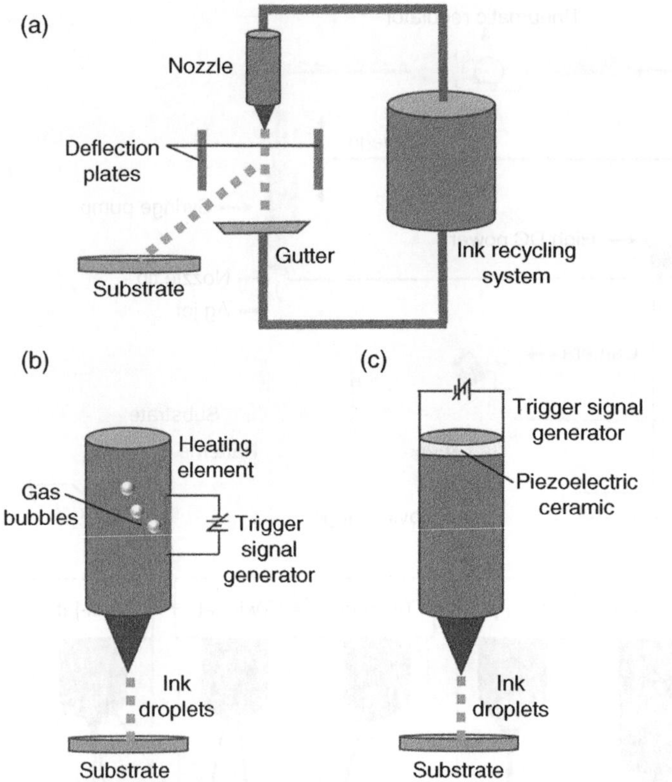

Figure 1.6 (a) Schematic representation of CIJ inkjet printing mechanism, (b) thermal actuator-based DOD printing, (c) piezoelectric actuator-based DOD printing. *Source:* Abdolmaleki et al. [49] / with permission of John Wiley & Sons.

while the bioink should have nonfibrous nature, low viscosity, and high rate of gelation [55]. Inkjet-based bioprinting was one of the first mechanisms for 2D cell printing due to advantages like high cell viability and droplet size [54].

The mechanism of inkjet printing involves ejection of ink droplets on a substrate from nozzles of various diameters [56]. Generally, two ejection modes are exhibited in inkjet printing, namely, drop on demand (DOD) and continuous inkjet (CIJ) (Figure 1.6). DOD printing involves ink droplet ejection on demand, and ejection from nozzle is governed by trigger signals through thermal or piezoelectric actuators present in the printhead. In thermal actuators, trigger signal passes through a heat layer, forming vapor bubbles, leading to drop ejection due to liquid expansion in nozzle. Whereas in piezoelectric actuators, trigger signal leads to piezoelectric element deformation, resulting in droplet ejection from nozzles, CIJ printing involves the ejection of continuous cylindrical ink column by means of an acoustic wave, fabricated using piezoelectric ceramic, in order to obtain minimum surface energy according to Rayleigh–Plateau instability rule. An electric field applied by two parallel deflector plates maneuvers drop deflection to the appropriate position on the substrate while the stray drops are collected in the gutter followed by recirculation into printhead nozzles [57]. Manipulation of optimum cell density, low droplet resolution due to the presence of biopolymers, clogging, and limited biomaterial compatibility resulted in reduced accuracy and mechanical instability [53, 57].

1.3.5 Stereolithography 3D Printing

Stereolithography apparatus (SLA) printing is the oldest printing technique, which uses a computer-controlled laser beam to photopolymerize liquid resin to print objects layer-by-layer. Charles Hull patented this method of fabrication in 1968 [59]. The fundamental elements of SLA printing include a computer-aided design (CAD) of tissue or organs of interest, a source of illumination, and a vertically moving stage in the resin tank. SLA 3DP is based on photopolymerization, wherein the photons from UV or visible light sources dissociate the photoinitiator into a high-energy radical state that interacts with monomers and produces polymeric chains. Biofabrication of 3D printed scaffold depends on the types and concentration of photoinitiator and the intensity of the light source [60].

SLA uses different approaches: (i) masked base writing, and (ii) direct laser writing [61]. SLA uses either a laser beam or projected light to photopolymerize the ink. Masked base writing, also called digital light processing-based SLA, uses digital micromirror technology to project an image of a predetermined pattern or a layer of a 3D CAD model by rerouting incoming light from a UV source. The schematic is shown in (Figure 1.7a). Direct laser writing (DLW) uses a laser beam to trace the 2D image on the surface of the resin (Figure 1.7b). This bottom-up approach is widely adopted wherein the fabrication platform is set just below the tank's surface and is filled with ink [64]. After the photopolymerization of one layer, the platform is lowered for the printing of the second layer. DLW with multiphoton polymerization prints scaffold with high resolution. Here, resin polymerization occurs via femtosecond light pulses from near-infrared laser and optical components to focus the laser beam. The movement of the focal plane of the laser leads to the fabrication of complex 3D structures with intricate features [65]. DLW is more time-consuming and expensive than the DLP technique.

Compared to the other bioprinting techniques, SLA is a faster technique with resolutions ranging from 5–100 μm [66, 67]. It is less time-consuming and independent of complexity or size as the entire image of the 2D slice is projected at once and gets photopolymerized upon irradiation. Owing to nozzle-free

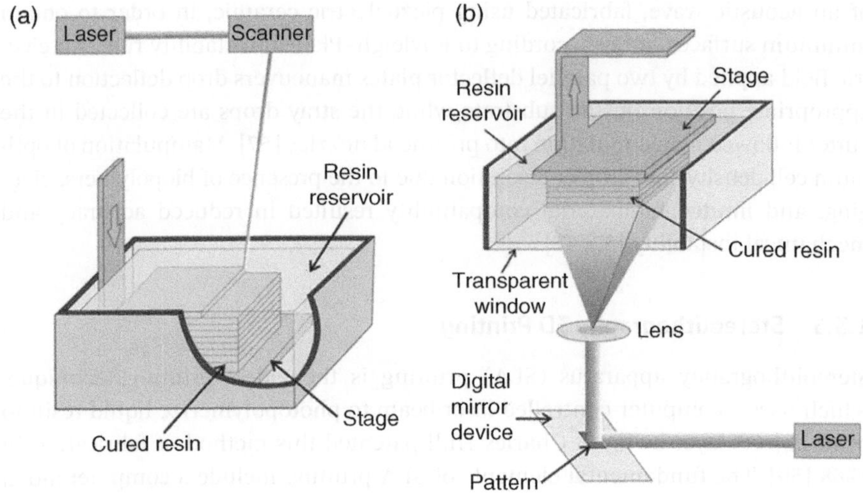

Figure 1.7 Schematic showing SLA based on (a) digital light processing and (b) direct laser writing [62]. Schematic representation of stereolithographic 3D printing. *Source:* Reproduced with permission Gross et al. [63] / American Chemical Society.

fabrication and shear stress independence, the cells could maintain up to 85–90% viability [68, 69]. Additionally, the temperature is low, which is also favorable for cells. Despite promising results in the fabrication of 3D scaffolds, it suffers from limitations. During post-processing, the structure undergoes shrinkage compromising the resolution [70]. The scarcity of biomaterials sensitive to photopolymerization, expensive equipment/photoinitiator, and cytotoxicity associated with uncured photoinitiator are other major drawbacks of the SLA approach.

1.4 Scaffold Modeling and G Coding

1.4.1 Scanning Technology

There are many 3D scanning technologies available for medical purposes, to collect and capture geometry data. Computed tomography (CT) and magnetic resonance imaging (MRI) are more favored in medical imaging, and the scanned data resembles the slice format for layer-by-layer RP.

1.4.2 CT Imaging

CT is a process of generating an image by the use of computer-assisted capturing of a series of X-ray images, through image acquisition at different angles, to produce cross-sectional views of scanned objects, especially soft tissue and bone of the human body through software combining all the slice images.

1.4.3 MRI Scanning

Magnetic resonance is a diagnostic tool with very high magnetic field (1.5 T) radio waves to generate a detailed image of the body's soft tissues and bones. The human body consists of organic compounds of carbon and hydrogen along with other important constituents. These atoms get aligned under the magnetic field. As soon as the radio waves are applied, these atoms deviate from their aligned position, and when the radio waves are turned off, the atoms try to realign to the magnetic field, releasing energy. This released energy signal is then traced with the probes available on the machine. The signal is then processed to generate the image. It should be noted that radio waves of different frequencies are applied while scanning to generate the image contrast.

1.4.4 Preferred Accuracy Parameters for Scanning

- Type of scanner axis movement mechanism
- Kind of scan – axial or helical
- Slice thickness 1.0 mm recommended
- Scan spacing 0.5 mm or half of the smallest dimension of interest
- Strength of signal
- Resolution
- *X–Y* dimensions of a single pixel
- Position of the object during scanning parallel to the bore
- Slice time – Two slides per second recommended

1.4.5 Biomodeling Process for RP

The CT/MRI-scanned medical images are typically in digital imaging and communications in medicine (DICOM) format. Scanning parameters, especially slice thickness and pixel size, are responsible for the clear and smooth virtual model. A 3D reconstruction technique called region growing is used for developing the 3D biomodels using software such as MIMICS, PowerDicom, etc. The reconstructed 3D biomodel is converted to editable file CAD, or directly to stereolithography STL format depending on the requirement. Since the STL file is generated from point cloud data, the model contour consists of planar triangular surfaces. The STL file is then sliced to develop a series of layers for RP fabrication. This sliced-layer file is uploaded to the 3D printer for printing objects.

1.5 Applications and Utility in Large-Scale Manufacturing

The quest of the human race to remain disease-free has led to the emergence of 3DP, which is a promising approach to bridging the gap between natural tissues and artificial tissue constructs. 3DP offers promise in a multitude of applications ranging from regenerative tissue construct models to the fabrication of artificial organs. These are again applied for diverse purposes like drug screening, toxicological purposes, tissue engineering, transplantation, and several clinical functions [71]. The present section highlights the applications and utilities of 3D bioprinting in diverse fields.

1.5.1 Bone

Bone is the hard connective tissue with an internal honeycomb-like matrix. Thus, the bioink for its 3DP should have tunable physicochemical properties. A wide variety of bioinks for bone may be explored by integrating lithographic techniques with others like extrusion-based or inkjet bioprinting. Naghieh et al. created osteoconductive scaffolds by fused deposition modeling with biochemical and mechanical features akin to bone, but the high temperature of the fabrication limited the impregnation of cells and other biological molecules [72]. Hence, predesigned, porous, and mechanically strong scaffolds made by laser-assisted bioprinting are appropriate for the attachment, proliferation, and differentiation of osteoblast cells [73]. Human bone marrow-derived mesenchymal stromal cells, grown on customized methacrylate hyaluronic acid bioprinted hydrogel scaffolds promoted bone formation with a cell survival rate of 64.4%. The combination of low-intensity pulsed ultrasound and 3D bioprinting of human MSCs and other biological components with PGA diacrylate bioink showed improvement in osteogenic cell function [74]. Amorphous magnesium phosphate-based bioink, containing nanofibers and dental pulp stem cells for the 3DP of craniofacial bone tissues, promotes osteogenic differentiation in regenerative dentistry [75]. $CaCl_2$ crosslinked alginate, agarose, and carboxymethyl-chitosan-based bioink containing induced pluripotent stem cells have been 3D printed into porous and stable constructs with the ability for their proliferation and differentiation [76].

The craniomaxillofacial bone defect has been rectified in a patient by constructing a tailor-made nanoengineered ionic covalent entanglement bioink. This ink forms ECM and has osteoinductive properties [77]. Patient-specific anatomical data can be exploited for making scaffolds for patients requiring alveolar and periodontal regeneration [78]. Also, data of patients from imaging techniques like MRI can be modeled to fabricate customized grafts.

1.5.2 Cartilage

Cartilage is one of the most common and simple 3D printed tissues. Bone and cartilage have been regenerated from bioprinted bone marrow stromal cells in alginate hydrogels, thereby paving the way for the application of cartilage 3DP in clinical settings [79]. In 3D bioprinted human nasal chondrocytes and bone marrow-derived mesenchymal stem cell-laden hydrogel constructs, in vivo chondrogenesis with proper mechanical strength was observed after 60 days of implantation in subcutaneous tissues of nude mice. The combination of both cells seemed to fetch more potency for cartilage regeneration [80]. Alginate hydrogel is a cell-compatible material, very frequently used in 3DP, with good flexibility and brittleness but somewhat poor mechanical properties. Cartilage regeneration was achieved by using two bioinks – nano-fibrillated cellulose with alginate and with hyaluronic acid for the delivery of human-derived induced pluripotent stem cells, where the ink with alginate proved to be more successful for cartilage regeneration [81]. 3DP ear cartilage has been produced by bioprinting, using human nasal chondrocytes with nano-fibrillated cellulose-alginate hydrogels indicating the material to be appropriate support for cartilage regeneration [82]. Inkjet bioprinting of MSCs with PCL demonstrated improved resolution of stratified articular cartilage, though with medium cell viability [83]. Though the conditions of the heating effect in SLS bioprinting are detrimental, the scaffolds generated by this method are often used in combination with cell-laden hydrogels. Thus, bioprinting of hydrogels within the SLS scaffolds can contribute strongly to scaffold remodeling and cell differentiation [84].

1.5.3 Skin

Skin is the largest connective tissue that has to undergo maximum trauma like wounds and burns. A 3D layered porous structure similar to native skin was produced by Ng et al. [85], by depositing keratinocytes, melanocytes, and fibroblasts within a suitable conducive microenvironment. Further growth conditions showed that it developed a layered epidermis with a continuous basement membrane protein layer. Similarly, Lee et al. cultured a multilayered biomimetic skin tissue made of human keratinocytes and skin fibroblasts using a printer based on pneumatic control, to produce skin transplants on demand [86]. A bioink of fibrous collagen gels, amniotic fluid stem cells, and bone marrow mesenchymal stem cells promoted wound healing by epithelial regeneration with microvessel and capillary formation [87]. Kim et al. created 3D printed, vascularized human skin equivalent, which comprised an epidermis derived from primary human epidermal keratinocytes, dermis derived from human dermal fibroblast-encapsulated skin-derived ECM with fibrinogen, and hypodermis from preadipocyte-embedded adipose-derived ECM with fibrinogen; HUVEC and thrombin contained 10% gelatin for vascularization. This full-thickness skin with vascularization reflected the actual complexity of native human skin rather than the skin models [88].

Researchers from the University of Toronto and Sunnybrook Health Sciences Center have recently developed a handheld bioprinter capable of printing skin sheets and their conformal delivery in a clinical setting, with clinical potential in the future. The bioink composition of enzymatically and thermally gelled fibrin, collagen, and human MSCs is directly dispensed as sheets on the wound bed, maintaining its topography and improving its re-epithelialization, dermal cell repopulation, and neovascularization. These sheets cover the wound and accelerate the healing process [89]. Citi Biotech has pioneered the world in producing immunized human skin by 3D bioprinting that contains immune cells.

1.5.4 Vascular Grafts

The vasculature system provides nutrients and growth factors to the cells. Tubular architecture can be printed around a solid mandrel, by extrusion of hydrogel in a vertical network of hollow tubules or by coaxial-printing. In the direct extrusion method, the construct is directly extruded in the form of a tubule. Hybrid bioink of HUVEC-encapsulated vascular-derived ECM and alginate has been used to create in vitro, perfusable, vascular models by coaxial cell printing. Endothelium maturation was followed by angiogenesis under the influence of angiogenic cues provided by proangiogenic factors like VEGF, and bFGF [90]. Thick vascular models could also be bioprinted by injection of vascular ink of Pluronic containing fibrinogen, thrombin, transglutaminase, human neonatal dermal fibroblasts, and human BMMSCs into vascular channels by direct extrusion method [91]. Luminal structures mimicking natural blood vessels could be formed by encapsulating endothelial cells and MSCs in a hydrogel, and by printing them by SLA bioprinting [92]. In vitro vascular chips could be fabricated out of human aortic smooth muscle cells and aortic endothelial cells via extrusion bioprinting. Protein expression as well as the contractile phenotype of smooth muscles pertaining to the vascular wall could be achieved [93]. Utilizing engineered elastin protein hydrogels with HUVECs as bioink for extrusion bioprinting, a vascularized chip platform was created for use in personalized medicine [94]. Bioprinting TPP with HUVECs and adipose-derived stem cells as supporting cells, a high-definition sub-micrometer range 3D vessel-on-chip platform was fabricated to facilitate angiogenesis [95].

In the indirect extrusion method, a tubular shape is printed using a sacrificial hydrogel which does not contain any cell and is printed either into or on a structural hydrogel. After the removal of the sacrificial hydrogel, the remaining network is perfused with media that contains vascular cells. After bioprinting, an in vitro cell-loaded chip system with embedded microfluidic channels within a 3D collagen matrix, and sacrificial gelatin templates were removed to form anatomical vasculature [96].

Classical extrusion printing demands complicated and fine preparation of bioink, requiring more extrusion print heads to maintain the location and orientation during the process. However, a hybrid printer with laser-based spray or droplet printing coupled to support cell-laden hydrogel printing might be suitable for printing vasculature [97].

1.5.5 Heart

Some cardiac diseases need immediate rescue, and 3DP cardiac tissue is the answer to cardiac disease modeling and heart engineering. Noor and coworkers pioneered bioprinting the heart in cell suspension using the patient's own myocardial cells and omental tissues. They separated somatic cells and matrix from self-omental tissue and converted the matrix into hydrogel into which cells were embedded. Somatic cells were induced towards cardiomyocyte and endothelial differentiation. Myocardial cells comprise the main body of the heart, and endothelial cells form the internal blood vessels. The heart was ~2 cm long with chambers and blood vessels inside. The heart could contract but not pump blood [98]. The 3DP of heart components fabricated by extrusion bioprinting by freeform reversible embedding of suspended hydrogels (FRESH) method bore a likeness to patient-specific anatomy of the heart, as evidenced by computed tomographic scans. Also, the personalized hydrogel was 3D printed after pH-induced gelation to yield thick, vascularized, and perfusable cardiac valves showing anatomical, cellular, biochemical, and immunological matching with that of the patient. ECM was regenerated from customized pluripotent stem cells. The valve showed synchronous contraction and performed directional potential action propagation [99]. Heart chips were constructed by DLP bioprinting of patterned neonatal mouse ventricular cardiomyocytes by photo-crosslinking method. It contracted synchronously with forced output [100]. Porcine valvular interstitial cells encapsulated within modified gelatin were crosslinked under UV light to create large-scale cardiac microtissues which might lead to the production of scalable organoids.

1.5.6 Lungs

A perfusable alveolar model with detailed internal structure and blood vessels was created by SLA bioprinting of photocurable hydrogel. Though very small in size, it was able to pump air into the airways, deliver oxygen, was strong enough to withstand inhalation and exhalation experiments, and allowed the perfusion of RBCs [101]. In vitro lung-on-a-chip was developed by Huh having suitable physiology alveolar–capillary interface, 3D microarchitecture, and dynamic mechanical activity. The model had two chambers separated by ~10 μm elastomeric and porous poly(dimethyl siloxane) membranes. Upper and lower chambers were seeded with human alveolar epithelial cells and pulmonary microvascular endothelial cells, respectively, and were allowed to adhere to ECM-coated membrane surfaces. Physiological breathing motions and the resulting deformation of alveolar–capillary barrier were emulated by applying cyclic vacuum suctions to both sides of the chambers [102]. Shrestha et al. also fabricated an in vitro lung-on-a-chip model on which Calu-3 cells were cultured [103]. In second generation lung-on-a-chip model, PDMS has been replaced with elastin and collagen, rendering it with stretchable, biodegradable, and biological properties with the added advantage of expression of alveolar epithelial cell markers, stretchable alveoli, and preservation of barrier [104].

1.5.7 Liver

The liver is the most vital organ for detoxification and hematopoiesis. 3D functional and metabolically active hepatic structures were printed using hepatic stellate cells and endothelial cells by a NovoGen MMX bioprinter TM. These structures possessed enzymatic activity, cell–cell interaction, and protein synthesis ability. In vitro, a liver chip was developed by combining human hepatic stellate cells and HepaRG cells on one hand, and perfusion and DLP bioprinting on the other. The result was higher cytochrome P450 and albumin content for liver [105].

A liver model was built by combining HUVECs, hiPSC-derived hepatic progenitor cells, and adipose stem cells with hydrogels via SLA. The model simulated a natural liver module, applicable for drug screening and disease modeling [106].

A 3D liver micro-organ, housing a 3D liver cell-encapsulated hydrogel-based tissue mimicking the natural hepatocyte microenvironment, could accomplish a biological liver function. It was developed as an in vitro drug metabolism model [107].

Survival of the Fah$^{-/-}$ Rag2$^{-/-}$ mouse model of liver injury improved upon transplanting their abdominal cavity with hepatorganoids composed of gelatin, alginate, and HepRG cells. These structures promoted the synthesis of liver-specific proteins, vasculature, and liver capacity [108].

1.5.8 Kidney and Urethra

A decellularized kidney can provide a blueprint for 3DP. Renal proximal tubules with lumen architecture were bioprinted using proximal tubule epithelial cells suspended in gelatin–fibrin hydrogel. These tubules had improved epithelial morphology [109]. A similar tubular structure of the urethra has been printed with PCL/poly (lactide-co-caprolactone) spiral scaffold, and cell-loaded fibrin hydrogel comprising of rabbit urothelial cells and smooth muscle cells [110].

1.5.9 Brain and Spinal Cord

Brain-like structures were developed by three-layered essential cortical neurons covered in RGD peptide-modified gellan gum hydrogel. The structures had three layers with the middle layer having axons. The structures are useful for understanding cell behavior, brain injuries, and neurodegenerative disorders. In vitro, brain matrix–mimetic microenvironment model, fabricated by using glial cells and hyaluronic acid via extrusion-based bioprinting method, simulated biological and mechanical properties of the human brain microenvironment [111].

Bioprinting can become handy for the creation of an interaction between the neural cells and ECM to impose positional control in the neural models, manufacture long-distance nerve-guided regeneration by customizable nerve guides, manufacture scaffolds with 3D bioprinted cells for nerve guides, and a filler for nerve guides, etc. [112]. Complex CNS structure for spinal cord regenerative medicine applications has been printed by microscale continuous projection printing method using neural progenitor cells to support axon regeneration. The establishment of synaptic transmission at the sites of complete spinal cord injury formed "neural relays." This study, done in rodents, is possible to be scaled up for humans too [113].

1.5.10 Cornea

Using an alginate/gelatin/collagen-based bioink, human corneal epithelial cells have been successfully bioprinted [114]. Human corneal keratocytes have been bioprinted by Isaacson et al. using collagen-based bioink by extrusion-based system showing cell viability of 87% till seven days [115]. Decellularized corneal ECM has been used as a bioink for corneal stroma bioprinting [116]. Bioprinting of full corneal replacements for clinical purposes is thus not far-fetched [117].

1.5.11 Therapeutics

The field of 3D bioprinting has immense potential in the field of therapeutics [118]. Some of these have been summarized.

1) *Drug printer*: Patient-centered dosages with control release properties can be delivered by 3DP. The first 3D printed tablet was developed by Aprecia Pharmaceuticals for the anti-epileptic medication levetiracetam, (Spritam) by powder bed, and inkjet 3DP technology known as ZipDose. The binder liquid is dropped on a powder bed by a printing head and the resultant pill contains a densely packed drug offering a high dose of medicine, thereby reducing the frequency of intake but providing steady release over time. It was approved by the FDA in 2015 [119, 120]. Extrusion printed, an antihypertensive drug for diabetics has been developed and is known as polypill, which has zero-order drug-release compartments of captopril and sustained drug-release compartments of glipizide and nifedipine. The porosity of this barrier layer can be controlled by 3DP. The quick disintegration of the barrier layer between the two compartments after consumption allows for the treatment of dyslipidemia and hypertension simultaneously with a single tablet [121]. 3DP paracetamol tablet, called printlet, has been prepared with Eudragit® L100–55 or Kollicoat® IR with high crushing strength, low friability, and pH-dependent release [122].
2) *RNA printer*: RNA printer has been developed by CureVac for rapid manufacture of clinic-ready vaccines as an outbreak response against several viral infections like MERS COVID-19, etc. [123, 124]. This technique was originally invented for the rabies vaccine [125].
3) *Cell delivery*: The dual antibiotic release 3DP scaffold containing rifampin and daptomycin, incorporated with viable macrophages in a hydrogel-based bioink, was fabricated for the prevention of craniotomy-associated infections in the mouse model. The presence of macrophages reduced *S. aureus* load and biofilm formation [126]. 3D printed dual-cell delivery porous titanium alloy scaffolds, filled with BMSCs and endothelial progenitor cells loaded in thermosensitive poloxamer 407 hydrogels, were constructed. The cell delivery promoted osteogenesis, and vascularization, and thus promised osseointegration in patients suffering from postoperative complications like loosening or dislocation of the prosthesis [127].
4) *Vaccine delivery*: Polymeric (PGA dimethacrylate), multifaceted microneedle arrays have been designed and 3D printed by continuous liquid interface production (CLIP) 3DP technology. The resulting patch yielded higher surface area, controlled geometries, high precision, and consistency in shape, size, spacing between every needle in the array, and higher surface coatings of vaccine components. Additionally, vaccine delivery by this method enhanced cargo retention in the skin, activated immune cells directly in the draining lymph nodes along with inducing more potent Th1-based cellular and humoral immune responses [128].

5) *Antibiotic delivery*: 3D scaffolds of rifampicin-loaded PCL have been designed and 3D printed at a temperature of 60 °C to prevent loss of antibacterial activity of antibiotic. Potential growth inhibition by this antibiotic-laden scaffold against *Escherichia coli* and *Staphylococcus aureus*, the representative causative organisms of osteomyelitis, was observed [129]. In a groundbreaking first, researchers have fabricated 3D scaffold implants at high temperatures, containing dual antibiotics – ciprofloxacin and gentamicin. The antibiotics are intercalated and covered within lamellar inorganic protectors – magnesium aluminum layered double hydroxides and α-zirconium phosphates. These scaffolds enable the sustained release of antibiotics through a polymer matrix to support bone regeneration during bone infections arising due to an injury or surgery [130].
6) *Cancer*: Cancer therapy calls for the manufacture of effective and reliable anti-cancer drugs, which can be tested on suitable models. Models of exocrine pancreas spheroid by laser-assisted 3DP, lung cancer tumors, and bladder cancer have been established by 3DP. These cell models show higher resistance and less sensitivity to chemotherapeutic drugs. Such miniaturized models recapitulate the cancer microenvironment, act as a drug screening platform, and throw light on the disease progression, and therapy strategies [131–133]. Gelatin-alginate-MatrigelTM mixed with patient-derived intrahepatic cholangiocarcinoma cells were used to bioprint a tumor construct. This tumor 3DP demonstrated a high proliferation rate, colony-forming ability, survival, high level of cancer stemness markers, epithelial to mesenchymal stem cell regulatory proteins, and matrix metalloproteinases, hence pointing towards invasiveness and metastatic cancer microenvironment [134].

1.6 Complications and Troubleshooting

The following are the notable complications and constraints during scaffold fabrication in specific bioprinting techniques.

1.6.1 Laser-Based Printing

- Material selection with required physical and chemical properties for any application is limited by the photosensitive polymer. Introducing photoresists in the material that may be capable of encoding different material properties or development of such material may be a suitable measure to address the issues.

- Material waste and cost expenses are a significant encumbrance in laser-based printing.
- Furthermore, sticking of material at the bed of the reservoir may damage the part and can be addressed by modifying the printing parameters. Additionally, the laser gun/source needs to be clean and well-maintained to get the desired focus, power, and wavelength.

1.6.2 Inkjet-Based Printing

- Inkjet printing requires bioink with low viscosity, which may lead to insufficient mechanical strength to withstand its weight. The problem can be solved by designing a bioink that maintains the stiffness of the scaffold [135, 136].
- Moreover, the gelation mechanism can also be designed to maintain bioink viscosity before and after printing.

1.6.3 Extrusion-Based Printing

- Nozzle clogging is the major problem for the extrusion printing strategy. Moreover, extrusion printing provides a comparatively low fabrication resolution (~100–200 μm). The problem of clogging can be addressed by cleaning the nozzle with a pin; however, frequent nozzle calibration would be required for the desired result [135–137].
- The material flow control is also a challenge as a result of delaying compressed gas volume in the extrusion system. One can tackle the issues with proper printing parameters.
- Additionally, extrusion-based printing has the lowest cell survival rate among all techniques due to the shear stress that arises during printing. Optimized bioink viscosity would solve the issue.

1.6.4 Droplet Printing

- The droplet printing is limited by high viscosity and cell density material. The problem can be solved by choosing materials with lower viscosity (<10 centipoise) and lover cell densities (<10 million cells/ml) without clogging the nozzle [138].
- The droplet printing technique requires an array of needles to hold the scaffold structure. The technique demands precise droplet size and sufficient viscosity to settle within the needle array. Optimization of printing parameters along with respect to biomaterial viscosity may address the technical glitch.

1.6.5 Stereolithography 3D Printing

- Hollow structures may contain remnants of precursor hydrogel that may lead to blockage of structures such as vessels, vasculature, or ducts. Optimization of the printing parameters and material properties may be required under such conditions [136, 138].
- Since the biomaterial required for the stereo lithography 3DP technique is UV/photocurable, therefore proper maintenance and cleaning of the light source would be essential in order to get quality scaffolds.

References

1 Pranzo, D., Larizza, P., Filippini, D., and Percoco, G. (2018). Extrusion-based 3D printing of microfluidic devices for chemical and biomedical applications: a topical review. *Micromachines* 9 (8): 374.

2 Jiang, Z., Diggle, B., Tan, M.L. et al. (2020). Extrusion 3D printing of polymeric materials with advanced properties. *Adv. Sci.* 7 (17): 1–32.

3 Thakar, C.M., Parkhe, S.S., Jain, A. et al. (2021). 3D printing: basic principles and applications. *Mater. Today Proc.* 51: 842–849. https://doi.org/10.1016/j.matpr.2021.06.272.

4 Wang, X., Jiang, M., Zhou, Z. et al. (2017). 3D printing of polymer matrix composites: a review and prospective. *Compos. Part B Eng.* 110: 442–458. https://doi.org/10.1016/j.compositesb.2016.11.034.

5 Malekpour, A. and Chen, X. (2022). Printability and cell viability in extrusion-based bioprinting from experimental, computational, and machine learning views. *J. Funct. Biomater.* 13: 40.

6 Schwab, A., Levato, R., D'Este, M. et al. (2020). Printability and shape fidelity of bioinks in 3D bioprinting. *Chem. Rev.* 120 (19): 11028–11055.

7 Engin, N., Sharda, V., Kunal, G. et al. (2021). From 3D printing to 3D bioprinting: the material properties of polymeric material and its derived bioink for achieving tissue specific architectures. *Cell Tissue Bank.* 0123456789. https://doi.org/10.1007/s10561-021-09975-z.

8 Vanaei, S., Parizi, M.S., Vanaei, S. et al. (2021). An overview on materials and techniques in 3D bioprinting toward biomedical application. *Eng. Regen.* November 2020 (2): 1–18. https://doi.org/10.1016/j.engreg.2020.12.001.

9 Samandari, M., Quint, J., Rodríguez-delarosa, A. et al. (2022). Bioinks and bioprinting strategies for skeletal muscle tissue engineering. *Adv. Mater.* 2105883 (34): 1–29.

10 Engler, A.J., Griffin, M.A., Sen, S. et al. (2004). Myotubes differentiate optimally on substrates with tissue-like stiffness: pathological implications for soft or stiff microenvironments. *J. Cell Biol.* 166 (6): 877–887.

11 Ning, L. and Chen, X. (2017). A brief review of extrusion-based tissue scaffold bio-printing. *Biotechnol. J.* 12 (1600671): 1–16.

12 Highley, C.B., Song, K.H., Daly, A.C., and Burdick, J.A. (2018). Jammed microgel inks for 3D printing applications. *Adv. Sci. News.* 6: 1801076.

13 Mouser, V.H.M., Melchels, F.P.W., Visser, J. et al. Yield stress determines bioprintability of hydrogels based on gelatin- methacryloyl and gellan gum for cartilage bioprinting. *Biofabrication* 8 (3): 1–13. https://doi.org/10.1088/1758-5090/8/3/035003.

14 Ouyang, L., Highley, C.B., Rodell, C.B. et al. (2016). 3D printing of shear-thinning hyaluronic acid hydrogels with secondary cross-linking. *ACS Appl. Mater. Interfaces* 2: 1743–1751.

15 Stella, S., Caballero, R., Saiz, E. et al. (2019). 3-D printing of chitosan-calcium phosphate inks: rheology, interactions and characterization. *J. Mater. Sci. Mater. Med.* https://doi.org/10.1007/s10856-018-6201-y.

16 Ji, S. and Guvendiren, M. (2017). Recent advances in bioink design for 3D bioprinting of tissues and organs. *Front. Bioeng. Biotechnol.* 5 (APR): 1–8.

17 Schwab, A., Levato, R., Este, M.D. et al. (2020). Printability and shape fidelity of bioinks in 3D bioprinting. *Chem. Rev.* 120: 11028.

18 Fu, Z., Naghieh, S., Xu, C. et al. (2021). Printability in extrusion bioprinting. *Biofabrication* 13: 033001.

19 Barki, A.M., Bocquet, L., and Stevenson, A. (2017). Linking rheology and printability for dense and strong ceramics by direct ink writing. *Sci. Rep.* 7 (January): 1–10.

20 Cooke, M.E. and Rosenzweig, D.H. (2021). The rheology of direct and suspended extrusion bioprinting The rheology of direct and suspended extrusion bioprinting. *APL Bioeng.* 5 (September 2020): 011502.

21 Materials Research Society (2018). 3D bioprinting of organs part 1 – MRS ondemand webminar. https://www.youtube.com/watch?v=dFXn-PCYRdQ (accessed 5 May 2022).

22 Placone, J.K. and Engler, A.J. (2019). Recent advances in extrusion-based 3D printing for biomedical applications. *Adv. Healthcare Mater.* 7 (8): 1–24.

23 Ding, S., Feng, L., Wu, J. et al. (2018). Bioprinting of stem cells: interplay of bioprinting process, bioinks, and stem cell properties. *ACS Biomater. Sci. Eng.* 4 (9): 3108–3124.

24 Zhang, J., Wehrle, E., Rubert, M., and Müller, R. (2021). 3d bioprinting of human tissues: Biofabrication, bioinks and bioreactors. *Int. J. Mol. Sci.* 22 (8).

25 Arslan-Yildiz, A., El Assal, R., Chen, P. et al. (2016). Towards artificial tissue models: past, present, and future of 3D bioprinting. *Biofabrication.* 8 (1).

26 Guillemot, F., Souquet, A., Catros, S., and Guillotin, B. (2010). Laser-assisted cell printing: principle, physical parameters versus cell fate and perspectives in tissue engineering. *Nanomedicine (London)* 5 (3): 507–515.

27 Li, J., Chen, M., Fan, X., and Zhou, H. (2016). Recent advances in bioprinting techniques: approaches, applications and future prospects. *J. Transl. Med.* 14 (1): 1.

28 Ozbolat, I.T. (2017). Laser-based bioprinting with minor contributions by Hemanth Gudapati, The Pennsylvania State University. In: *3D Bioprinting* (ed. I.T. Ozbolat), 165–197. Academic Press.

29 Wang, S., Zhou, Z., Li, B. et al. (2021). Progresses on new generation laser direct writing technique. *Mater. Today Nano.* 16: 100142.

30 Jia, Y.C., Wang, S.X., and Chen, F. (2020). Femtosecond laser direct writing of flexibly configured waveguide geometries in optical crystals: fabrication and application. *Opto-Electronic Adv.* 3 (10): 1–12.

31 Zennifer, A., Subramanian, A., and Sethuraman, S. (2022). Design considerations of bioinks for laser bioprinting technique towards tissue regenerative applications, e00205. *Bioprinting* 27: https://doi.org/10.1016/j.bprint.2022.e00205.

32 Odde, D.J. and Renn, M.J. (1999). Laser-guided direct writing for applications in biotechnology. *Trends Biotechnol.* 17 (10): 385–389.

33 Odde, D.J. and Renn, M.J. (2000). Laser-guided direct writing of living cells. *Biotechnol. Bioeng.* 67: 312–318.

34 Chrisey, D.B., Piqué, A., McGill, R.A. et al. (2003). Laser deposition of polymer and biomaterial films. *Chem. Rev.* 103 (2): 553–576.

35 Fitz-Gerald, J.M., Wu, H.D., Pique, A. et al. (2000). Maple direct write: a new approach to fabricate ferroelectric thin film devices in air at room temperature. *Integr. Ferroelectr.* 28 (1–4): 13–28.

36 Piqué, A. (2011). The matrix-assisted pulsed laser evaporation (MAPLE) process: origins and future directions. *Appl. Phys. A Mater. Sci. Process.* 105 (3): 517–528.

37 Papavlu, A.P., Dinca, V., Filipescu, M., and Dinescu, M. (2017). Matrix-assisted pulsed laser evaporation of organic thin films: applications in biology and chemical sensors. In: *Laser Ablation – From Fundamentals to Applications* IntechOpen. https://doi.org/10.5772/intechopen.70676.

38 Schiele, N.R., Chrisey, D.B., and Corr, D.T. (2011). Gelatin-based laser direct-write technique for the precise spatial patterning of cells. *Tissue Eng – Part C Methods.* 17 (3): 289–298.

39 Devillard, R., Pagès, E., Correa, M.M. et al. (2014). Cell patterning by laser-assisted bioprinting. *Methods Cell Biol.* 119: 159–174.

40 Guillemot, F., Souquet, A., Catros, S. et al. (2010). High-throughput laser printing of cells and biomaterials for tissue engineering. *Acta Biomater.* 6 (7): 2494–2500.

41 Unger, C., Gruene, M., Koch, L. et al. (2011). Time-resolved imaging of hydrogel printing via laser-induced forward transfer. *Appl. Phys. A* 103 (2): 271–277.

42 Choi, J.W., Kim, H.C., and Wicker, R. (2011). Multi-material stereolithography. *J. Mater. Process. Technol.* 3 (211): 318–328.

43 Zhou, C., Chen, Y., Yang, Z., and Khoshnevis, B. (2013). Digital material fabrication using mask-image-projection-based stereolithography. *Rapid Prototyp. J.* 19 (3): 153–165.

44 Jakus, A.E., Rutz, A.L., and Shah, R.N. (2016). Advancing the field of 3D biomaterial printing. *Biomed. Mater.* 11: 014102. https://doi.org/10.1088/1748-6041/11/1/014102.

45 Hu, F., Mikolajczyk, T., Pimenov, D.Y., and Gupta, M.K. (2021). Extrusion-based 3d printing of ceramic pastes: mathematical modeling and in situ shaping retention approach. *Materials (Basel)* 14 (5): 1–22.

46 Davoodi, E., Sarikhani, E., Montazerian, H. et al. (2020). Extrusion and microfluidic-based bioprinting to fabricate biomimetic tissues and organs. *Adv. Mater. Technol.* 5 (8): 1901044.

47 Chung, J.H.Y., Yue, Z., Kapsa, R. et al. (2013). Bio-ink properties and printability for extrusion printing living cells. *Biomater. Sci.* 1. https://doi.org/10.1039/C3BM00012E.

48 Hospodiuk, M., Dey, M., Sosnoski, D., and Ozbolat, I.T. (2017). The bioink: a comprehensive review on bioprintable materials. *Biotechnol. Adv.* 35 (2): 217–239. https://doi.org/10.1016/j.biotechadv.2016.12.006.

49 Abdolmaleki, H., Kidmose, P., and Agarwala, S. (2021). Droplet-based techniques for printing of functional inks for flexible physical sensors. *Adv. Mater.* 33: 2006792.

50 Can, T.T.T., Nguyen, T.C., and Choi, W.S. (2019). Patterning of high-viscosity silver paste by an electrohydrodynamic-jet printer for use in TFT applications. *Sci. Rep.* 9 (1): 1–8.

51 Lee, A., Jin, H., Dang, H.W. et al. (2013). Optimization of experimental parameters to determine the jetting regimes in electrohydrodynamic printing. *Langmuir* 29 (44): 13630–13639.

52 Kwon, H., Hong, J., Nam, S.Y. et al. (2021). Overview of recent progress in electrohydrodynamic jet printing in practical printed electronics: focus on the variety of printable materials for each component. *Mater. Adv.* 2: 5593.

53 Arslan-Yildiz, A., El Assal, R., Chen, P. et al. (2016). Towards artificial tissue models: past, present, and future of 3D bioprinting. *Biofabrication.* 8 (1).

54 Xu, T., Jin, J., Gregory, C. et al. (2005). Inkjet printing of viable mammalian cells. *Biomaterials* 26: 93–99.

55 Roth, E.A., Xu, T., Das, M. et al. (2004). Inkjet printing for high-throughput cell patterning. *Biomaterials* 25: 3707–3715.

56 Smith, M., Choi, Y.S., Boughey, C., and Kar-Narayan, S. (2017). Controlling and assessing the quality of aerosol jet printed features for large area and flexible electronics. *Flexible Printed Electron.* 2 (1): 015004.

57 Daly, R., Harrington, T.S., Martin, G.D., and Hutchings, I.M. (2015). Inkjet printing for pharmaceutics – a review of research and manufacturing. *Int. J. Pharm.* 494 (2): 554–567. https://doi.org/10.1016/j.ijpharm.2015.03.017.

58 Kim, J.D., Choi, J.S., Kim, B.S. et al. (2010). Piezoelectric inkjet printing of polymers: stem cell patterning on polymer substrates. *Polymer (Guildf)* 51 (10): 2147–2154. https://doi.org/10.1016/j.polymer.2010.03.038.

59 Hull, C. W. (1986). Apparatus for production of three-dimensional objects by stereolithography. Vol. 4,575,330, US Patent.

60 Arcaute, K., Mann, B.K., and Wicker, R.B. (2006). Stereolithography of three-dimensional bioactive poly(ethylene glycol) constructs with encapsulated cells. *Ann. Biomed. Eng.* 34 (9): 1429–1441.

61 Kumar, H. and Kim, K. (2020). Stereolithography 3D bioprinting. In: *3D Bioprinting: Principles and Protocols* (ed. J.M. Crook), 93–108. New York, NY: Springer US https://doi.org/10.1007/978-1-0716-0520-2_6.

62 Mouzakis, D.E. (2018). Advanced technologies in manufacturing 3D-layered structures for defense and aerospace. In: *Lamination – Theory and Application* (ed Osheku, C.A.). IntechOpen. https://doi.org/10.5772/intechopen.74331.

63 Gross, B.C., Erkal, J.L., Lockwood, S.Y. et al. (2014). Evaluation of 3D printing and its potential impact on biotechnology and the chemical sciences. *Anal. Chem.* 86 (7): 3240–3253.

64 Lee, S.J., Kang, H.W., Park, J.K. et al. (2008). Application of microstereolithography in the development of three-dimensional cartilage regeneration scaffolds. *Biomed. Microdevices* 10 (2): 233–241.

65 Lee, K.S., Kim, R.H., Yang, D.Y., and Park, S.H. (2008). Advances in 3D nano/microfabrication using two-photon initiated polymerization. *Prog. Polym. Sci.* 33 (6): 631–681.

66 Gauvin, R., Chen, Y.C., Lee, J.W. et al. (2012). Microfabrication of complex porous tissue engineering scaffolds using 3D projection stereolithography. *Biomaterials* 33 (15): 3824–3834. https://doi.org/10.1016/j.biomaterials.2012.01.048.

67 Raman, R., Bhaduri, B., Mir, M. et al. (2016). High-resolution projection microstereolithography for patterning of neovasculature. *Adv. Healthcare Mater.* 5 (5): 610–619.

68 Lin, H., Zhang, D., Alexander, P.G. et al. (2013). Application of visible light-based projection stereolithography for live cell-scaffold fabrication with designed architecture. *Biomaterials* 34 (2): 331–339. https://doi.org/10.1016/j.biomaterials.2012.09.048.

69 Wang, Z., Abdulla, R., Parker, B. et al. (2015). A simple and high-resolution stereolithography-based 3D bioprinting system using visible light crosslinkable bioinks. *Biofabrication* 7 (4): 45009. https://doi.org/10.1088/1758-5090/7/4/045009.

70 Konstantinou, G., Kakkava, E., Hagelüken, L. et al. (2020). Additive micro-manufacturing of crack-free PDCs by two-photon polymerization of a single, low-shrinkage preceramic resin. *Addit. Manuf.* 35 (May): 101343. https://doi.org/10.1016/j.addma.2020.101343.

71 Tan, B., Gan, S., Wang, X. et al. (2021). Applications of 3D bioprinting in tissue engineering: advantages, deficiencies, improvements, and future perspectives. *J. Mater. Chem B. R. Soc. Chem.* 9: 5385–5413.

72 Naghieh, S., Ravari, M.R.K., Badrossamay, M. et al. (2016). Numerical investigation of the mechanical properties of the additive manufactured bone scaffolds fabricated by FDM: the effect of layer penetration and post-heating. *J. Mech. Behav. Biomed. Mater.* https://doi.org/10.1016/j.jmbbm.2016.01.031.

73 Keriquel, V., Oliveira, H., Rémy, M. et al. (2017). In situ printing of mesenchymal stromal cells, by laser-assisted bioprinting, for in vivo bone regeneration applications. *Sci. Rep.* 7 (1): 1778.

74 Zhou, X., Castro, N.J., Zhu, W. et al. (2016). Improved human bone marrow mesenchymal stem cell osteogenesis in 3D bioprinted tissue scaffolds with low intensity pulsed ultrasound stimulation. *Nat. Publ. Gr.* (September): 6, 1–12.

75 Dubey, N., Ferreira, J.A., Malda, J. et al. (2020). Extracellular matrix/amorphous magnesium phosphate bioink for 3D bioprinting of craniomaxillofacial bone tissue. *ACS Appl. Mater. Interfaces* 12 (21): 23752–23763. https://doi.org/10.1021/acsami.0c05311.

76 Gu, Q., Tomaskovic-Crook, E., Wallace, G.G., and Crook, J.M. (2017). 3D bioprinting human induced pluripotent stem cell constructs for in situ cell proliferation and successive multilineage differentiation. *Adv. Healthcare Mater.* 6 (17): 1–11.

77 Chimene, D., Miller, L., Cross, L.M. et al. (2020). Nanoengineered osteoinductive bioink for 3D bioprinting bone tissue. *ACS Appl. Mater. Interfaces* 12 (14): 15976–15988.

78 Yu, N., Nguyen, T., Cho, Y.D. et al. (2019). Personalized scaffolding technologies for alveolar bone regenerative medicine. *Orthod. Craniofacial Res.* 22 (S1): 69–75.

79 Fedorovich, N.E., Weeren, V., Ph, D. et al. (2012). Biofabrication of osteochondral tissue equivalents by printing topologically defined, cell-laden hydrogel scaffolds. *Tissue Eng. Part C Methods* 18 (1): 33.

80 Möller, T., Amoroso, M., Hägg, D. et al. (2017). In vivo chondrogenesis in 3D bioprinted human cell-laden hydrogel constructs. *Plast. Reconstr. Surg. Glob. Open* 5 (2): 1–7.

81 Nguyen, D., Hgg, D.A., Forsman, A. et al. (2017). Cartilage tissue engineering by the 3D bioprinting of iPS cells in a nanocellulose/alginate bioink. *Sci. Rep.* 7 (1): 1–10.

82 Markstedt, K., Mantas, A., Tournier, I. et al. (2015). 3D bioprinting human chondrocytes with nanocellulose-alginate bioink for cartilage tissue engineering applications. *Biomacromolecules* 16: 1489–1496.

83 Daly, A.C., Cunniffe, G.M., Sathy, B.N. et al. (2016). 3D bioprinting of developmentally inspired templates for whole bone organ engineering. *Adv. Healthcare Mater.* 5 (18): 2353–2362.

84 Gupta, S., Bissoyi, A., and Bit, A. (2018). A review on 3D printable techniques for tissue engineering. *BioNanoScience* 8 (3): 868–883.

85 Ng, W.L., Qi, J.T.Z., Yeong, W.Y., and Naing, M.W. (2018). Proof-of-concept: 3D bioprinting of pigmented human skin constructs. *Biofabrication* 10 (2): 025005.

86 Skardal, A., Mack, D., and Soker, S. (2012). Oprinted amniotic fluid-derived stem cells accelerate healing of large skin wounds. *Stem Cells Transl. Med.* 1 (792): 792–802.

87 Skardal, A. (2015). Bioprinting essentials of cell and protein viability. In: *Essentials of 3D Biofabrication and Translation* (eds Anthony A. and James J.Y.), 1–17. Elsevier, Academic Press, http://dx.doi.org/10.1016/B978-0-12-800972-7/00001-3.

88 Kim, B.S., Gao, G., Kim, J.Y., and Cho, D.W. (2019). 3D cell printing of perfusable vascularized human skin equivalent composed of epidermis, dermis, and hypodermis for better structural recapitulation of native skin. *Adv. Healthcare Mater.* 8 (7): 1–11.

89 Cheng, R.Y., Eylert, G., Gariepy, J. et al. (2020). Handheld instrument for wound-conformal delivery of skin precursor sheets improves healing in full-thickness burns Handheld instrument for wound-conformal delivery of skin precursor sheets improves healing in full-thickness burns. *Biofabrication* 12 (2): 025002.

90 Gao, Q., He, Y., Fu, J. et al. (2015). Biomaterials coaxial nozzle-assisted 3D bioprinting with built-in microchannels for nutrients delivery. *Biomaterials* 61: 203–215. https://doi.org/10.1016/j.biomaterials.2015.05.031.

91 Kolesky, D.B., Homan, K.A., Skylar-Scott, M.A., and Lewis, J.A. (2016). Three-dimensional bioprinting of thick vascularized tissues. *Proc. Natl. Acad. Sci. USA* 113 (12): 3179–3184.

92 Zhu, W., Qu, X., Zhu, J. et al. (2017). Biomaterials direct 3D bioprinting of prevascularized tissue constructs with complex microarchitecture. *Biomaterials* 124: 106–115. https://doi.org/10.1016/j.biomaterials.2017.01.042.

93 Abudupataer, M., Chen, N., Yan, S. et al. (2020). Bioprinting a 3D vascular construct for engineering a vessel-on-a-chip. *Biomed. Microdevices* 22: 1–10.

94 Duarte Campos, D.F., Lindsay, C.D., Roth, J.G. et al. (2020). Bioprinting cell- and spheroid-laden protein-engineered hydrogels as tissue-on-chip platforms. *Front. Bioeng. Biotechnol.* 8 (April): 1–13.

95 Dobos, A., Gantner, F., Markovic, M. et al. (2020). On-chip high-definition bioprinting of microvascular structures. *Biofabrication* 13 (1): 015016.

96 Wan, L., Skoko, J., Yu, J. et al. (2017). Mimicking embedded vasculature structure for 3D cancer on a chip approaches through micromilling. *Sci. Rep.* 7 (1): 1–8.

97 Chen, E.P., Toksoy, Z., Davis, B.A., and Geibel, J.P. (2021). 3D bioprinting of vascularized tissues for in vitro and in vivo applications. *Front. Bioeng. Biotechnol.* 9: 664188.

98 Noor, N., Shapira, A., Edri, R. et al. (2019). 3D printing of personalized thick and perfusable cardiac patches and hearts. *Adv. Sci.* 6: 1900344.

99 Lee, A., Hudson, A.R., Shiwarski, D.J. et al. (2019). 3D bioprinting of collagen to rebuild components of the human heart. *Science (80-)* 365 (6452): 482–487.

100 Liu, J., Miller, K., Ma, X. et al. (2020). Biomaterials direct 3D bioprinting of cardiac micro-tissues mimicking native myocardium. *Biomaterials* 256: 120204. https://doi.org/10.1016/j.biomaterials.2020.120204.

101 Grigoryan, B., Paulsen, S.J., Corbett, D.C. et al. (2019). Multivascular networks and functional intravascular topologies within biocompatible hydrogels. *Science (80-)* 364 (6439): 458–464.

102 Huh, D.D. (2015). A human breathing lung-on-a-chip. *Ann. Am. Thorac. Soc.* 12: S42.

103 Shrestha, J., Ghadiri, M., Shanmugavel, M., and Razavi, S. (2020). Organs-on-a-Chip A rapidly prototyped lung-on-a-chip model using 3D-printed molds. *Organs-on-a-Chip* 1: 100001. https://doi.org/10.1016/j.ooc.2020.100001.

104 Zamprogno, P., Wüthrich, S., Achenbach, S. et al. (2021). Second-generation lung-on-a-chip with an array of stretchable alveoli made with a biological membrane. *Commun. Biol.* 1–10. https://doi.org/10.1038/s42003-021-01695-0.

105 Grix, T., Ruppelt, A., Thomas, A. et al. (2018). Bioprinting perfusion-enabled liver equivalents for advanced organ-on-a-chip applications. *Genes (Basel)* 9 (4): 176.

106 Ma, X., Qu, X., Zhu, W. et al. (2016). Deterministically patterned biomimetic human iPSC-derived hepatic model via rapid 3D bioprinting. *Proc. Natl. Acad. Sci. USA* 113 (8): 2206–2211.

107 Chang, R., Emami, K., Wu, H., and Sun, W. (2010). Biofabrication of a three-dimensional liver micro-organ as an in vitro drug metabolism model. *Biofabrication* 2 (4): 045004.

108 Yang, H., Sun, L., Pang, Y. et al. (2021). Three-dimensional bioprinted hepatorganoids prolong survival of mice with liver failure. *Gut* 70 (3): 567–574.

109 Homan, K.A., Kolesky, D.B., Skylar-scott, M.A. et al. (2016). Bioprinting of 3D convoluted renal proximal tubules on perfusable chips. *Nat. Publ. Gr.* 1–13. https://doi.org/10.1038/srep34845.

110 Zhang, K., Fu, Q., Yoo, J. et al. (2016). 3D bioprinting of urethra with PCL/PLCL blend and dual autologous cells in fibrin hydrogel: an in vitro evaluation of biomimetic mechanical property and cell growth environment. *Acta Biomater.* https://doi.org/10.1016/j.actbio.2016.12.008.

111 Lozano, R., Stevens, L., Thompson, B.C. et al. (2015). Biomaterials 3D printing of layered brain-like structures using peptide modi fi ed gellan gum substrates. *Biomaterials* 67: 264–273. https://doi.org/10.1016/j.biomaterials.2015.07.022.

112 Li, X.Y., Hu, H.L., Fei, J.R. et al. (2015). One-stage human acellular nerve allograft reconstruction for digital nerve defects. *Neural Regen. Res.* 10 (1): 95–98.

113 Koffler, J., Zhu, W., Qu, X. et al. (2019). Biomimetic 3D-printed scaffolds for spinal cord injury repair. *Nat. Med.* 25 (2): 263–269. https://doi.org/10.1038/s41591-018-0296-z.

114 Zhang, W., Chen, J., Backman, L.J. et al. (2017). Surface topography and mechanical strain promote keratocyte phenotype and extracellular matrix formation in a biomimetic 3D corneal model. *Adv. Healthcare Mater.* 6 (5): 1601238.

115 Isaacson, A., Swioklo, S., and Connon, C.J. (2018). 3D bioprinting of a corneal stroma equivalent. *Exp. Eye Res.* 173 (March): 188–193. https://doi.org/0.1016/j.exer.2018.05.010.

116 Wu, Z., Su, X., Xu, Y. et al. (2016). Bioprinting three-dimensional cell-laden tissue constructs with controllable degradation. *Nat Publ Gr* (March): 1–10. https://doi.org/10.1038/srep24474.

117 Kim, H., Jang, J., Kim, H.K. et al. (2019). 3D cell printed corneal stromal analogues for corneal tissue engineering. *2018 IEEE International Conference on Cyborg and Bionic Systems (CBS)*, Shenzhen, China, 2018, pp. 191–194. https://doi.org/10.1109/CBS.2018.8612218.

118 Yi, H., Kim, H., Kwon, J. et al. (2021). Application of 3D bioprinting in the prevention and the therapy for human diseases. *Signal Transduction Targeted Ther.* https://doi.org/10.1038/s41392-021-00566-8.

119 West, J.T. (2019). *3D and 4D Printing in Biomedical Applications*. Wiley.

120 Agrawal, A. and Gupta, A.K. (2019). 3D printing technology in pharmaceuticals and biomedical: a review. *J. Drug Deliv. Ther.* 9: 1–4.

121 Khaled, S.A., Burley, J.C., Alexander, M.R., and Roberts, C.J. (2014). Desktop 3D printing of controlled release pharmaceutical bilayer tablets. *Int. J. Pharm.* 461 (1–2): 105–111. https://doi.org/10.1016/j.ijpharm.2013.11.021.

122 Fina, F., Goyanes, A., Gaisford, S., and Basit, A.W. (2017). Selective laser sintering (SLS) 3D printing of medicines. *Int. J. Pharm.* 529 (1–2): 285–293. https://doi.org/10.1016/j.ijpharm.2017.06.082.

123 Khoury, M., Cuenca, J., Cruz, F.F. et al. (2020). Current status of cell-based therapies for respiratory virus infections: applicability to COVID-19. *Eur. Respir. J.* 55: 2000858.

124 Gouglas, D., Christodoulou, M., Plotkin, S.A., and Hatchett, R. (2019). Cepi: driving progress toward epidemic preparedness and response. *Epidemiol. Rev.* 41: 28–33.

125 Armbruster, N., Jasny, E., and Petsch, B. (2019). Advances in RNA vaccines for preventive indications: a case study of a vaccine against rabies. *Vaccines* 7: 132.

126 Aldrich, A., Kuss, M.A., Duan, B., and Kielian, T. (2019). 3D bioprinted scaffolds containing viable macrophages and antibiotics promote clearance of staphylococcus aureus craniotomy-associated biofilm infection. *ACS Appl. Mater. Interfaces* 11: 12298–12307.

127 Zhao, H., Shen, S., Zhao, L. et al. (2021). 3D printing of dual-cell delivery titanium alloy scaffolds for improving osseointegration through enhancing angiogenesis and osteogenesis. *BMC Musculoskelet. Disord.* 22 (1): 1–12.

128 Caudill, C., Perry, J.L., Iliadis, K. et al. (2021). Transdermal vaccination via 3D-printed microneedles induces potent humoral and cellular immunity. *Proc. Natl. Acad. Sci. USA* 118 (39): 1–8.

129 Lee, J., Baik, J., Young-soo, Y. et al. (2020). Development of a heat labile antibiotic eluting 3D printed scaffold for the treatment of osteomyelitis. *Sci. Rep.* 1–8. https://doi.org/10.1038/s41598-020-64573-5.

130 Cámara-Torres, M., Duarte, S., Sinha, R. et al. (2021). 3D additive manufactured composite scaffolds with antibiotic-loaded lamellar fillers for bone infection prevention and tissue regeneration. *Bioact. Mater.* 6 (4): 1073–1082.

131 Hakobyan, D., Médina, C., Dusserre, N. et al. (2020). Laser-assisted 3D bioprinting of exocrine pancreas spheroid models for cancer initiation study. *Biofabrication* 12 (3): 035001.

132 Wang, X., Zhang, X., Dai, X. et al. (2018). Tumor-like lung cancer model based on 3D bioprinting. *3 Biotech.* 8 (12): 1–9. https://doi.org/10.1007/s13205-018-1519-1.

133 Kim, M.J., Chi, B.H., Yoo, J.J. et al. (2019). Structure establishment of three-dimensional (3D) cell culture printing model for bladder cancer. *PLoS One* 14: e0223689.

134 Mao, S., He, J., Zhao, Y. et al. (2020). Bioprinting of patient-derived in vitro intrahepatic cholangiocarcinoma tumor model: establishment, evaluation and anti-cancer drug testing. *Biofabrication* 12 (4): 045014.

135 Zhu, W., Ma, X., Gou, M. et al. (2016). 3D printing of functional biomaterials for tissue engineering. *Curr. Opin. Biotechnol.* 40: 103–112.

136 Mahfouzi, S.H., Safiabadi Tali, S.H., and Amoabediny, G. (2021). 3D bioprinting for lung and tracheal tissue engineering: criteria, advances, challenges, and future directions. *Bioprinting* 21 (4): e00124.

137 Seoane-Viaño, I., Januskaite, P., Alvarez-Lorenzo, C. et al. (2021). Semi-solid extrusion 3D printing in drug delivery and biomedicine: personalised solutions for healthcare challenges. *J. Control. Release* (332): 367–389.

138 Kačarević, Ž.P., Rider, P.M., Alkildani, S. et al. (2018). An introduction to 3D bioprinting: possibilities, challenges and future aspects. *Materials (Basel)* 11 (11): 2199.

CCC RightsLink

Home Help Live Chat Sign in Create Account

ACS Publications
Most Trusted. Most Cited. Most Read.

Optimization of Experimental Parameters To Determine the Jetting Regimes in Electrohydrodynamic Printing

Author: Ayoung Lee, Howon Jin, Hyun-Woo Dang, et al
Publication: Langmuir
Publisher: American Chemical Society
Date: Nov 1, 2013

PERMISSION/LICENSE IS GRANTED FOR YOUR ORDER AT NO CHARGE

This type of permission/license, instead of the standard Terms and Conditions, is sent to you because no fee is being charged for your order. Please note the following:

- Permission is granted for your request in both print and electronic formats, and translations.
- If figures and/or tables were requested, they may be adapted or used in part.
- Please print this page for your records and send a copy of it to your publisher/graduate school.
- Appropriate credit for the requested material should be given as follows: "Reprinted (adapted) with permission from (COMPLETE REFERENCE CITATION). Copyright (YEAR) American Chemical Society." Insert appropriate information in place of the capitalized words.
- One-time permission is granted only for the use specified in your RightsLink request. No additional uses are granted (such as derivative works or other editions). For any uses, please submit a new request.

If credit is given to another source for the material you requested from RightsLink, permission must be obtained from that source.

BACK CLOSE WINDOW

CCC RightsLink

Home Help Live Chat Sign in Create Account

ACS Publications
Most Trusted. Most Cited. Most Read.

Evaluation of 3D Printing and Its Potential Impact on Biotechnology and the Chemical Sciences

Author: Bethany C. Gross, Jayda L. Erkal, Sarah Y. Lockwood, et al
Publication: Analytical Chemistry
Publisher: American Chemical Society
Date: Apr 1, 2014

PERMISSION/LICENSE IS GRANTED FOR YOUR ORDER AT NO CHARGE

This type of permission/license, instead of the standard Terms and Conditions, is sent to you because no fee is being charged for your order. Please note the following:

- Permission is granted for your request in both print and electronic formats, and translations.
- If figures and/or tables were requested, they may be adapted or used in part.
- Please print this page for your records and send a copy of it to your publisher/graduate school.
- Appropriate credit for the requested material should be given as follows: "Reprinted (adapted) with permission from (COMPLETE REFERENCE CITATION). Copyright (YEAR) American Chemical Society." Insert appropriate information in place of the capitalized words.
- One-time permission is granted only for the use specified in your RightsLink request. No additional uses are granted (such as derivative works or other editions). For any uses, please submit a new request.

If credit is given to another source for the material you requested from RightsLink, permission must be obtained from that source.

BACK CLOSE WINDOW

2

Cellular Requirements and Preparation for Bioprinting

Shalini Dasgupta, Vriti Sharma, and Ananya Barui

Centre for Healthcare Science and Technology, Indian Institute of Engineering Science and Technology, Shibpur, India

2.1 Introduction

The tissue engineering sector has progressed quite a lot in the 21st century and is making steady improvements year by year. Artificial organs were once considered an impossible pipe dream but have now been converted into reality by the miracles associated with 3D bioprinting. Since the first bioprinting was demonstrated in 1988 by Klebe, after Charles Hull developed the first 3D printer using the stereolithography method [1], modern bioprinting has come a long way, with different kinds of biocompatible inks, loaded with cells printed with absolute precision in a sterile environment. Bioprinted scaffolds have not only demonstrated the potential of developing viable artificial organs for transplantation and regeneration but also opened a new path for developing in vitro models for disease research [2]. The current generation of bioprinted constructs includes cells already embedded within the prepared scaffolds. Based upon the different parameters associated with bioprinting, or the material property of the composition, the cells proliferate in a definite lineage and develop into tissues and vessels as required within the construct. Rather than seeding cells on the surface and waiting for cell infiltration, bioprinting live cells, along with the fabricated scaffolds, allow greater control over the cell density in the required area as well as better cell interaction with the biomaterial environment.

While bioink composition has always been optimized depending upon different parameters, the quantity and quality of cells being mixed have also been a major

3D Bioprinting from Lab to Industry, First Edition. Edited by Prosenjit Saha, Sabu Thomas, Jinku Kim, and Manojit Ghosh.

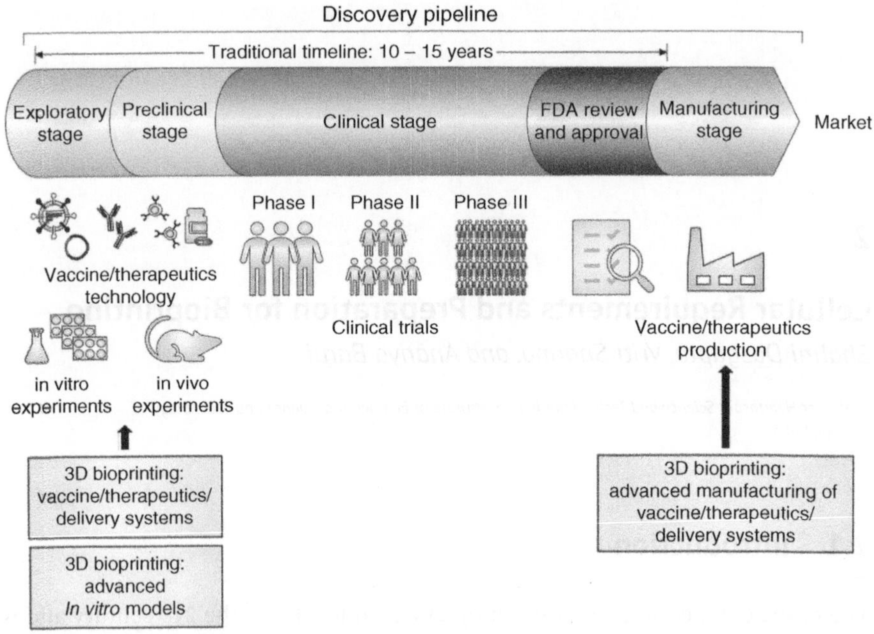

Figure 2.1 Potential of bioprinting different organs. *Source:* Yi et al. [3] / Springer Nature / Licensed Under CC BY 4.0.

factor integral to the success of bioprinted grafts as seen in Figure 2.1 [3]. Every 3D bioprinted experimental setup includes a strenuous and extensive optimization process with proper sterilization, nozzle pressure, bioink density, etc., after which we can obtain the optimum results required for our grafts.

Bioprinting a cell-embedded ink into a definite organ is no mean feat and requires close oversight over a multitude of parameters. Beginning from maintaining sterile conditions to maintaining cell viability, the type of mechanical forces used, or the kind of printing used, providing the required stimuli, differentiation, or potency, many different aspects govern the successful preparation of a 3D printed construct [4]. Here, in this chapter, we shall review some of the parameters associated with cellular requirements for developing bioprinted constructs.

2.2 Types of Bioprinting

Using a similar concept to a printing press, the 3D bioprinting technique uses a methodology that creates discrete layers of different dimensions [5]. The procedure, in three simple parts, is organized according to a generic outline: the preparatory phase in which bioink is being made, the printing phase in which soft tissues

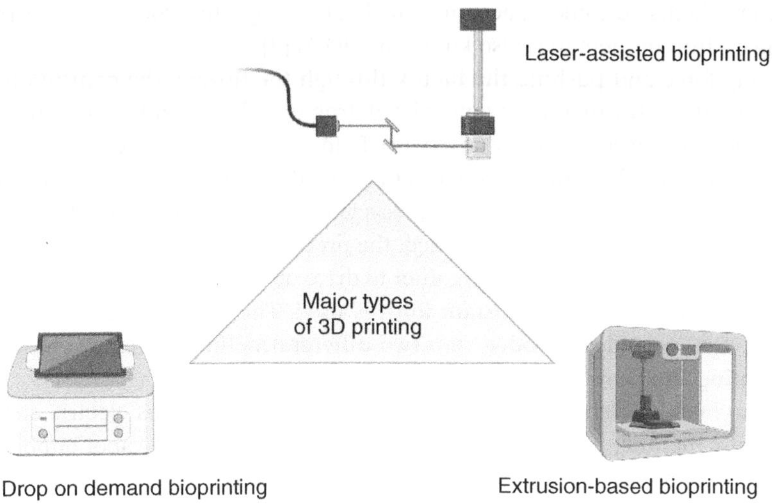

Figure 2.2 Types of bioprinting. *Source:* with Permission of biorender.

are being printed with the help of bioprinters, and the post-handling phase where the embedded cells present begin to grow to develop into a specific tissue [6]. Computer-aided design (CAD) is employed for fabricating the target tissue in the step of reprocessing. To build the specific organ model, various kinds of healthcare-associated imaging methodologies like computer tomography as well as magnetic resonance imaging (MRI) are utilized. A 3D structure that defines the composition and distribution of the materials and cells is then created from the blueprint. By breaking down the model designs into a series of 2D single-layered units, a complex multidimensional tissue model is recreated. Cells and biomaterials have to be added one layer on top of the other during the printing process by utilizing algorithm-assisted precise printing technologies. Post processing methods involve incubating the bioprinted models in a sterile environment [7].

For the purpose of selectively patterning cells and biomaterials to create functional tissue constructions as seen in Figure 2.2, many different 3D printing methodologies have been created, such as inkjet-assisted printing, extrusion-associated bioprinting, and laser-assisted printing.

2.2.1 Inkjet-Assisted Printing

This approach makes use of "bioink," containing a suspension of cells along with a biomaterial having low viscosity that could be placed over a "biopaper" like a scaffold, Petri dish, modeled tissue, etc. [6]. The abovementioned method is a form of noncontact automated manufacturing (AM) printing technique, which is digitally controlled, and could be accomplished in two ways: in a continuous form of printing or drop-on-demand (also known as DOD) [8].

By exerting force and pushing the bioink through the nozzle, the continuous printing method creates an uninterrupted jet of droplets. This extruded ink is then redirected towards the construct by an electric field that is applied next. The extra droplets that do not follow the necessary pattern are diverted in the direction of a disposal tray, where they are collected and recycled. With the exception of the fact that droplets are only formed when needed, the process used to create droplets is identical to continuous inkjet (CIJ). In order to drive out the droplets, a pressure-associated pulse rather than a constant force is used. The printing technologies used by the DOD could be divided into two different methods: pressure-based electric printing and heat-based inkjet printing [6].

In order to heat an element, the thermal-based printer generates electric current via a pulse method. As a result of the pressure generated by the vapor bubble, the heat created rapidly turns the bioink into vapor inside the chamber. The vaporized bioink droplets are then forced onto the substrate through the nozzle aperture. The cells are only exposed to brief bursts of high temperature; thus, their temperature does not rise significantly above the surrounding environment, allowing them to continue to function. Acoustic waves are also used by piezoelectric-based equipment to expel the bioink [9]. This process restricts the usage of extremely dense bioinks because their viscous nature affects the applied pressure given and prevents droplet release [10].

The primary benefits of DOD technology include inexpensive prices, as the devices used are comparably cheaper than commercially available equipment, faster printability, because of the parallel operation of printheads, and good cell viability [11]. However, the DOD approach has drawbacks such as limited options for specific materials, temperature-associated changes during printing, and recurrent nozzle blockage. Hybrid cell printing techniques are now being researched as a means of overcoming such constraints [12]. The use of inkjet-associated printing created 3D models of cartilaginous tissue [13], fabricated neural tissue-like structures [14, 15], blood–brain barrier models [16], renal capillaries [17], muscle models, dermal constructs, lab-on-chip models, vascular models, hepatic tissue [18], and other similar tissue-engineered models [19]. Also, inkjet-associated printing might help with processes like wound healing because it provides a tiered application of various kinds of cell populations according to the depth of the lesion [20].

2.2.2 Extruder-Assisted Printing

Extruder-assisted printing is the most economical, versatile, and extensively utilized method, as well as the most practical additive printing approach. Pneumatic, piston, and screw-driven bioprinting are all included in this technique [1, 21]. In pressure-based methodologies, the release force has been obtained from pneumatic mechanisms, and due to power-based systems which include pistons as well as screw drives, the bioprinting approach has been controlled with perpendicular and rotating forces. The flow of bioink can be directly controlled by a piston-driven system, and screw-associated mechanisms offer better precision-based control and could be implemented to release bioink at higher viscosity [22]. Bioprinters may differ in the direction of layer deposition. The majority of the time, the bioink cartridge is usually bound with a mechanical appendage traveling in the z and y axis above the collection area and traversing in the x direction, allowing for the production of multidimensional printed models [23].

The fundamental benefit of the comparatively modest extrusion speed and pressure is that it avoids the severe circumstances that the cells may experience in other bioprinting techniques (shear, shock, heat, etc.). Pressure extrusion has other advantages, including the usage of different viscous and density gradients for polymer-based bioinks, with a lot of embedded cells, and various cell concentrations [24, 25]. The drawbacks of this approach, however, are the mechanical changes in hydrogels, comparatively limited precision, possible nozzle jamming, and mortality of implanted tissue, mostly because of the produced force applied inside the nozzle [23]. 3D tissues and biological constructs have been created using this bioprinting technique, including kidney [26], liver [27], vasculature [28], 3D printed musculature [29], intestinal constructs, adipogenic tissue models, trachea, vascular models, dermal tissue scaffolds, brain, cartilage constructs, bone, as well as other engineered structures [30, 31].

2.2.3 Laser-Assisted Bioprinting

This technique employs laser beams in a pulsed manner that operates like a ribbon, primarily constructed with glass, with layers of biological material made of cells or hydrogel and layers of energy uptake material (gold or titanium), which are subsequently collected on a substrate facing the ribbon. This methodology helps to print the ribbon gold layer, which is then gathered on the substrate, to produce a high-pressure bubble [32]. High cell viabilities (>95) and a wide range of resolutions are both characteristics of this scaffold-free approach [33]. Since there was no documented change in gene expression even after retaining the culture for a few days, the LAB laser pulse caused no impact on the embedded mesenchymal stem cells (MSCs) [34]. Compared to conventional printing techniques, which are hampered by clogging issues brought on by cells or biomaterials, laser-assisted bioprinting does not require nozzles [11].

Another advantage of the laser-assisted printing methodology is the possibility of printing with biomaterials that have increased viscosity and cell densities. It permits nozzle or needle processing while preventing the material's shear stress for high-resolution printing [35]. The lack of accuracy in the geometry of the microscale structure in other bioprinting technologies can be remedied by laser-assisted bioprinting [36]. Compared to extrusion- and inkjet-based bioprinting, this approach demonstrated greater cell viability [37]. These bioprinters can fabricate a variety of 3D tissue grafts, including skin, bone, and hollow tubular tissue structures [38]. These bioprinters were successfully used to produce the 3D shape of a human liver that mirrored the lobule structure [36].

For laser-based stereolithography, to cure high-speed photopolymers, a UV light source is employed. Liquid polymers become solid when the energy from the UV light forms a covalent bond amid nearby polymer chains [39]. This laser-based bioprinting technique involves four steps: (i) A thin (0.10 mm) layer of photopolymer is applied to the platform, which has been covered and perforated over the organ that needs to be printed. (ii) UV exposure soon causes the fluid bioink to solidify, creating the 3D printed construct's first layer. (iii) Once the initial layer has developed, the loading takes place to reveal the newly added top layer of resin polymer. (iv) The layer of the organ becomes painted as a result of repetitive UV light exposure, which develops links between the prior layers. It takes several iterations of these steps to complete the organ.

2.3 Features Required for Bioprinting with Cells

There are quite a few parameters associated with the successful implementation of bioprinting. Printing fidelity as well as cell function retention are important factors in determining how well 3D printed tissue structures and constructs perform in their intended applications [40]. These crucial traits are influenced by a variety of parameters like sterility, pressure applied, printing speed, culture conditions, viscosity, cell-laden hydrogels, pH, etc. which result in successful bioprinting and make the accelerated development of 3D bioprinting conceivable as observed in Figure 2.3 below [41, 42].

2.3.1 Sterility Parameters

To successfully shift from research to applied clinical practice, sterilization of the used polymers is crucial throughout the initial stages of producing bioinks. One of the best conventionally advised methods of sterilization is treatment with 70% ethanol for disinfection in biofabrication [43]. Yet, as a noncompendial approach, it has little clinical translation potential. Implant-related infections associated with greater patient morbidity and deaths severely demonstrate the necessity of sterility in biomaterials and implants [44, 45]. UV light irradiation is another disinfectant that is used to maintain sterility. This method is also considered

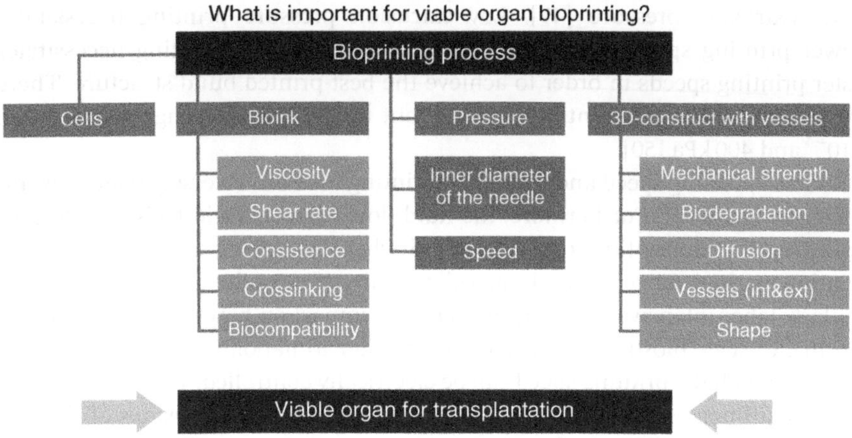

Figure 2.3 Some of the parameters associated with viable 3D bioprinting. *Source:* Klak et al. [41] / MDPI / Licensed Under CC BY 4.0.

Table 2.1 Some techniques for sterilization and disinfection.

S. no.	Name	Method	References
1.	UV (ultraviolet radiation)	UV-C irradiation at 254 nm.	[47, 48]
2.	Filtration under sterile conditions (filtration + Lyophilization)	Membrane filtration (≤0.22 μm)	[46]
3.	Autoclave (solution, autoclaved) (powder)	15 min of sterilizing in saturated steam at 121 °C.	[46]

noncompendial. Autoclaving the solutions is a necessity to attain a contamination-free solution. Lyophilization after sterile filtering at low polymer concentrations yields both sterile and printable bioinks [46]. Autoclaving (liquid and powder forms) and UV light disinfection, two noncompendial methods of sterilization, are compared for efficacy (Table 2.1).

2.3.2 Printing Speed and Pressure

For the efficient management of living cells during bioprinting, these parameters are essential. In terms of cell viability, printing with increased extrusion force results in a lower percentage of viable cells as higher extrusion pressure increases shear pressure at the nozzle, but weakens the membrane cell's integrity and encourages cell death. In order to print live cells, it is imperative to fabricate at a lesser extrusion pressure [49]. Low-extrusion pressure printing necessitates slower printing speeds, whereas high-extrusion pressure printing necessitates faster printing speeds in order to achieve the best-printed build structure. There are varying ranges for printing pressure, but the best-suited range lies between $5 \cdot 10^{-4}$ and 400 kPa [50].

Because printing speed and overall bioprinting time are inversely related, printing speed is crucial. Furthermore, the fluid flow (filament width) is mainly controlled by the bioprinting force and speed while employing an extrusion-based printing method. Since maintaining cell viability after prolonged printing is particularly labor intensive, printing speed has an impact on how well millimeter- or centimeter-scale biostructures are built. Picoliter to nanoliter per minute is the range at which the printing speed can be electrically controlled. The robot motors' mobility influences the total printing time as well as the dimensions of the finished filament or droplet [40].

2.3.3 pH and Osmotic Condition

Culture media are frequently used as a hydrogel solvent in place of water to maintain osmotic equilibrium during the creation of bioink. To avoid osmotic stress (OS), a cell culture medium's ionic concentration, or that of any physiological buffer used to break down hydrogels, is typically adjusted to 290–320 mM [51]. It is possible, nonetheless, for the culture media to react unexpectedly with hydrogels and produce materials or sediments that are bad for the cells. Examples include the development of gel particles when alginate is dissolved in a culture media with calcium ions [52], or the development of precipitates when Carbopol reacts with divalent and trivalent cations in the medium or buffer [53]. To shield cells against osmotic pressure, covering them with gels has been used in addition to the use of an ionic-balanced medium [54]. The additional gel coating gives cells greater rigidity under hypotonic conditions, stopping the cell membrane from rupturing and protecting the inner processes. The application of chemically inert osmotic regulators, like sucrose, during the creation of the bioink, is another method for shielding cells from osmotic pressure [55]. Comparing the bioink control groups, where the hydrogel is dissolved in physiological buffer solutions, it has been demonstrated that proper modulation of the sucrose concentration in the bioink can assist in preserving and promoting cell viability.

Different pH values are relevant for 3D bioprinting based on specific requirements of the bioink, cell type as well as target tissue being produced. The literature indicates that a pH range between 7.38 and 7.87 is ideal for cell development [56]. The pH of the bioink solution, which contains the cells and biomaterials, needs to be carefully controlled to ensure cell viability and bioink stability. The ideal pH for cell survival, maturation, or function may not always be the same as the ideal pH for optimal bioprinting [57]. Although the physiological pH 7.4 is believed to promote sustained cell growth since it is physiologically compatible with most mammalian cells, some bioprinting procedures benefit from an ink with a wider pH range. To illustrate, it has been demonstrated that the optimal bioprinting capabilities of collagen-based bioinks are extremely pH-dependent and are maintained at pH values between 8–9.5 [58]. To offer structural stability, some bioinks go through gelation after printing. It is possible for the pH conditions during gelation and the physical features of the printed object to be altered [59]. For instance, calcium ions are frequently utilized to induce gelation in the case of bioinks made from alginate. The availability of calcium ions and the ionization of alginate are both influenced by pH, which has an impact on gelation [60]. Alginate gelation often requires a pH that is somewhat acidic. After printing, the pH conditions in the post-printing phase are crucial for tissue maturation and functionality. Maintaining a physiological pH range (around 7.2–7.4) is typically important for promoting cell proliferation, ECM production, and overall tissue development during the maturation phase.

2.3.4 Hydrogel Generation

It could be challenging to select the ideal biomaterials for usage as bioink because they need to meet a number of specific requirements, including bioactivity and compatibility with the cells they are meant to be used with, the ability to be deposited (either by extrusion or inkjet), the capacity to conform to predefined forms, and more mechanical qualities that support the printed architecture. These materials may be of natural or synthetic origin [61]. Due to their high water-based content and physio-mechanical qualities, which allow the creation of the ECM like that of the original tissue, hydrogels are the biomaterials that are most commonly investigated in bioprinting [62]. Hydrogels were prepared from a crosslinked polymeric network that can be classed as either natural or synthetic, depending on the origin of its backbone polymeric material [63].

The bioprinting method, tissue type, and chosen cells are all closely tied to the choice of hydrogel (natural or synthetic) as a bioink. Additionally, its formulation needs to meet biological and rheological standards. In the post-printing process, there are three primary forms of crosslinking that are used: heat-based, chemical crosslinkers, or physical (using UV radiation, etc.). The rheological properties and crosslinker quantity should also be optimized [50]. Other aspects like cytotoxic effects, printable resolution [64], mechanical properties [65], in vitro or in vivo degradation capability as well as the impact of byproducts in the growth media, are critical for the replication of organ constructs [66].

The origin of polymers also plays a big role in their usage and preparation. Alginate, collagen, gelatin, and fibrin are examples of natural hydrogels that come

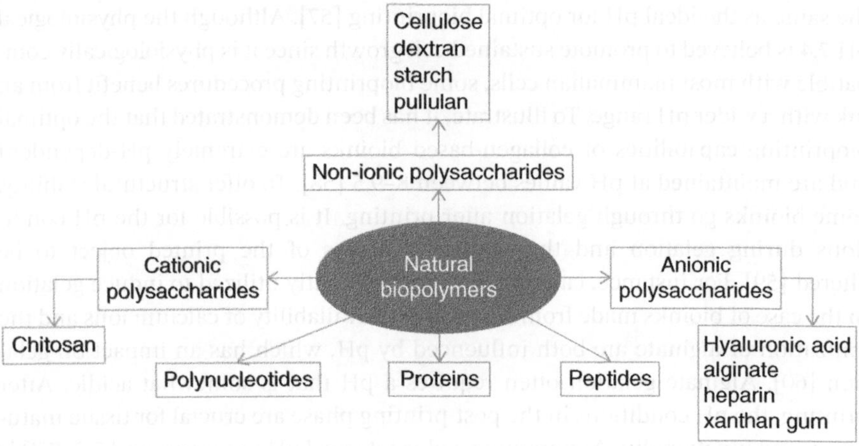

Figure 2.4 Different classifications of polymers. *Source:* Atanase et al. [68] / MDPI / Licensed Under CC BY 4.0.

from both mammalian and non-mammalian sources [67]. Mammalian-derived hydrogels often exhibit high levels of biocompatibility and a human-like ECM, supporting cellular bioactivities. This is mostly because these hydrogels include natural ligands that can perform many tasks, including promoting cell attachment. The origin of polymers based upon their different classification methods has been explained in detail in Figure 2.4. However, the instability, poor mechanical properties, high breakdown rates, and limited tunability of natural hydrogels limit their utility [69].

2.3.4.1 Natural Polymers

Natural polymers have performed numerous crucial roles in a variety of cellular and biomolecular interactions and bioartificial organ generations and maturations, both prior to and following the 3D bioprinting process. These vital functions include providing adequate spaces for cell-based and biomolecular function (such as proliferation, locomotion, segregation, and lineage-based differentiation), sufficient room for ECM-like patterns (such as formation, growth factor release, and direction), biophysical and other chemical signals for tissue and organ morphologies (such as development and layer-wise modeling), and stepwise vasculogenic, neurogenic, and immune-associated development [70, 71]. Such naturally available polymers for usage in additive printing stand out for three reasons: they are highly biocompatible, have low mechanical strength, and degrade quickly. Here are a few examples of usable biomaterial components utilized for 3D bioprinting.

Hyaluronic Acid

The extracellular matrix and the majority of connective tissues contain hyaluronic acid, which is a straight glycosaminoglycan without sulfate bonds [72, 73]. The naturally available hydrogel demonstrates exceptional biocompatible nature and hydrophilic behavior during cell growth. Its weak mechanical performance demands binding with other compounds to meet the physicochemical criteria of additive printing [74]. Enhancing hyaluronic acid's mechanical strength is a good substitute for photopolymerization, which crosslinks materials in the presence of ultraviolet (UV) radiation. For developing a pentanoate-based hyaluronic acid (PHA) solution for 3D bioprinting, Kiyotake et al. [75] utilized the UV-based crosslinking procedure.

Collagen

Type I collagen is another naturally available hydrogel that is widely used. It is noteworthy because it is needed by other tissues to create their extracellular matrix and is a component of musculoskeletal tissue. Collagen also provides properties that are crucial for 3D bioprinting, including biocompatibility [76], biodegradability, and cell adhesion. The photopolymerization (UV) process improves collagen's low mechanical properties, which are similar to those of hyaluronic acids, biodegradability, and cell adhesion. Collagen's low mechanical property, like that of hyaluronic acid, is enhanced by photopolymerization (UV) [77]. But only collagen-based bioink has limited mechanical properties and insufficient stability after 3D printing because of the slow crosslinking generated due to a decrease in temperature to body temperature (37 °C) [78]. To get around this issue, collagen solutions are usually blended with a stronger kind of polymeric material, such as HA or alginate [79, 80].

Gelatin

Collagen denaturation produces the polypeptide known as gelatin [81]. It was investigated for the creation of the additive ink because of its biocompatible properties, including biodegradability, affordability, manufacturing ease, as well as cell attachment [82]. Additionally, gelatin's mechanical properties can be enhanced by crosslinking it with additional compounds like methacrylic anhydride [83]. In an effort to enhance material qualities, many research initiatives have been implemented to develop a bioink that contains an active form of gelatin, known as gelatin-methacryloyl (GelMA). Gelatin could also be homogenized with other materials like silk fibroin [84]. In a study conducted on silk fibroin gelatin, the prepared bioink had excellent print fidelity, which indicated that it had a good chance of working to regenerate cartilage tissue [85].

Agarose

Agarose is a natural polymeric material consisting of the sugars D-galactose, along with 3,6-anhydro-L-galactose [86]. An agarose scaffold limits cell multiplication; however, when mixed with other hydrogels, it can be utilized to have higher cell compatibility. Additionally, it serves as a model for composite 3D tissue cultures [87]. Fan et al. [88] prepared an excellent rheologically combined agarose-matrigel technology for 3D bioprinting. The results demonstrated the advantages of agarose for maintaining the bioprinted construct along with the biocompatibility of matrigel for cell proliferation [88]. Another study in 2018 indicated the physical properties of agarose-combined hydrogels. Alginate as well as many other agarose-associated materials displayed good cytocompatibility, proving their possible usage as bioinks [89].

Alginate

Alginate is also a naturally available polymer made up of the following units – D-mannuronic acid along with L-glucuronic acid [90, 91]. It was extensively utilized in 3D additive printing due to its biocompatible nature, print fidelity, low cost, and adaptability [92]. Its gelation ability due to the divalent cations (such as Ca^{2+}) also improves the stability of the scaffold and lessens the effects of shear force on cell growth, favoring its utilization in extrusion as well as inkjet bioprinting [93]. Alginate's physical properties should be thoroughly evaluated before it is employed in 3D printing as the bioink viscosity of the hydrogel is highly related to its density, molecular weight, phenotypic traits, and cell density [60, 94]. Since pure alginate has low mechanical properties, it can be a challenge to promote cell development in it. These shortcomings can be changed by integrating them with other materials [95]. For use in the extrusion bioprinting procedure, an alginate-cellulose bioink was developed, which demonstrated that the material had excellent shear characteristics, the construct's geometry was maintained, and there had been no cellular damage [96]. Decellularized methacrylated extracellular matrix (dECM), produced from bone extracts, was created and added to an alginate-based bioink by Lee et al. in 2020, which indicated that the material could print 3D cellular structures while maintaining cell viability [95].

Fibrin

The formation of fibrin, which results from the thrombin-mediated crosslinking of fibrinogen, is essential for the maintenance of tissue homeostasis and wound rehabilitation [97]. Thrombin-catalyzed polycondensation processes allow fibrin networks to swiftly self-crosslink and produce comparably stiffer gel matrixes than other natural hydrogels. The thrombin concentrations can be changed to affect the solidification period and the strength of the resultant fibrin hydrogels [98]. However, fibrin gels normally dissolve quickly in 15 days, in line with their function in reversible clot formation. Fibrin has been blended with other biomaterials (such as polyurethane (PU), chitosan, or alginate), which were observed to lengthen the shelf life of the prepared construct in order to catch up to the desired regenerative growth rate [99]. Using extrusion-based 3D bioprinting, Piard et al. [100] proposed an alternate technique to imitate the patterned development of human umbilical vein endothelial cells (HUVECs) along with human MSCs (hMSCs) [101].

Chitosan

Chitin, the primary structural element of crab exoskeletons, is deacylated to produce chitosan, a positively charged polysaccharide [102]. Chitosan hydrogels can be created through covalent crosslinking with glutaraldehyde or formaldehyde, ionic complexation tripolyphosphate, or sodium salts of polyaspartic acid as well as physical crosslinking (although these hydrogels typically have weak hydrogen bonds and can be extremely pH-sensitive) [103]. The most widely employed covalent crosslinkers, however, can pose serious toxicity problems if the remaining crosslinking agent is not completely eliminated from the fabricated structure gel after usage. Contrarily, pure chitosan is not dissoluble at neutral pH, necessitating the usage of different acidic solvents to produce scaffolds that could be harmful to cells. Even though chitosan is not naturally biodegradable, with time, oxidative breakdown of the glycosidic linkages may cause it to degrade back into oligomeric sugars [104]. The solubility issue can be resolved by using tweaked chitosan derivatives, like carboxymethyl chitosan; however, doing so may somewhat offset some of the chitosan's advantages with regard to cell adhesion or antibacterial capabilities. It has been known for some time that chitosan can help improve bone development by inducing more alkaline phosphatase activity as well as increased mineralized calcium presence in certain microenvironments, but this information has just recently come to light [105].

Silk

A naturally occurring protein-based polymer, silk is derived from a variety of Lepidoptera larvae, including silkworms and spiders. The strongest natural fiber, silk fiber, is the ideal combination of high strength, low weight, outstanding durability, and suppleness. Sericin, a glue-like protein with 18 different types of amino acids [106], and silk fibroin, which is generated from the fibers of the filament, are the two unique major proteins that make up silk. In addition to its exceptional mechanical qualities, silk has biocompatibility, thermostability (up to 250 °C), and an array of processing temperatures. Silk fibroin is one of the commonly known materials that are perfect for osteogenic tissue construct designs due to its distinct physical characteristics, higher control over degradability than many other polymers, and strong bioactive properties [107].

Apart from the above natural polymers, there are quite a few synthetic polymers being used extensively for the development of tissue-engineered constructs.

2.3.4.2 Synthetic Polymers

Biomaterial inks can already be made using a wide variety of synthetic polymers. Synthetic polymers, however, typically lack strong biocompatibility but have outstanding mechanical qualities. So, among synthetic polymers, PU, poly(l-lactic) acid, polyethylene glycol, etc. are still the most often utilized biomaterials [108], which are often modified to match the specific tissues' properties. However, due to a number of issues (usage of hazardous solutions, with higher melting points compared to body temperature, and issues with encasing cells), they only constitute roughly ten percent of current biological additive printing technologies. Biological cues that are present in natural ECM and are necessary for promoting cell growth and lineage transition, along with sites for cell-based receptor recognition, are often absent from synthetic polymers. Despite the fact that adding functionalities to synthetically developed bioinks can augment their biological features, tailoring a scaffold's mechanobiological properties requires adaptable side groups [101].

Polyethylene Glycol

Polyethylene glycol (PEG) is a synthetically developed polymer that is generally regarded as being quite safe and whose physicochemical characteristics, for example, polymer chain length as well as structure, are relatively simple to control. In addition, it possesses high mechanical qualities. It is a biomaterial that is not immunogenic and is not toxic to cells (depending on molecular weight) [109]. PEG, along with its derived hydrogels, is perhaps the most researched artificial polymer utilized in soft tissue healing. PEG-based scaffolds by themselves are unable to provide the ideal environment for cell attachment as well as tissue development because they do not have RGD motifs. However, the functionalized hydroxyl chains associated with PEG-diol could be transformed into additional bonds for attachments (such as acrylate, thiol, or carboxyl) that can be adjusted for hydrogel generation or for coupling with biomolecules via physical, ionic, or covalent crosslinking [110].

Polysiloxane

Silicone, also known as polysiloxane, has found extensive use in the therapeutic category due to its biocompatible nature and structural rigidity [111]. It is simple to engineer polysiloxane processing and casting at higher molecular weights to generate solid objects by combining polysiloxane with a curing agent (such as platinum) [112]. The fabrication of a hydrogel structure using silicone and a UV-curable system that has been methacrylated or thiolated is quite comparable to how other polymers are used. Such a structure is excellent for biomedical applications like bioink since it dries quickly and exhibits less toxicity. In order to create 3D objects with good surface characteristics and fidelity, polysiloxane can readily control photo-crosslinking reactions [113, 114].

Polycaprolactone

PCL is a bioinert and biodegradable poly-alpha-hydroxyester polymer. But because it is highly hydrophobic and crystalline, it degrades gradually [115]. Nowicki et al. employed fused deposition printing to develop a composite osteochondral construct with PCL-based biomaterial [116]. PCL is currently employed extensively in the biomedical domain and has strong biocompatibility and a low rate of biodegradation. However, because PCL is hydrophobic, the bioscaffolds that are produced typically have comparatively less viability [117, 118]. The properties of PLA include high ductility, rigidity, biocompatibility, and increased biodegradation. Sun et al. developed two stereoisomeric polymers of PLA in order to alter the stiffness of the scaffolds as well as broaden the applications of PLA [119].

Polyurethane

With its heat-curing property, good biocompatible nature, and physical stiffness, PU is considered to be a potential biodegradable elastomer. The type of solvent it employs allows it to be separated from water-based PU and conventional PU. In the latter, water is utilized for solubilizing purposes, whereas different organic solvents are used in the former [120, 121]. In SLA and DLP printing, PU is employed, and PLA can be used to evaluate how well it degrades. High printing resolution and superior cytocompatibility strengthen the advantages of 3D bioprinting. Chondrocytes are desired for cartilage tissue engineering, and the creation of bone, tendon-based, and neurogenic constructs. It has a distinct, elastomer-like feature that can withstand repeated tensile stress, making it a fantastic implant for musculoskeletal growth [121, 122].

Polylactic Acid

The widely used polymeric bioink PLA is an aliphatic polyester. It has printing capabilities, biocompatibility, and degradability [123, 124]. PLA is used most frequently as a precursor polymer in the FDM process. In order to replace ligaments and nonbiodegradable fibers, filaments generated from muscle and skeletal tissue engineering may enable the utilization of PLA. Because of the acidic byproducts created when PLA breaks down, which inflames tissues and kills cells, PLA's long-term biocompatibility is compromised [125, 126]. Additionally, PLA's fragile behavior limits its application because it lowers its overall strength relative to bone. The restriction can be overcome by mixing it with affordable ceramic components, like calcium phosphate. It lessens acid production and strengthens bone scaffolds [125, 127].

Polyvinyl Alcohol

Due to its superior mechanical strength compared to PEG, and significant water retention, PVA has demonstrated its capability as a stiffer scaffolding material for tissue engineering-based applications [128]. PVA has only a few applications in tissue regeneration because it is a polymer that cannot degrade in a biological environment and does not naturally help in cell attachment. Furthermore, PVA's hydrogen bonding-based crosslinking technique makes it more difficult to modify other polymers chemically to overcome these shortcomings. To get around this issue, composite PVA hydrogels have been created by physically encasing natural ECM elements like collagen or gelatin [129]. In the later stage, a polyvinyl alcohol-polymer functionalized with tyramine was produced, which was photopolymerized for crosslinking and could be degraded via ester hydrolysis, which connects the PVA–Tyr groups. Gelatin can also be added through tyrosine-based bonding [130].

2.3.5 Culture Duration and Conditions

The cell culture conditions need to be optimized in a sterile manner to sustain the 3D bioprinted product for in vitro testing. Unlike traditional 3D printing materials, bioinks must have the following characteristics: crosslinking substances made of living organisms that are safe and can be modified by cells after printing and temperatures that are not higher than what is considered as physiological circumstances of weak gelation [93]. Different cell culture types require different amounts of time for proper cell viability and cell development. The biological niche has an extensive range of stimuli, including physico-chemical and biological properties, to govern cell attachment, proliferative growth, and differentiate into different lineages [131]. In vitro, cell expansion and differentiation begin with cell isolation from the tissue sample. Depending on the application, the type of cells employed may be stem cells or primary cells tailored to a patient's or an organ's demands. The bioink is then generated using the separated cells, growth factors, and bioprinting ingredients in accordance with the physiological temperature, pH, and printing structure criteria [33]. By specifically planning the surface morphology of scaffolds, effects on cell behavior that were obvious have been proven [132].

Incubators or bioreactors are frequently used to keep the printed structure during the culture phase, which offers the best conditions for cell proliferation and tissue formation. Maintaining a certain temperature, humidity, oxygenation, nutrition delivery, and mechanical stimulation are a few examples of these circumstances. The printed structure's cells can multiply, differentiate, and arrange themselves into useful tissue during the culture period. Additionally, it gives the cells time to interact, create extracellular matrix, and develop functions unique to different tissues. The time frame is frequently chosen in accordance with the desired tissue characteristics, such as mechanical strength, cell density, or functioning, which may take a certain period of time to develop.

2.3.6 Rheological Properties

The key rheological factors that affect how 3D printed tissue structures turn out include fluid behavior, viscous nature, shear-based force, and elastomeric properties [133]. Hydrogels' rheological properties, which demonstrate their resistance to deformation-based forces, are influenced due to the relationship in the middle of shear stress as well as shear rate. Typically, based on the flow, flow behavior is categorized as either Newtonian or non-Newtonian [134]. For 3D bioprinting, it is essential to characterize the bioink's flow behavior. The anticipated bioinks were said to have shear-thinning behavior, which made it possible for them to flow freely without clogging up along with enhancing the printability and steadfast nature of bioprinted scaffolds [135, 136]. Most hydrogel-based bioinks exhibit non-Newtonian flow [134, 137].

The viability of the cells may be compromised by higher viscosity, whereas lower viscosity would provide cells with a more hospitable environment but make printing more challenging. Because excessive viscosity might lead to nozzle blockage, the flow can also be controlled depending on the nozzle tip diameter. Viscosity for different compositions of bioink could be regulated by adjusting the density, concentration of different polymer components, additive mass, heat, and crosslinkers present [138].

The viscous nature of the bioinks helps to identify the shear stress induced by the 3D printing operations, which may affect cell survival and proliferation. This is owing to the likelihood of cell damage at higher shear stress levels [139]. Hydrogel-based bioinks with less viscosity at medium pressure levels are favored because they enable optimal printability as well as the capacity to sustain viable cells both during in-lab and clinical trial-based testing [134, 139–141].

The kind of hydrogel, molarity, and crosslinking properties can have a significant impact on viscoelasticity, which can, in turn, have a significant impact on properties including cell-scaffold-based crosstalk, porous nature, and the degradability of 3D printed scaffolds. Moreover, it controls cell growth and differentiation, as well as the structural stability and integrity of hydrogels [142]. It is significant that bioinks with higher storage moduli have greater gelation behavior, offering structural stability but at the risk of clogging and filament breakage. Contrarily, while easier to work with, hydrogel-based bioinks with larger loss moduli may result in the development of less stable 3D structures [143].

2.4 Bioprinting Methodologies for Cell Expansion and Proliferation

To create bioidentical tissue for varied uses, 3D bioprinting accurately layers cells, biological scaffolds, and growth stimulants [144]. The scaffold, which is a highly porous 3D substrate, is a vital component in the biomedical field. Cells have been grown in in vitro conditions before incorporating them into the scaffold. The scaffold can offer a biocompatible surface on which cells can attach, develop, and survive while forming the biological niche, which is made up of various functional proteins and saccharides, that form the bioprinted organ [70]. The

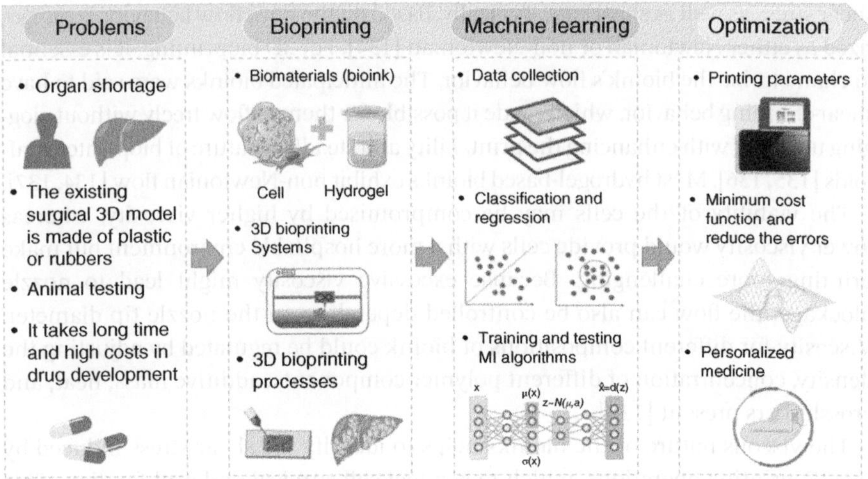

Figure 2.5 Development of organs from the 3D bioprinting process. *Source:* Shin et al. [146] / MDPI / Licensed Under CC BY 4.0.

inclusion of biomimetic components into a bioprinted construct dynamically impacts cellular attachment, motility, proliferative growth, and function. The development of growing and differentiating cells in a construct is made possible by materials, which have a substantial impact on cell attachment, cell size, and cell shape [33]. Numerous biological processes, including cell division and proliferation, are significantly influenced by hydrostatic pressure (HP), OS, and many other factors [145]. Figure 2.5 provides a concise explanation for the development and fabrication of organs from the 3D bioprinting methods.

In its most fundamental state, bioink is the composite material that has been applied in different tiers throughout the bioprinting process. The essential elements of bioink are a supporting structure, additives (growth stimulants, signaling chemicals, etc.), and biological components. The scaffold's characteristics are imperative in sustaining the physical requirements of the printing methodology. In addition to being protected from the mechanical and thermal stresses of printing, cells must be able to adhere firmly to bioink scaffolds. Furthermore, they must encourage cellular development and proliferation while avoiding altering the cell phenotype until absolutely necessary [147, 148]. The niche of the native tissue is frequently replicated in additive manufacturing applications that demand scaffold-promoting lineage transition. This mimicry extends to mechanical, chemical, and structural characteristics [149]. The kind and order of integrin-binding proteins are two of the most crucial ECM structural factors that affect stem cell fate. Prowse et al. [150], in their experimental studies, found that the form and configuration of integrins present in a scaffold can affect stem cell development [150]. In general, soft, pliable constructs (0.1–5 kPa) induce the growth of neurogenic and adipogenic tissue, while robust, rigid scaffolds (8–30 kPa) improve the development of tendons, chondrogenic, and osteogenic tissue [23, 151–154]. Bioprinting techniques include inkjet 3D bioprinting which prints several droplets of bioink on a surface using heat and pressure-based or magnetic forces. Extrusion-based bioprinting is a technique wherebioink is released via a nozzle by physical or pressure-based forces whereas in stereolithography, bioink is cured layer-by-layer using digital light. On the other hand, laser-assisted 3D bioprinting propels bioink and cells to a receptive substrate by vaporizing them on the bottom of a ribbon. The proportion of viable cells with the maximum rate of expansion and proliferation while employing laser-assisted 3D bioprinting has been identified [144]. Each bioprinting technique has a distinctive impact on the bioink properties. Every printing method has intrinsic stressors that can affect stem cell differentiation. Mechanical pressure, which is prominent in inkjet-associated printing, was shown to affect MSCs' ability to develop into cartilage and bone [155]. It has been demonstrated that the shear stresses present during extrusion printing affect the development of cells towards bone and vascular tissue [156]. Alternately, a comparatively better force-independent bioprinting technique, such as laser-associated printing, may be used to maintain the multipotent nature and make use of other strategies to encourage cell specification.

2.5 The Impact of Bioprinting Process Conditions on Phenotype Alterations

Bioprinting technologies have a much more significant impact on the morphological and physiological aspects of cell growth as compared to what was known before. Depending upon the parameters required to be kept in check during the process, and even the type of bioprinting used, phenotypic alterations and lineage differentiation could be induced in various types of stem cells [23]. Just as stem cells within the body differentiate and develop into other specialized cells due to various cues, in vitro stem cell differentiation can also be induced by various bioengineering strategies. These strategies can include the type of bioink used, the pressure and stiffness induced, and the porosity provided in the prepared construct, along with the hormones and other additives provided. Bioprinting can include the development of a bioinert scaffold designed specifically to maintain the stemness of the cells or the preparation of a bioactive scaffold that guides a specific lineage from the stem cells [157].

It has already been well established that phenotype alterations come from stem cells differentiating towards a specific type of cell. Primary cells can also change their phenotype depending upon the internal stimuli provided by various biochemical and biomechanical cues (mostly for cancer). For example, various tumor models have been prepared by 3D printing, transitioning from cell type to differentiated cell type over the course of any additive provided or mechanical cues present [158]. Various tissue-engineered grafts have also been developed where, by controlling the pressure, shear stress, or viscosity of the bioink, a stem cell could be induced to develop into any kind of cell as required. The parameters of the bioink, from its chemical to rheological properties, can also have a major influence on the alteration of phenotypes. Many research groups have utilized different bioprinting approaches made of innovative polymers and materials to induce osteogenic or chondrogenic phenotypes [93, 159, 160]. The major reasons for differentiating stem cells in the graft rather than directly using the primary cells themselves are because of the robust nature of the stem cells, easier growth, and tolerance in vitro as well as the immune-tolerated properties of undifferentiated MSCs in vivo [161, 162]. This allows the cell-laden grafts to differentiate into the chosen cells when placed in the affected region without fear of immunogenic rejection. Depending on the different kinds of factors utilized, bioprinting has an important impact on differentiating cell lineages and the development of different phenotypes.

2.5.1 Bioprinting Techniques for Stem Cell Differentiation

One of the major factors associated with bioprinting that can cause differentiation is shear stress and mechanical stress [23]. Depending upon the type of stress and the amount of stress caused by the bioprinting technique, pluripotent stem cells could be induced toward a specific lineage of choice as per the requirements of the prepared tissue. As mentioned previously, specific kinds of shear stress could induce endothelial and osteogenic differentiation, while mechanical stress could enforce chondrogenic and, to an extent, osteogenic differentiation as well [163]. This has been observed even more in the 3D printed scaffolds prepared by extrusion-based bioprinting, as compared to less force utilizing techniques such as laser-assisted and inkjet bioprinting. Adipogenic differentiation for scaffolds has been observed via the addition of various growth factors that are temperature- and shear-sensitive and thus can be induced by the other two techniques as well [164]. In order to mitigate and reduce the effect of pressure as well as shear force on the stem cells, bioinks of different viscosities are carefully calibrated and chosen out of shear-thinning polymers. Depending upon the requirements of the artificial tissue graft being prepared, the cell-laden scaffolds can be fabricated in a controlled environment, only differentiating when needed. The microenvironmental cues required for cell growth, its morphological shape, and its physiological behavior can all be controlled by modulating various parameters of bioprinting.

2.5.1.1 Bioprinting Strategies for Cellular Environment Alterations

There have been many different methodologies used for bioprinting in order to promote a biocompatible microenvironment for proliferative growth and tissue formation. They can include the usage of various natural and synthetic biomaterials that promote cell growth, maintain stemness, and promote lineage transition. Different approaches for the alteration of the cellular microenvironment towards a more biocompatible and proliferation-friendly aspect include the viscosity and chemical composition of bioinks, pH, ionic properties, thermal properties, crosslinking techniques, etc. [159]. The topography, micropatterns, or pores present on the scaffold surface and the various mechanobiological cues associated with the scaffold can influence the microenvironment of the cell-seeded graft as well. The microenvironment of an artificial graft is also influenced by different cytokines, small molecules as well as other additives present in the bioink itself, whether encapsulated or loaded. In the following Table 2.2, the impact of the above cues present with bioprinting on the microenvironment of the cells is observed and understood.

2.5.1.2 Bioprinting Strategies for Cell Behavior Modulation

Many different bioprinting strategies have been utilized in recent times in order to influence the morphological and physiological behavior of cells. From techniques improving cell adhesion to the spatial arrangement of cells in the bioinks for specific cell–cell and cell–substrate interaction for differentiation and tissue development [181], quite a few bioprinting methodologies have been utilized to successfully achieve the objectives required for the specified graft. Cell morphology could be controlled as per the needs associated with the tissue and can be modified by the presence of various external and internal cues provided by bioprinting. Just like how various additives [3] and physicochemical changes can be implemented within the bioink to alter the microenvironment for stem cell differentiation, various other methodologies can be utilized that would directly impact the physiological behavior of cells themselves. It could be due to the biochemical receptors interacting directly with the cells or the mechanical and physical shear stress affecting the cells themselves and directing them toward a specified lineage [182]. The directed transition of stem cells (towards a specific phenotype) seeded on a scaffold over time may impact the construct's shape and other properties, matching the theory of 4D bioprinting.

Many different bioprinting techniques often provide controlled stimulation so as to not disturb the potency of the undifferentiated cells. Inkjet as well as laser-assisted bioprinting techniques utilize comparatively less shear and mechanical stress and have been considered for use in printing immunotolerant scaffolds with stem cells maintaining their potency as compared to unprinted cells [131]. The addition of several chemicals to the bioinks has been observed to directly impact

Table 2.2 Mechanotransduction cues caused by different factors and their impact on microenvironment.

S. no.	Types of cues	Category	Impact on microenvironment	Reference
1.	Biochemical additives included with bioink	Small molecules	Small molecules can be used to target specific steps in the signal transduction pathway to affect cell differentiation. Their presence in ECM may support in continuing their self-renewal or provide a definite directive in terms of lineage differentiation.	[165]
		Growth factors	Includes a variety of cytokines and trophic hormones. Involves impact on growth, proliferation, inflammatory response, repair, etc. Affects various pathways and may influence differentiation in a controlled manner.	[166]
		Synthetic peptide molecules	Peptide-based biomolecules are known to mimic physiological conditions for specific biomaterials to provide the required impact – either self-renewal or proliferation.	[167]
		Microvesicles and exosomes	Single membrane vesicles that help in cell–cell communication. Helps in protein, small molecules, and miRNA transport. Recently coated on biomaterial scaffolds to help support stem cell differentiation and growth.	[168]
2.	Nucleotide-based additives included with bioink	DNA	Various DNA functionalized hydrogels, DNA-loaded scaffolds; and polypeptide DNA scaffolds have been prepared which have influenced the microenvironment for easier cell viability, potency, and lineage transition.	[169]
		Micro RNAs	Encapsulated mi-RNAs via exosomes or other microcarriers have been utilized to deliver to the ECM of the bioprinted graft to ensure cells follow a directed cycle of differentiation or maintain self-renewal.	[170]
3.	Bioink	Natural polymers	Natural biomaterials like collagen, gelatin, hyaluronic acid, impact stem cell development and are components of the ECM themselves.	[171]
		Synthetically created polymers	Artificial biomaterials like PLA, PGA, PCL, etc. often have additional RGD motifs embedded to provide a biocompatible microenvironment to help in stem cell adhesion and proliferation.	[172]
		Bioglass	Bioglasses are made of different ceramic and ionic components that have been known to induce angiogenesis in bioprinted grafts by directing stem cell fate via their ECM.	[173]

		Bioceramics	Various materials like hydroxyapatite added with other bioink additives can influence osteogenic differentiation and help in providing an odontological environment.	[174]
		Microcarriers	Microcarriers support adhesive cell types for being carried in the bioinks. They are made of bioinert chemicals and can be used to efficiently transport cells. They can be chemically modified to direct toward a particular differentiation strategy or to carry hormones and other growth factors to be added to the microenvironment.	[175]
4.	Topography		Specifically added ridges, controlled pore formation, micropatterns and grooves added by different bioprinting techniques form a crucial part of the cell microenvironment and may influence differentiation and tissue as well as capillary formation. Changes in substrate topography radically affect the ECM and impact stem cell morphology and physiology.	[176]
5.	Viscosity		Changes in the viscosity of bioinks can affect the ECM environment and stem cell lineage and differentiation. Low viscosity bioinks often require crosslinking mechanisms which may have cytotoxic side effects. High viscosity can often lead to increased shear stress and frictional force affecting the cell fluid interface.	[35]
6.	Printing method	Extrusion	High-pressure force-based printing. Often can lead to the differentiation of cells based upon the mechanical and shear stress given to be microenvironment.	[177]
		Inkjet	Comparatively gentler method. This can lead to the maintenance of stem cell potency and self-renewal.	[28]
		Laser-assisted	Same as inkjet bioprinting. However, easier to print in different micropattern designs which form a crucial part of the microenvironment.	[178]
7.	Nozzle shape		Change in nozzle shape contributes to change in strain rate and thus affects the biophysical microenvironment.	[179]
8.	Rheology		Most bioinks are prepared to have shear thinning properties to have a consistent flow rate without clogging and to be able to print high-fidelity shapes after deposition.	[180]

cell lineage and behavior. The presence of β-glycerophosphate, dexamethasone, and insulin has been utilized to provide osteogenic differentiation for stem cells, whereas the addition of indomethacin, isobutyl methylxanthine, and dexamethasone leads to adipogenic differentiation [183]. If a DNA-methylating agent like 5-azacytidine is added with the bioinks, stem cells seeded on the scaffolds can move towards a cardiomyocyte-like phenotype [184]. Many other bioprinting techniques, like extrusion bioprinting, were used where osteogenic stem cells mixed with methylacrylamide gelatin scaffold showed osteogenic development as compared to unprinted cells kept in the same media [185]. Other kinds of cells like embryo-extracted stem cells, were bioprinted on an RGD-coupled alginate scaffold, where it was observed that due to the subsequent printing technique and further incubation in an appropriate medium, the cells showed stages of differentiation. Alginate-gelatin bioink, mixed with mouse planta dermis extract, was found to promote epidermal progenitor lineage from stem cells, inducing the development of sweat glands. They did not affect already-differentiated cells [23]. The physical characterization of the prepared construct was observed to resemble that of mouse dermis and thus was seen to affect stem cell behavior. Certain cells in a particular bioprinted construct had an elongated shape with a tendency towards higher proliferation as compared to cells added to the scaffolds without bioprinting.

2.5.1.3 Bioprinting Strategies for Genetic Modulation and Transcriptomics Variation

Various bioprinting strategies include providing a simulated environment with added pressure, shear, and mechanical stress or with biochemical additives present in the bioink. The viscosity, type of crosslinker present, temperature, and rheological properties of the bioink were all seen to have a large impact on the genetic and transcriptomics variation of the developed tissue [64]. Many different bioprinted constructs have been used as modes of efficient gene delivery to the cells for gene modulation as well as providing a specified pathway for differentiation. Different bioprinting parameters can be modified to induce changes at the genomic level and have been utilized to promote vascularization, osteogenesis, tissue engineering, cancer, etc. [186]. New methodologies have been implemented to print gene-encapsulated 3D printed nanostructures on the surface of a construct in order to promote a genetic transition of the stem cells toward a specific lineage or pathway. These methodologies may include the traditional techniques of gene therapy like splicing, silencing, editing, controlling, or inactivating certain genes in order to obtain a specific result. There are two major methods of gene delivery to any target cell: (i) viral vectors and (ii) nonviral carriers [187]. Many new research groups have experimented with bioprinting these nonviral carriers along with the scaffolds in order to effectively invade the target cell and modify the specified gene.

One of the major issues with gene therapy is the efficiency of the release mechanisms governing the delivery of these targets to the cells in a timely manner [188]. Thanks to 3D bioprinting, a lot of issues surrounding the kinetics of gene delivery have been solved to quite an extent. Bioprinting provides a stable, personalized approach with fully controlled parameters associated with temperature, stability, viscosity, biocompatibility, and cell growth [1]. It helps to improve conventional dosage issues and allows for the incorporation of nucleic acids, growth factors, etc. within the scaffolds, allowing direct access to the embedded cells present.

Additive bioprinting has already demonstrated its possibilities in a recent study where a construct created specifically to treat bone defects had embedded specific nucleic acids bioprinted within to help promote osteogenic differentiation, proliferation, and mineralization in the stem cells [160]. Bioprinting is also extensively utilized to study cancer and develop lab-based tumor constructs in order to understand which particular pathways can be modulated via genetic variations to reduce invasive growth.

A prime example is when fabricated PLA scaffolds were applied with rhodamine-associated PAMAM dendrimers and one miRNA gene (FAM-mir-503) was embedded within these scaffolds as seen in Figure 2.6 [189]. The researchers were capable of studying the impact on cell growth thanks to the internalization of the fluorescent dendrimers, which acted as nontoxic delivery vectors for their miRNA cargo. Many studies have utilized 3D bioprinting for increased efficiency in

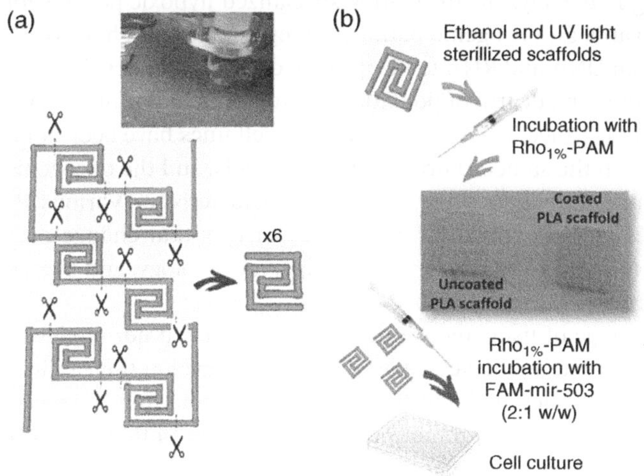

Figure 2.6 Bioprinted PLA applied with PAMAM (showing fluorescence) and embedded miRNA [189] where (a) denotes the bioprinted design, (b) denotes UV and ethanol sterilization followed by PAMAM dendrimer after which cell culture has been conducted. *Source:* Paolini et al. [189] / Springer Nature / CC BY 4.0.

transfection as compared to standard in vitro protocols. A research group found that porcine aortic endothelial cells had over 90% post-transfection cell viability in their 3D bioprinted scaffolds, whereas the in vitro transfection efficiency of unprinted cells was close to 10% [190]. Thus, gene modification and subsequent transcriptomic changes can be observed and obtained in a controlled manner via different bioprinting strategies.

2.5.2 Bioprinting Techniques for Tumorigenic Differentiation

In vitro, carcinogenic constructs have been utilized for understanding the pathology of cancer for quite some time now. 2D models often depict a very simplistic version of cancerous growth and are unable to recreate all the aspects that are associated with a cancer microenvironment. Cancerous growth is not only affected by the proliferation of oncogenic cells but also by changes in the ECM and the dynamics of cell–cell adhesion [191]. Tumor formation often leads to localized hypoxia-like pockets surrounded by dense tissue, which in turn increase proliferation and cause uncontrolled growth of the cells. Various aspects of tumorigenic growth provide specialized environments via different substrate biomaterials to activate different pathways for oncogenic development. As bioprinting has already shown its mettle in developing artificial tissues and even organs, creating specialized 3D and even 4D tumor models would be highly possible via the abovementioned methods [22]. From providing an ECM with the desired parameters of porosity, tensile strength, density, etc. to creating specialized hypoxic pockets for cells to proliferate towards an oncogenic pathway, to even providing hormones, growth factors, and even DNA and RNA to incorporate construct for developing a tumorigenic tissue, there is no dearth of potential for preparing tumor models via bioprinting [192]. Various cancer cell lines and healthy cell lines have been compared by seeding them on the same 3D printed tumor models, and the oncogenic cells developed into tumor-like tissue with all its noted characteristics. Various 4D bioprinting techniques include growing cell-laden constructs that change over time to form realistic tumor tissue, changing the shape, morphology, and overall structure of the construct [193].

Another major advantage of using bioprinting techniques for tumor models is the ability to print and create blood vessels to sustain and develop the carcinogenic tissue [194]. Tumor tissues are identifiable by their tortuous and leaky vessels, which are necessary for their continuous growth. Therefore, with the help of bioprinting, it is possible to recreate a carcinogenic environment to improve our understanding of the pathways and to observe the mechanism of action for the chemotherapeutics being developed without having to deal with patient risks. With the different kinds of bioprinting techniques present, it is very much possible to control the spatial properties of the matrix, along with manipulating its density and biocompatibility, as well as printing "biomolecular payloads" to promote uncontrolled cell proliferation if necessary. Within a specifically designed 3D environment, it is easy to observe the different integrin and laminin-associated interactions and how they have been affected by cancer progression. Thus, a more comprehensive in vitro evaluation of the tumor tissue can be conducted, and it can be implanted in vivo to recreate the impact of the said tissue in the organism instead of inducing and destroying healthy tissue.

2.5.2.1 Bioprinting Strategies for Oncogenic Cell Growth

Based on the many different techniques present in the bioprinting strategies, quite a few are borrowed from basic tissue engineering approaches. To understand the biomechanics and physiological processes associated with cancer, we have to first understand how cancer cells begin to form and spread via metastasis [195]. Most cancer studies focus on evaluating the metastatic nature of tumor models and the invasive migration patterns of cancer cells. Tumor formation is a very complex process, and its development can depend on a plethora of various pathways and physiological factors all occurring at the same time. Figure 2.7 demonstrates how the cell microenvironment consisting of ECM undergoes various changes during metastasis in cancer.

Cancer cells invade the ECM and the underlying tissue via changes in the integrin and laminin dynamics, but also through extra collagen and ECM deposition in inaccurate locations, twisted and leaky blood vessels, and increased interstitial pressure on the tumor tissue [191]. 2D cancer models can be used to detect cell invasion and molecular dynamics to a certain extent, but they cannot accurately predict drug efficacy or range of action. 3D models have been utilized to a greater

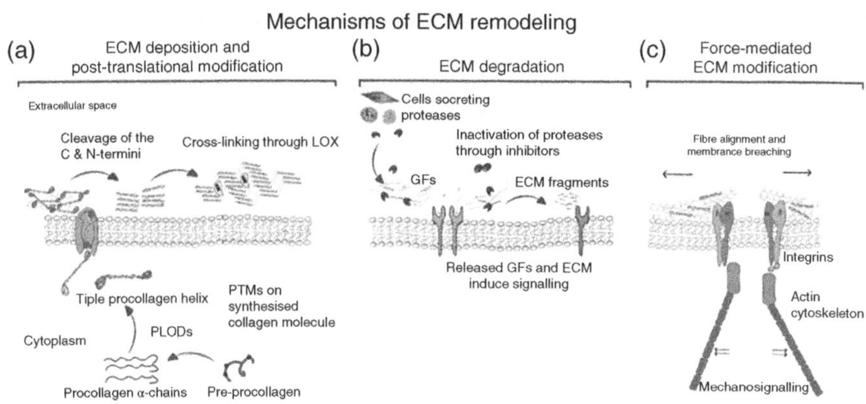

Figure 2.7 Mechanism of ECM remodeling during cancer. (a) denotes the process of ECM deposition and crosslinking for different proteins like collagen. (b) denotes degradarion of ECM via protease interactions. (c) denotes ECM modification due to mehanosignaling. *Source:* Winkler et al. (2020) / Springer Nature / CC BY 4.0.

extent and have been able to provide a lot of insight in terms of the complex dynamics of ECM and how it affects oncogenic cultures [196], along with the changes affecting drug delivery,. Inbred tumors in animal models are still the gold standard for obtaining a complete understanding of malignant formation as well as the challenges associated with the proposed drug mechanisms.

However, thanks to the strides made in the field of bioprinting, oncogenic cell growth may be observed in a proper in vitro carcinogenic environment with the added physical and biochemical cues required for tumor progression. Bioprinting can be used to mimic accurate cancer microenvironments with precision-controlled cell position and modified hydrogel parameters. Bioprinting could also revolutionize cancer research with its ease of reproducibility, thus allowing us to get tunable cancer models that can provide us with reproducible and safe results from understanding the molecular pathophysiology involved to test out potential anti-cancer therapeutics with results comparable to in vivo assays.

Bioprinting can also be utilized to create precise microfluidic channels on the surface of printed ECM-like substrates for vascular channel-like development for oncogenic cell growth. Jung et al. [197] prepared a high throughput cancer metastasis model to observe cell migration and invasion by adding a combination of cancer cell lines to PEG maleimide hydrogel crosslinked with thiol activators [197]. The addition of different breast cancer and ovarian cancer cell lines printed with the 3D printed scaffolds was evaluated for migration distance depending upon the different elastic moduli of different sample constructs, along with immunostaining and gene expression assays. Thanks to precise bioprinting technology, different cell lines indicated different kinds of results on hydrogels with different physical parameters. This allowed us to estimate the kind of tumor model that can be designed based on the type of cell line in mind. Breast cancer cells printed within the construct allowed for a very viable in vitro drug screening assay where cell migration was impeded with different experimental drugs and the results were observed. The cells printed within these constructs can also be retrieved quite easily using trypsin or any company-specific standardized solution and analyzed further for any alteration in their molecular dynamics.

Oncogenic cell growth and differentiation were also observed in quite a few more studies thanks to the development of biocompatible bioprinted constructs. One of the major features of metastatic growth is the recruitment of adipocytes and fibroblasts. Adipogenic stem cells often are recruited by the tumor microenvironment, where they cross-talk with the carcinogenic cells to form cancer-associated adipocytes (CAA). These adipocytes release fatty acids to be utilized by the cancer cell's mitochondria as energy for growth. Horder et al. [198] fabricated a 3D printed breast cancer construct including stem cells and observed a subsequent transition towards CAA, which increased oncogenic cell growth, remodeled the ECM, and contributed to the growth of tumor tissue [198]. Fibroblasts are often recruited by the cancer cells to become cancer-associated fibroblasts (CAF) and promote specific cell–cell, and cell–ECM dynamics that instigate tumor progression. Bioprinted a lung carcinoma xenograft model with fibroblasts and observed the CAF-promoted resistance to anti-PD1 and anti-PDL-1 as well as CTLA-4 immunotherapy [199]. Thus, oncogenic cell growth and differentiation as well as their impact on other cell lines, can be studied properly via bioprinted models in a much more distinct manner as compared to 2D cancer models.

2.5.2.2 Bioprinting Strategies for the Development of Tumor Models

As observed previously with oncogenic growth, bioprinting has taken on a major role, from simply helping to observe cancer cell interaction to recreating the cancer environment with as much accuracy as possible. With the help of the development of spheroids and cancer stem cell niches, bioprinting has taken tumor development to a whole new level. Development of a fully functional tumor tissue included not only the invasive migration of stem cells but also anastomosis and extravasation, along with the presence of hypoxic and necrotic sites. All this can be created or observed in progression via bioprinted models, which replicate the malignant microenvironment down to the deepest molecular level. Zervantonakis et al. [200] allowed the entire phenomenon of extravasation caused by tumor endothelial cells to be observed in a breast adenocarcinoma bioprinted model by affecting the endothelial barrier [200]. Cancer stem cell niches and their subsequent migration patterns had to be studied in low oxygen and low adherence suspension-like culture models, which were expensive, often impractical, and time-consuming. However, Suzuka et al. [201] demonstrated that bioprinted double PEG hydrogels could be utilized to convert tumor-differentiated cells to cancer stem cells in the 24-hour timeframe for many major oncogenic cultures [201]. An overall increase in stemness as well as an increase in proliferation were observed in the above study.

Spheroid cultures have also been utilized in the last ten years to develop a tumorigenic microenvironment. They develop in wells of non-adherent substrates, where they start to aggregate and form a necrotic core. However, long-term maintenance of spheroid cells is rather difficult and has its own challenges. Bioprinting has been able to create biocompatible and porous hydrogel-like structures where spheroids can be created with hypoxic pores and get access to nutrients just like in the cancer microenvironment in vivo, allowing the tumor structures to develop and become a more realistic interpretation of the malignancy [202]. They have been observed to be superior in terms of vascularized growth, development, and proliferation in comparison to both monocultures and multidimensional coculture models.

2.6 Discussion

Bioprinting has thus proven as a useful methodology in the biomedical field. It has provided a scalable method for developing specialized artificial tissue with a multitude of layers and as many cells as are required to be embedded within the tissue. Bioprinting has also allowed for a variety of designs, channels, and other patterns to be added to the artificial construct as per demand, allowing a unique topography to develop as required for the tissue or organ in question. From developing disease models to preparing artificial tissues for replacement or repair for injuries, bioprinting has revolutionized our in vitro studies, providing accurate results comparable to in vivo assessments. The different types of bioprinting have a very important part in the artificial tissue or in vitro model that is being created. Each technique has its own pros and cons and cost-effectiveness, and the mechanical parameters and method of deposition also change depending on the kind of method. However, a lot of parameters, both physicochemical and biological, have to be kept within a limited range for a bioprinted model to function properly. These include various mechanical variables such as pressure, temperature, pH, osmotic conditions, etc. For bioprinting to occur properly, it is imperative that the cells combined with bioink or added separately do not face more undue stress than necessary. Depending upon the variation in the above parameters, the same kind of cells on the same type of bioink substrate could develop into a different lineage entirely.

The maintenance of cells growing within a bioprinted construct is another major factor to contend with when it comes to artificial tissue development. Contamination can become a very common issue with all the different sensitive biological cells and biocompatible materials involved. Depending on the bioink requirements, sterility techniques utilized as per ISO norms may not be useful or sufficient for preventing contamination. Other mechanobiological factors can affect the production of bioprinted constructs, such as the viscoelastic properties of the bioink used, etc. Ultimately, it all comes down to the end goal for the construct prepared, whether it will be used as a biocompatible scaffold, a full tissue-engineered construct, or as a viable in vitro disease model. Depending on the microenvironment required to form, the bioprinting parameters may change drastically. Many complex structures depend upon the resolution of the bioink, the rigidity of the support matrix, along with the usage of sacrificial layers to form ridges as well as microchannels as required. Other features, such as porosity, can also be controlled during the printing process, allowing researchers an easier level of precision over the different kinds of artificial grafts formed.

Bioprinting could make artificial organ systems a reality, reducing our dependency on transplants. However, quite a lot more extensive research is required. The longevity of the artificial constructs being produced and their efficacy for the purpose for which they were created were the major limitations currently being dealt with in the regenerative medicine sector. The crosstalk between various cells and layers of scaffolds and their overall impact on lab and clinical trials is crucial for researchers to provide a more inclusive understanding of the functioning of organ systems. The greater the number of variables involved in bioprinting, the greater the precision required in order to replicate a particular microenvironment or tissue. However, understanding the cellular requirements associated with bioprinting allows us to prepare and fabricate more biocompatible scaffolds that may support increased growth, proliferation, and differentiation in the direction needed and allow us to form even more complex constructs that could be utilized to replicate as much of the human body as possible.

2.7 Conclusion

Bioprinting is a field with a multitude of potentials and possibilities. From developing artificial organs to accurate in vitro models for drug delivery and disease pathology, it could be utilized to recreate a biological environment without having to worry about ethics or model organism availability. However, a lot more progress is required for fine-tuning the process to develop an intricate tissue system with all its components. Commendable efforts have been made in this area, allowing us a deeper insight into how the cell-ECM interaction affects various pathways within a certain region and how to provide angiogenic growth to an affected portion of the body. Still, a lot of progress leaves much to be desired in terms of the specialization of bioinks and cell printing technologies. Even so, bioprinting could very well change the landscape of tissue engineering if it can provide replicable, scalable tissue models or artificial constructs developed exactly as per the myriad requirements of the biomedical sector within a sustainable cost and a compact time frame. Understanding the cellular requirements associated with bioprinting could help us prepare a more effective and biocompatible bioink that does not provide too much stress to the cells other than what is required.

2.8 Future Prospects

Cell-based bioprinting is paving the way for the future where two, three, or even four types of cells are printed on similar constructs for different layers to replicate segregated tissue. 4D bioprinting is the new method being utilized in the 21st century where, due to interaction with the cells as well as the microenvironment, the form of the printed construct changes and begins to evolve over time, developing into the intended model. From treating chronic injuries beyond the normal range of repair to developing fully functional models capable of surviving on their own within a controlled experimental setup, bioprinting could be very beneficial to the medical field and to mankind in the future. And with an increase in precision related to the cellular aspects and further improved control over the cell parameters associated with bioprinting, developing complex multilayered constructs perfectly mimicking a tissue or organ can become a reality.

References

1 Gu, Z., Fu, J., Lin, H., and He, Y. (2020). Development of 3D bioprinting: from printing methods to biomedical applications. *Asian J. Pharm. Sci.* 15 (5): 529–557.

2 Fransen, M.F., Addario, G., Bouten, C.V. et al. (2021). Bioprinting of kidney in vitro models: cells, biomaterials, and manufacturing techniques. *Essays Biochem.* 65 (3): 587–602.

3 Yi, H.G., Kim, H., Kwon, J. et al. (2021). Application of 3D bioprinting in the prevention and the therapy for human diseases. *Signal Transduct. Target. Ther.* 6: 177. https://doi.org/10.1038/s41392-021-00566-8.

4 Jain, P., Kathuria, H., and Dubey, N. (2022). Advances in 3D bioprinting of tissues/organs for regenerative medicine and in-vitro models. *Biomaterials* 287: 121639.

5 Birla, R.K. and Williams, S.K. (2020). 3D bioprinting and its potential impact on cardiac failure treatment: an industry perspective. *APL Bioeng.* 4 (1): 010903.

6 Agarwal, S., Saha, S., Balla, V.K. et al. (2020). Current developments in 3D bioprinting for tissue and organ regeneration–a review. *Front. Mech. Eng.* 6: 589171.

7 He, Y., Xie, M., Gao, Q., and Fu, J. (2019). *Why Choose 3D Bioprinting? Part I: A Brief Introduction of 3D Bioprinting for the Beginners*, 221–224. Springer.

8 Moldovan, N.I., Hibino, N., and Nakayama, K. (2017). Principles of the Kenzan method for robotic cell spheroid-based three-dimensional bioprinting. *Tissue Eng. B Rev.* 23 (3): 237–244.

9 Gupta, D. and Negi, N.P. (2022). 3D bioprinting: printing the future and recent advances. *Bioprinting* 27: e00211.

10 Tekin, E., Smith, P.J., and Schubert, U.S. (2008). Inkjet printing as a deposition and patterning tool for polymers and inorganic particles. *Soft Matter* 4 (4): 703–713.

11 Roseti, L., Cavallo, C., Desando, G. et al. (2018). Three-dimensional bioprinting of cartilage by the use of stem cells: a strategy to improve regeneration. *Materials* 11 (9): 1749.

12 Mandrycky, C., Wang, Z., Kim, K., and Kim, D.-H. (2016). 3D bioprinting for engineering complex tissues. *Biotechnol. Adv.* 34 (4): 422–434.

13 Kwon, H., Paschos, N.K., Hu, J.C., and Athanasiou, K. (2016). Articular cartilage tissue engineering: the role of signaling molecules. *Cell. Mol. Life Sci.* 73: 1173–1194.

14 Raveendran, N.T., Vaquette, C., Meinert, C. et al. (2019). Optimization of 3D bioprinting of periodontal ligament cells. *Dent. Mater.* 35 (12): 1683–1694.

15 Abelseth, E., Abelseth, L., De la Vega, L. et al. (2018). 3D printing of neural tissues derived from human induced pluripotent stem cells using a fibrin-based bioink. *ACS Biomater. Sci. Eng.* 5 (1): 234–243.

16 Lee, C., Abelseth, E., De La Vega, L., and Willerth, S. (2019). Bioprinting a novel glioblastoma tumor model using a fibrin-based bioink for drug screening. *Mater. Today Chem.* 12: 78–84.

17 Addario, G., Djudjaj, S., Fare, S. et al. (2020). Microfluidic bioprinting towards a renal in vitro model. *Bioprinting* 20: e00108.

18 Nguyen, D.G., Funk, J., Robbins, J.B. et al. (2016). Bioprinted 3D primary liver tissues allow assessment of organ-level response to clinical drug induced toxicity in vitro. *PLoS One* 11 (7): e0158674.

19 Xu, T., Zhao, W., Zhu, J.-M. et al. (2013). Complex heterogeneous tissue constructs containing multiple cell types prepared by inkjet printing technology. *Biomaterials* 34 (1): 130–139.

20 Chang, C.C., Boland, E.D., Williams, S.K., and Hoying, J.B. (2011). Direct-write bioprinting three-dimensional biohybrid systems for future regenerative therapies. *J. Biomed. Mater. Res. B Appl. Biomater.* 98 (1): 160–170.

21 Vijayavenkataraman, S., Yan, W.-C., Lu, W.F. et al. (2018). 3D bioprinting of tissues and organs for regenerative medicine. *Adv. Drug Deliv. Rev.* 132: 296–332.

22 Derakhshanfar, S., Mbeleck, R., Xu, K. et al. (2018). 3D bioprinting for biomedical devices and tissue engineering: a review of recent trends and advances. *Bioact. Mater.* 3 (2): 144–156.

23 Irvine, S.A. and Venkatraman, S.S. (2016). Bioprinting and differentiation of stem cells. *Molecules* 21 (9): 1188.

24 Colosi, C., Shin, S.R., Manoharan, V. et al. (2016). Microfluidic bioprinting of heterogeneous 3D tissue constructs using low-viscosity bioink. *Adv. Mater.* 28 (4): 677–684.

25 Faulkner-Jones, A., Fyfe, C., Cornelissen, D.-J. et al. (2015). Bioprinting of human pluripotent stem cells and their directed differentiation into hepatocyte-like cells for the generation of mini-livers in 3D. *Biofabrication* 7 (4): 044102.

26 Lawlor, K.T., Vanslambrouck, J.M., Higgins, J.W. et al. (2021). Cellular extrusion bioprinting improves kidney organoid reproducibility and conformation. *Nat. Mater.* 20 (2): 260–271.

27 Norona, L.M., Nguyen, D.G., Gerber, D.A. et al. (2019). Bioprinted liver provides early insight into the role of Kupffer cells in TGF-β1 and methotrexate-induced fibrogenesis. *PLoS One* 14 (1): e0208958.

28 Zhou, X., Nowicki, M., Sun, H. et al. (2020). 3D bioprinting-tunable small-diameter blood vessels with biomimetic biphasic cell layers. *ACS Appl. Mater. Interfaces* 12 (41): 45904–45915.

29 Bour, R., Sharma, P., Turner, J. et al. (2020). Bioprinting on sheet-based scaffolds applied to the creation of implantable tissue-engineered constructs with potentially diverse clinical applications: tissue-Engineered Muscle Repair (TEMR) as a representative testbed. *Connect. Tissue Res.* 61 (2): 216–228.

30 Kundu, R., Mahada, P., Chhirang, B., and Das, B. (2022). Cellulose hydrogels: green and sustainable soft biomaterials. *Curr. Res. Green Sustainable Chem.* 5: 100252.

31 Yang, Q., Peng, J., Xiao, H. et al. (2022). Polysaccharide hydrogels: functionalization, construction and served as scaffold for tissue engineering. *Carbohydr. Polym.* 278: 118952.

32 Guillotin, B., Souquet, A., Catros, S. et al. (2010). Laser assisted bioprinting of engineered tissue with high cell density and microscale organization. *Biomaterials* 31 (28): 7250–7256.

33 Murphy, S.V. and Atala, A. (2014). 3D bioprinting of tissues and organs. *Nat. Biotechnol.* 32 (8): 773–785.

34 Gruene, M., Deiwick, A., Koch, L. et al. (2011). Laser printing of stem cells for biofabrication of scaffold-free autologous grafts. *Tissue Eng. Part C Methods* 17 (1): 79–87.

35 Hölzl, K., Lin, S., Tytgat, L. et al. (2016). Bioink properties before, during and after 3D bioprinting. *Biofabrication* 8 (3): 032002.

36 Ma, X., Qu, X., Zhu, W. et al. (2016). Deterministically patterned biomimetic human iPSC-derived hepatic model via rapid 3D bioprinting. *Proc. Natl. Acad. Sci.* 113 (8): 2206–2211.

37 Choudhury, D., Anand, S., and Naing, M.W. (2018). The arrival of commercial bioprinters–towards 3D bioprinting revolution! *Inter. J. Bioprinting* 4 (2): 139.

38 Gruene, M., Pflaum, M., Deiwick, A. et al. (2011). Adipogenic differentiation of laser-printed 3D tissue grafts consisting of human adipose-derived stem cells. *Biofabrication* 3 (1): 015005.

39 Choi, J.S., Kang, H.-W., Lee, I.H. et al. (2009). Development of micro-stereolithography technology using a UV lamp and optical fiber. *Int. J. Adv. Manuf. Technol.* 41: 281–286.

40 Fatimi, A., Okoro, O.V., Podstawczyk, D. et al. (2022). Natural hydrogel-based bio-inks for 3D bioprinting in tissue engineering: a review. *Gels* 8 (3): 179.

41 Klak, M., Bryniarski, T., Kowalska, P. et al. (2020). Novel strategies in artificial organ development: what is the future of medicine? *Micromachines* 11 (7): 646.

42 Dey, M. and Ozbolat, I.T. (2020). 3D bioprinting of cells, tissues and organs. *Sci. Rep.* 10 (1): 14023.

43 Rutz, A.L., Lewis, P.L., and Shah, R.N. (2017). Toward next-generation bioinks: tuning material properties pre-and post-printing to optimize cell viability. *MRS Bull.* 42 (8): 563–570.

44 Galante, R., Pinto, T.J., Colaco, R., and Serro, A.P. (2018). Sterilization of hydrogels for biomedical applications: a review. *J. Biomed. Mater. Res. B Appl. Biomater.* 106 (6): 2472–2492.

45 Rutala, W. and Weber, D. (1999). Infection control: the role of disinfection and sterilization. *J. Hosp. Infect.* 43: S43–S55.

46 Lorson, T., Ruopp, M., Nadernezhad, A. et al. (2020). Sterilization methods and their influence on physicochemical properties and bioprinting of alginate as a bioink component. *ACS Omega* 5 (12): 6481–6486.

47 Gupta, A., Avci, P., Dai, T. et al. (2013). Ultraviolet radiation in wound care: sterilization and stimulation. *Adv. Wound Care* 2 (8): 422–437.

48 Lah, E.F.C., Musa, R.N.A.R., and Ming, H.T. (2012). Effect of germicidal UV-C light (254 nm) on eggs and adult of house dustmites, Dermatophagoides pteronyssinus and Dermatophagoides farinae (Astigmata: Pyroglyhidae). *Asian Pac. J. Trop. Biomed.* 2 (9): 679–683.

49 Fakhruddin, K., Hamzah, M.S.A., and Razak, S.I.A. editors. (2018) Effects of extrusion pressure and printing speed of 3D bioprinted construct on the fibroblast cells viability. IOP Conference Series: Materials Science and Engineering, June 25, 2018. Terrengganu, Malaysia: IOP Publishing.

50 Mancha Sánchez, E., Gómez-Blanco, J.C., López Nieto, E. et al. (2020). Hydrogels for bioprinting: a systematic review of hydrogels synthesis, bioprinting parameters, and bioprinted structures behavior. *Front. Bioeng. Biotechnol.* 8: 776.

51 Gagne, F. (2014). *Biochemical Ecotoxicology: Principles and Methods.* Elsevier.

52 Ning, L., Guillemot, A., Zhao, J. et al. (2016). Influence of flow behavior of alginate–cell suspensions on cell viability and proliferation. *Tissue Eng. Part C Methods* 22 (7): 652–662.

53 Jin, Y., Compaan, A., Bhattacharjee, T., and Huang, Y. (2016). Granular gel support-enabled extrusion of three-dimensional alginate and cellular structures. *Biofabrication* 8 (2): 025016.

54 Cahall, C.F., Kaur, A.P., Davis, K.A. et al. (2020). Cell death persists in rapid extrusion of lysis-resistant coated cardiac myoblasts. *Bioprinting* 18: e00072.

55 Carvalho, A., Gasperini, L., Ribeiro, R. et al. (2018). Control of osmotic pressure to improve cell viability in cell-laden tissue engineering constructs. *J. Tissue Eng. Regen. Med.* 12 (2): e1063–e1067.

56 Mackenzie, C.G., Mackenzie, J.B., and Beck, P. (1961). The effect of pH on growth, protein synthesis, and lipid-rich particles of cultured mammalian cells. *J. Cell Biol.* 9 (1): 141–156.

57 Hu, J.B., Tomov, M.L., Buikema, J.W. et al. (2018). Cardiovascular tissue bioprinting: physical and chemical processes. *Appl. Phys. Rev.* 5 (4): 041106.

58 Diamantides, N., Wang, L., Pruiksma, T. et al. (2017). Correlating rheological properties and printability of collagen bioinks: the effects of riboflavin photocrosslinking and pH. *Biofabrication* 9 (3): 034102.

59 Lin, D., Kelly, A.L., and Miao, S. (2022). The impact of pH on mechanical properties, storage stability and digestion of alginate-based and soy protein isolate-stabilized emulsion gel beads with encapsulated lycopene. *Food Chem.* 372: 131262.

60 Axpe, E. and Oyen, M.L. (2016). Applications of alginate-based bioinks in 3D bioprinting. *Int. J. Mol. Sci.* 17 (12): 1976.

61 Mehrban, N., Teoh, G.Z., and Birchall, M.A. (2016). 3D bioprinting for tissue engineering: stem cells in hydrogels. *Inter. J. Bioprinting* 2 (1): 214.

62 Lee, K.Y. and Mooney, D.J. (2001). Hydrogels for tissue engineering. *Chem. Rev.* 101 (7): 1869–1880.

63 Jin, R. and Dijkstra, P.J. (2010). Hydrogels for tissue engineering applications. In: *Biomedical Applications of Hydrogels Handbook* (ed. R.M. Ottenbrite and T. Okano), 203–225. Springer.

64 Lee, S.C., Gillispie, G., Prim, P., and Lee, S.J. (2020). Physical and chemical factors influencing the printability of hydrogel-based extrusion bioinks. *Chem. Rev.* 120 (19): 10834–10886.

65 Dzobo, K., Motaung, K.S.C.M., and Adesida, A. (2019). Recent trends in decellularized extracellular matrix bioinks for 3D printing: an updated review. *Int. J. Mol. Sci.* 20 (18): 4628.

66 Naghieh, S. and Chen, X. (2021). Printability–a key issue in extrusion-based bioprinting. *J. Pharm. Anal.* 11 (5): 564–579.

67 Jang, J., Park, J.Y., Gao, G., and Cho, D.-W. (2018). Biomaterials-based 3D cell printing for next-generation therapeutics and diagnostics. *Biomaterials* 156: 88–106.

68 Atanase, L.I. (2021). Micellar drug delivery systems based on natural biopolymers. *Polymers* 13 (3): 477.

69 Bajaj, P., Schweller, R.M., Khademhosseini, A. et al. (2014). 3D biofabrication strategies for tissue engineering and regenerative medicine. *Annu. Rev. Biomed. Eng.* 16: 247–276.

70 Derby, B. (2012). Printing and prototyping of tissues and scaffolds. *Science* 338 (6109): 921–926.

71 Toh, W.S. and Loh, X.J. (2014). Advances in hydrogel delivery systems for tissue regeneration. *Mater. Sci. Eng. C* 45: 690–697.

72 Samadian, H., Maleki, H., Allahyari, Z., and Jaymand, M. (2020). Natural polymers-based light-induced hydrogels: promising biomaterials for biomedical applications. *Coord. Chem. Rev.* 420: 213432.

73 Vasvani, S., Kulkarni, P., and Rawtani, D. (2020). Hyaluronic acid: a review on its biology, aspects of drug delivery, route of administrations and a special emphasis on its approved marketed products and recent clinical studies. *Int. J. Biol. Macromol.* 151: 1012–1029.

74 Hospodiuk, M., Dey, M., Sosnoski, D., and Ozbolat, I.T. (2017). The bioink: a comprehensive review on bioprintable materials. *Biotechnol. Adv.* 35 (2): 217–239.

75 Kiyotake, E.A., Douglas, A.W., Thomas, E.E. et al. (2019). Development and quantitative characterization of the precursor rheology of hyaluronic acid hydrogels for bioprinting. *Acta Biomater.* 95: 176–187.

76 Ahn, S., Yoon, H., Kim, G. et al. (2010). Designed three-dimensional collagen scaffolds for skin tissue regeneration. *Tissue Eng. Part C Methods* 16 (5): 813–820.

77 Osidak, E.O., Kozhukhov, V.I., Osidak, M.S., and Domogatsky, S.P. (2020). Collagen as bioink for bioprinting: a comprehensive review. *Int. J. Bioprinting* 6 (3): 1137.

78 Shim, J.-H., Kim, J.Y., Park, M. et al. (2011). Development of a hybrid scaffold with synthetic biomaterials and hydrogel using solid freeform fabrication technology. *Biofabrication* 3 (3): 034102.

79 Park, J.Y., Choi, J.-C., Shim, J.-H. et al. (2014). A comparative study on collagen type I and hyaluronic acid dependent cell behavior for osteochondral tissue bioprinting. *Biofabrication* 6 (3): 035004.

80 Yeo, M., Lee, J.-S., Chun, W., and Kim, G.H. (2016). An innovative collagen-based cell-printing method for obtaining human adipose stem cell-laden structures consisting of core–sheath structures for tissue engineering. *Biomacromolecules* 17 (4): 1365–1375.

81 Shah, P.P., Shah, H.B., Maniar, K.K., and Özel, T. (2020). Extrusion-based 3D bioprinting of alginate-based tissue constructs. *Procedia CIRP* 95: 143–148.

82 Benwood, C., Chrenek, J., Kirsch, R.L. et al. (2021). Natural biomaterials and their use as bioinks for printing tissues. *Bioengineering* 8 (2): 27.

83 Perez-Valle, A., Del Amo, C., and Andia, I. (2020). Overview of current advances in extrusion bioprinting for skin applications. *Int. J. Mol. Sci.* 21 (18): 6679.

84 Kim, S.H., Yeon, Y.K., Lee, J.M. et al. (2018). Precisely printable and biocompatible silk fibroin bioink for digital light processing 3D printing. *Nat. Commun.* 9 (1): 1620.

85 Singh, Y.P., Bandyopadhyay, A., and Mandal, B.B. (2019). 3D bioprinting using cross-linker-free silk–gelatin bioink for cartilage tissue engineering. *ACS Appl. Mater. Interfaces* 11 (37): 33684–33696.

86 Lima, T.P., Canelas, C.A.A., Concha, V.O. et al. (2022). 3D bioprinting technology and hydrogels used in the process. *J. Funct. Biomater.* 13 (4): 214.

87 Gopinathan, J. and Noh, I. (2018). Recent trends in bioinks for 3D printing. *Biomater. Res.* 22: 1–15.

88 Fan, R., Piou, M., Darling, E. et al. (2016). Bio-printing cell-laden Matrigel–agarose constructs. *J. Biomater. Appl.* 31 (5): 684–692.

89 López-Marcial, G.R., Zeng, A.Y., Osuna, C. et al. (2018). Agarose-based hydrogels as suitable bioprinting materials for tissue engineering. *ACS Biomater. Sci. Eng.* 4 (10): 3610–3616.

90 Malikmammadov, E., Tanir, T.E., Kiziltay, A. et al. (2018). PCL and PCL-based materials in biomedical applications. *J. Biomater. Sci. Polym. Ed.* 29 (7–9): 863–893.

91 Jovic, T.H., Kungwengwe, G., Mills, A.C., and Whitaker, I.S. (2019). Plant-derived biomaterials: a review of 3D bioprinting and biomedical applications. *Front. Mech. Eng.* 5: 19.

92 Mobaraki, M., Ghaffari, M., Yazdanpanah, A. et al. (2020). Bioinks and bioprinting: a focused review. *Bioprinting* 18: e00080.

93 Gungor-Ozkerim, P.S., Inci, I., Zhang, Y.S. et al. (2018). Bioinks for 3D bioprinting: an overview. *Biomater. Sci.* 6 (5): 915–946.

94 Ching, S.H., Bansal, N., and Bhandari, B. (2017). Alginate gel particles–a review of production techniques and physical properties. *Crit. Rev. Food Sci. Nutr.* 57 (6): 1133–1152.

95 Lee, J., Hong, J., Kim, W., and Kim, G.H. (2020). Bone-derived dECM/alginate bioink for fabricating a 3D cell-laden mesh structure for bone tissue engineering. *Carbohydr. Polym.* 250: 116914.

96 Wu, Y., Lin, Z.Y.W., Wenger, A.C. et al. (2018). 3D bioprinting of liver-mimetic construct with alginate/cellulose nanocrystal hybrid bioink. *Bioprinting* 9: 1–6.

97 Ehrbar, M., Rizzi, S.C., Hlushchuk, R. et al. (2007). Enzymatic formation of modular cell-instructive fibrin analogs for tissue engineering. *Biomaterials* 28 (26): 3856–3866.

98 Janmey, P.A., Winer, J.P., and Weisel, J.W. (2009). Fibrin gels and their clinical and bioengineering applications. *J. R. Soc. Interface* 6 (30): 1–10.

99 Xu, F., Dawson, C., Lamb, M. et al. (2022). Hydrogels for tissue engineering: addressing key design needs toward clinical translation. *Front. Bioeng. Biotechnol.* 10: 849831.

100 Piard, C., Baker, H., Kamalitdinov, T., and Fisher, J. (2019). Bioprinted osteon-like scaffolds enhance in vivo neovascularization. *Biofabrication* 11 (2): 025013. https://doi.org/10.1088/1758-5090/ab078a. PMID: 30769337; PMCID: PMC7195919.

101 Khoeini, R., Nosrati, H., Akbarzadeh, A. et al. (2021). Natural and synthetic bioinks for 3D bioprinting. *Adv. NanoBiomed Res.* 1 (8): 2000097.

102 Islam, M., Shahruzzaman, M., Biswas, S. et al. (2020). Chitosan based bioactive materials in tissue engineering applications–a review. *Bioact. Mater.* 5 (1): 164–183.

103 Nicodemus, G.D. and Bryant, S.J. (2008). Cell encapsulation in biodegradable hydrogels for tissue engineering applications. *Tissue Eng. B Rev.* 14 (2): 149–165.

104 Einbu, A., Grasdalen, H., and Vårum, K.M. (2007). Kinetics of hydrolysis of chitin/chitosan oligomers in concentrated hydrochloric acid. *Carbohydr. Res.* 342 (8): 1055–1062.

105 Wang, L. and Stegemann, J.P. (2010). Thermogelling chitosan and collagen composite hydrogels initiated with β-glycerophosphate for bone tissue engineering. *Biomaterials* 31 (14): 3976–3985.

106 Choi, J.H., Kim, D.K., Song, J.E. et al. (2018). Silk fibroin-based scaffold for bone tissue engineering. In: *Novel Biomaterials for Regenerative Medicine* (ed. Wim E. Crusio, Haidong Dong, Heinfried H. Radeke, Nima Rezaei, Ortrud Steinlein, and Junjie Xiao), 371–387. Springer.

107 Reddy, M.S.B., Ponnamma, D., Choudhary, R., and Sadasivuni, K.K. (2021). A comparative review of natural and synthetic biopolymer composite scaffolds. *Polymers* 13 (7): 1105.

108 Yang, P., Ju, Y., Hu, Y. et al. (2023). Emerging 3D bioprinting applications in plastic surgery. *Biomater. Res.* 27 (1): 1–27.

109 Choi, Y.-J., Park, H., Ha, D.-H. et al. (2021). 3D bioprinting of in vitro models using hydrogel-based bioinks. *Polymers* 13 (3): 366.

110 Shan Wong, Y., Yong Tay, C., Wen, F. et al. (2012). Engineered polymeric biomaterials for tissue engineering. *Curr. Tissue Eng.* 1 (1): 41–53.

111 Rahimi, A. and Mashak, A. (2013). Review on rubbers in medicine: natural, silicone and polyurethane rubbers. *Plast., Rubber Compos.* 42 (6): 223–230.

112 Lambert, J.M. (2006). The nature of platinum in silicones for biomedical and healthcare use. *J. Biomed. Mater. Res. Part B: Appl. Biomater.* 78 (1): 167–180.

113 Yang, X., Chen, Q., Bao, H. et al. (2018). Preparation and performance of ultraviolet curable silicone resins used for ultraviolet cured coating and ultraviolet-assisted 3D printing materials. *OSA Continuum* 1 (2): 542–552.

114 Xiang, H., Wang, X., Ou, Z. et al. (2019). UV-curable, 3D printable and biocompatible silicone elastomers. *Prog. Org. Coat.* 137: 105372.

115 Aoki, K. and Saito, N. (2020). Biodegradable polymers as drug delivery systems for bone regeneration. *Pharmaceutics* 12 (2): 95.

116 Nowicki, M., Zhu, W., Sarkar, K. et al. (2020). 3D printing multiphasic osteochondral tissue constructs with nano to micro features via PCL based bioink. *Bioprinting* 17: e00066.

117 Cheng, Y., Yang, H., Yang, Y. et al. (2018). Progress in TiO2 nanotube coatings for biomedical applications: a review. *J. Mater. Chem. B* 6 (13): 1862–1886.

118 Kolan, K.C., Semon, J.A., Bromet, B. et al. (2019). Bioprinting with human stem cell-laden alginate-gelatin bioink and bioactive glass for tissue engineering. *Inter. J. Bioprinting* 5 (2.2): 204.

119 Magin, C.M., Alge, D.L., and Anseth, K.S. (2016). Bio-inspired 3D microenvironments: a new dimension in tissue engineering. *Biomed. Mater.* 11 (2): 022001.

120 Shie, M.-Y., Chang, W.-C., Wei, L.-J. et al. (2017). 3D printing of cytocompatible water-based light-cured polyurethane with hyaluronic acid for cartilage tissue engineering applications. *Materials* 10 (2): 136.

121 Hung, K.C., Tseng, C.S., and Hsu, S. (2014). Synthesis and 3D printing of biodegradable polyurethane elastomer by a water-based process for cartilage tissue engineering applications. *Adv. Healthcare Mater.* 3 (10): 1578–1587.

122 Merceron, T.K., Burt, M., Seol, Y.-J. et al. (2015). A 3D bioprinted complex structure for engineering the muscle–tendon unit. *Biofabrication* 7 (3): 035003.

123 Garlotta, D. (2001). A literature review of poly (lactic acid). *J. Polym. Environ.* 9: 63–84.

124 Serra, T., Mateos-Timoneda, M.A., Planell, J.A., and Navarro, M. (2013). 3D printed PLA-based scaffolds: a versatile tool in regenerative medicine. *Organogenesis* 9 (4): 239–244.

125 Guvendiren, M., Molde, J., Soares, R.M., and Kohn, J. (2016). Designing biomaterials for 3D printing. *ACS Biomater. Sci. Eng.* 2 (10): 1679–1693.

126 Asti, A. and Gioglio, L. (2014). Natural and synthetic biodegradable polymers: different scaffolds for cell expansion and tissue formation. *Int. J. Artif. Organs* 37 (3): 187–205.

127 Serra, T., Planell, J.A., and Navarro, M. (2013). High-resolution PLA-based composite scaffolds via 3-D printing technology. *Acta Biomater.* 9 (3): 5521–5530.

128 Schmedlen, R.H., Masters, K.S., and West, J.L. (2002). Photocrosslinkable polyvinyl alcohol hydrogels that can be modified with cell adhesion peptides for use in tissue engineering. *Biomaterials* 23 (22): 4325–4332.

129 Wu, Z., Kong, B., Liu, R. et al. (2018). Engineering of corneal tissue through an aligned PVA/collagen composite nanofibrous electrospun scaffold. *Nanomaterials* 8 (2): 124.

130 Lim, K.S., Alves, M.H., Poole-Warren, L.A., and Martens, P.J. (2013). Covalent incorporation of non-chemically modified gelatin into degradable PVA-tyramine hydrogels. *Biomaterials* 34 (29): 7097–7105.

131 Li, J., Chen, M., Fan, X., and Zhou, H. (2016). Recent advances in bioprinting techniques: approaches, applications and future prospects. *J. Transl. Med.* 14 (1): 1–15.

132 Roth, E.A., Xu, T., Das, M. et al. (2004). Inkjet printing for high-throughput cell patterning. *Biomaterials* 25 (17): 3707–3715.

133 Ribeiro, A., Blokzijl, M.M., Levato, R. et al. (2017). Assessing bioink shape fidelity to aid material development in 3D bioprinting. *Biofabrication* 10 (1): 014102.

134 Ning, L., Sun, H., Lelong, T. et al. (2018). 3D bioprinting of scaffolds with living Schwann cells for potential nerve tissue engineering applications. *Biofabrication* 10 (3): 035014.

135 Paxton, N., Smolan, W., Böck, T. et al. (2017). Proposal to assess printability of bioinks for extrusion-based bioprinting and evaluation of rheological properties governing bioprintability. *Biofabrication* 9 (4): 044107.

136 Ning, L., Zhu, N., Mohabatpour, F. et al. (2019). Bioprinting Schwann cell-laden scaffolds from low-viscosity hydrogel compositions. *J. Mater. Chem. B* 7 (29): 4538–4551.

137 Mouser, V.H., Melchels, F.P., Visser, J. et al. (2016). Yield stress determines bioprintability of hydrogels based on gelatin-methacryloyl and gellan gum for cartilage bioprinting. *Biofabrication* 8 (3): 035003.

138 Rutz, A.L., Hyland, K.E., Jakus, A.E. et al. (2015). A multimaterial bioink method for 3D printing tunable, cell-compatible hydrogels. *Adv. Mater.* 27 (9): 1607–1614.

139 Blaeser, A., Duarte Campos, D.F., Puster, U. et al. (2016). Controlling shear stress in 3D bioprinting is a key factor to balance printing resolution and stem cell integrity. *Adv. Healthcare Mater.* 5 (3): 326–333.

140 Forget, A., Blaeser, A., Miessmer, F. et al. (2017). Mechanically tunable bioink for 3D bioprinting of human cells. *Adv. Healthcare Mater.* 6 (20): 1700255.

141 Piedra-Cascón, W., Krishnamurthy, V.R., Att, W., and Revilla-León, M. (2021). 3D printing parameters, supporting structures, slicing, and post-processing procedures of vat-polymerization additive manufacturing technologies: a narrative review. *J. Dent.* 109: 103630.

142 Chaudhuri, O. (2017). Viscoelastic hydrogels for 3D cell culture. *Biomater. Sci.* 5 (8): 1480–1490.

143 Schwab, A., Levato, R., D'Este, M. et al. (2020). Printability and shape fidelity of bioinks in 3D bioprinting. *Chem. Rev.* 120 (19): 11028–11055.

144 Bishop, E.S., Mostafa, S., Pakvasa, M. et al. (2017). 3-D bioprinting technologies in tissue engineering and regenerative medicine: current and future trends. *Genes Dis.* 4 (4): 185–195.

145 Kourouklis, A.P., Wahlsten, A., Stracuzzi, A. et al. (2023). Control of hydrostatic pressure and osmotic stress in 3D cell culture for mechanobiological studies. *Biomater. Adv.* 145: 213241.

146 Shin, J., Lee, Y., Li, Z. et al. (2022). Optimized 3D bioprinting technology based on machine learning: a review of recent trends and advances. *Micromachines* 13: 363. https://doi.org/10.3390/mi13030363.

147 Chimene, D., Lennox, K.K., Kaunas, R.R., and Gaharwar, A.K. (2016). Advanced bioinks for 3D printing: a materials science perspective. *Ann. Biomed. Eng.* 44: 2090–2102.

148 Slaughter, B.V., Khurshid, S.S., Fisher, O.Z. et al. (2009). Hydrogels in regenerative medicine. *Adv. Mater.* 21 (32–33): 3307–3329.

149 Lane, S.W., Williams, D.A., and Watt, F.M. (2014). Modulating the stem cell niche for tissue regeneration. *Nat. Biotechnol.* 32 (8): 795–803.

150 Prowse, A.B., Chong, F., Gray, P.P., and Munro, T.P. (2011). Stem cell integrins: implications for ex-vivo culture and cellular therapies. *Stem Cell Res.* 6 (1): 1–12.

151 Gao, G., Schilling, A.F., Hubbell, K. et al. (2015). Improved properties of bone and cartilage tissue from 3D inkjet-bioprinted human mesenchymal stem cells by simultaneous deposition and photocrosslinking in PEG-GelMA. *Biotechnol. Lett.* 37: 2349–2355.

152 Huebsch, N., Arany, P.R., Mao, A.S. et al. (2010). Harnessing traction-mediated manipulation of the cell/matrix interface to control stem-cell fate. *Nat. Mater.* 9 (6): 518–526.

153 Engler, A.J., Sen, S., Sweeney, H.L., and Discher, D.E. (2006). Matrix elasticity directs stem cell lineage specification. *Cell* 126 (4): 677–689.

154 Engler, A., Sweeney, H., Discher, D., and Schwarzbauer, J.E. (2007). Extracellular matrix elasticity directs stem cell differentiation. *J. Musculoskelet. Nueronal Interact.* 7 (4): 335.

155 Shav, D. and Einav, S. (2010). The effect of mechanical loads in the differentiation of precursor cells into mature cells. *Ann. N. Y. Acad. Sci.* 1188 (1): 25–31.

156 Stolberg, S. and McCloskey, K.E. (2009). Can shear stress direct stem cell fate? *Biotechnol. Prog.* 25 (1): 10–19.

157 Lopez, J., Mouw, J., and Weaver, V. (2008). Biomechanical regulation of cell orientation and fate. *Oncogene* 27 (55): 6981–6993.

158 Belgodere, J.A., King, C.T., Bursavich, J.B. et al. (2018). Engineering breast cancer microenvironments and 3D bioprinting. *Front. Bioeng. Biotechnol.* 6: 66.

159 Vanaei, S., Parizi, M., Salemizadehparizi, F., and Vanaei, H. (2021). An overview on materials and techniques in 3D bioprinting toward biomedical application. *Eng. Regener.* 2: 1–18.

160 Tripathi, S., Mandal, S.S., Bauri, S., and Maiti, P. (2023). 3D bioprinting and its innovative approach for biomedical applications. *MedComm.* 4 (1): e194.

161 Han, Y., Li, X., Zhang, Y. et al. (2019). Mesenchymal stem cells for regenerative medicine. *Cells* 8 (8): 886.

162 Ullah, I., Subbarao, R.B., and Rho, G.J. (2015). Human mesenchymal stem cells-current trends and future prospective. *Biosci. Rep.* 35 (2): e00191.

163 Köpf, M., Nasehi, R., Kreimendahl, F. et al. (2022). Bioprinting-associated shear stress and hydrostatic pressure affect the angiogenic potential of human umbilical vein endothelial cells. *Int. J. Bioprinting* 8 (4): 606.

164 Choi, J.H., Gimble, J.M., Lee, K. et al. (2010). Adipose tissue engineering for soft tissue regeneration. *Tissue Eng. B Rev.* 16 (4): 413–426.

165 Maan, Z., Masri, N.Z., and Willerth, S.M. (2022). Smart bioinks for the printing of human tissue models. *Biomolecules* 12 (1): 141.

166 Kim, J., Choi, H.S., Kim, Y.M., and Song, S.C. (2023). Thermo-responsive nanocomposite bioink with growth-factor holding and its application to bone regeneration. *Small* 19 (9): 2203464.

167 Firipis, K., Nisbet, D.R., Franks, S.J. et al. (2021). Enhancing peptide biomaterials for biofabrication. *Polymers* 13 (16): 2590.

168 Chen, P., Zheng, L., Wang, Y. et al. (2019). Desktop-stereolithography 3D printing of a radially oriented extracellular matrix/mesenchymal stem cell exosome bioink for osteochondral defect regeneration. *Theranostics* 9 (9): 2439.

169 Dobres, S., Mula, G., Sauer, J., and Zhu, D. (2022). Applications of 3D printed chimeric DNA biomaterials. *Eng. Regener.* 3 (1): 13–23.

170 Pan, T., Song, W., Xin, H. et al. (2022). MicroRNA-activated hydrogel scaffold generated by 3D printing accelerates bone regeneration. *Bioact. Mater.* 10: 1–14.

171 Liu, F., Chen, Q., Liu, C. et al. (2018). Natural polymers for organ 3D bioprinting. *Polymers* 10 (11): 1278.

172 Kumar, S. (2021). Synthetic polymer-derived single-network inks/bioinks for extrusion-based 3D printing towards bioapplications. *Mater. Adv.* 2 (21): 6928–6941.

173 Kajave, N.S., Schmitt, T., Nguyen, T.-U. et al. (2021). Bioglass incorporated methacrylated collagen bioactive ink for 3D printing of bone tissue. *Biomed. Mater.* 16 (3): 035003.

174 Kim, W. and Kim, G. (2019). Collagen/bioceramic-based composite bioink to fabricate a porous 3D hASCs-laden structure for bone tissue regeneration. *Biofabrication* 12 (1): 015007.

175 Levato, R., Visser, J., Planell, J.A. et al. (2014). Biofabrication of tissue constructs by 3D bioprinting of cell-laden microcarriers. *Biofabrication* 6 (3): 035020.

176 Gu, Y., Forget, A., and Shastri, V.P. (2022). Biobridge: an outlook on translational bioinks for 3D bioprinting. *Adv. Sci.* 9 (3): 2103469.

177 Zhang, T., Zhao, W., Xiahou, Z. et al. (2021). Bioink design for extrusion-based bioprinting. *Appl. Mater. Today* 25: 101227.

178 Zhang, Z., Xu, C., Xiong, R. et al. (2017). Effects of living cells on the bioink printability during laser printing. *Biomicrofluidics* 11 (3): 034120.

179 Reina-Romo, E., Mandal, S., Amorim, P. et al. (2021). Towards the experimentally-informed in silico nozzle design optimization for extrusion-based bioprinting of shear-thinning hydrogels. *Front. Bioeng. Biotechnol.* 9: 701778.

180 Wang, Q., Backman, O., Nuopponen, M. et al. (2021). Rheological and printability assessments on biomaterial inks of nanocellulose/photo-crosslinkable biopolymer in light-aided 3D printing. *Front. Chem. Eng.* 3: 723429.

181 Ashammakhi, N., Ahadian, S., Xu, C. et al. (2019). Bioinks and bioprinting technologies to make heterogeneous and biomimetic tissue constructs. *Mater. Today Bio* 1: 100008.

182 Chen, Y., Ju, L., Rushdi, M. et al. (2017). Receptor-mediated cell mechanosensing. *Mol. Biol. Cell* 28 (23): 3134–3155.

183 Langenbach, F. and Handschel, J. (2013). Effects of dexamethasone, ascorbic acid and β-glycerophosphate on the osteogenic differentiation of stem cells in vitro. *Stem Cell Res. Ther.* 4 (5): 1–7.

184 Zhou, G.-S., Zhang, X.-L., Wu, J.-P. et al. (2009). 5-Azacytidine facilitates osteogenic gene expression and differentiation of mesenchymal stem cells by alteration in DNA methylation. *Cytotechnology* 60: 11–22.

185 Tharakan, S., Khondkar, S., and Ilyas, A. (2021). Bioprinting of stem cells in multimaterial scaffolds and their applications in bone tissue engineering. *Sensors* 21 (22): 7477.

186 Chen, E.P., Toksoy, Z., Davis, B.A., and Geibel, J.P. (2021). 3D bioprinting of vascularized tissues for in vitro and in vivo applications. *Front. Bioeng. Biotechnol.* 9: 664188.

187 Goverdhana, S., Puntel, M., Xiong, W. et al. (2005). Regulatable gene expression systems for gene therapy applications: progress and future challenges. *Mol. Ther.* 12 (2): 189–211.

188 Pan, X., Veroniaina, H., Su, N. et al. (2021). Applications and developments of gene therapy drug delivery systems for genetic diseases. *Asian J. Pharm. Sci.* 16 (6): 687–703.

189 Paolini, A., Leoni, L., Giannicchi, I. et al. (2018). MicroRNAs delivery into human cells grown on 3D-printed PLA scaffolds coated with a novel fluorescent PAMAM dendrimer for biomedical applications. *Sci. Rep.* 8 (1): 1–11.

190 Shende, P. and Trivedi, R. (2021). 3D printed bioconstructs: regenerative modulation for genetic expression. *Stem Cell Rev. Rep.* 14: 1–12.

191 Walker, C., Mojares, E., and del Río, H.A. (2018). Role of extracellular matrix in development and cancer progression. *Int. J. Mol. Sci.* 19 (10): 3028.

192 Lamhamedi-Cherradi, S.-E., Santoro, M., Ramammoorthy, V. et al. (2014). 3D tissue-engineered model of Ewing's sarcoma. *Adv. Drug Deliv. Rev.* 79: 155–171.

193 Arif, Z.U., Khalid, M.Y., Zolfagharian, A., and Bodaghi, M. (2022). 4D bioprinting of smart polymers for biomedical applications: recent progress, challenges, and future perspectives. *React. Funct. Polym.* 179: 105374.

194 Datta, P., Dey, M., Ataie, Z. et al. (2020). 3D bioprinting for reconstituting the cancer microenvironment. *npj Precis. Oncol.* 4 (1): 18.

195 Liu, Q., Zhang, H., Jiang, X. et al. (2017). Factors involved in cancer metastasis: a better understanding to "seed and soil" hypothesis. *Mol. Cancer* 16: 1–19.

196 Winkler, J., Abisoye-Ogunniyan, A., Metcalf, K.J., and Werb, Z. (2020). Concepts of extracellular matrix remodelling in tumour progression and metastasis. *Nat. Commun.* 11 (1): 5120.

197 Jung, M., Skhinas, J.N., Du, E.Y. et al. (2022). A high-throughput 3D bioprinted cancer cell migration and invasion model with versatile and broad biological applicability. *Biomater. Sci.* 10 (20): 5876–5887.

198 Horder, H., Guaza Lasheras, M., Grummel, N. et al. (2021). Bioprinting and differentiation of adipose-derived stromal cell spheroids for a 3D breast cancer-adipose tissue model. *Cells* 10 (4): 803.

199 Germain, N., Dhayer, M., Dekiouk, S., and Marchetti, P. (2022). Current advances in 3D bioprinting for cancer modeling and personalized medicine. *Int. J. Mol. Sci.* 23 (7): 3432.

200 Zervantonakis, I.K., Hughes-Alford, S.K., Charest, J.L. et al. (2012). Three-dimensional microfluidic model for tumor cell intravasation and endothelial barrier function. *Proc. Natl. Acad. Sci.* 109 (34): 13515–13520.

201 Suzuka, J., Tsuda, M., Wang, L. et al. (2021). Rapid reprogramming of tumour cells into cancer stem cells on double-network hydrogels. *Nat. Biomed. Eng.* 5 (8): 914–925.

202 Samadian, H., Jafari, S., Sepand, M. et al. (2021). 3D bioprinting technology to mimic the tumor microenvironment: tumor-on-a-chip concept. *Mater. Today Adv.* 12: 100160.

3

3D Bioprinting: Materials for Bioprinting Bioinks Selection

Mona Moaness and Mostafa Mabrouk

Refractories, Ceramics and Building Materials Department, Advanced Materials, Technology and Mineral Resources Research Institute, National Research Centre, Cairo, Egypt

3.1 Introduction

Like 3D printing, 3D bioprinting is an additional biofabrication technique that constructs an object, layer-by-layer, using a digital file as a template. Bioprinters use living cells, bioactive molecules, and biomaterials to fabricate organs having microstructures, which allows living cell proliferation. Despite being a relatively new technology, bioprinting has the potential to fill many important gaps in medical research, including uses in drug discovery, regenerative medicine, functional organ replacement, and cosmetics testing [1–3]. To produce the necessary tissue construct, a variety of materials, techniques, and cells can be used, depending on the application. In organ and tissue 3D bioprinting, cultured cells are combined with different biocompatible materials to make bioinks, which then can be 3D bioprinted into functional tissue constructs that need to be treated. Despite the availability of various 3D bioprinting approaches, such as extrusion-based bioprinting, inkjet bioprinting/droplet bioprinting, and laser-assisted bioprinting, as well as the use of multi-head deposition systems (MHDSs), the application of these methods has been limited. Their limitation is primarily due to a lack of suitable bioinks that both need to meet the requirements for functional bioprinting and to have proper biocompatibilities and bioactivities [4–6].

Figure 3.1 shows the characteristics of standardized bioinks that should mimic the extracellular matrix environment structurally and mechanically, promoting

3D Bioprinting from Lab to Industry, First Edition. Edited by Prosenjit Saha, Sabu Thomas, Jinku Kim, and Manojit Ghosh.

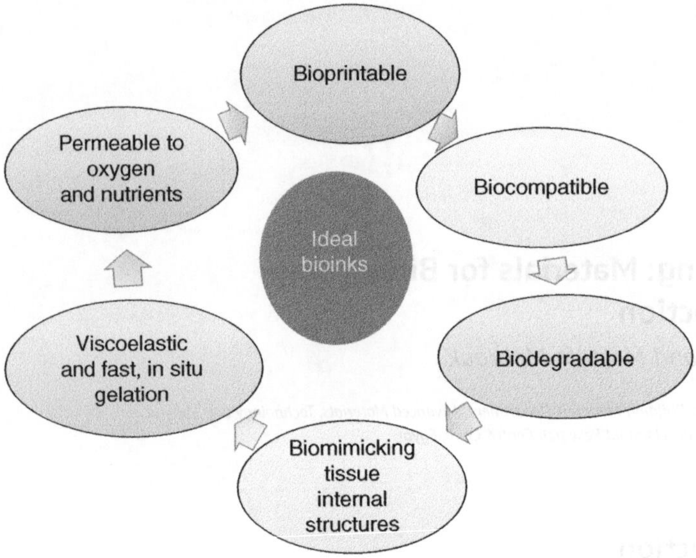

Figure 3.1 The characteristics of standard bioinks.

cell adhesion, proliferation, and differentiation. These bioinks should possess adjustable gelation and stabilization with high shape fidelity, adequate biodegradability to imitate the targeted tissues' natural microenvironment, flexibility for chemical modifications that fulfill tissue-specific needs, flowability,deformability, and potential for large-scale production with high accuracy. Such bioinksare greatly needed for their use in different bioprinting applications [7].

Typically, natural or synthetic biomaterials, or a combination of these materials, are used to create bioinks. Polycaprolactone (PCL), polyvinyl pyrrolidone (PVP), pluronic, and polyethylene glycol (PEG) are synthetic polymers, while hyaluronic acid, gelatin, collagen, and matrigel are natural polymers. To create the necessary tissue-like constructions, 3D bioprinting most frequently employs extrusion-based techniques that involve integrating cells into a fluid matrix that solidifies after printing. However, these techniques are known to reduce cell viability because they introduce shear forces during extrusion [8, 9].

The goal of this book chapter is to classify bioprinting materials (bioinks) according to constituting material type, cell dependence, bioprinting methods according to bioink type, bioinks selection according to biomedical applications and tissue type to be fabricated, and then explain the term of multicomponent bioinks. Finally, we turn to future recommendations associated with developing current bioinks to meet the requirements of desired functional tissue to be bioprinted [10, 11].

3.2 Bioprinting Materials

3.2.1 Biomaterials

Biomaterials are a diverse class of materials made up of crosslinking polymers, metals, or ceramics, physically and/or chemically. Recent advances in biology and materials technology have increased the variety of biomaterials utilized in tissue engineering applications [12]. Polymeric materials are classified as natural (extracted naturally) or synthetic (laboratory fabricated). In terms of tissue engineering, natural polymers are used because they mimicthe nature of the extracellular matrix of tissues,while synthetic ones can be modified by altering molecular weight to produce desirable mechanical properties. The biomaterial has to stimulate cell attachment and retain cellular functionality for the needed tissue's proliferation, differentiation, and maturation. A specific biomaterial's rheological and mechanical qualities are required to enable both printability and mechanical stability in order to retain construct integrity after printing and throughout culture.

To obtain desirable functional, mechanical, and supporting qualities, complex combinations or gradients are modified or created, in order to assist bioprinter deposition while also displaying required post-printing qualities. Tissue engineering is one important application of biomaterials in which they are utilized to produce 3D structures that imitate the design and function of natural tissues and organs. This is accomplished by the use of specialized printers to layer bioinks into a specific shape, a process known as bioprinting. Bioinks are typically made up of a hydrogel matrix that gives structure and support, and they can be loaded with living cells, growth factors, and other biomolecules to form functioning tissue. Natural polymers like collagen and hyaluronic acid, synthetic polymers like PEG and PCL, and hydrogels like alginate and gelatin can all be used to make bioinks. These materials are chosen for their biocompatibility, mechanical qualities, and capacity to facilitate cell growth and differentiation [13]. Overall, using biomaterials and bioinks in tissue engineering has the ability to develop medicine through modification of complex and functional tissues and organs for transplantation, drug testing, and other applications. In addition, bioinks are an important component of biomaterials used in tissue engineering and regenerative medicine, as they enable the creation of complex and functional tissues and organs [14].

3.2.2 Cells

Because of its major influence on mechanical and biological qualities, the choice of biomaterials/cells is crucial in scaffold architecture. As building blocks, cells assist in the basic function of target tissues. Recent developments in cell biology have made it possible to use a variety of cell sources, including cell lines, primary cell lines (from either human or animal sources), stem cells, and stem cell-derived cells. The majority of these cell sources are now available commercially, allowing for the use of a diverse spectrum of these cells in scaffold construction [15]. The selection of cells for tissue or organ manufacturing is important to the operation of the resulting structure. Tissues and organs contain several cell types that perform specific and important biological functions that must be duplicated in the implanted tissue. These cells have many functions like supporting, structural, and barrier, and are engaged in vascularization. Manufacturing cell-loaded scaffolds entails either injecting many initial cell types into patterns that accurately reproduce natural tissue or depositing stem cells that can multiply and develop into required cell types [16].

Any cell type chosen for bioprinting should be differentiated and proliferated to amount enough for printing. Controlling cell growth precisely in vitro and in vivo is critical for bioprinting. Insufficient proliferation may result in the transplanted viability failure, whereas overgrowing proliferation may result in death or hyperplasia. Controlling cell proliferation in the transplanted scaffold is critical for attaining functional and supportive cells. Furthermore, the timing of cellular growth is critical. A high cellular proliferation rate may be beneficial at first to populate the construct, but in the long run, proliferation must be maintained at a rate appropriate for tissue homeostasis, but without hyperplasia [16]. Transfection of viruses [17] and small molecules [18, 19] are expected solutions for cell growth stimulation. Bioprinted tissue must achieve and maintain cellular homeostasis, self-renewal, and tissue's proper response to damage or injury for long-term functionality after transplantation.

Employing an autologous source of cells, or tolerance-induction procedures are possible solutions to overcome host's immune system rejection of bioprinted implant. Biopsies are taken from autologous cell sources in healthy patients, as cells of genetically abnormal or ill patients will be nonfunctional [20].Stem cells are an appealing cell type for tissue-engineering applications due to their ability to replicate in an unidentified but powerful phase (self-renewal) and generate a broad range of functional tissue-specific cell characteristics. Embryonic stem cells and pluripotent stem cells that have been induced have been shown to self-renew continuously and to be undivided for an excess of eighty passages [21]. The capacity of these stem cells to generate massive amounts of cells demonstrates their promise for bioprinting and additional therapeutic applications. More research to ensure the safety of these cells would be immensely valuable to the field.

Well-differentiated stem cells, originating from bone marrow [22] and fat [23] as well as perinatal ones from amniotic fluid [24] or placenta [25], are thought to be safer for clinical transplantation. Mesenchymal stromal cells (MSCs) may be a suitable cell source for bioprinted structures with proven techniques for isolation, growth, and differentiation. MSCs have been successfully cultivated in vitro for clinical trials in clinically meaningful numbers, and advances in cell culture techniques are likely to make use of other stem cell types for bioprinting therapeutic applications. The cells undergoing the bioprinting process must be able to withstand physiological stresses, like pressure and shear stress, once transplanted, as well as biological ones like toxin presence, enzymes, and nonphysiological pH. Fibroblasts or transformed cell lines underwent much bioprinting research as they are well-known for their ability to proliferate and be particularly resilient. Bioprinting technology may need to be modified to awide range of cells that can tolerate factors such as culture conditions and shear stress in addition to the time required to produce the construct.

3.2.3 Biomolecules or Additive Molecules

Biomolecules are required to fine-tune and regulate cell functionality in addition to the bioprinting of 3D structures made of various materials and cells [26, 27]. As a result, structures with features that release biomolecules have been created [28]. Different therapeutic substances, including growth factors and medications, can be released in a controlled fashion using hydrogels, both spatially and temporally. Hydrogels are employed as a stable substrate for various encapsulating medicines that are used to control release of drugs due to their tunable physical features and programmable degradability [29]. Diverse biomolecular gradients were successfully produced using bioinks, and it was later shown that they could be used to direct cell differentiation and function in 3D bioprinted constructions [30]. One popular technique is to physically or chemically couple hydrogels to biomolecules with gradient concentrations, such as growth factors. For vascularized bone tissue, Byambaa et al. created a bioactive bioink based on GelMA, having gradient vascular endothelial growth factor (VEGF), and then bioprinted bone structures with various VEGF distributions by chemically conjugating VEGF with gradient concentrations to GelMA prepolymer [30].As shown in Figure 3.2, the incorporation of bioactive molecules is a wise choice, and many additive molecules were used for this purpose and differ according to type of tissue to bebioprinted, for e.g., bone morphogenetic proteins vascular, endothelial growth factor, fibroblast growth factor, transforming growth factor, stromal cell-derived factor, platelet-rich plasma, stem cell secretomes, peptide motifs, and other molecules [31].

3.2.4 Hydrogels

Because of their desirable properties, hydrogels are mostly employed as bioink material in scaffold-based bioprinting. Their biodegradability, biocompatibility, and the existence of cell-binding sites for cell attachment, proliferation, or differentiation are the primary reasons [32].Hydrogels are the most suitable material type for cell support, yet, there are significant bioprinting limitations. First, hydrogels do not contain ECM proteins unique to distinct cell types. As a result, they are unable to provide a natural environment. Second, hydrogels enclose and constrain

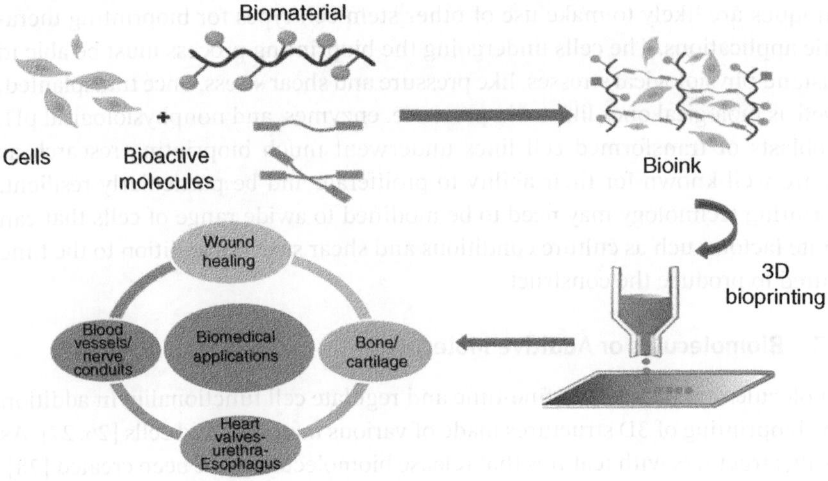

Figure 3.2 Components and applications of bioprinting materials in the biomedical field.

cells, thereby restricting their interactions. Furthermore, achieving the same high-cell density as in natural tissues is difficult. The concentration of hydrogels is also crucial because higher concentrations improve mechanical qualities while limiting biological activity. Furthermore, stiffer hydrogels necessitate higher pressure levels for extrusion or droplet production. A higher concentration of hydrogel reduces cell mobility, resulting in reduced cell proliferation and ECM protein deposition. The cellular microenvironment provided by hydrogels can vary greatly [33]. The cellular environment offered by hydrogels may be drastically different from that of real tissues. Because of the lack of enzymes, immunological reactions, and blood flow, hydrogels disintegrate significantly more slowly in vitro. Hydrogels, on the other hand, must be designed to breakdown in vivo in a synchronized manner with respect to the development of new tissue. Another significant flaw is the toxicity of breakdown products, which can be damaging to cells or newly produced tissue. Another significant constraint that may emerge at various phases of the bioprinting process is instability. Because most hydrogels are exceedingly soft, they spread and lose structural integrity throughout the bioprinting process.

3.3 Bioinks Selectivity Guide

3.3.1 Printability of Materials

Printability is one of the important considerations in bioinks selection. It refers to the ability of the ink to be processed for a particular bioprinting application, followed by the generation of high-quality biological scaffolds [34]. Shape accuracy, resolution, and cell survival of the 3Dprinted object are all factors that influence printability. These parameters can be changed by adjusting material qualities including surface tension, viscosity, and mechanism of crosslinking. In addition, altering bioink viscosity by changing biomaterial content, temperature, or the number of loaded cells has all been demonstrated to affect shape fidelity. Duan et al. recently developed bioinks based on methacrylated hyaluronic acid (Me-HA) and methacrylated gelatin (Me-Gel) to bioprint heart valve conduits with encapsulated human aortic valvular interstitial cells (HAVIC) [35]. The researchers noticed that changing the Me-Gel content changed the viscosity of the bioink. Likewise, printing with an elevated viscosity bioink resulted in structures of superior quality, with excellent shape accuracy, whilst printing with an inferior viscosity bioink resulted in fluid-like, spongy structures that lost their shape following printing. Surface tension is another key factor in affecting the printability of a bioink. Surface tension is created by the adhesion forces that are present between the components in bioink and has been shown to affect print quality, resolution, and object size [36]. In order to build constructs with well-defined three-dimensional features, surface tension is essential to maintain the skeletal strength of the bioink as printed [37].

Xu et al. demonstrated that increasing cell population in alginate bioinks was associated with a decrease in surface tension through inkjet printing [38]. This also reduced the size and droplet shape, limiting the resolution and dimensions of the printed material. The degree of crosslinking within the bioink substance is another consideration for printability since it typically determines the final shape and size of the printed 3D scaffold. Bioinks typically use materials with excellent crosslinking capacities because they provide good shape fidelity following bioprinting. Rutz et al., for example, created a bioink consisting of gelatin methacrylate and multiarmed PEG, with PEG injected to loosely crosslink gelatin methacrylate, to achieve the needed viscosity for bioprinting [39]. Their method allowed them to fine-tune the bioink's chemical and physical properties. The resulting constructs exhibit excellent form and structural fidelity while preserving cytocompatibility [39]. This opens up a lot of possibilities for establishing customized platforms for researching cell–cell signaling and tissue morphogenesis in 3D, as well as fabricating more customized, functional, and biomimetic 3D-printed tissue structures.

3.3.2 Material Biocompatibility

The purpose of biocompatibility has shifted with the advancement of tissue engineering. Instead of implanted materials coexisting with endogenous tissue without generating any negative local or systemic consequences in the host, it is now envisioned that those substances will either involuntarily enable or actively induce beneficial outcomes in the host [40]. Biocompatibility in the overall scheme of bioprinting refers to the assumption of a predictable and effective influence on the construct's biological and functional elements. Engaging with the immune system and/or local tissues, fostering optimal cellular activity, and supporting molecular or mechanical signaling pathways are all important for transplantation to succeed and perform.

3.3.3 Structural Properties

It is widely understood that the mechanical strength of a scaffold should be comparable to that of the repaired tissue/organs. Notably, the mechanical strength of scaffolds varies over time during the tissue restoration process. On the one hand, the mechanical strength of scaffold polymers decreases over time, on the other hand, cell development and tissue regeneration occur and thereby add mechanical strength to the scaffold's composite build. A material's mechanical qualities primarily refer to mechanical strength and are defined by the connection between applied force and resulting deformation. The printability of a biomaterial and, consequently, the scaffold structure that results are strongly influenced by its mechanical properties. To maintain the structural integrity of printed scaffolds throughout manufacture, and subsequently to provide the mechanical support necessary for cell growth and tissue regeneration, a biomaterial must have sufficient mechanical strength. If a substance is necessary for a 3D structure's uninterrupted function, such as resisting or creating specific forces, or acting as a fixing point for mechanical influence, the material's attributes must be maintained. Materials must be carefully selected depending on the mechanical characteristics of the build, and different demands on structure will be required for different tissue types such as skin [41, 42], liver [43], and bone [44]. One possibility for bypassing this drawback is to use transitory materials that can provide the required structural and mechanical features for a set duration of time. This disposable component might be used throughout printing to make space for appropriate crosslinking in the structure [45, 46], or it could be incorporated into the structure itself.

3.3.4 Materials Degradation

The attached cells produce proteases and, as a result, proteins from the ECM that shape the new tissue as the scaffold dissolves [47]. The dynamics of material degradation need to be investigated and governed. There are several degradation parameters that should be considered. The first is the ability to govern deterioration, ideally by matching rates of degradation to the ability of cells to substitute damaged materials through their own ECM proteins. This is tricky because materials with acceptable functional and mechanical qualities for a certain tissue may not be capable of replacing the material following dissolution. Degradation metabolites are of special importance because they impact the biocompatibility of biodegradable materials. The byproducts of metabolism should be safe, easily metabolized, and rapidly eliminated from the body. Toxic products may consist of minute proteins and chemicals, as well as nonphysiological pH, temperature, and other variables that might impair cell viability and function. Some high-molecular-weight polymers in general, for example, can be broken down into oligomers or monomers that cells can recognize and use to cause inflammation and other harmful effects. The materials' swelling and elastic properties are very significant in the design of tissue-engineering devices. Over-swelling components may take up fluid from tissues nearby, whereas compression can close pores or capillaries needed for cell migration and nutritional transport. Additionally, knowing these behaviors is crucial when using substances with dissimilar swelling or tensile behavior, since this might end up in layer coherence loss or final construction deformation.

3.3.5 Biomimicry

The significance of biomimicry in biocompatibility has just lately been investigated. The capacity to load biomimetic materials into a bioprinted scaffold can have a positive effect on both endogenous and exogenous cell adhesion, migration, proliferation, and function. Materials have been proven to possess a significant impact on cell adhesion, and also on cell shape and dimensions [48–51], and those concepts may be helpful in regulating the growth and differentiation of cells in a substrate. Topical agents applied to a substance can increase cell adherence and proliferation on its surface [52, 53]. The cellular components are lysed and removed during this procedure. Tissue decellularization challenges include establishing a compromise between the total removal of cellular components and the preservation of fine vasculature and other tissue features. Furthermore, when cells are cultured on decellularized tissue scaffolds, considerable toxicity has been seen, possibly due to the retention of the decellularization detergent inside the ECM [54]. In mammals, there are around 300 ECM protein molecules, alongside various ECM-modifying enzymes, ECM-binding growth factors, and other ECM-associated proteins [55]. The most prevalent and extensively researched proteins are collagens, proteoglycans, and glycoproteins. The development of techniques for incorporating these materials into structures utilizing bioprinting technique, whilst making certain the components have proper degradation times and waste products, along with easily comprehended and adjustable physical and biological reactions in the construct, is one challenge.

3.4 Classification of Bioprinting Materials

3.4.1 According to Material Type

The following features should be present in an ideal bioink for it to be termed biofunctional: (i) appropriate biological properties, and (ii) physico-mechanical properties [4]. Additionally, the bioink needs to be printable and should be able to preserve cell viability. A mixture of two or more biomaterials is typically required to achieve all of these properties, especially for extrusion-based bioprinting [3]. Multicomponent bioinks are frequently superior to single ones because only one component bioinkhas low biocompatibility and needs exceptional functional and mechanical requirements to produce biomimetic tissues [56].

3.4.1.1 Polymers

Natural

Agaroseis a naturally occurring carbohydrate polymer generated from red seaweed, which is composed of repeated D-galactose and 3, 6-anhydro-L-galactopyranose units of disaccharide [57]. Due to its biocompatibility, Agarose is often used in biomedical applications and thermoreversible gelling mechanisms [57, 58]. During bioprinting, agarose is employed as bioink and has been designed to auto-degrade in order to form microchannels [59], to help in creating vascularized tissue constructs. Oxygen and nutrient diffusion across this microstructure provide a support structure for cell proliferation [58]. Agarose is traditionally employed as bioink for applications of 3D printing to build neural mini-tissue constructs. Agarose bioinks are used to bioprint cartilage tissue [58, 60]. To improve agarose's ability to maintain cell survival, pure agarose itself does not interact with other biological environments, so it is frequently combined with other natural biocompatible polymers like chitosan, gelatin, and collagen to overcome this problem.

Alginate is a brown algae-derived biocompatible anionic block copolymer and is widely employed in medical applications such as tissue engineering, drug transport, and wound healing due to its inexpensive cost and biocompatibility [61, 62]. It is considered a bioink as it has the capacity to produce hydrogels with similar characteristics to the ECM of tissues. Alginates possess long-term stability in vivo as mammals do not produce enzymes that cause their degradation, so modification through oxidization by periodate makes it prone to hydrolytic degradation [61–63].Pure alginate has low viscosity, zero shear viscosity [64], and minimal bioactivity [64–66], which, in turn, decreases the proliferation of cells chances. These obstacles can be modified by blending with other hydrogels like nanocellulose to improve the printability of alginate-based bioinks, through the improvement of rheological and mechanical properties [60].

Gelatin is defined as a polypeptide produced by denaturing collagen [67]. Because of its unique features, such as biocompatibility [68], low cost, biodegradability, ease of processing, and cell affinity [69], it has been researched for the production of bioink. Methacrylic anhydride addition is used to crosslink gelatin with other materials that can in turn improve its mechanical properties [67]. Several studies have been conducted to improve material properties by using this functionalized gelatin in print (Gelatin-methacryloyl/GelMA) [70–72].

Hyaluronic acid is a nonsulfated linear glycosaminoglycan found in the extracellular matrix and a majority of connective tissues [73]. This natural hydrogel exhibits great hydrophilicity and biocompatibility during cell development [74]. To meet the physicochemical requirements of 3D bioprinting, it crosslinked with other polymers to overcome low mechanical properties [75]. Another effective technique for improving the mechanical feature of hyaluronic acid is to crosslink it with ultraviolet (UV) [31]. Hyaluronic acid and polylactic acid (PLA) were copolymerized by Antich et al. [76] to develop a novel type of bioink to be employed in the 3D bioprinting of tissues and cartilage [76]. In order to generate a pentanoate-functionalized hyaluronic acid (PHA) bioink for 3D bioprinting, Kiyotake et al. [77] used photocrosslinking (UV) [77].

Collagen Type I collagen is one of the most interesting and extensively utilized natural hydrogels. It is significant because it is a part of the extracellular matrix in other tissues and is a component of musculoskeletal tissue [78]. Additionally, collagen has features that are essential to 3D bioprinting, such as biocompatibility, biodegradability, and cell attachment [79]. Collagen's poor mechanical property, similar to that of hyaluronic acid, is enhanced by photopolymerization (UV) [80]. Shi et al. [81] paper gives proof of this. Employing photopolymerization of collagen and methacrylated gelatin hydrogel (GelMa), he generated a novel type of bioink that could be a wonderful choice for 3D bioprinting of tissues to heal damaged tissue [81].

Silk Fibroin FDA-approved silk fibroin (SF) is a fascinating material [82] for the fabrication of bioinks. The 3D bioprinting industry has taken notice of this material, which is made from the silkworm, Bombyx mori (B. mori), for its excellent biodegradability [83], biocompatibility [83], and processing in a variety of formats (hydrogels, films, membranes, etc.). The viability of the growth of cells must also be taken into consideration when producing bioinks [84]. In this regard, SF is unique due to its great vitality for various cell lines, good cell and bioactive molecule encapsulation, and high stability. However, due to rheological limitations, natural polymers may need to be combined with other polymeric materials [85–87].

Decellularized Extracellular Matrix Decellularization of tissues employing a number of chemical and physical techniques, such as cycles of freezing and thawing, chemicals, or enzymatic agents, results in dECM [88]. Tissue decellularization strives to eliminate all of the tissue's cellular components while retaining almost all of the ECM's framework, even if some ECM disruption is unavoidable during this technique [88]. Two benefits of preserving the original structure of the ECM for use as a bioink material include the ability to infuse tissue-specific qualities into printed structures by choosing the tissue source of the dECM and avoiding crosslinker employing [89–92]. A range of tissue sources, including the heart [89], skin, liver [90], intestines [91], cornea [90], bones [92], and tendons [93, 94] have been used to create dECM bioinks for extrusion printing. For cardiovascular applications, significant dECM analysis has been carried out. Cardiomyocyte growth, viability, and maturation have all been shown to be supported by decellularized cardiac tissue [92, 93]. Furthermore, printing human cardiac progenitor cells with heart dECM bioinks is provided. Jang et al. used heart dECM bioink to print pre-vascularized stem cell patches that were injected with cardiac progenitor cells and MSCs [95]. Patches created of the two cell types boosted vascularization, preserved the survival of cells, and diminished cardiac reconfiguration and fibrosis in mice [95].

Synthetic

PEGA common pharmaceutical adjuvant is PEG. With FDA permission, gels based on PEG can be created through physical or covalent crosslinking and utilized internally [96]. A PEG derivative known as PEGDA, which has been acrylated, can be photopolymerized to create a gel with improved mechanical characteristics. PEGDA is a simple material to work with for creating 3D scaffolds for extrusion-based bioprinting. PEG-based gels are often not printed alongside cells because cells cannot effectively move and multiply on printed PEGDA scaffolds [97]. Instead, they are more commonly used as vehicles for dispensing bioactive substances in order to increase mechanical strength and preserve the 3D structure [98]. Peptide modification is an additional technique for increasing PEG's bioactivity. For instance, human mesenchymal stem cells were printed using PEG-based microgels that had been altered with the cell adhesive peptide to create 3D constructs that encourage cell spreading and proliferation.

PCL Excellent bioink qualities like stiffness, biocompatibility, and degradability can be found in PCL, a less expensive polymer. PCL is a nontoxic polymer that has exceptional stability. The stability has a biological half-life of three years and normally lasts for six months [99]. The porous structure that encourages connection, the rough surface, and the compactness that resembles bones, all contribute to bone regeneration and cell ingrowth in SLS-printed PCL scaffolds. However, with scaffolds intended for uses other than bone tissue creation, the prolonged biological half-life of the scaffold creates an additional hurdle. Additionally, the substance's higher hydrophobicity leads to reduced bioactivity, which decreases cell growth and tissue [100].

PVP is an amorphous nonionic polymer that is nontoxic, water-soluble, and highly soluble in polar solvents and widely used in pharmaceuticals, tissue engineering and cosmetics. Izgordu et al. [101], investigate the effect of PVP on the biological and mechanical performance of 3D printed cartilage-like constructs. In this investigation, low molecular weight CS, alginic acid sodium salt, and HA were employed independently in the polymer matrix, along with 3D constructs of PCL and PVP as the matrix polymers. Evaluations were done on the biocompatibility, degradability, swelling, chemical, morphological, thermal, and mechanical characteristics of 3D constructions. With the addition of fillers, the melting temperature of the matrix rose, with PCL/3 wt% PVP/1 wt% HA having the maximum melting temperature [101].

Pluronic, a kind of poloxamer, is used in bioprinting as a substance to create sacrifice structures. It is a potential candidate for usage in bioinks because of its good printability and temperature-responsive gelation. As a result, pluronic may be easily removed by washing it away after printing, if necessary, because it liquefies at 4 °C or lower. Despite the fact that pluronic has been used extensively as a sacrificial bioink, its biocompatibility prevents it from being used directly as a standard bioink for cell culture since it is insufficient to support long-term cell survival. In a recent work. Muller et al. described a technique called nanostructuring that made it possible to improve pluronic's biocompatibility by combining acrylated and unmodified pluronic and removing unmodified molecules after UV crosslinking. They were able to significantly extend the viability of encapsulated chondrocytes up to two weeks in culture. This was made possible by the enhanced porosity that resulted from the removal of unmodified molecules. However, it was discovered that the construct had poor mechanical strength, which was successfully enhanced by the inclusion of HAMA. These findings indicated that pluronic might be coupled with different polymers and used for bioprinting of various tissue constructions [8].

3.4.1.2 Nanocomposites

The incorporation of micro- and nanofillers is one of the most effective approaches to offer biofunctionality to 3D bioprinted scaffolds. The adaptability of the incorporated organic and/or inorganic NPs, as well as the specific needs of the target tissues and organs, may be completely understood, allowing the nanocomposite bioinks to be adjusted to build 3D-printed scaffolds with the necessary qualities. As a result, nanofillers can actively assist the generated scaffolds in achieving the intended goals, while also passively improving the printability and exact shape fidelity of the bioprinted constructs [98]. Maintaining or supplying a cell niche by simulating the microenvironment of the tissue or organ of interest, or enabling real-time monitoring of the cellular processes during the maturation, may involve either influencing cell proliferation or differentiation into the intended cell type.

It has been found to be effective to add ceramic NPs to bone TE scaffolds to strengthen their toughness and stiffness, along with utilizing electrically-responsive NPs, such as nanotubes made of carbon, to provide hydrogels the electrical properties needed to create muscle tissue in the skeleton, neuronal, or heart tissue [102].

3.4.1.3 Nanoparticles

Over the past few years, the field of biomedical engineering has quickly increased its interest in nanomaterials [103, 104]. Due to their large specific surface area, nanoparticles of any concentration can have a considerable effect on the characteristics of a hydrogel network [103]. Numerous biological applications, including imaging, medication delivery, cancer therapy, and the creation of biosensors, have taken advantage of the special features of nanomaterials. Bioinks have been given new functions by using nanomaterials, such as improved electrical conductivity, stimuli responsiveness, control over cell behavior, printability enhancements, and mechanical reinforcement. According to the nature and properties of the included nanoparticles for biomedical applications, they were divided into four groups. Ceramic NPs, NPs with biological components conjugated to them, NPs that respond to stimuli, and NPs for real-time monitoring [105–111]. The use of nanofillers in fabricating light-responsive hydrogels has also been investigated, with graphite–graphene oxide structures being among the most frequently used in this creation process [68, 112]. Graphite-derived composites exhibit the previously indicated electrical response in addition to being light-sensitive, which enables the development of hydrogels with customizable properties by using external light stimuli when using them as reinforcement materials [113]. In order to increase the generated nanocomposite hydrogels' photothermal stimulation responsively, graphite oxide nanostructures' infrared light absorption characteristics were employed. In Figure 3.3, we showed different types of 3D bioprinted scaffolds from different sources [114–118].

3.4.2 According to Cell Dependence

3.4.2.1 Cell-based Bioinks

To put it simply, a bioink is "a formulation of cells suitable for processing by a computerized biofabrication equipment, which may also contain biologically active substances and biomaterials." Therefore, presence of cells is mandatory. These cells have many forms, single, coated, or aggregated, or they can be combined with other substances like microcarriers, microgels, physical hydrogels, or hydrogel precursors. Scaffold-based bioinks is the first approach where cells and/or biomolecules are printed within a biomaterial matrix, such as hydrogels in the scaffold-based technique, to promote tissue repair and regeneration along with

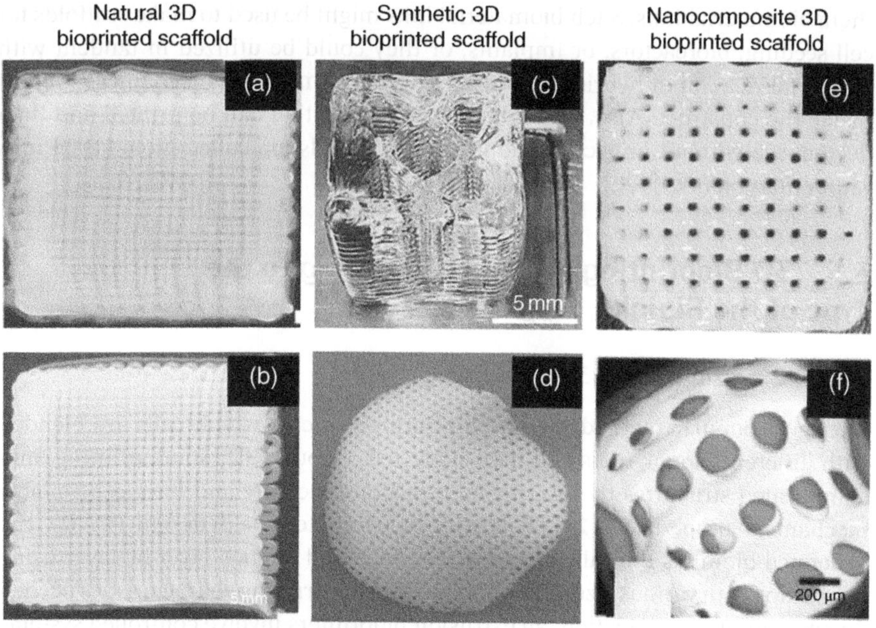

Figure 3.3 Different types of 3D bioprinted scaffold from different sources (a) and (b) *Source:* From Kamdem et al. [114] / MDPI, (c) *Source:* Chen et al. [115] / MDPI, (d) *Source:* Santis et al. [116], (e) *Source:* Chen et al. [117] / MDPI and (f) *Source:* Kakarla et al. [118] / MDPI.

three-dimensional growth [113, 118]. Because of their biocompatibility, low cytotoxicity, and high water contact, hydrogels have been widely employed in 3D bioprinting to load cells for tissue engineering [113, 118].

Another cell-based bioinks approach is called the scaffold-free bioinks. In which, the cell pellets are maintained in a printed mold structure, so they can consolidate and splash ECM components to hold them together [119]. In actuality, scaffold-free methods do not have the same negative consequences as scaffold materials, which can contain toxic chemicals, by-products of degrading processes, and high processing temperatures that reduce cell viability or result in inflammatory reactions when implanted [119]. Furthermore, because of improved cell–cell communication and interaction, scaffold-free methods give cells a natural environment in which to produce more ECM.

3.4.2.2 Cell-Free Bioinks (Biomaterial Inks)

Biomaterials that can be printed and then seeded with cells after printing, but are not directly formed with cells, do not qualify as a bioink. It is suggested to call them biomaterial inks. Such biomaterial inks might be used to make scaffolds for cell seeding, bioreactors, or implants, or they could be utilized in tandem with bioink fabrication in hybrid techniques to generate intrinsic mechanical support inside the construct. Also, biodegradable materials that can be printed and then decompose without impacting the life of living cells are called biomaterial inks rather than bioinks [120].

3.5 3D Bioprinting Methods According to the Type of the Bioinks

3.5.1 Extrusion-Based 3D Bioprinting

The most popular method for 3D bioprintingis one in which cells are blended with different biopolymers and then subjected, through 3D bioink printing, into complicated structures that mirror biological architecture and have controllable mechanical properties [121]. Following a computer-aided design (CAD), the generated bioink is typically stored in a syringe and applied layer-by-layer using nozzle tips with various sizes. To accommodate the printing characteristics needed for each unique composition, all extrusion bioprinters involve controlled systems of pressure and temperature [121]. Because the bioink is continually ejected, strengthening the mechanical hardness of the scaffolds, this technology is appropriate for the mass manufacture of medical devices. By raising the pressure or temperature throughout the printing process, extrusion 3D bioprinting excels at overcoming viscosity problems. However, bioprinting quality and cell viability may be compromised by the slow increase of extrusion pressure inside the walls of the dispensing nozzle [122, 123]. Due to increased shear pressures, viscosity of nanocomposite bioinks increased, impairing cell survival in extrusion-based 3D bioprinting as well as printability with various approaches [124].

3.5.2 Inkjet 3D Bioprinting

3D inkjet bioprinting is a technique for creating 3D objects. It works by using several pressures, including piezoelectric, thermal, and electromagnetic, to propel the bioink into precise 3D structures. This method, which enables the parallel 3D bioprinting of heterogeneous solutions, like drug and gene carriers containing bioinks, promotes the interweaving of various biological constituents because it does not require direct contact with the deposition surface in order to function [125, 126]. This then makes printing faster, more accurate, and with no effect on the viability of the enclosed cells [127]. Contrarily, inkjet bioprinting is hindered by the requirement for low-viscosity solutions, making its applications for bioinks with a high gelling rate very impracticable.

3.5.3 Stereolithography 3D Bioprinting

A liquid photopolymerizable solution is used as a precursor in the stereolithography-based bioprinting process. This precursor is poured into a mold with the appropriate structure. The liquid polymer is then exposed to UV light or laser to begin the photocrosslinking process, which drives the solidification of the bioink as it begins to take on the shape of the previously designed volume. The stereolithography-based bioprinter uses a CAD to produce the final structure, which is repeated until it solidifies into a hydrogel [128]. Despite the high printing resolution that this bioprinting technique offers, the changing cycles of the bioink injection and irradiation curing phases make it quite slow. Additionally, UV and laser radiation can damage cell viability by disrupting cellular components and their physiological functions [129, 130].

3.5.4 Laser-Based 3D Bioprinting

In theory, laser-based bioprinting and inkjet bioprinting are relatively comparable because both do not have direct contact with the deposition surface. The bioink droplet is heated by a laser beam that is sent at it through multiple different angled mirrors, making it easier for it to move toward a surface target. The aforementioned procedure is then repeated numerous times in a layer-by-layer fashion until the final bioprinted scaffold is established by the stacking of cohesive layers created by nearby droplets [37]. In exchange, this method allows for precise, high-speed printing and can significantly increase the maximum number of cells that can be included in the bioink, making it appropriate for applications that call for large starting cell numbers [131]. However, the laser beam supply can have a negative impact on the initial cell survival due to the droplet's heat absorption [132].

3.5.5 Bioplotting

Bioplotting is a 3D printing technique used in tissue biofabrication [133]. Bioplotting is based on the ability to eject material in tube or spheroid forms. These printers can use different cell types and multiple syringes, thus allowing for the production of different bioengineered soft tissues. The methodology of this technique involves layers being stacked one on one another, which are then cured through UV radiation or a chemical reaction [134] Choosing appropriate materials for extrusion is the main obstacle in bioplotting, as they should be viscous, cell-supporting, and should produce a useful cellular microenvironment. The bioplotting method frequently makes use of materials including thermoset resins, polymer melts, pastes with high filler contents, cement, polymer solutions, and even macromolecules like proteins [133]. Bioplotting is the ideal method,

Table 3.1 The benefits and drawbacks of 3D bioprinting techniques.

Printing technique	Benefits	Drawbacks	Cell compatibility
Extrusion-based	• Highly controllable printing speed and high fidelity	• Cell viability could be affected by shear stress	Compatible
Inkjet-based	• High printing speed • Low cost	• Failed with high-viscosity materials	Compatible
Stereolithography (SLA)	• High resolution	• Large amounts of material needed • Long operation time • Long operational times decrease cell viability	Compatible
Laser-based	• Can be used with viscous materials • Highly accurate	• Laser can affect cell viability	Compatible
Bioplotting	• Can be used with high filler contents and macromolecules and multiple contents	• Small outcome • Cost effective	Compatible

according to recent studies, for producing co-cultured scaffolds and tissues that do not need high-resolution details [134]. Table 3.1 provides an overview of the benefits and drawbacks of the bioprinting processes.

3.6 Bioinks Selection According to Biomedical Application

While there are still not enough donors, the need for tissue regeneration and organ replacement has increased rapidly in recent years. The traditional animal model has been applied frequently and produces meaningful findings. However, a recent study demonstrated that the disease phenotype and manifestations are not the same in mice and humans. The reason medications fail to attain therapeutic safety and efficacy in clinical trials is clear: many animals cannot precisely imitate the human underlying molecular pathways, and this issue is gradually worsening. Costly animal studies, lengthy research processes, and a limited ability to completely see occurrences at the cellular level are some further downsides. Additionally, the main goal of any biomedical study is to provide more efficient methods of diagnosis and treatment to treat human illnesses in a better way.

Unmet requirements in pharmaceutical research and tissue and organ regeneration can be addressed by 3D bioprinting. The goal of tissue engineering, a subfield of biomedical engineering, is to create biological replacements and tissues outside the body, implant them at the site of the damage, and then heal, cultivate, and nurture the damaged organs or tissues. Tissue engineering applications for various bioprinting technology types are numerous.

Bone 3D bioprinting 3D bioprinting with hydroxyapatite nanoparticles as a substrate has enabled the creation of tissue. However, large, complicated faults occurred, necessitating more biological and technical advances. Furthermore, research is being conducted on the development of bone implants using cutting-edge bioresorbable materials such as poly(propylene carbonate) in order to avoid the toxic byproducts of the breakdown process. Synthetic clays are another type of nanoparticle that has sparked a lot of interest in 3D bioprinting of bones. The 2D coin-shaped physical structures, which have shear thinning capabilities and unique charge distribution, form strong noncovalent binds with natural and synthetic polymers [135, 136]. Another example is that artificial nanosilicates and GelMA enhanced viscosity at shear rates lower than those required for 3D bioprinting [137]. Hong et al. observed a comparable improvement in the mechanical and biological properties of PEGDA/alginate composites by incorporating silicate nanoparticles [138]. These findings offer light on a recently proposed approach that employs nanoparticles.

Cartilage 3D bioprinting Small concentrations of nanoparticles can be introduced to polymeric hydrogels to affect their physical and chemical properties, such as stiffness, shear-thinning, and resistance to degradation in physiological conditions [139]. The existence of shear-thinning properties is critical to the success of soft tissue engineering. At room temperature, for example, alginate hydrogels containing nanofibrillated cellulose were used to generate anatomically precise reproductions of cartilage tissues such as the meniscus and ear [139]. The natural tissue imitated stiffness when different alginate concentrations and cationic crosslinkers were applied. Chondrocytes from human cartilage multiplied and had a viability rate of more than 70% inside the bioinks. It is still necessary to optimize how the nanoparticles affect the mechanical properties of the bioink. Most currently available bioprinted tissues do not function as well as genuine tissue does. They can keep their shape and serve as the tissue's mechanical support system. However, bioinks' ability to restore the tissue's activity is entirely dependent on the cells. Mannoor et al. have created a special bioink for ears and cartilage tissue engineering in order to overcome these problems. The combination of cell-seeded hydrogel bioinks and an interwoven conducting polymer containing silver nanoparticles allowed the reading of inductively coupled signals from cochlea-shaped electrodes [140]. This approach combined biologic and nanoelectronic properties with 3D bioprinting to produce functional tissue.

Cardiovascular Urethral and Esophageal Regeneration Cardiovascular disease is a significant global problem that causes 20 million mortalities annually [141]. The best method to deal with these issues is through replacement, which employs entirely distinct, load-bearing biological and physical systems. Modern synthetic constructs, xenografts, and allografts all have specific disadvantages and are not useful in the long run. Therefore, breakthroughs in strong tissue engineering are urgently needed to develop cardiovascular implants. The main study areas include 3D bioprinting of cardiac repairing patches and heart valves. An earlier work employed laser-assisted bioprinting to regenerate cardiac patches and extrusion-based bioprinting to construct a trileaflet heart valve conduit [142]. Numerous hydrogels, including those made of hyaluronic acid, gelatin, and alginate, have been used because of their outstanding biocompatibility. For 3D bioprinting of blood vessels, carbon nanotubes (CNTs) were explored in an alginate matrix [143]. By examining in vitro measurements of the perfusion and permeability, cell survival, production of the extracellular matrix, and tissue histology, the bioinks with nanoparticles were discovered to be mechanically desired vessels. These CNTs can be changed out for natural protein nanofibers to further integrate these conduits in the mass production of tissues. By introducing magnetically controlled nanoparticles, the position of the vessels within tissues was controlled by a magnetic field [144].

Treatment for urethral structure brought on by urethral injury cannot be adequately met by conventional urethral tissue engineering and therapy techniques. A novel method for urology treatment is provided by 3D bioprinted urethra, which may imitate the mechanical properties and structure of human urethral tissue [145]. Although therapy for many esophageal illnesses, including esophageal stenosis, cancer, atresia, etc., has a specific impact, it may also have a number of negative effects [146]. With the use of 3D bioprinting, replacements for these esophageal abnormalities can be created, which nearly mirror the intrinsic makeup and original esophageal structure [147]. It is without a doubt that the development of 3D bioprinting will open up a variety of opportunities for tissue engineering and regenerative medicine.

3D Bioprinted Wound Dressing Skin injury that does not heal naturally can be brought on by external pathogenic germs, mishaps, or physical wounds. Clinically, autologous grafts and other skin tissue substitutes are preferred for solving this issue, but their availability is limited, and their use may have unintended negative effects on the body [148]. Serious immunological rejection may also occur after an allogeneic transplant. New ideas on how to repair and regenerate skin tissues are provided by the creation of tissue-engineered skin substitutes. Among the different biomanufacturing methods, 3D bioprinting has received a lot of interest because of its strong ability to control complex 3D structures and produce useful 3D tissue scaffolds [149]. The 3D bioprinting of bionic medical dressings is a common method for wound healing. The purpose of this dressing is to move and release different cell types and vital biomolecules that are required for wound healing, accelerating the healing process. Researchers have created a special PCL fiber wound dressing that is suitable for use in hospitals. It is created using 3D bioprinting, contains human gingival mesenchymal stem cells, and has anisotropic mechanical characteristics. In the rat injury model, this dressing group's wound healing rate is considerably faster than that of the other groups, and the full-thickness excision wound scar development is reduced [150]. The development of skin tissue substitutes was mandatory since simple bandages were ineffective in treating patients with serious skin injuries. With the use of 3D bioprinting and fresh bionic bioinks, a novel method for producing functional living skin (FLS) was developed. FLS can significantly aid in the growth of skin appendages like hair follicles and sebaceous glands. The procedure to regenerate skin without scars seems promising. Due to its mechanical and bioadhesive properties, FLS, which has the physiological structure of normal skin and may promote skin and blood vessel regeneration, offers new hope for the treatment of large-area skin abnormalities in the future [151]. The usage of 3D bioprinting dressings also extends to other tissues, such as the nasal mucosa and oral cavity, and is not just limited to the restoration and regrowth of skin tissues.

Neural Tissue Regeneration Peripheral and central nerve regeneration has been limited due to substantial abnormalities, severe injury, and poor functional outcomes [152]. Brain tissue engineering with 3D bioprinting is still in its early phases. The key barriers to 3D bioprinting of brain cells are the bioinks' insufficient mechanical flexibility and electrical conductivity. CNTs have acquired prominence as flexible and electrically conductive materials due to their outstanding and new properties [153]. The in vivo results suggest that electroconductive nanoparticles could be used at the interface between peripheral neural cells and the nerve conduit. Other nano/microfabrication techniques, such as self-assembly and electrospinning, are more favorable for brain regeneration due to the aggregation of nanoparticles in bioinks. For the regenerative process of peripheral and central nerves, both synthetic and natural nanofibrous scaffolds such as electrospun PCL-based constructions and self-assembled peptide nanostructures have shown considerable promise [154].

3D Bioprinting in Cancer Research Cancer research is one area where experts hope 3D bioprinting can be valuable. The tumor microenvironment, or the biological structures and elements that surround the tumor rather than just the cells, is an important aspect of cancer behavior that is still poorly understood [155, 156].Experimental animals and 2D cell cultures have been used with some success, but they do not accurately represent the behavior of human cancer cells in vivo, which consists of a complex interaction between tumor cells and the cellular environment in 3D space along with other cells. According to researcher findings, 3D models better explain tumor cell behavior, and bioprinting is an appropriate choice that is being examined. One advantage of 3D bioprinting over other methods is that cells with extracellular and structural elements can be directly built into a 3D model [155, 156].

3.7 Multicomponent Bioinks

Multicomponent bioinks are described as a combination of two or more biomaterials. In this manner, they can reinforce the limited printability and biological performance of single-component bioinks [157]. For example, blending different polymers, and improving the rheological properties of the bioinks during and after printing, in order to obtain high printability and structural stability. Furthermore, the biological performance of single-component bioinks can be improved by the addition of bioactive or extracellular components [157]. Therefore, it is desirable to further develop bioinks to provide a more in vivo-like microenvironment, while ensuring high printability [158]. Multicomponent bioinks are often divided into four groups, according to Cui et al. [157]. Multimaterial bioinks contain two or more types of biomaterials that are capable of covalent crosslinking with one another. In contrast, interpenetrating network bioinks consist of biomaterials that create distinct networks based on independent crosslinking mechanisms. In the printed construct, the networks are intertwined without the need for covalent connections [33]. As a result, by increasing the number of polymers involved in the production of the crosslinking networks in the resulting hydrogel, the mechanical stability of the printed structure can be enhanced [157–158].

For instance, when printing, the ionic or photocrosslinking of alginate initially takes place to protect the printed shape. The secondary crosslinking procedure used on the remaining materials can also help to increase the construct's shape stability [31]. Nanoparticles or nanostructures found in nanocomposite bioinks can make printed structures stiffer and shear-thinning behavior better. Additionally, through boosting bioactivity, nanocomposites can modify cellular behavior. In particular, tissues that require electric impulses, including muscles and nerves, are created using nanocomposite bioinks that incorporate conductive nanobiomaterials. According to the developed stress, supramolecular bioinks build and break reversible links between functional groups connected to a biomaterial via a guest–host bond [31]. These bioinks use the supramolecular crosslinking technique. This capacity for self-healing not only results in shear-thinning behavior but also permits the creation of a network devoid of any chemical agents and external stimuli that can compromise cell viability.

3.8 Future Prospects

Due to their static state after printing, 3D bioprinted transplants may not be able to adapt to a dynamic physiological environment. In contrast, normal human tissues constantly go through physiological dynamic changes brought on by external stimuli in order to carry out their complicated and specialized functional activities. To overcome these 3D bioprinting limitations, it is imperative to create structures that mimic the physiological responses of natural human tissues under a variety of external stimuli. Recently, four-dimensional bioprinting has emerged, where advanced multicomponent bioink materials are capable of being bioprinted and changing their structure in response to external stimuli (such as water, temperature, pH, light, ions concentrations, magnetic field, and electric field). Additionally, thanks to advancements in artificial intelligence technological advances, surgical robots may quickly deposit improved bioinks in real-time through visual examination (*in situ*) and acquire the three- or four-dimensional structure of abnormalities, considerably reducing operation time and patient discomfort.

References

1 Tibbitt, M.W., Rodell, C.B., Burdick, J.A., and Anseth, K.S. (2015). Progress in material design for biomedical applications. *Proc. Natl. Acad. Sci.* 112 (47): 14444–14451.

2 Datta, P., Barui, A., Wu, Y. et al. (2018). Essential steps in bioprinting: from pre- to post-bioprinting. *Biotechnol. Adv.* 36 (5): 1481–1504. https://doi.org/10.1016/j.biotechadv.2018.06.003. (Elsevier Inc., 01 Sept 2018).

3 Gungor-Ozkerim, P.S., Inci, I., Zhang, Y.S. et al. (2018). Bioinks for 3D bioprinting: an overview. *Biomater. Sci.* 6 (5): 915–946.

4 Williams, D., Thayer, P., Martinez, H. et al. (2018). A perspective on the physical, mechanical and biological specifications of bioinks and the development of functional tissues in 3D bioprinting. *Bioprinting* 9: 19–36.

5 Ng, W.L., Yeong, W.Y., and Naing, M.W. (2017). Polyvinylpyrrolidone-based bioink improves cell viability and homogeneity during drop-on-demand printing. *Materials (Basel)* 10 (2): 190.

6 Dababneh, A.B. and Ozbolat, I.T. (2014). Bioprinting technology: a current state-of-the-art review. *J. Manuf. Sci. Eng.* 136 (6): 061016.

7 Xu, T., Binder, K.W., Albanna, M.Z. et al. (2012). Hybrid printing of mechanically and biologically improved constructs for cartilage tissue engineering applications. *Biofabrication* 5 (1): 015001.

8 Müller, M., Becher, J., Schnabelrauch, M., and Zenobi-Wong, M. (2015). Nanostructured Pluronic hydrogels as bioinks for 3D bioprinting. *Biofabrication* 7 (3): 035006.

9 Kolesky, D.B., Truby, R.L., Gladman, A.S. et al. (2014). 3D bioprinting of vascularized, heterogeneous cell -laden tissue constructs. *Adv. Mater.* 26 (19): 3124–3130.

10 O'Connell, C., Ren, J., Pope, L. et al. (2020). Characterizing bioinks for extrusion bioprinting: printability and rheology. *Methods in Molecular Biology* 111–133.

11 Liu, F., Chen, Q., Liu, C. et al. (2018). Natural polymers for organ 3D bioprinting. *Polymers (Basel)* 10 (11): 1–26.

12 Pugliese, R., Beltrami, B., Regondi, S., and Lunetta, C. (2021). Polymeric biomaterials for 3D printing in medicine: an overview. *Ann. 3D Print. Med.* 2: 100011.

13 Luo, Z., Mu, X., and Zhang, Y.S. (2022). Biomaterials for bioprinting. In: *Bioprinting*, 51–86. London: Academic, Elsevier.

14 Tripathi, S., Mandal, S.S., Bauri, S., and Maiti, P. (2023). 3D bioprinting and its innovative approach for biomedical applications. *MedComm. 4* (1): e194.

15 Murphy, S.V. and Atala, A. (2014). 3D bioprinting of tissues and organs. *Nat. Biotechnol. 32* (8): 773–785.

16 Ambesi-Impiombato, F.S., Parks, L.A., and Coon, H.G. (1980). Culture of hormone-dependent functional epithelial cells from rat thyroids. *Proc. Natl. Acad. Sci. USA* 77: 3455–3459.

17 Hamm, A., Krott, N., Breibach, I. et al. (2002). Efficient transfection method for primary cells. *Tissue Eng.* 8: 235–245.

18 Okumura, N., Ueno, M., Koizumi, N. et al. (2009). Enhancement on primate corneal endothelial cell survival in vitro by a ROCK inhibitor. *Invest. Ophthalmol. Vis. Sci.* 50: 3680–3687.

19 Yu, Z., Liu, M., Fu, P. et al. (2012). ROCK inhibition with Y27632 promotes the proliferation and cell cycle progression of cultured astrocyte from spinal cord. *Neurochem. Int.* 61: 1114–1120.

20 Dimri, G.P., Lee, X., Basile, G. et al. (1995). A biomarker that identifies senescent human cells in culture and in aging skin in vivo. *Proc. Natl. Acad. Sci. USA* 92: 9363–9367.

21 Reubinoff, B.E., Pera, M.F., Fong, C.Y. et al. (2000). Embryonic stem cell lines from human blastocysts: somatic differentiation in vitro. *Nat. Biotechnol.* **18**: 399–404.

22 Dominici, M., Le Blanc, K., Mueller, I. et al. (2006). Minimal criteria for defining multipotent mesenchymal stromal cells. The International Society for Cellular Therapy position statement. *Cytotherapy* **8**: 315–317.

23 Zuk, P.A., Zhu, M., Ashjian, P. et al. (2002). Human adipose tissue is a source of multipotent stem cells. *Mol. Biol. Cell* **13**: 4279–4295.

24 De Coppi, P., Bartsch, G., Siddiqui, M.M. et al. (2007). Isolation of amniotic stem cell lines with potential for therapy. *Nat. Biotechnol.* 25: 100–106.

25 Murphy, S. et al. (2010). Amnion epithelial cell isolation and characterization for clinical use. *Curr. Protoc. Stem Cell Biol.* 13: 1E–6E.

26 Lin, C.-C. and Anseth, K.S. (2009). PEG hydrogels for the controlled release of biomolecules inregenerative medicine. *Pharm. Res.* 26 (3): 631–643.

27 Koons, G.L. and Mikos, A.G. (2019). Progress in three-dimensional printing with growthfactors. *J. Control. Release* 295: 50–59.

28 Ionescu, L.C., Lee, G.C., Sennett, B.J. et al. (2010). An anisotropicnanofiber/microsphere composite with controlled release of biomolecules for fibrous tissue engineering. *Biomaterials* 31 (14): 4113–4120.

29 Caliceti, P., Salmaso, S., Lante, A. et al. (2001). Controlled release of biomolecules from temperature-sensitivehydrogels prepared by radiation polymerization. *J. Control. Release* 75 (1–2): 173–181.

30 Byambaa, B., Annabi, N., Yue, K. et al. (2017). Bioprinted osteogenic and vasculogenic patterns for engineering 3D bone tissue. *Adv. Healthcare Mater.* 6 (16): 1700015.

31 Ashammakhi, N., Ahadian, S., Xu, C. et al. (2019). Bioinks and bioprinting technologies to make heterogeneous and biomimetic tissue constructs. *Mater. Today Bio* 1: 100008.

32 Daszek, J., Costantini, M., Karlsen, T.A. et al. (2019). 3D bioprinting of hydrogel constructs with cell and material gradients for the regeneration of full-thickness chondral defectusing a microfluidic printing head. *Biofabrication* 11: 044101.

33 Choi, Y.J., Park, H., Ha, D.H. et al. (2021). 3D bioprinting of in vitro models using hydrogel-based bioinks. *Polymers* 13: 366.

34 Ji, S. and Guvendiren, M. (2017). Recent advances in bioink design for 3D bioprinting of tissues and organs. *Front. Bioeng. Biotechnol.* 5: 1–8. https://doi.org/10.3389/fbioe.2017.00023.

35 Duan, B., Kapetanovic, E., Hockaday, L.A., and Butcher, J.T. (2014). Three-dimensional printed trileaflet valve conduits using biological hydrogels and human valve interstitial cells. *Acta Biomater.* 10: 1836–1846. https://doi.org/10.1016/j.actbio.2013.12.005.

36 Vafaei, S., Tuck, C., Ashcroft, I., and Wildman, R. (2016). Surface microstructuring to modify wettability for 3D printing of nano-filled inks. *Chem. Eng. Res. Des.* 109: 414–420. https://doi.org/10.1016/j.cherd.2016.02.004.

37 Mandrycky, C., Wang, Z., Kim, K., and Kim, D.H. (2016). 3D bioprinting for engineering complex tis-sues. *Biotechnol. Adv.* 34: 422–434. https://doi.org/10.1016/j.biotechadv.2015.12.011.

38 Xu, C., Zhang, M., Huang, Y. et al. (2014). Study of droplet formation process during drop-on-demand inkjetting of living cell-laden bioink. *Langmuir* 30: 9130–9138. https://doi.org/10.1021/la501430x.

39 Rutz, A.L., Hyland, K.E., Jakus, A.E. et al. (2015). A multimaterial bioink method for 3D printing tunable, cell-compatible hydrogels. *Adv. Mater.* 27: 1607–1614.

40 Williams, D.F. (2008). On the mechanisms of biocompatibility. *Biomaterials* 29: 2941–2953.

41 Murphy, S.V., Skardal, A., and Atala, A. (2013). Evaluation of hydrogels for bio-printing applications. *J. Biomed. Mater. Res. A* 101: 272–284.

42 Ringeisen, B.R., Kim, H., Barron, J.A. et al. (2004). Laser printing of pluripotent embryonal carcinoma cells. *Tissue Eng.* 10: 483–491.

43 Ananthanarayanan, A., Narmada, B.C., Mo, X. et al. (2011). Purpose-driven biomaterials research in liver-tissue engineering. *Trends Biotechnol.* 29: 110–118.

44 Hutmacher, D.W. (2000). Scaffolds in tissue engineering bone and cartilage. *Biomaterials* 21: 2529–2543.

45 Limpanuphap, S. and Derby, B. (2002). Manufacture of biomaterials by a novel printing process. *J. Mater. Sci. Mater. Med.* 13: 1163–1166.

46 Miller, J.S., Stevens, K.R., Yang, M.T. et al. (2012). Rapid casting of patterned vascular networks for perfusable engineered three-dimensional tissues. *Nat. Mater.* 11: 768–774.

47 Wade, R.J., Bassin, E.J., Rodell, C.B., and Burdick, J.A. (2015). Protease degradable electrospun fibrous hydrogels. *Nat. Commun.* 6: 6639.

48 Hopp, B., Smausz, T., Kresz, N. et al. (2005). Survival and proliferative ability of various living cell types after laser-induced forward transfer. *Tissue Eng.* 11: 1817–1823.

49 Zhang, S., Holmes, T.C., DiPersio, C. et al. (1995). Self-complementary oligopeptide matrices support mammalian cell attachment. *Biomaterials* 16: 1385–1393.

50 Hersel, U., Dahmen, C., and Kessler, H. (2003). RGD modified polymers: biomaterials for stimulated cell adhesion and beyond. *Biomaterials* 24: 4385–4415.

51 Teixeira, A.I., Nealey, P.F., and Murphy, C.J. (2004). Responses of human keratocytes to micro- and nanostructured substrates. *J. Biomed. Mater. Res. A* 71: 369–376.

52 Karp, J.M., Yeh, J., Eng, G. et al. (2007). Controlling size, shape and homogeneity of embryoid bodies using poly(ethylene glycol) microwells. *Lab Chip* 7: 786–794.

53 Price, R.L., Haberstroh, K.M., and Webster, T.J. (2003). Enhanced functions of osteoblasts on nanostructured surfaces of carbon and alumina. *Med. Biol. Eng. Comput.* 41: 372–375.

54 Sullivan, D.C., Mirmalek-Sani, S.H., Deegan, D.B. et al. (2012). Decellularization methods of porcine kidneys for whole organ engineering using a high-throughput system. *Biomaterials* 33: 7756–7764.

55 Hynes, R.O. and Naba, A. (2012). Overview of the matrisome–an inventory of extracellular matrix constituents and functions. *Cold Spring Harb. Perspect. Biol.* 4: a004903.

56 Kim, D., Kim, M., Lee, J., and Jang, J. (2022). Review on multicomponent hydrogel bioinks based on natural biomaterials for bioprinting 3D liver tissues. *Front. Bioeng. Biotechnol.* 14 (10): 764682. https://doi.org/10.3389/fbioe.2022.764682.
57 Lee, K. and Cha, C. (2020). Advanced polymer-based bioink technology for printing soft biomaterials. *Macromol. Res.* 28: 689–702.
58 Zarrintaj, P., Manouchehri, S., Ahmadi, Z. et al. (2018). Agarose-based biomaterials for tissue engineering. *Carbohydr. Polym.* 187: 66–84.
59 Aydin, L., Kucuk, S., and Kenar, H. (2020). A universal self-eroding sacrificial bioink that enables bioprinting at room temperature. *Polym. Adv. Technol.* 31: 1634–1647.
60 Daly, A.C., Critchley, S.E., Rencsok, E.M., and Kelly, D.J. (2016). A comparison of different bioinks for 3D bioprinting of fibrocartilage and hyaline cartilage. *Biofabrication* 8: 045002.
61 Lee, K.Y. and Mooney, D.J. (2012). Alginate: properties and biomedical applications. *Prog. Polym. Sci.* 37: 106–126.
62 Emami, Z., Ehsani, M., Zandi, M., and Foudazi, R. (2018). Controlling alginate oxidation conditions for making alginate-gelatin hydrogels. *Carbohydr. Polym.* 198: 509–517.
63 Bouhadir, K.H., Lee, K.Y., Alsberg, E. et al. (2001). Degradation of partially oxidized alginate and its potential application for tissue engineering. *Biotechnol. Prog.* 17: 945–950.
64 Lee, J., Hong, J., Kim, W., and Kim, G.H. (2020). Bone-derived dECM/alginate bioink for fabricating a 3D cell-laden mesh structure for bone tissue engineering. *Carbohydr. Polym.* 250: 116914. [CrossRef] [PubMed].
65 Müller, M., Öztürk, E., Arlov, Ø. et al. (2017). Alginate sulfate–nanocellulose bioinks for cartilage bioprinting applications. *Ann. Biomed. Eng.* 45: 210–223.
66 Jia, J., Richards, D.J., Pollard, S. et al. (2014). Engineering alginate as bioink for bioprinting. *Acta Biomater.* 10: 4323–4331.
67 Rajabi, N., Rezaei, A., Kharaziha, M. et al. (2021). Recent advances on bioprinted gelatin methacrylate-based hydrogels for tissue repair. *Tissue Eng.-Part A* 27: 679–702.
68 Zhang, L., Zhang, X., Li, L. et al. (2020). Fabrication of photothermally responsive nanocomposite hydrogel through 3D printing. *Macromol. Mater. Eng.* 305 (2): 1900718.
69 Lee, A.R.H.A., Hudson, A.R., Shiwarski, D.J. et al. (2019). 3D bioprinting of collagen to rebuild components of the human heart. *Science* 365 (6452): 482–487.
70 Jain, T., Baker, H.B., Gipsov, A. et al. (2021). Impact of cell density on the bioprinting of gelatin methacrylate (GelMA) bioinks. *Bioprinting* 22: e00131.
71 Kim, S.H., Yeon, Y.K., Lee, J.M. et al. (2018). Precisely printable and biocompatible silk fibroin bioink for digital light processing 3D printing. *Nat. Commun.* 9: 1–14.

72 Singh, Y.P., Bandyopadhyay, A., and Mandal, B.B. (2019). 3D bioprinting using cross-linker-free silk-gelatin bioink for cartilage tissue engineering. *ACS Appl. Mater. Interfaces* 11: 33684–33696.

73 Samadian, H., Maleki, H., Allahyari, Z., and Jaymand, M. (2020). Natural polymers-based light-induced hydrogels: promising biomaterials for biomedical applications. *Coord. Chem. Rev.* 420: 213432.

74 Vasvani, S., Kulkarni, P., and Rawtani, D. (2020). Hyaluronic acid: a review on its biology, aspects of drug delivery, route of administrations and a special emphasis on its approved marketed products and recent clinical studies. *Int. J. Biol. Macromol.* 151: 1012–1029.

75 Hospodiuk, M., Dey, M., Sosnoski, D., and Ozbolat, I.T. (2017). The bioink: a comprehensive review on bioprintable materials. *Biotechnol. Adv.* 35 (2): 217–239.

76 Antich, C., de Vicente, J., Jiménez, G. et al. (2020). Bio-inspired hydrogel composed of hyaluronic acid and alginate as a potential bioink for 3D bioprinting of articular cartilage engineering constructs. *Acta Biomater.* 106: 114–123.

77 Kiyotake, E.A., Douglas, A.W., Thomas, E.E. et al. (2019). Development and quantitative characterization of the precursor rheology of hyaluronic acid hydrogels for bioprinting. *Acta Biomater.* 95: 176–187.

78 Guvendiren, M., Molde, J., Soares, R.M.D., and Kohn, J. (2016). Designing biomaterials for 3D printing. *ACS Biomater. Sci. Eng.* 2: 1679–1693.

79 Ahn, S., Yoon, H., Kim, G. et al. (2010). Designed three-dimensional collagen scaffolds. *Tissue Eng. Part C Methods* 16: 813–820.

80 Chan, W.W., Chen, D., Yeo, L. et al. (2020). Additive biomanufacturing with collagen inks. *Bioengineering* 7: 66.

81 Shi, Y., Xing, T.L., Zhang, H.B. et al. (2018). Tyrosinase-doped bioink for 3D bioprinting of living skin constructs. *Biomed. Mater.* 13 (3): 035008.

82 Wang, H.Y., Zhang, Y.Q., and Wei, Z.G. (2021). Characterization of undegraded and degraded silk fibroin and its significant impact on the properties of the resulting silk biomaterials. *Int. J. Biol. Macromol.* 176: 578–588.

83 Chakraborty, J. and Ghosh, S. (2020). Cellular proliferation, self-assembly, and modulation of signaling pathways in silk fibroin gelatin-based 3D bioprinted constructs. *ACS Appl. Bio Mater.* 3: 8309–8320.

84 Wang, Q., Han, G., Yan, S., and Zhang, Q. (2019). 3D printing of silk fibroin for biomedical applications. *Materials* 12: 504.

85 Oliveira, J.M. (2020). Current and future trends of silk fibroin-based bioinks in 3D printing. *J. 3D Print. Med.* 4: 69–73.

86 Wei, L., Wu, S., Kuss, M. et al. (2019). 3D printing of silk fibroin-based hybrid scaffold treated with platelet rich plasma for bone tissue engineering. *Bioact. Mater.* 4: 256–260.

87 Kim, E., Seok, J.M., Bae, S.B. et al. (2021). Silk fibroin enhances cytocompatibilty and dimensional stability of alginate hydrogels for light-based three-dimensional bioprinting. *Biomacromolecules* 22: 1921–1931.

88 Keane, T.J., Swinehart, I.T., and Badylak, S.F. (2015). Methods of tissue decellularization used for preparation of biologic scaffolds and in vivo relevance. *Methods* 84: 25–34.

89 Pati, F., Jang, J., Ha, D.H. et al. (2014). Printing three-dimensional tissue analogues with decellularized extracellular matrix bioink. *Nat. Commun.* 5: 3935.

90 Han, W., Singh, N.K., Kim, J.J. et al. (2019). Directed differential behaviors of multipotent adult stem cells from decellularized tissue/organ extracellular matrix bioinks. *Biomaterials* 224: 119496.

91 Shi, L., Hu, Y., Ullah, M.W. et al. (2019). Cryogenic free-form extrusion bioprinting of decellularized small intestinal submucosa for potential applications in skin tissue engineering. *Biofabrication* 11: 035023.

92 Kim, J.-Y., Ahn, G., Kim, C. et al. (2018). Synergistic effects of beta tri-calcium phosphate and porcine-derived decellularized bone extracellular matrix in 3D-printed polycaprolactone scaffold on bone regeneration. *Macromol. Biosci.* 18: 1800025.

93 Toprakhisar, B., Nadernezhad, A., Bakirci, E. et al. (2018). Development of bioink from decellularized tendon extracellular matrix for 3D bioprinting. *Macromol. Biosci.* 18: e1800024.

94 Zhao, F., Cheng, J., Sun, M. et al. (2020). Digestion degree is a key factor to regulate the printability of pure tendon decellularized extracellular matrix bio-ink in extrusion-based 3D cell printing. *Biofabrication* 12: 045011.

95 Jang, J., Park, H.J., Kim, S.W. et al. (2017). 3D printed complex tissue construct using stem cell-laden decellularized extracellular matrix bioinks for cardiac repair. *Biomaterials* 112: 264–274.

96 Veronese, F.M. and Pasut, G. (2005). PEGylation, successful approach to drug delivery. *Drug Discov. Today* 10 (21): 1451–1458.

97 Ozbolat, I.T. and Hospodiuk, M. (2016). Current advances and future perspectives in extrusion-based bioprinting. *Biomaterials* 76: 321–343.

98 Bakht, S.M., Pardo, A., Gómez-Florit, M. et al. (2021). Engineering next-generation bioinks with nanoparticles: moving from reinforcement fillers to multifunctional nanoelements. *J. Mater. Chem. B* 9 (25): 5025–5038.

99 Aoki, K. and Saito, N. (2020). Biodegradable polymers as drug delivery systems for bone regeneration. *Pharmaceutics* 12 (2): 95.

100 Nowicki, M., Zhu, W., Sarkar, K. et al. (2020). 3D printing multiphasic osteochondral tissue constructs with nano to micro features via PCL based bioink. *Bioprinting* 17: e00066.

101 Izgordu, M.S., Uzgur, E.I., Ulag, S. et al. (2021). Investigation of 3D-printed polycaprolactone-/polyvinylpyrrolidone-based constructs. *Cartilage* 13 (2_suppl): 626S–635S.
102 Du, X., Fu, S., and Zhu, Y. (2018). 3D printing of ceramic-based scaffolds for bone tissue engineering: an overview. *J. Mater. Chem. B* 6 (27): 4397–4412.
103 Tamay, D.G., Usal, T.D., Alagoz, A.S. et al. (2019). 3D and 4D printing of polymers for tissue engineering applications. *Front. Bioeng. Biotechnol.* 7: 164.
104 Spangenberg, J., Kilian, D., Czichy, C. et al. (2021). Bioprinting of magnetically deformable scaffolds. *ACS Biomater. Sci. Eng.* 7 (2): 648–662.
105 Gaharwar, A.K., Peppas, N.A., and Khademhosseini, A. (2014). Nanocomposite hydrogels for biomedical applications. *Biotechnol. Bioeng.* 111 (3): 441–453.
106 Tadic, D. and Epple, M. (2004). A thorough physicochemical characterisation of 14 calcium phosphate-based bone substitution materials in comparison to natural bone. *Biomaterials* 25 (6): 987–994.
107 Wang, X., Schro¨der, H.C., Wiens, M. et al. (2012). Bio-silica and bio-polyphosphate: applications in biomedicine (bone formation). *Curr. Opin. Biotechnol.* 23 (4): 570–578.
108 Poldervaart, M.T., Gremmels, H., Van Deventer, K. et al. (2014). Prolonged presence of VEGF promotes vascularization in 3D bioprinted scaffolds with defined architecture. *J. Control. Release* 184: 58–66.
109 Gonzalez-Fernandez, T., Rathan, S., Hobbs, C. et al. (2019). Pore-forming bioinks to enable spatio-temporally defined gene delivery in bioprinted tissues. *J. Control. Release* 301: 13–27.
110 Deng, Z., Hu, T., Lei, Q. et al. (2019). Stimuli-responsive conductive nanocomposite hydrogels with high stretchability, self-healing, adhesiveness, and 3D printability for human motion sensing. *Mater. Interfaces* 11 (7): 6796–6808.
111 De Sanctis, A., Mehew, J.D., Craciun, M.F., and Russo, S. (2018). Graphene-based light sensing: fabrication, characterisation, physical properties and performance. *Materials* 11 (9): 1762. https://doi.org/10.3390/ma11091762.
112 Mu, J., Hou, C., Wang, H. et al. (2015). Origami-inspired active graphene-based paper for programmable instant self-folding walking devices. *Sci. Adv.* 1 (10): e1500533.
113 Pardo, A., Pelaz, B., Gallo, J. et al. (2020). Synthesis, characterization, and evaluation of superparamagnetic doped ferrites as potential therapeutic nanotools. *ACS Nano* 32 (6): 2220–2231.
114 Kamdem Tamo, A., Doench, I., Walter, L. et al. (2021). Development of bioinspired functional chitosan/cellulose nanofiber 3D hydrogel constructs by 3D printing for application in the engineering of mechanically demanding tissues. *Polymers* 13 (10): 1663.
115 Chen, S.Y., Cho, Y.C., Yang, T.S. et al. (2021). A tailored biomimetic hydrogel as potential bioink to print a cell scaffold for tissue engineering applications: printability and cell viability evaluation. *Appl. Sci.* 11 (2): 829.

116 De Santis, R., Russo, T., Rau, J.V. et al. (2021). Design of 3D additively manufactured hybrid structures for cranioplasty. *Materials* 14 (1): 181.

117 Chen, Y., Gong, Y., Shan, L. et al. (2023). Research on cartilage 3D printing technology based on SA-GA-HA. *Materials* 16 (15): 5312.

118 Kakarla, A.B., Kong, I., Kong, C., and Irving, H. (2022). Extrusion-based bioprinted boron nitride nanotubes reinforced alginate scaffolds: mechanical, printability and cell viability evaluation. *Polymers* 14 (3): 486.

119 Zheng, Z., Wu, J., Liu, M. et al. (2018). 3D bioprinting of self-standing silk-based bioink. *Adv. Healthcare Mater.* 7 (6): 1701026.

120 Khoshnood, N. and Zamaniain, A. (2020). A comprehensive review on scaffold-free bioinks for bioprinting. *Bioprinting* 19: e00088.

121 Xiaorui, L., Fuyin, Z., Xudong, W. et al. (2023). 1Biomaterial inks for extrusion-based 3D bioprinting: Property, classification, modification, and selection. *Int. J. Bioprinting* 9 (2): 649.

122 Ooi, H.W., Mota, C., Ten Cate, A.T. et al. (2018). Thiol–Ene alginate hydrogels as versatile bioinks for bioprinting. *Biomacromolecules* 19: 3390–3400. https://doi.org/10.1021/acs.biomac.8b00696.

123 Choudhury, D., Anand, S., and Naing, M.W. (2018). The arrival of commercial bioprinters-Towards 3D bioprinting revolution! *Int. J. Bioprinting* 4: 139.

124 Emmermacher, J., Spura, D., Cziommer, J. et al. (2020). Engineering considerations on extrusion-based bioprinting: interactions of material behavior, mechanical forces and cells in the printing needle. *Biofabrication* 12: 025022.

125 Bedell, M.L., Torres, A.L., Hogan, K.J. et al. (2022). Human gelatin-based composite hydrogels for osteochondral tissue engineering and their adaptation into bioinks for extrusion, inkjet, and digital light processing bioprinting. *Biofabrication* 14: 045012. https://doi.org/10.1088/1758-5090/ac8768.

126 Li, X., Liu, B., Pei, B. et al. (2020). Inkjet bioprinting of biomaterials. *Chem. Rev.* 120: 10793–10833.

127 Gudapati, H. and Ozbolat, I.T. (2020). The role of concentration on drop formation and breakup of collagen, fibrinogen, and thrombin solutions during inkjet bioprinting. *bioRxiv* 36: 15373–15385. https://doi.org/10.1021/acs.langmuir.0c02926.

128 Dufour, A., Gallostra, X.B., O'Keeffe, C. et al. (2022). Integrating melt electrowriting and inkjet bioprinting for engineering structurally organized articular cartilage. *Biomaterials* 283: 121405. https://doi.org/10.1016/j.biomaterials.2022.121405.

129 Kumar, H. and Kim, K. (2020). *Stereolithography 3D Bioprinting*, vol. 2140, 93–108. Germany: Springer; Berlin/Heidelberg.

130 Lam, T., Dehne, T., Krüger, J.P. et al. (2019). Photopolymerizable gelatin and hyaluronic acid for stereolithographic 3D bioprinting of tissue-engineered cartilage. *J. Biomed. Mater. Res. Part B Appl. Biomater.* 107: 2649–2657.

131 Yang, H., Yang, K.-H., Narayan, R.J., and Ma, S. (2021). Laser-based bioprinting for multilayer cell patterning in tissue engineering and cancer research. *Essays Biochem.* 65: 409–416.

132 Sorkio, A., Koch, L., Koivusalo, L. et al. (2018). Human stem cell based corneal tissue mimicking structures using laser-assisted 3D bioprinting and functional bioinks. *Biomaterials* 171: 57–71. https://doi.org/10.1016/j.biomaterials.2018.04.034.

133 Dou, C., Perez, V., Qu, J. et al. (2021). A state-of-the-art review of laser-assisted bioprinting and its future research trends. *ChemBioEng Rev.* 8: 517–534. https://doi.org/10.1002/cben.202000037.

134 Pfister, A., Landers, R., Laib, A. et al. (2004). Biofunctional rapid prototyping for tissue-engineering applications: 3D bioplotting versus 3D printing. *J. Polym. Sci. Part A Polym. Chem.* 42 (3): 624–638.

135 O'Brien, C.M., Holmes, B., Faucett, S., and Zhang, L.G. (2014). Three-dimensional printing of nanomaterial scaffolds for complex tissue regeneration. *Tissue Eng. B Rev.* 21 (1): 103–114.

136 Chimene, D., Alge, D.L., and Gaharwar, A.K. (2015). Two-dimensional nanomaterials for biomedical applications: emerging trends and future prospects. *Adv. Mater.* 27 (45): 7261e84.

137 Kang, H.-W., Lee, S.J., Ko, I.K. et al. (2016). A 3D bioprinting system to produce human-scale tissue constructswith structural integrity. *Nat. Biotechnol.* 34 (3): 312.

138 Xavier, J.R., Thakur, T., Desai, P. et al. (2015). Bioactive nanoengineered hydrogels for bone tissue engineering: agrowth-factor-free approach. *ACS Nano* 9 (3): 3109e18.

139 Hong, S., Sycks, D., Chan, H.F. et al. (2015). 3D printing of highly stretchable and tough hydrogels into complex,cellularized structures. *Adv. Mater.* 27 (27): 4035e40.

140 Markstedt, K., Mantas, A., Tournier, I. et al. (2015). 3D bioprinting human chondrocytes with nanocelluloseealginate bioink for cartilage tissue engineering applications. *Biomacromolecules* 16 (5): 1489e96.

141 Mannoor, M.S., Jiang, Z., James, T. et al. (2013). 3D printed bionic ears. *Nano Lett.* 13 (6): 2634e9.

142 Mosadegh, B., Xiong, G., Dunham, S. et al. (2015). Current progress in 3D printing for cardiovascular tissue engineering. *Biomed. Mater.* 10 (3): 034002.

143 Duan, B., Kapetanovic, E., Hockaday, L.A. et al. (2014). Three-dimensional printed trileaflet valve conduits using biologicalhydrogels and human valve interstitial cells. *Acta Biomater.* 10 (5): 1836e46.

144 Dolati, F., Yu, Y., Zhang, Y. et al. (2014). In vitro evaluation of carbon-nanotube-reinforced bioprintablevascular conduits. *Nanotechnology* 25 (14): 145101.

145 Talelli, M., Rijcken, C.J., Lammers, T. et al. (2009). Superparamagnetic iron oxide nanoparticles encapsulated in biodegradable thermosensitive polymeric micelles: toward a targeted nanomedicine suitable for image-guided drug delivery. *Langmuir* 25 (4): 2060e7.

146 Agung, N.P., Nadhif, M.H., Irdam, G.A., and Mochtar, C.A. (2021). The role of 3D-printed phantoms and devices for organ-specified appliances in urology. *Int. J. Bioprinting* 7: 333. https://doi.org/10.18063/ijb.v7i2.333.

147 Farhat, W., Chatelain, F., Marret, A. et al. (2021). Trends in 3D bioprinting for esophageal tissue repair and reconstruction. *Biomaterials* 267: 120465.

148 Nam, H., Jeong, H.-J., Jo, Y. et al. (2020). Multilayered free-form 3D cell-printed tubular construct with decellularized inner and outer esophageal tissue-derived bioinks. *Sci. Rep.* 10: 7255.

149 Hu, H. and Xu, F.-J. (2020). Rational design and latest advances of polysaccharide-based hydrogels for wound healing. *Biomater. Sci.* 8: 2084–2101.

150 Cubo, N., Garcia, M., del Cañizo, J.F. et al. (2016). 3D bioprinting of functional human skin: production and in vivo analysis. *Biofabrication* 9: 015006.

151 Shafiee, A., Cavalcanti, A.S., Saidy, N.T. et al. (2021). Convergence of 3D printed biomimetic wound dressings and adult stem cell therapy. *Biomaterials* 268: 120558.

152 Zhou, F., Hong, Y., Liang, R. et al. (2020). Rapid printing of bio-inspired 3D tissue constructs for skin regeneration. *Biomaterials* 258: 120287.

153 Chan, K.M., Gordon, T., Zochodne, D.W., and Power, H.A. (2014). Improving peripheral nerve regeneration: from molecular mechanisms to potential therapeutic targets. *Exp. Neurol.* 261: 826e35.

154 Ahn, H.-S., Hwang, J.Y., Kim, M.S. et al. (2015). Carbon-nanotube-interfaced glass fiber scaffold for regeneration of transected sciatic nerve. *Acta Biomater.* 13: 324e34.

155 Zhu, W., O'Brien, C., O'Brien, J.R. et al. (2014). 3D nano/microfabrication techniques and nanobiomaterials for neuraltissue regeneration. *Nanomedicine* 9 (6): 859e75.

156 Ozbolat, I.T., Peng, W., and Ozbolat, V. (2016). Application areas of 3D bioprinting. *Drug Discov. Today* 21 (8): 1257–1271.

157 Novak, C.M., Horst, E.N., Taylor, C.C. et al. (2019). Fluid shear stress stimulates breast cancer cells to display invasive and chemoresistant phenotypes while upreg- ulating PLAU in a 3D bioreactor. *Biotechnol. Bioeng.* 116 (11): 3084–3097.

158 Cui, X., Li, J., Hartanto, Y. et al. (2020). Advances in extrusion 3D bioprinting: a focus on multicomponent hydrogel-based bioinks. *Adv. Healthc. Mater.* 9: 1901648.

4

Printed Scaffolds in Tissue Engineering

Thara Tom[1,3], Samanta Sam[2], Josmin P. Jose[1], M.S. Sreekala[3], and Sabu Thomas[4,5,6]

[1] *Mar Thoma College, Tiruvalla, Kerala, India*
[2] *School of Energy Materials, Mahatma Gandhi University, Kottayam, Kerala, India*
[3] *School of Chemical Sciences, Mahatma Gandhi University, Kottayam, Kerala, India*
[4] *School of Energy Materials, School of Nanoscience and Nanotechnology, School of Polymer Science and Technology, School of Chemical Science and International and Inter University Centre for Nanoscience and Technology (IIUCNN), Mahatma Gandhi University, Kottayam, Kerala, India*
[5] *Department of Chemical Sciences, University of Johannesburg, Doornfontein, Johannesburg, South Africa*
[6] *TrEST Research Park, Sreekariyam, Trivandrum, Kerala, India*

4.1 Introduction

3 dimensional (3D) printing, which is officially known as additive manufacturing (AM), has emerged as a tremendous tool for the manufacturing of complex objects in a less time-consuming and cost-effective manner over the past few decades [1]. Charles Hull is known as the father of 3D printing as this technique was first patented by him in 1986 [2]. AM helps to manufacture objects that are difficult to manufacture by conventional manufacturing techniques, in a technically and economically feasible way. 3D printing is a rapid prototyping technique that involves the production of 3D physical models by layer-by-layer deposition of material using 3D computer models [3]. Nowadays, 3D printing technology is widely utilized in the world for the mass customization and production of materials with suitable designs in various fields like agriculture, automotive industries, health care units, and aerospace industries [4]. Its utilization in healthcare systems, like in drug delivery [5], scaffolds [6], and tissue engineering [7], is increasing tremendously because of the advantages it offers. 3D printing helps to replicate the cellular and intricate architecture of tissues and organs. So, it is a potential technique in biomedical research and tissue engineering. Materials like polymers, ceramics, metals, etc. can be used for

3D Bioprinting from Lab to Industry, First Edition. Edited by Prosenjit Saha, Sabu Thomas, Jinku Kim, and Manojit Ghosh.

Table 4.1 Method, specifications, and materials for different 3D printing technologies.

3D printing method	Specifications	Materials used	Reference
Stereolithography (SLA)	Uses UV light to focus material matrix and print.	Photo-curable polymers and resins	[8]
Fused deposition modelling (FDM)	Material is deposited through a heated nozzle and printing takes place in a layered fashion	Thermoplastics like polylactic acid (PLA)	[9]
Selective laser sintering (SLS)	Uses a high-energy laser source and focuses on the granular powder matrix for printing.	Ceramics and metals	[10]
Electron Beam Melting (EBM)	Uses high energy electron beam on powder layers	Ceramic metal powders Co-Cr alloy	[11]
Extrusion based printing	Uses mechanical/pneumatic forces for extruding materials for printing	Hydrogels polymers	[12]
3D bioprinting	Layer-wise assembly of living cells, bioink for model fabrication	Hydrogels bio-glasses	[13]

AM using different printing technologies. Table 4.1 shows different 3 dimensional methods and materials used for various 3D printing technologies.

AM has various potential applications in automobile, aerospace, electronic, and mechanical engineering areas for the construction of even complex geometries within a short period of time. Among all of these applications in various fields, biomedical fabrications like scaffolds and implants have more dominance due to their large necessity all over the world. Even during the COVID pandemic, 3D constructs played an important role in the well curing. This chapter focuses on the biomedical application of 3D printing technology in tissue engineering. It briefly explains the various 3D printing techniques like fused deposition modeling (FDM), stereolithography (SLA), and various others for the 3D printing of scaffolds and implants. It also focuses on the wide applicability of 3D printing during the COVID-19 pandemic and the advantages of 3D printed scaffolds in various tissue engineering applications.

4.2 Biomedical Application of 3D Printing

3D printing technology has evolved as one of the leading manufacturing techniques in the healthcare sector since it helps in the synthesis of patient-specific implants and instruments that are a perfect match with the individual anatomy [14]. 3D printed materials find a wide range of medical applications in dentistry, scaffolds, drug formulation, tissue engineering, medical devices, regenerative medicine, etc. In order to understand the true potential of these 3D printed devices, it is necessary to know about different techniques used for its preparation.

There are different types of 3D printing technologies for biomedical applications [11]. This includes powder-based printing, material deposition, SLA, and sheet lamination. Some examples of such techniques and the material used in it and their biomedical application is shown in Table 4.2. Powder-based technique includes powder bed fusion (PBF) and binder jetting. Material deposition includes direct energy deposition (DED), material extrusion, material jetting (MJ), and bioprinting [21]. PBF involves the melting of the material using laser beams and a new layer of powder is deposited on the top and the process continues. Binder jetting, which is also known as Powder bed binder jet printing (BJP), is a low-cost and fast technique for the process of complex structures in which the powder is deposited layer-by-layer and is bonded with the binder selectively [22]. Both these methods use metal powders as raw material. DED adds material which is either metal powder or wire alongside the heat input (such as laser, plasma arc, and electron beam) simultaneously [23]. The powder DED system uses single or multiple nozzles for the ejection of metal powders. Extrusion-based 3D printing is one of the most common printing methodologies and uses thermoplastic as the material, which is melted to a semiliquid state and gets deposited layer-by-layer one over the other [24].

Table 4.2 3D printing techniques, materials used, and their biomedical application.

3D printing technique	Material used for printing	Biomedical applications	Reference
Binder jetting	Silk powder	Sutures for wound healing	[15]
Inkjet	Magnesium oxide (MgO)/ Zinc oxide (ZnO)-Tricalcium phosphate (TCP)	Bone repair and regeneration applications.	[16]
Binder jetting	Bioactive glass (BG)-TCP	Bone graft and regeneration application	[17]
Stereolithography (SLA)	nanocrystalline hydroxyapatite (nHA)-poly(lactic-co-glycolic) acid (PLGA) nanospheres encapsulated with transforming growth factor β1 (TGF-β1)	Mechanically stable osteochondral scaffolds	[18]
Stereolithography (SLA)	cartilage extracellular matrix (ECM)/gelatin methacrylate (GelMA)/exosome	Osteochondral defect regeneration	[19]
Powder bed fusion (PBF)	Ti-6Al-4V alloy powder	Orthopedic application	[20]

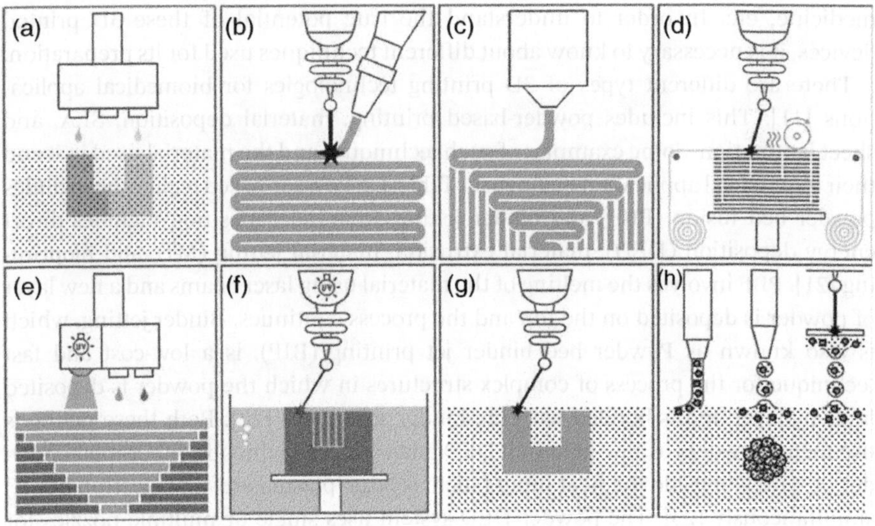

Figure 4.1 Types of 3D printing (a) binder jetting (b) directed energy deposition (c) material extrusion (d) sheet lamination (e) material jetting (MJ) (f) stereolithography (g) powder bed fusion (h) extrusion bioprinting (left), Inkjet bioprinting (center), Laser-assisted bioprinting (right) [21].

This technique can be used with pastes and hydrogels for coprinting with living cells. Extrusion and melting methods are considered slow methods when compared with SLA because of the low resolution and adhesion properties shown by the former. SLA uses light as a source of energy and is one of the most commonly used 3D printing methods because of its accuracy and resolution. SLA 3DP mechanism involves the computer-spatially-controlled solidification of a liquid resin by a process called photopolymerization [2]. As the photopolymerized layer becomes solidified on exposure to light, the process then continues. Another method used is 3D bioprinting. This method involves the 3D printing of organs and tissues layer-by-layer [25]. The ink used for bioprinting is known as bioink. It consists of polymer composition along with living cells where the latter is absent in printing scaffolds otherwise present during the 3D printing of tissues and organs. Figure 4.1 shows various types of 3D printing techniques [21].

Table 4.2 discusses 3D printing methods and applications based on different materials.

4.2.1 Implants and Scaffolds

3D printing is at the forefront of current medical technology in creating scaffolds for medical implants [26]. 3D printing helps in the manufacture of implants that are perfect in size and shape for the recipient and this can be done by computed topography (CT) and magnetic resonance imaging (MRI). 3D porous architecture of the scaffold-based strategies offers great promise for regenerative medicine and tissue engineering since it helps in tissue regeneration [27]. One such 3D printed scaffold for bone tissue regeneration was synthesized by Khalyfa et al. using tetracalcium phosphate (TTCP) as a reactive component and β-tricalcium phosphate (β-TCP) or calcium sulfate as a biodegradable filler [28]. Calcium phosphate is similar to bone minerals and is a biocompatible material that helps in tissue regeneration. Engineered scaffolds and implants can be regarded as an extracellular matrix (ECM) analog which helps in cell integration and its growth [29]. Two calcium phosphate (CaP) based scaffolds with hydroxyapatite (HA), that is, HA and HA-TCP, were manufactured by SLA 3D printing technology to study the intra-pore bone formation and osseointegration for further intra-oral bone regeneration application [30]. This scaffold was found to be biocompatible and showed osseointegration and non-cytotoxicity properties. Another 3D printed scaffold for bone tissue engineering was fabricated by Radhakrishnan et al. This was a 3D printed polycaprolactone/silver ion nanoparticle (PCL/Ag NPs) filament using the FDM technique [31]. Customized fabrication of this scaffold was done by encapsulating Ag NPs into the PCL matrix. Figure 4.2 shows the 3D printed scaffolds with neat PCL and the Ag NPs encapsulated PCL scaffold; the brown color is due to the Ag nanoparticles. 1Ag2, 3Ag2 in which 1 and 3 indicate two different percentages of silver nitrate solution (1% and 3% w/v) added to the PCL matrix. The in vivo studies indicated that the scaffolds displayed less cytotoxicity, indicating that it is a potential candidate to be used in bone tissue engineering applications. Its antibacterial activity against *E. Coli* was conducted and both of them showed an inhibition zone against the bacteria but the inhibition zone was more profound for the 3Ag2 scaffold because of the increased Ag ion content in it when compared with the 1Ag2 scaffold. The study showed that the synthesized scaffold is biocompatible and has antimicrobial properties which is an essential criterion for a scaffold for these tissue engineering purposes.

The use of alumina-based ceramic implant specimens is on the rise due to their high compression strength and wear resistance properties. These qualities make

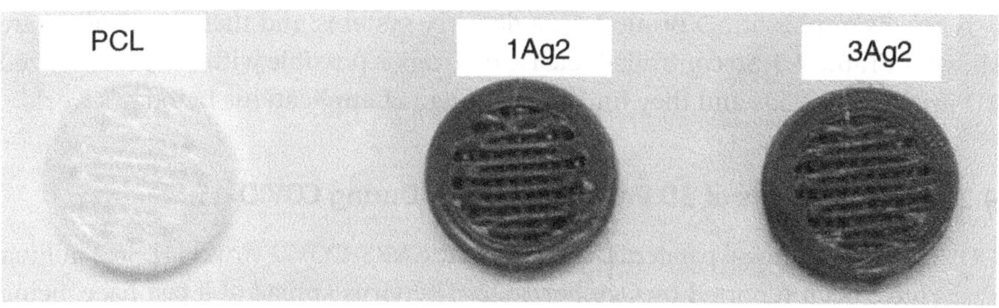

Figure 4.2 Photo of the 3D printed scaffolds (1 cm is denoted by scale bar). *Source:* Radhakrishnan et al. [31] / with permission from Elsevier.

them particularly useful in stomatology and hip-joint surgery [32]. They are biologically inert and can be used as a good implant. Safonov et al. synthesized an alumina-based ceramic implant in the shape of a trabecular bone as a bone substitute using the 3D stereolithographic technique [33]. Other studies were conducted to fabricate stereolithographic 3D printed bioceramic scaffold Ca2.5Na(PO4)2, which is found to be noncytotoxic, biocompatible, and showed high human fibroblast activity [34]. There are a wide variety of projects working on the synthesis of biocompatible 3D printed scaffolds with different techniques and bioinks.

4.2.2 Drug Delivery/Drug Modeling Application

3D printed drug delivery systems are way more promising than conventional drug delivery systems [35]. This is because it helps in the fabrication of personalized drug delivery devices, specific for each and every individual, which is otherwise impossible with the conventional systems as it is common to each one of them and can cause side effects due to overdose or underdose. 3D printed drug delivery systems facilitate sustained and controlled release of drugs. Goyanes et al. synthesized anti-acne patches/masks that match the anatomy of the patient by using fused deposition modeling (FDM) and SLA, in which an anti-acne agent, salicylic acid, was incorporated [36]. Among them, SLA printing is found to be more efficient in the sustained release of drugs, which is the salicylic acid to the specific affected areas. Economidou et al. synthesized 3D printed microneedles (MN) in pyramid and sphere shapes for insulin delivery by using the SLA method [37]. It is a transdermal drug delivery in which the drug will get pierced through the human skin. The in vitro release of insulin was studied and it was found that 3D printed pyramid showed a faster release of insulin that is approximately 80% within two minutes and up to 92% within eight minutes, whereas sphere-shaped MNs showed slower insulin release of 62–70% within two minutes and 81–84% within eight minutes, which can be observed in Figure 4.3. In order to preserve the insulin activity before the deposition of insulin sugar films on the MN surface, carriers like mannitol, trehalose, and xylitol were used. Overall, a rapid release of insulin can be observed. So, the synthesized MNs can be used for drug delivery applications.

A few examples of 3D printed drug delivery systems and their application are shown in Table 4.3. So controlled release of drugs is possible with the personalized 3D printed scaffolds and they find a wide range of applications in today's world.

4.2.3 Applications of 3D Printed Scaffolds During COVID-19

COVID-19 was a global pandemic caused by the SARS-COV-2 virus [44]. 36.2 million cases have been reported by October 2022. The virus spread at a fast pace, being transmitted among individuals when they came into contact with infected persons through the droplets released while sneezing or coughing. Many deaths were

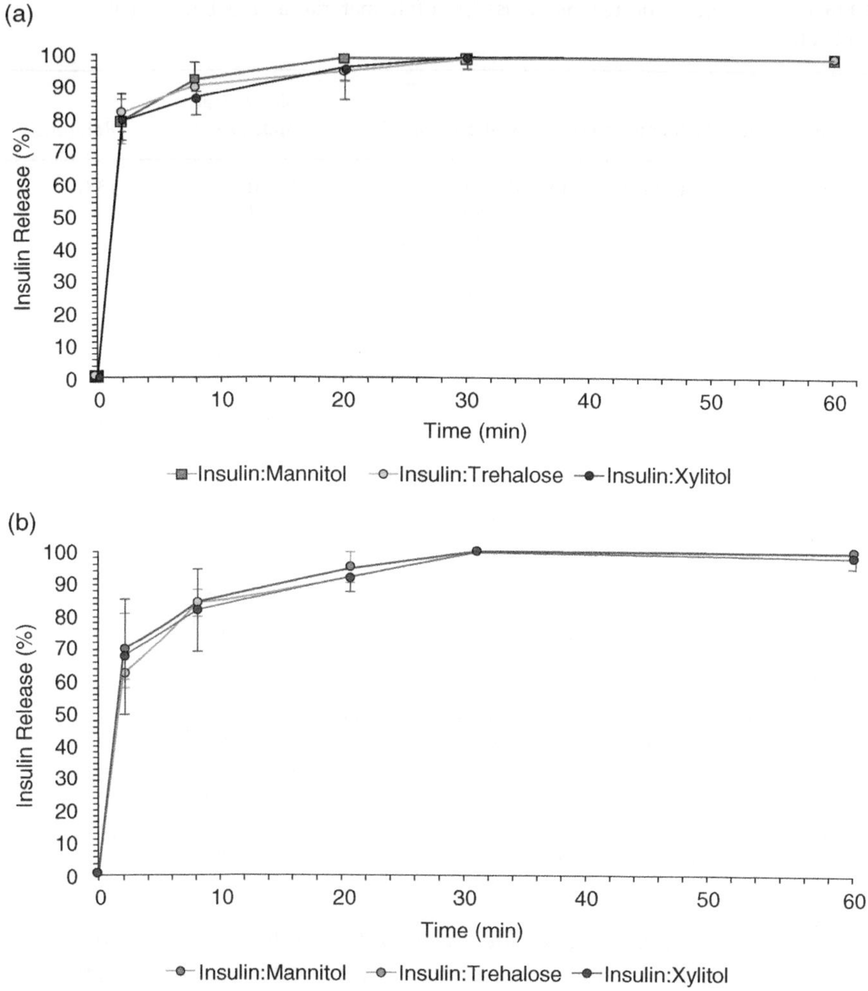

Figure 4.3 In vitro release of insulin through porcine skin from (a) pyramid and (b) spear-shaped MN for all drug carriers studied here. *Source:* Economidou et al. [37] / Elsevier.

reported during this period and strict lockdowns along with social distancing were enforced by the World Health Organization [45]. As a result, many companies had to shut down and this had created a shortage of medical and basic emergency need supplies. So, administrators had to find a way to increase the production of medical equipment during the COVID-19 pandemic. That is where the significance of 3D printing technology came, which enabled the customization of personalized protective equipment. Figure 4.4 shows different 3D printed devices synthesized during COVID-19.

Table 4.3 Targeted drug, technique used, scaffold material, and its biomedical applications.

Targeted drug	Technique used	Scaffold material	Biomedical Application	Reference
Cisplatin	Stereolithography (SLA)	Polyvinyl caprolactam-polyvinyl acetate-polyethylene glycol polymer	For treatment of A-431 epidermoid skin carcinoma cells.	[38]
Gentamicin	Fused deposition modeling (FDM)	Polylactic acid (PLA)/akermanite	Bone tissue engineering application	[39]
IGF-1	Fused deposition modeling (FDM)	Polylactic glycolic acid (PLGA)-polydopamine (PDA)-polycaprolactone (PCL)	Cartilage tissue regeneration	[40]
Doxorubicin, cisplatin, ifosfamide, methotrexate	Stereolithography (SLA)	Poly-L-lactic acid (PLLA)	Treatment of osteosarcoma	[41]
5-fluorouracil (5-FU)	Semi-solid extrusion	Calcium phosphate cement (CPC)	Bone graft material for bone cancer treatment	[42]
Cisplatin	Fused deposition modeling (FDM)	Gelatin	Cancer treatment	[43]

3D printing allowed the easy and customizable synthesis of protection equipment [46]. Spake et al. fabricated 3D printed face masks to be used during the pandemic [47]. Vankova et al. fabricated a polylactic (PLA) face mask using FDM technique [48]. PLA is a biodegradable polymer and is a promising candidate in 3D printing technology. PLA can be incorporated with graphene where the latter shows enhanced antimicrobial activity and inhibition toward SARS-COV-2 attachment [49]. PLA-G-based objects were found to be a game changer against pandemic spreading. Another 3D printed equipment is self-sanitizing gloves, in which it has embedded channels through which the sanitizing solution runs, and this was fabricated after measuring the palm size of the user [50]. Figure 4.5 shows the glove made using the common molding method and 3D printing method. 3D printing helps in the systematic and uniform coverage of channels over the entire glove surface. The glove is entirely porous for the uniform spreading of the sanitizing solution. This glove can be used by people working in the health sectors, schools, delivery units, etc.

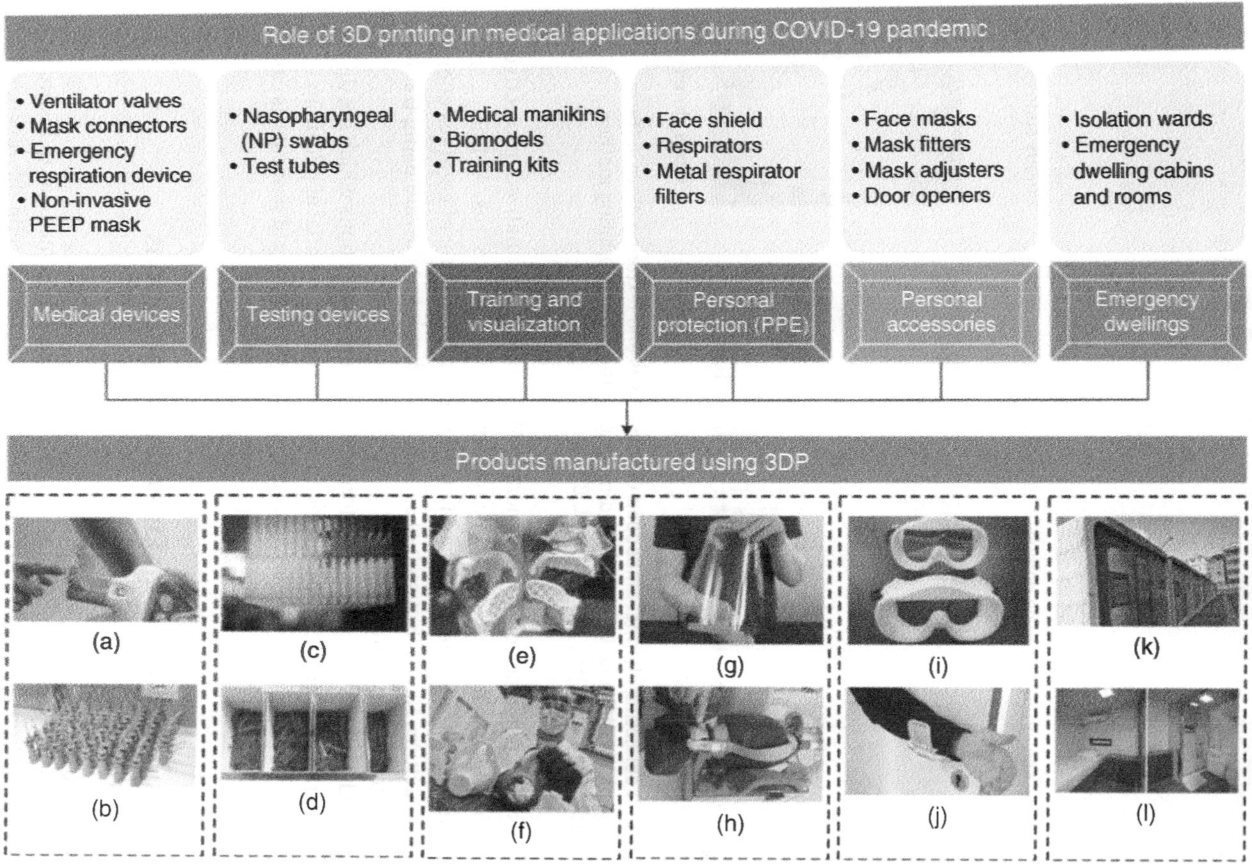

Figure 4.4 Applications of 3D printed devices (a) Valves used to convert the snorkeling face masks into ventilators (b) Valves for respiratory devices (c) Nasopharyngeal swabs (d) Syringes (e) Medical manikins for swabs (f) Silicon masks (g) Protective face shields (h) Emergency respiratory equipment (i) Safety goggles (j) Contact-free door handles. *Source:* Wang et al. [45] / with permission from Elsevier.

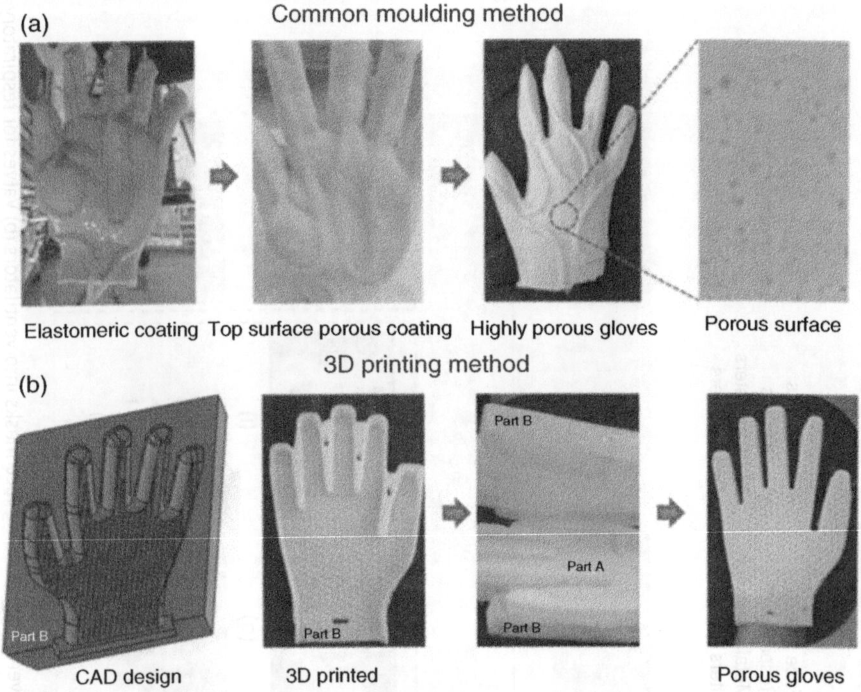

Figure 4.5 Fabricating channel embedded elastic gloves and texturing glove surface with porous morphology. (a) Common mold-making method (b) 3D-printing method. *Source:* Sadasivuni et al. [50] / Elsevier / CC BY 4.0.

Various research was done to generate 3D printed personalized protective equipment (PPE) to defend against the transmission of coronavirus. AM plays an important role in the medical field and facilitates the fabrication of complex engineering structures like PPE which is somewhat difficult to fabricate using traditional manufacturing techniques [51].

So, printed scaffolds and 3D printed forms have many applications which are discussed in detail. Among all, tissue engineering is one of the potential applications of 3D printing currently. Regeneration and further growth of damaged tissue parts can be actively carried out with the help of printed scaffolds by AM.

4.3 Tissue Engineering: Emerging Applications by 3D Printing

Tissue engineering is one of the most rapidly emerging biomedical fields that use the applications of 3D printing currently. It mainly focuses on the regeneration and repair of defective tissues in different parts of the body. For bone defect repair, there are some traditional methods like autografts, allografts, and the replacement of defective bone with artificial ones, etc. For scaffold preparation, there are different methods like freeze-drying, gas forming, and phase separation, but it is difficult to tune the pore size of fabricated scaffolds by these techniques. 3D printing is an advanced method introduced for developing scaffolds with excellent pore sizes and bioactivity, which can replace defective tissues and promote further growth of new tissues effectively [52]. Tissue engineering mainly focuses on patients who suffer from skin tumors, organ failure, bone defects, heart disease, etc. By imitating features of extracellular matrices of these damaged tissues, it is possible to control cell behavior and regeneration of damaged tissues. Using a combination of stem cells, engineering methods, and scaffold fabrications, cartilage regeneration possibilities can be developed [53]. Table 4.4 shows the different materials and the specific properties of the scaffolds by that material.

Table 4.4 Scaffold materials and properties for tissue engineering applications.

Scaffold material	Properties of scaffolds in tissue engineering	Reference
Alginate	Anionic polysaccharide, non-toxic, biocompatibility, and they mainly stimulate the regeneration of nerve tissues.	[54]
Chitosan	Low-cost, biodegradable, and biocompatible scaffold having mechanical strength. Even complex structures can print at room temperature.	[55]
Gelatin and GeIMA	Scaffolds with methacylic acid gelatin (GeIMA) having cell viability >97%. Porous scaffolds with excellent flexibility and tensile strength.	[56]
Polyvinyl-alcohol (PVA) hydrogel	Scaffolds by PVA do not involve any toxic crosslinking agents and show support for cell viability and extracellular matrix proteins. Mainly for cartilage tissue engineering applications.	[57]
Polyesters like PCL, PLA, PLGE, etc.	Good compatibility. Scaffolds with guidance channels can develop. Ability to sustain cell growth and attachment over scaffolds. Mainly for neural tissue engineering scaffolds.	[58]
Carbon-based nanomaterials	Scaffold can be fabricated with hydrogel and polyester having high mechanical, flexibility, thermal, and electrical properties; mainly supports neural tissue engineering. Promotes cell adhesion, proliferation, and neurogenic differentiation.	[59]

4.3.1 Cartilage Tissue Engineering by Printed Scaffolds

Many of the reports suggested that 3D printed cartilage scaffolds show excellent mechanical strength, pore distribution, and accuracy. In cartilage tissue engineering, the scaffolds which are made up of hydrogels have a promoting effect on cartilage repair [60]. Sadeghianmaryan et al. developed chitosan-based printed scaffolds (Figure 4.4) for cartilage tissue engineering, and, in vitro, characterizations were conducted [61]. Figure 4.6a, b shows the fabricated scaffolds and its SEM images with varying chitosan concentration. Its flexibility demonstration and cell attachment profile shown in Figure 4.6c, d.

Scaffolds with varying concentrations (w/v) of chitosan were developed using different concentrations (8%, 10%, 12%) of acetic acid solvent through extrusion-based 3D printing technology. Subsequently, scanning electron microscopy (SEM), mechanical testing, cell attachment studies, and cytotoxicity studies were conducted. Porous cartilage scaffolds with 2–4 mm pore size are formed, and they facilitate cell migration, ingrowth, nutrient flow, etc. over it. These scaffolds show 84% of ATDC5 cell attachment and are properly biocompatible for these cells. Above this, 10% w/v chitosan scaffold shows optimum results overall [61]. Yang et al. developed bioink for scaffold fabrication in cartilage tissue engineering using a collagen–alginate system. Using this bioink, it is possible to develop scaffolds that are biocompatible, low cytotoxicity, and water content for tissue regeneration. Since both the raw materials had high stiffness, the corresponding scaffold showed good mechanical strength and resistance to deformation [62]. Maihemuti et al. developed an extrusion-based 3D printed cartilage scaffold with gelatin and sodium alginate for knee cartilage defects. In vivo studies show the excellent bioactivity of the scaffolds, and fluorescence microscope images detected the cell growth of C28 cells. The results showed excellent biocompatibility. Implantation of the fabricated scaffold into rats was carried out and observed for months using the staining method for the biological activity of scaffolds. This study shows that the scaffold replaced the injured meniscus and, at the same time, provided nutrition for the injured knee joints [63].

4.3.2 Liver Tissue Engineering by Printed Scaffolds

Damage to liver tissues takes place due to viral infection, poisoning from drugs, etc. Shortage of donors is one of the major issues in the field of liver transplantation. Liver tissue engineering includes the fabrication of scaffolds for mimicking the microstructure of the liver and for maintaining hepatic function [64]. An ideal liver tissue scaffold should possess the following characteristics: (i) should have a suitable microenvironment for mimicking the native liver architecture, (ii) should contain a pre-established vascular bed and enhance waste removal from the body

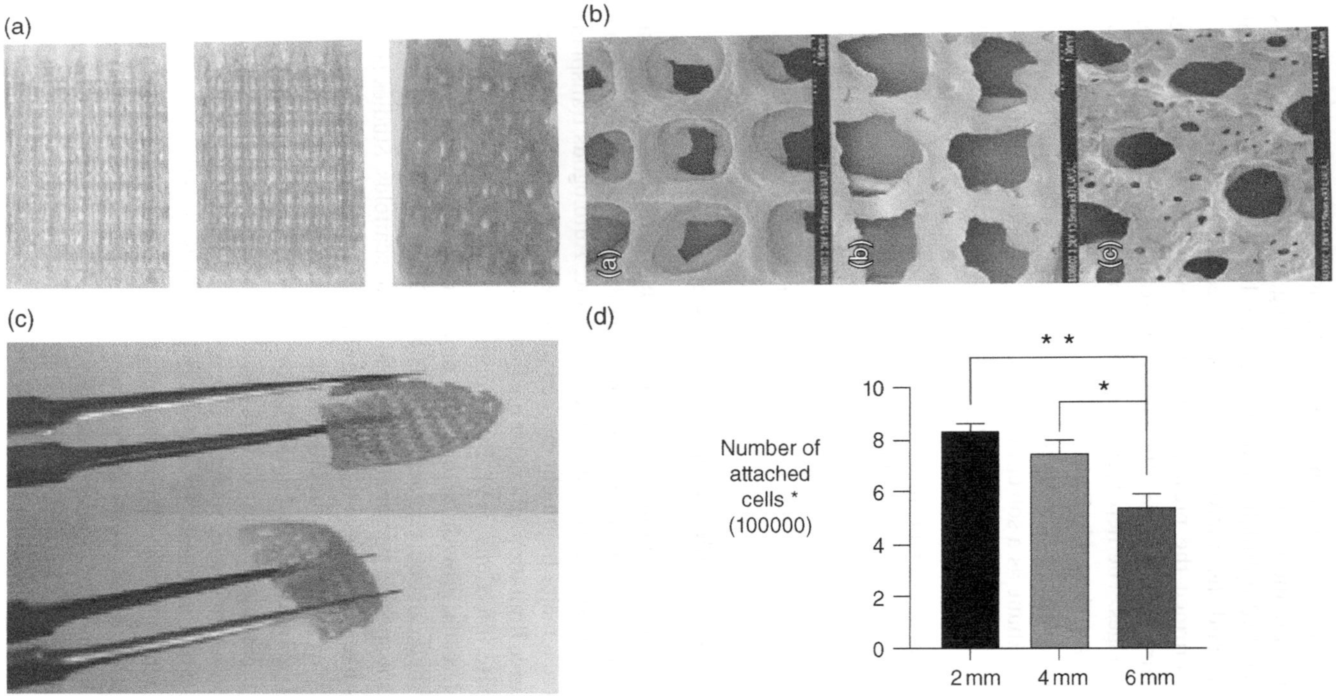

Figure 4.6 (a) images of 3D printed scaffold (i) 8% (ii) 10% (iii) 12% (w/v) chitosan. (b) SEM images of the scaffolds (a) 8% (b) 10% (c) 12% scaffold. (c) Flexibility demonstration of the fabricated scaffold. (d) Number of attached cells to scaffolds with different pore sizes of 2, 4, and 6 mm. *Source:* Sadeghianmaryan et al. [61] / with permission from Elsevier.

with time for nutrient and oxygen delivery scaffolds, (iii) should possess highly porous scaffold surface [65]. A chitosan–gelatin hybrid scaffold developed with a pore size of 100 μm and porosity >90%. Swelling and degradation studies show that it possesses hydrophilicity and biodegradability. The growth of hepatic cell culture over it, and the analysis of scanning electron microscopy results after seven days confirmed the presence of a large number of colonies over scaffolds [66]. Figure 4.7a, b shows the fabricated scaffolds with hepatic chambers and its SEM images. The hepatocyte proliferation and degradation studies over the scaffold surface shown in Figure 4.7c, d.

Meng et al. developed a 3D printed scaffold using polylactic acid (PLA) filament with dichloromethane as a solvent for liver tissue engineering applications [67]. The prepared scaffold by the FDM method showed a large thickness and better mechanical and thermal conductivity. The pores are uniformly distributed and have a diameter of 0.53 ± 0.06 mm. Surface morphological studies showed that the scaffold exhibited a suitable surface for tissue support and adhesion. 3D bioprinted polyvinyl alcohol (PVA)/chitosan scaffold shows bioadhesive and mechanical strength characteristics. They also facilitate uniform growth of hepatic cells over the scaffold as part of better liver tissue regeneration [68].

4.3.3 Nerve Tissue Engineering by Printed Scaffolds

Controlling and regulation of the human body is mainly done by the nervous system, which is the complex network in the body. Damage to the nervous system leads to paralysis, sensory loss, weakness, etc. Autologous transplantation was considered a major treatment method in the early stages, but it meets drawbacks like source limitation, diameter mismatch between the donor nerve recipient and disease infections. [69]. By combining nerve substituents along with transplantable cells, growth factors, and biopolymers, it is possible to develop scaffolds for nerve tissue regeneration. Using 3D bioprinting, it is possible to develop patient-specific personalized neural tissue scaffolds. Studies show that 2D materials like graphene and GO have unique properties to act as bioink formulations as part of nerve tissue engineering. Derivatives of graphene have nerve cell adhesion properties [70]. Another 3D printed PLA/chitosan scaffold for neural tissue engineering as a part of Parkinson's disease was developed by Saylam et al. [71]. This is a neurodegenerative disease that can be treated by neural tissue support and differentiation. The scaffold that was fabricated had a pore size of 100–200 μm, which allows cell penetration and growth. The synthesized scaffold showed a sustained and controlled release of levodopa drug and the biocompatibility was confirmed by fluorescence microscopy. The nervous system plays an important role in the electric signal transmission in the body. So, an ideal nerve scaffold would be better to have electric properties for enhancing the

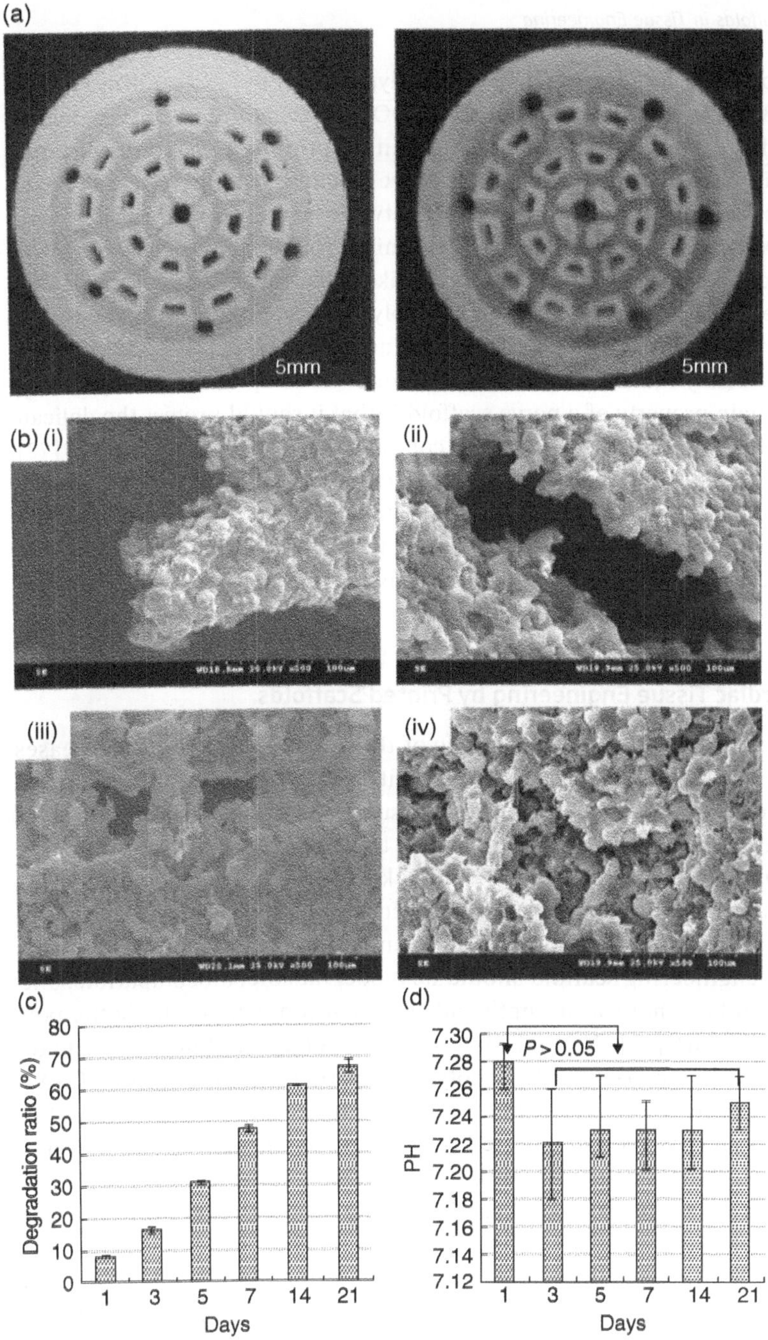

Figure 4.7 (a) resin mold of chitosan–gelatin scaffold and scaffold with hepatic chambers and channel network (b) SEM images of hepatocyte proliferation in the hepatic chambers of chitosan–gelatin scaffolds after being cultured for one day (i), three days (ii), five days (iii), and seven days (iv) (c) degradation ratio of scaffolds (d) pH of the resultant solution. *Source:* Jiankang et al. [66] / with permission from Elsevier.

proliferation and migration of nerve cells. Polypyrrole (PPy), polyaniline (PANI), and poly (3, 4-ethylenedioxy thiophene) PEDOT are some promising conductive polymers that can act as the scaffold matrix in 3D printing [72]. To repair large-gap-nerve injuries, the best alternative is nerve guidance conduits (NGCs). NGCs are tubular structures that can be placed between nerve gaps as a bridge, and they will axillarate axon regeneration. Customized NGCs can be developed by 3D printing technology using natural materials like collagen, chitosan, agarose, etc., and synthetic materials like polyglycolide polyhydroxybutyrate, etc. The tensile strength limit of normal sciatic and acellular nerves in rats is 1.4–2.7 MPa. Better scaffolds normally meet the mechanical strength value up to the corresponding level. The main property of a nerve scaffold is that it should protect the delicate nerves and have the ability to resist compression and shearing force [73]. Development of conductive scaffolds is very important for nerve tissue engineering applications. Conductive 3D printed PPy nanoparticle-containing scaffolds are developed by extrusion-based 3D printing technology. A porous scaffold with uniformly distributed pores having good mechanical strength supports nerve tissue engineering applications [74].

4.3.4 Cardiac Tissue Engineering by Printed Scaffolds

Cardiovascular diseases (CVDs) are one of the most life-threatening diseases prevailing worldwide in terms of mortality. The difficulties in electrical impulses and oxygen as well as nutrient supply over heart tissues, lead to heart attacks. Studies show that almost one billion cardiomyocytes (CMs) are damaged and lost during even a minor attack. However, there is no chance for auto-regeneration for the corresponding lost tissues. Scaffolds containing ECM and cardiomyocytes can promote regeneration of cardiac tissues. A cardiac tissue engineering scaffold should be biodegradable, biocompatible, and have sufficient mechanical strength and electrical conductance for better performance of heart [75]. Natural materials like collagen, chitosan, alginate, and synthetic polymers like PCL, poly D, L-lactic-co-glycolic acid (PLGA), and poly(lactic acid) (PLA) can be used for cardiac scaffold Fabrication [76]. Extrusion-based 3D shows increased cell density using alginate bioink. This highly viscous bioink develops a uniformly porous cardiac scaffold [77]. Also, 3D printed alginate bioink-based scaffold using droplet-inkjet printing technology for fabricating cardiac branched tubes. They exhibited structural integrity and cell encapsulation for better cardiac functioning [78]. 3D printed scaffolds as cardiac patches and branched tubes show high-resolution geometry and cell viability using PCL and polyethylene glycol diacrylate (PEGDA) using laser-assisted printing [79]. Ruther et al. developed a 3D scaffold using poly (glycol sebacate) (PGS)- zein for potential cardiac tissue engineering

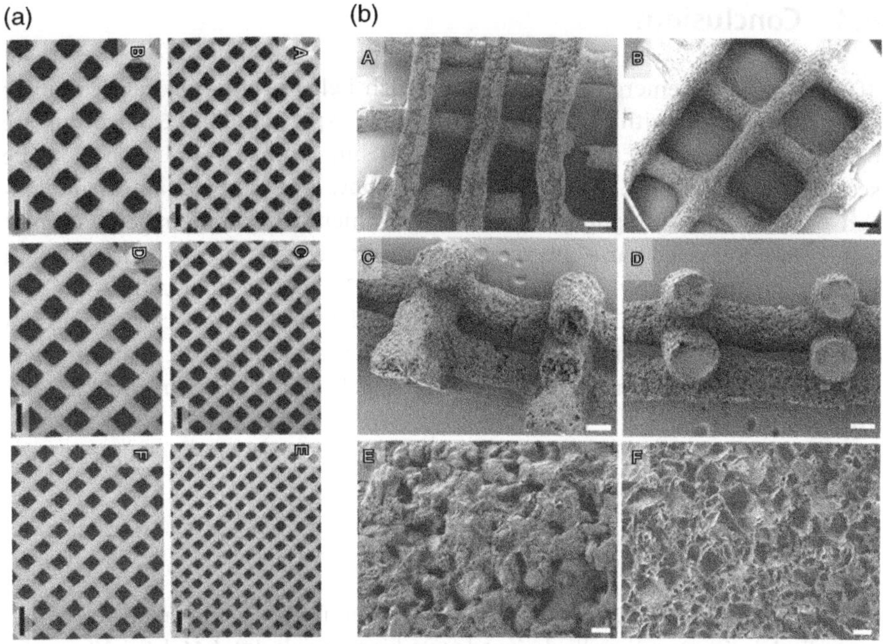

Figure 4.8 (a) Optical images of 3D printed PGS-zein constructs after crosslinking and salt leaching. Top view of hierarchical 50PGS10Zein3Gly constructs with a strand distance of (A, B) 1.75 mm as well as 50PGS30Zein10Gly with a strand distance of (C, D) 1.75 mm, and (E, F) 1.25 mm. (b) SEM micrographs of crosslinked and sat-leached 50PGS10Zein3Gly and (B, D) 50PGS30ZeinGly scaffolds at different magnifications in (A, B) top view and (C, D) cross-sectional view (E, F). Close-up of a 50PGS30Zein10Gly strut before and after leaching of the salt particles. *Source:* Ruther et al. [80] / John Wiley & Sons / CC BY 4.0.

applications. Biological analysis shows that no cytotoxic effect on C2C12 cardiac cells and surface morphological studies revealed the high degree of porosity and surface area of scaffold. Microscopic results give the mean value of pore diameter as $15 \pm 7\,\mu m$ and exhibit hydrophilic character helping for the passage of nutrient growth factors over the surface [80]. Figure 4.8a (A–F) shows the optical images of scaffolds with varying strand distance and its SEM images, b (A–F).

An electroconductive 3D printed poycaprolactone (PCL) scaffolds for cardial tissue engineering, which is decorated with gold nanoparticles (GNPs), was developed. PCL/0.5 wt% GNPs showed optimized values for appropriate myocardium regeneration with improved mechanical and electrical conductivity. Gold nanoparticles in scaffolds enhance the tissue proliferation and differentiation to a large extent [81].

4.4 Conclusions

3D printing is an emerging technology, which helps in the easy fabrication of complex structures within a short period of time. Applications of 3D printing are extended over scientific and engineering fields, but dominant applications can be seen in biomedical area. 3D printed scaffolds show efficient mechanical and bioactive character with highly porous surfaces, which helps in the regeneration of tissues over damaged parts effectively. Polymeric, ceramic, metal, hydrogels, and a combination of materials can be suitable for 3D scaffold fabrications for biomedical applications. Various tissue engineering like nerve, cardiac, cartilage, and liver use 3D printed scaffolds for regeneration and repair purposes. In future, new treatment methods with advanced printed models will lead the medical field to great success.

References

1 Zadpoor, A.A. (2017). Design for additive bio-manufacturing: from patient-specific medical devices to rationally designed meta-biomaterials. *Int. J. Mol. Sci.* 18 (8): 1607.

2 Robles Martinez, P., Basit, A.W., and Gaisford, S. (2018). The history, developments and opportunities of stereolithography. *AAPS Adv. Pharm. Sci. Ser.* 31: 55–79.

3 Rengier, F., Mehndiratta, A., Von Tengg-Kobligk, H. et al. (2010). 3D printing based on imaging data: review of medical applications. *Int. J. Comput. Assist. Radiol. Surg.* 5 (4): 335–341.

4 Shahrubudin, N., Lee, T.C., and Ramlan, R. (2019). An overview on 3D printing technology: technological, materials, and applications. *Procedia Manuf.* 35: 1286–1296.

5 Norman, J., Madurawe, R.D., Moore, C.M.V. et al. (2017). A new chapter in pharmaceutical manufacturing: 3D-printed drug products. *Adv. Drug Deliv. Rev.* 108: 39–50.

6 Ahangar, P., Akoury, E., Luna, A.S.R.G. et al. (2018). Nanoporous 3D-printed scaffolds for local doxorubicin delivery in bone metastases secondary to prostate cancer. *Materials (Basel)* 11 (9): 1485.

7 Rosenzweig, D.H., Carelli, E., Steffen, T. et al. (2015). 3D-printed ABS and PLA scaffolds for cartilage and nucleus pulposustissue regeneration. *Int. J. Mol. Sci.* 16 (7): 15118–15135.

8 Chaput, C., De, T.C., Céramiques, D.T., and Technopole, E. (2020). Fabrication of ceramics by stereolithography. *R Tejournal - Forum für Rapid Technologie* 9 (1): 1–15.

9 Masood, S.H. and Song, W.Q. (2004). Development of new metal/polymer materials for rapid tooling using fused deposition modelling. *Mater. Des.* 25 (7): 587–594.

10 Cansiz, E., Arslan, Y.Z., Turan, F., and Atalay, B. (2014). Computer-assisted design of patient-specific sagittal split osteotomy guide and soft tissue retractor. *J. Med. Biol. Eng.* 34 (4): 363–367.

11 Gokuldoss, P.K., Kolla, S., and Eckert, J. (2017). Additive manufacturing processes: selective laser melting, electron beam melting and binder jetting-selection guidelines. *Materials (Basel)* 10 (6).

12 Panwar, A. and Tan, L.P. (2016). Current status of bioinks for micro-extrusion-based 3D bioprinting. *Molecules* 21 (6): 685.

13 Murphy, S.V. and Atala, A. (2014). 3D bioprinting of tissues and organs. *Nat. Biotechnol.* 32 (8): 773–785.

14 Anadioti, E., Kane, B., and Soulas, E. (2018). Current and emerging applications of 3D printing in restorative dentistry. *Curr. Oral Health Rep.* 5 (2): 133–139.

15 Zhang, J., Allardyce, B.J., Rajkhowa, R. et al. (2021). 3D printing of silk powder by Binder Jetting technique. *Addit. Manuf.* 38: 101820.

16 Ke, D. and Bose, S. (2018). Effects of pore distribution and chemistry on physical, mechanical, and biological properties of tricalcium phosphate scaffolds by binder-jet 3D printing. *Addit. Manuf.* 22: 111–117.

17 Bose, S., Bhattacharjee, A., Banerjee, D. et al. (2021). Influence of random and designed porosities on 3D printed tricalcium phosphate-bioactive glass scaffolds. *Addit. Manuf.* 40: 101895.

18 Castro, N.J., O'Brien, J., and Zhang, L.G. (2015). Integrating biologically inspired nanomaterials and table-top stereolithography for 3D printed biomimetic osteochondral scaffolds. *Nanoscale* 7 (33): 14010–14022.

19 Chen, P., Zheng, L., Wang, Y. et al. (2019). Desktop-stereolithography 3D printing of a radially oriented extracellular matrix/mesenchymal stem cell exosome bioink for osteochondral defect regeneration. *Theranostics* 9 (9): 2439–2459.

20 Lee, Y., Jung, A., Heo, S.J. et al. (2023). Influences of surface topography of porous titanium scaffolds manufactured by powder bed fusion on osteogenesis. *J. Mater. Res. Technol.* 23: 2784–2797.

21 Ahangar, P., Cooke, M.E., Weber, M.H., and Rosenzweig, D.H. (2019). Current biomedical applications of 3D printing and additive manufacturing. *Appl. Sci.* 9 (8): 1713.

22 Mostafaei, A., Toman, J., Stevens, E.L. et al. (2017). Microstructural evolution and mechanical properties of differently heat-treated binder jet printed samples from gas- and water-atomized alloy 625 powders. *Acta Mater.* 124: 280–289.

23 Sing, S.L., Tey, C.F., Tan, J.H.K. et al. (2019). 3D printing of metals in rapid prototyping of biomaterials: techniques in additive manufacturing. In: *Rapid Prototyping of Biomaterials: Techniques in Additive Manufacturing*, 2e, 17–40. Woodhead Publishing Series in Biomaterials.

24 Placone, J.K. and Engler, A.J. (2018). Recent advances in extrusion-based 3D printing for biomedical applications. *Adv. Healthcare Mater.* 7 (8): 1–11.

25 Prabhakar, P., Sen, R.K., Dwivedi, N. et al. (2021). 3D-printed microfluidics and potential biomedical applications. *Front. Nanotechnol.* 3 (March): 1–16.

26 Tan, X.P., Tan, Y.J., Chow, C.S.L. et al. (2017). Metallic powder-bed based 3D printing of cellular scaffolds for orthopaedic implants: a state-of-the-art review on manufacturing, topological design, mechanical properties and biocompatibility. *Mater. Sci. Eng. C* 76: 1328–1343.

27 Li, J., Wu, C., Chu, P.K., and Gelinsky, M. (2020). 3D printing of hydrogels: rational design strategies and emerging biomedical applications. *Mater. Sci. Eng. R. Rep.* 140 (January): 100543.

28 Khalyfa, A., Vogt, S., Weisser, J. et al. (2007). Development of a new calcium phosphate powder-binder system for the 3D printing of patient specific implants. *J. Mater. Sci. Mater. Med.* 18 (5): 909–916.

29 Wu, L., Virdee, J., Maughan, E. et al. (2018). Stiffness memory nanohybrid scaffolds generated by indirect 3D printing for biologically responsive soft implants. *Acta Biomater.* 80: 188–202.

30 Le Guéhennec, L., Van Hede, D., Plougonven, E. et al. (2020). In vitro and in vivo biocompatibility of calcium-phosphate scaffolds three-dimensional printed by stereolithography for bone regeneration. *J. Biomed. Mater. Res. – Part A* 108 (3): 412–425.

31 Radhakrishnan, S., Nagarajan, S., Belaid, H. et al. (2021). Fabrication of 3D printed antimicrobial polycaprolactone scaffolds for tissue engineering applications. *Mater. Sci. Eng. C* 118: 111525.

32 Antoniac, I.V. (2016). *Handbook of Bioceramics and Biocomposites*, 1–1386. Springer.

33 Safonov, A., Maltsev, E., Chugunov, S. et al. (2020). Design and fabrication of complex-shaped ceramic bone implants via 3d printing based on laser stereolithography. *Appl. Sci.* 10 (20): 1–18.

34 Putlyaev, V.I., Yevdokimov, P.V., Mamonov, S.A. et al. (2019). Stereolithographic 3D printing of bioceramic scaffolds of a given shape and architecture for bone tissue regeneration. *Inorg. Mater. Appl. Res.* 10 (5): 1101–1108.

35 Wang, J., Zhang, Y., Aghda, N.H. et al. (2021). Emerging 3D printing technologies for drug delivery devices: current status and future perspective. *Adv. Drug Deliv. Rev.* 174: 294–316.

36 Goyanes, A., Det-Amornrat, U., Wang, J. et al. (2016). 3D scanning and 3D printing as innovative technologies for fabricating personalized topical drug delivery systems. *J. Control. Release* 234: 41–48.

37 Economidou, S.N., Pere, C.P.P., Reid, A. et al. (2019). 3D printed microneedle patches using stereolithography (SLA)for intradermal insulin delivery. *Mater. Sci. Eng. C* 102: 743–755.

38 Uddin, M.J., Scoutaris, N., Economidou, S.N. et al. (2020). 3D printed microneedles for anticancer therapy of skin tumours. *Mater. Sci. Eng. C* 107: 110248.

39 Arastouei, M., Khodaei, M., Atyabi, S.M., and Jafari, N.M. (2021). The in-vitro biological properties of 3D printed poly lactic acid/akermanite composite porous scaffold for bone tissue engineering. *Mater. Today Commun.* 27: 102176.

40 Wei, P., Xu, Y., Gu, Y. et al. (2020). IGF-1-releasing PLGA nanoparticles modified 3D printed PCL scaffolds for cartilage tissue engineering. *Drug Deliv.* 27 (1): 1106–1114.

41 Wang, Y., Sun, L., Mei, Z. et al. (2020). 3D printed biodegradable implants as an individualized drug delivery system for local chemotherapy of osteosarcoma. *Mater. Des.* 186: 108336.

42 Wu, Y., Woodbine, L., Carr, A.M. et al. (2020). 3d printed calcium phosphate cement (CPC) scaffolds for anti-cancer drug delivery. *Pharmaceutics* 12 (11): 1–15.

43 Jiang, Y., Pan, X., Yao, M. et al. (2021). Bioinspired adhesive and tumor microenvironment responsive nanoMOFs assembled 3D-printed scaffold for anti-tumor therapy and bone regeneration. *Nano Today* 39: 101182.

44 Kantaros, A., Laskaris, N., Piromalis, D., and Ganetsos, T. (2021). Correction to: manufacturing zero- waste COVID-19 personal protection equipment: a case study of utilizing 3D printing while employing waste material recycling. *Circ. Econ. Sustain.* 1 (3): 871–871.

45 Wang, Y., Ahmed, A., Azam, A. et al. (2021). Applications of additive manufacturing (AM) in sustainable energy generation and battle against COVID-19 pandemic: the knowledge evolution of 3D printing. *J. Manuf. Syst.* 60 (August): 709–733.

46 François, P.M., Bonnet, X., Kosior, J. et al. (2021). 3D-printed contact-free devices designed and dispatched against the COVID-19 pandemic: the 3D COVID initiative. *J. Stomatol. Oral Maxillofac. Surg.* 122 (4): 381–385.

47 Spake, C.S.L., Carruthers, T.N., Crozier, J.W. et al. (2021). 3D printed N-95 masks during the COVID-19 pandemic: lessons learned. *Ann. Biomed. Eng.* 49 (12): 3666–3675.

48 Vaňková, E., Kašparová, P., Khun, J. et al. (2020). Polylactic acid as a suitable material for 3D printing of protective masks in times of COVID-19 pandemic. *PeerJ.* 8: 1–20.

49 De Maio, F., Rosa, E., Perini, G. et al. (2022). 3D-printed graphene polylactic acid devices resistant to SARS-CoV-2: Sunlight-mediated sterilization of additive manufactured objects. *Carbon N. Y.* 194: 34–41.

50 Sadasivuni, K.K., Maurya, M.R., Houkan, M.T. et al. (2023). Self-sanitizing reusable glove via 3D-printing and common mold making method. *Mater. Today Proc.* https://doi.org/10.1016/j.matpr.2023.03.232.

51 Tarfaoui, M., Nachtane, M., Goda, I. et al. (2020). Additive manufacturing in fighting against novel coronavirus COVID-19. *Int. J. Adv. Manuf. Technol.* 110 (11–12): 2913–2927.

52 Cao, S., Zhao, Y., Hu, Y. et al. (2020). New perspectives: in-situ tissue engineering for bone repair scaffold. *Compos. Part B Eng.* 202 (August): 108445.

53 Choi, J.R., Yong, K.W., and Choi, J.Y. (2018). Effects of mechanical loading on human mesenchymal stem cells for cartilage tissue engineering. *J. Cell. Physiol.* 233 (3): 1913–1928.

54 Liu, J., Willför, S., and Xu, C. (2015). A review of bioactive plant polysaccharides: biological activities, functionalization, and biomedical applications. *Bioact. Carbohydr. Dietary Fibre* 5 (1): 31–61.

55 Li, G., Xue, C., Wang, H. et al. (2018). Spatially featured porous chitosan conduits with micropatterned inner wall and seamless sidewall for bridging peripheral nerve regeneration. *Carbohydr. Polym.* 194: 225–235.

56 Van Den Bulcke AI, Bogdanov B, De Rooze N, Schacht EH, Cornelissen M, Berghmans H. 甲基丙烯酰胺改性明胶水凝胶的结构与流变性能 *Biomacromolecules* 2000;1(1):31–8.

57 Meng, Y., Cao, J., Chen, Y. et al. (2020). 3D printing of a poly(vinyl alcohol)-based nano- composite hydrogel as an artificial cartilage replacement and the improvement mechanism of printing accuracy. *J. Mater. Chem. B* 8 (4): 677–690.

58 Jahromi, H.K., Farzin, A., Hasanzadeh, E. et al. (2020). Enhanced sciatic nerve regeneration by poly-L-lactic acid/multi-wall carbon nanotube neural guidance conduit containing Schwann cells and curcumin encapsulated chitosan nanoparticles in rat. *Mater. Sci. Eng. C* 109: 110564.

59 Fraczek-Szczypta, A. (2014). Carbon nanomaterials for nerve tissue stimulation and regeneration. *Mater. Sci. Eng. C* 34 (1): 35–49.

60 Jung, W., Jung, Y.H., Pikhitsa, P.V. et al. (2021). Three-dimensional nanoprinting via charged aerosol jets. *Nature* 592 (7852): 54–59.

61 Sadeghianmaryan, A., Naghieh, S., Alizadeh Sardroud, H. et al. (2020). Extrusion-based printing of chitosan scaffolds and their in vitro characterization for cartilage tissue engineering. *Int. J. Biol. Macromol.* 164: 3179–3192.

62 Yang, X., Lu, Z., Wu, H. et al. (2018). Collagen-alginate as bioink for three-dimensional (3D) cell printing based cartilage tissue engineering. *Mater. Sci. Eng. C* 83: 195–201.

63 Maihemuti, A., Zhang, H., Lin, X. et al. 3D-printed fish gelatin scaffolds for cartilage tissue engineering. *Bioact. Mater.* 2023 (26): 77–87.

64 Wang, J., Huang, D., Yu, H. et al. (2022). Developing tissue engineering strategies for liver regeneration. *Eng. Regen.* 3 (1): 80–91.

65 Shin, M., King, K., Matsuda, K. et al. (2004). Endothelialized networks with a vascular geometry in microfabricated poly(dimethyl siloxane). *Biomed. Microdevices* 6: 269–278.

66 Jiankang, H., Dichen, L., Yaxiong, L. et al. (2009). et al, Preparation of chitosan-gelatin hybrid scaffolds with well-organized microstructures for hepatic tissue engineering. *Acta Biomater.* 5 (1): 453–461.

67 Meng, C., Zhao, J., Yin, Y. et al. (2020). Preparation and characterization of PLA film/3D printing composite scaffold for tissue engineering application. *Fibers Polym.* 21 (4): 709–716.

68 Supriya Bhatt S, Thakur G, Nune M. Preparation and characterization of PVA/Chitosan cross-linked 3D scaffolds for liver tissue engineering. *Mater. Today Proc.*2023https://doi.org/10.1016/j.matpr.2023.02.251.

69 Yan, J., Wu, R., Liao, S. et al. (2020). Applications of polydopamine-modified scaffolds in the peripheral nerve tissue engineering. *Front. Bioeng. Biotechnol.* 8 (October): 1–7.

70 Jung Jo, H., Sung Kang, M., Jeong Jang, H. et al. (2023). Advanced approaches with combination of 2D nanomaterials and 3D printing for exquisite neural tissue engineering. *Mater. Sci. Addit. Manuf.* 2 (2): 1.

71 Saylam, E., Akkaya, Y., Ilhan, E. et al. (2021). Levodopa-loaded 3d- printed poly (Lactic) acid/chitosan neural tissue scaffold as a promising drug delivery system for the treatment of parkinson's disease. *Appl. Sci.* 11 (22).

72 Zhang, L.G. and Kaplan, D.L. (2016). *Neural Engineering: From Advanced Biomaterials to 3D Fabrication Techniques*, 1–306. Springer.

73 Drury, J.L. and Mooney, D.J. (2003). Hydrogels for tissue engineering: scaffold design variables and applications. *Biomaterials* 24 (24): 4337–4351.

74 Ma, C., Jiang, L., Wang, Y. et al. (2019). 3D printing of conductive tissue engineering scaffolds containing polypyrrole nanoparticles with different morphologies and concentrations. *Materials (Basel)* 12 (15): 2491.

75 Xu, Y., Patnaik, S., Guo, X. et al. (2014). Cardiac differentiation of cardiosphere-derived cells in scaffolds mimicking morphology of the cardiac extracellular matrix. *Acta Biomater.* 10 (8): 3449–3462.

76 Liu, Q., Tian, S., Zhao, C. et al. (2015). Porous nanofibrous poly(l-lactic acid) scaffolds supporting cardiovascular progenitor cells for cardiac tissue engineering. *Acta Biomater.* 26: 105–114.

77 Christensen, K., Xu, C., Chai, W. et al. (2015). Freeform inkjet printing of cellular structures with bifurcations. *Biotechnol. Bioeng.* 112 (5): 1047–1055.

78 Xu, T., Baicu, C., Aho, M. et al. (2009). Fabrication and characterization of bio- engineered cardiac pseudo tissues. *Biofabrication* 1 (3): 035001.

79 Zhu, W., Qu, X., Zhu, J. et al. (2017). Direct 3D bioprinting of prevascularized tissue constructs with complex microarchitecture. *Biomaterials* 124: 106–115.

80 Ruther, F., Roether, J.A., and Boccaccini, A.R. (2022). 3D printing of mechanically resistant poly (Glycerol Sebacate) (PGS)-zein scaffolds for potential cardiac tissue engineering applications. *Adv. Eng. Mater.* 24 (9): 2101768.

81 Ghaziof, S., Shojaei, S., Mehdikhani, M. et al. (2022). Electro- conductive 3D printed polycaprolactone/gold nanoparticles nanocomposite scaffolds for myocardial tissue engineering. *J. Mech. Behav. Biomed. Mater.* 132 (May): 105271.

5

Printability and Shape Fidelity in Different Bioprinting Processes

Prajisha Prabhakar[1], *Aiswarya Sathian*[1,2], *and Sabu Thomas*[3,4,5]

[1] *International and Inter University Centre for Nanoscience and Nanotechnology, Mahatma Gandhi University Kottayam, Kottayam, Kerala, India*
[2] *College of Science and Engineering, James Cook University, Townsville, Queensland, Australia*
[3] *School of Energy Materials, School of Nanoscience and Nanotechnology, School of Polymer Science and Technology, School of Chemical Science and International and Inter University Centre for Nanoscience and Technology (IIUCNN), Mahatma Gandhi University, Kottayam, Kerala, India*
[4] *Department of Chemical Sciences, University of Johannesburg, Doornfontein, Johannesburg, South Africa*
[5] *TrEST Research Park, Sreekariyam, Trivandrum, Kerala, India*

5.1 Introduction

Bioprinting, an advanced amalgamation of 3D printing and biology with tissue engineering, has sparked a revolution in regenerative medicine and biomedical research. This groundbreaking method enables the meticulous creation of intricate biological structures, such as tissues and organs, by precisely layering bioinks enriched with living cells and biomaterials. As bioprinting advances, printability, and shape fidelity have emerged as pivotal determinants for its widespread implementation and effectiveness. These two factors heavily influence the quality and accuracy of the bioprinted constructs, propelling the technology toward transformative applications in personalized medicine and regenerative therapies.

Printability pertains to the capacity of a bioprinting process to consistently generate precise and top-notch prints, featuring intricate geometries and exact dimensions. This concept encompasses the mastery over diverse printing parameters, including layer height, printing speed, nozzle size, and material viscosity, all aimed at achieving optimal outcomes. The printability of a bioprinting process acts as a decisive element in influencing the structural integrity, functionality, and biological effectiveness of the produced printed structures. Conversely, shape fidelity assumes significant importance as it assesses the degree to which a bioprinted

3D Bioprinting from Lab to Industry, First Edition. Edited by Prosenjit Saha, Sabu Thomas, Jinku Kim, and Manojit Ghosh.

object faithfully replicates the intended design or digital model. Attaining shape fidelity requires meticulous alignment of the physical output with the original 3D model, encompassing dimensions, surface smoothness, and intricate details. In the context of bioprinting applications in regenerative medicine and personalized healthcare, shape fidelity stands as a paramount factor, directly influencing the functional performance and viability of the printed tissues and organs [1].

Effective tissue rejuvenation through the use of 3D bioprinted scaffolds necessitates fulfilling particular conditions, such as fostering cell migration, proliferation, and differentiation, removing waste components, and ultimately promoting vascularization. The careful choice of biomaterials and the rheological and mechanical characteristics they possess can directly impact the formation of the scaffold's structure, thereby enabling its intended functionality. Moreover, the deliberate control of pore size and porosity, as well as the connectivity of pores within the scaffold architecture, aids in emulating the in vivo microenvironment, thereby promoting tissue growth. Studies indicate that alterations in pore size and configuration influence cellular activity within the support framework. Nevertheless, the extrusion-based bioprinting (EBB) technique frequently exhibits noticeable differences between the intended design and the actual fabricated part, primarily stemming from variations in both material and process characteristics. Hence, attaining shape fidelity, ensuring biocompatibility, and preserving the mechanical integrity of the scaffold framework, pose challenges, making the identification of suitable biomaterials for fabricating the 3D regulated porous arrangement created through additive manufacturing a vibrant area of ongoing research [2].

In this chapter, we delve into the fundamentals of printability and shape fidelity in different bioprinting processes. We investigate the techniques applied in extrusion-based bioprinting, inkjet-based bioprinting, and stereolithography-based bioprinting, analyzing their capabilities and challenges concerning printability and shape fidelity. Furthermore, we explore the struggles experienced while optimizing critical aspects. Gaining a comprehensive understanding of the intricate relationship between printability and shape fidelity holds utmost importance in propelling the field of bioprinting forward and harnessing its complete potential in regenerative medicine, drug development, and tissue engineering. By grasping the underlying factors that impact the quality and precision of bioprinted constructs, researchers and practitioners can pave the path for groundbreaking applications, personalized therapies, and enhanced patient outcomes within the realm of bioprinting.

5.2 Fundamentals of Printability

The 3D printing of living body tissues using the bioinks fabricated by the innovative technique refers to bioprinting. Bioprinting entails a number of important factors, one of which is printability, which refers to the accuracy and precision with which a bioink is deposited during the 3D printing process in a controlled manner, to obtain the best results. The viability and functionality of the fabricated tissue should be considered. It is essential that inks suitable for bioprinting meet a number of key requirements so that they can be used with a 3D bioprinter. For them to be useful as regenerative medicine and tissue engineering products, they need to meet all the criteria dictated by the applications they are designed for. The main challenge is that they need to be printable. To ensure successful printing and structural stability of a bioink, it needs to meet specific requirements. These include having the right viscosity, exhibiting shear-thinning behavior, and possessing appropriate mechanical properties. These characteristics enable the bioink to flow smoothly through the printer nozzle and maintain its structure during the printing process.

Scientists are always searching for ways to improve the printability of bioinks and achieve more intricate and complex tissue structures. They do this by experimenting with various formulations and printing parameters to find the optimal combination. The field is constantly advancing as researchers strive to achieve the best possible printability. The ultimate goal is to generate tissues and organs capable of being transplanted and used functionally in the host's body.

Often referred to as print accuracy, shape fidelity describes the structural permanence of the distinct strands during extrusion and the printed structure when compared to the initial computer design showed in Figure 5.1 [1].

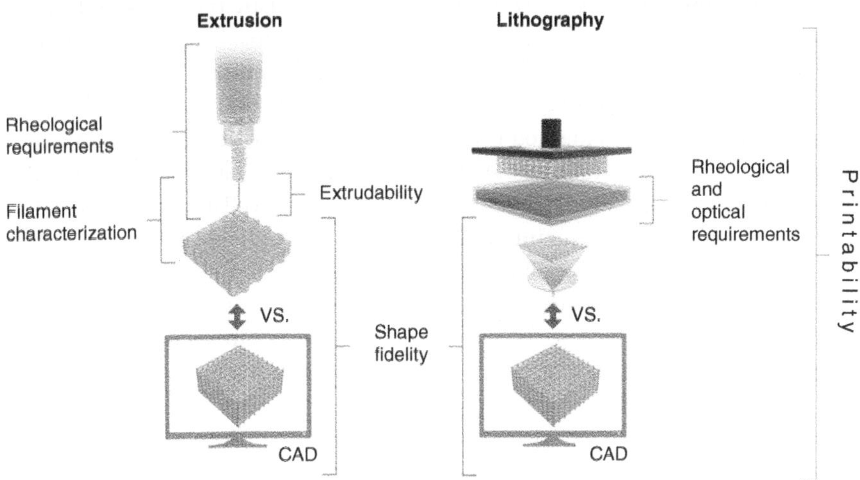

Figure 5.1 Shows bioprinting using extrusion and lithography: key aspects to assess printability. *Source:* Andrea Schwab et al. (2020) / with permission of American Chemical Society.

5.3 Bioprinting Techniques and Printability

3D printing in conjunction with biology is emerging as a transformative technology with enormous potential in regenerative as well as tissue engineering. Bioprinting techniques and printability are explored, including the diverse methodologies for fabricating living tissues and organs. However, it is the printability of the bioprinting process that ensures the faithful translation of complex digital designs into biological constructs that are tangible and functional. By understanding the fundamentals of bioprinting techniques, and key factors that influence printability, researchers, and clinicians can unlock new frontiers in personalized medicine and revolutionize healthcare as we know it. We offer a comprehensive journey into the art and science of bioprinting, uncovering the cutting-edge innovations that promise to reshape the future of regenerative therapies. The 3D bioprinting approach, based on extrusion, comprehends the filament deposition (which is also mentioned as strands or struts) filled within the cartridge of printable cell-containing formulation [1].

5.3.1 Extrusion-Based Bioprinting

The method is a commonly employed 3D printing approach that has spurred significant advancements in tissue engineering, allowing for the production of biocompatible scaffolds with diverse complexities, suitable for transplantation. Extrusion-based systems come in piston, pneumatic, or screw variations. The screw system is particularly well-suited for extruding high-viscosity materials and is primarily employed for the 3D printing of thermoplastic polymers. Extruding apparatus come in piston, pneumatic, or screw variations. The screw system is particularly well-suited for extruding high-viscosity materials and is primarily employed for the 3D printing of thermoplastic polymers. The primary consequence of employing this method is to manufacture or create structures incorporating living cells in several ways. According to extrusion-based bioprinting, the bioink filament perpetually extrudes through a nozzle as well as permits the deposition. The bioink needs to maintain low viscosity throughout the extrusion process, so that it hinders all possible clogging of the extrusion needle, furthermore protecting cells from excessive fluid shear stress [4]. In order to maintain the shape of the deposited bioink, the bioink should solidify rapidly once it is deposited on the printer bed (Huang, [5, 6]. Extrusion bioprinting normally has a resolution of 50–1000 μm [4].

There are three types of extrusion-based systems: piston, pneumatic, and screw-driven in Figure 5.2a – Extrusion bioprinting arranges filaments using a hydrogel-based bioink, and the defined shape becomes permanent upon crosslinking the hydrogel precursor; Figure 5.2b, which may include pH and

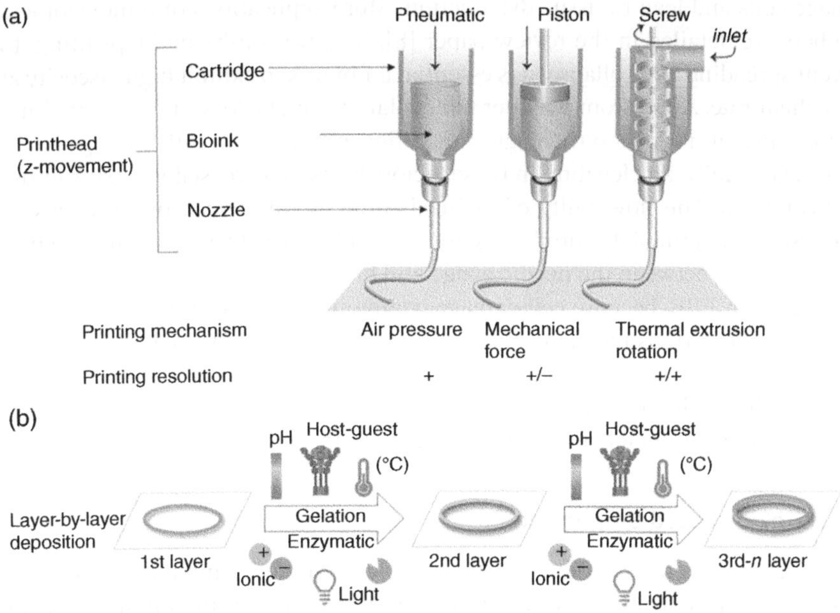

Figure 5.2 (a) extrusion-based bioprinting system, (b) bottom-up method in a layer-by-layer deposition of bioink. *Source:* Schwab et al. [1] / with permission of American Chemical Society.

temperature changes, ionic crosslinking, photochemical reactions, enzyme crosslinking, or guest–host interactions [1].

For describing printability, dimensionless numbers or ratios can be used. It is useful to measure inertial, viscous, capillary, and elastic forces on material extrusion using rheological dimensionless numbers like Reynolds, Weber, and Weissenberg numbers [8]. Among the Reynolds numbers, the Weber number measures the ratio between inertial and capillary forces, and the Weissenberg number measures the ratio between elastic and viscous forces. In the context of extrusion bioprinting, printability refers to the capability of creating and maintaining consistent bioink-infused 3D structures through bioprinting methods. It has a direct impact on the structure of the printed scaffolds, influencing both their mechanical and biological characteristics. The significance of the printability concept lies in the fact that deviations between a printed framework and an ideal design can alter mechanical and biological attributes, such as strength and cell function. In the context of extrusion-based techniques, printability can influence the shape accuracy of biofabricated scaffolds and, consequently, impact cell performance. The essence of printability is crucial as the scaffold's structure governs the morphology and growth of cells post-printing [9, 10]. Poor printability can damage cells and lead to tissue dysfunction. More applications of dimensionless numbers are detailed in the review paper [8]. In extrusion-based bioprinting, to prevent spreading and collapse, it is essential for bioink to exhibit high viscosity at a low shear rate. Apart from polymer molecular weight, factors such as branching degree, concentration, and rheological modifiers play a role in influencing viscosity [12]. Generally, an elevation in these factors leads to increased viscosity across all shear rates. The flowability of bioink is directly influenced by its viscosity. Conversely, an optimal, frictionless system can enhance extrusion by minimizing surface tension between the needle gauge and bioink.

However, in EBB, because cell damage commonly occurs during the extrusion process due to various mechanical actions applied to the cells, including tensile, compressive, shear, and extensional stresses. This occurrence is influenced by factors such as cell types, biocompatibility, bioprinting techniques, and bioprinting parameters, ultimately resulting in a decrease in cell viability [13].

5.3.2 Inkjet-Based Bioprinting

The inkjet approach also belongs to the additive manufacturing method for replicating materials or images, or accurately ejecting and steering ink in the form of droplets to predetermined points. Typically, the droplets are transferred from the printhead to the receiving substrates in an inkjet-based bioprinting system without direct contact, employing gravity, pressure, and fluidic mechanisms. This technology boasts numerous benefits, including cost-effectiveness, the ability to be programmed, high precision, rapid production, swift printing speed, and adaptability to a variety of biogenic substances. Additionally, by adjusting drop densities or sizes, intricate 3D structures can be generated by introducing concentration gradients that include cells, materials, or growth factors. The diagrammatic representation of classification of INKJET printing system is shown in Figure 5.3. Originally utilized within office printers and microengineering for printing digital boards, this technology had its inaugural exploration in biomedical applications by Klebe in 1988. Using HP inkjet printers, he generated synthetic tissues in both 2D and 3D formats by depositing collagen and fibronectin. Subsequently, numerous research has explored diverse inkjet printers applied in the field of biomedicine, encompassing regenerative medicine, toxicology, disease modeling, and pharmaceutical science [11, 14].

Bioinks suitable for inkjet printing technologies need to satisfy not just the criteria of biocompatibility and biodegradability but also conform to the physical limitations of bioprinters, including resolution, orifice size, and viscosity. These challenges would inevitably impact the selection of appropriate bioinks. It is crucial to highlight that for diverse bioprinting applications, bioinks must either remain in a liquid state or have substantial water content [15].

To achieve stable and effective printed arrangements, meeting solid, biological, and environmental criteria is imperative. The behavior and effectiveness of the

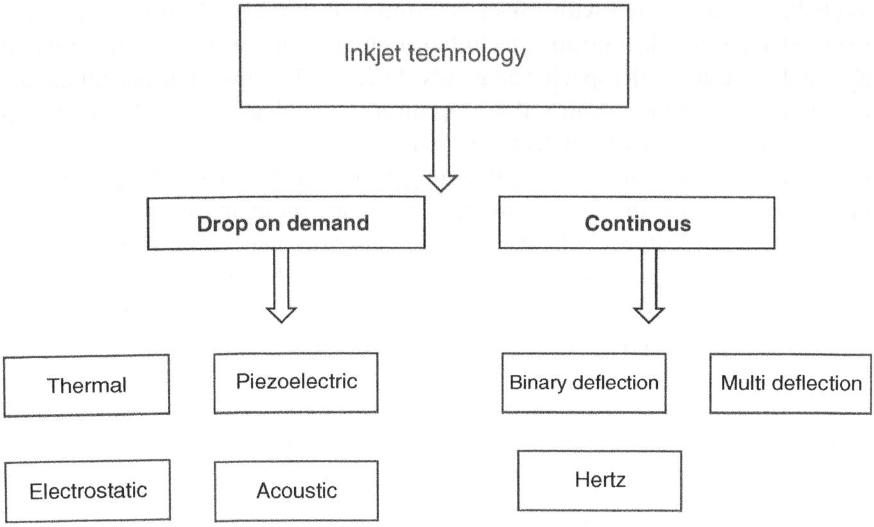

Figure 5.3 Diagrammatic classification of INKJET printing system [11].

printed designs are impacted by tangible factors like orifice dimensions, mechanical vibrations, matrix composition, ink viscosity, and temperature. Those variables significantly impact the droplet size, resolution, and sharpness of the printed configuration. Among the essential characteristics in advancing novel printing technology are print resolution and the excellence of the printed design. This bioprinting technology encompasses ejecting an array of droplets onto substrates, which blend to construct the printed patterns and lines. The clearness of the printed motifs is altered by the deposition of the inks drop-by-drop. Improved printing quality is achieved by reducing the size of droplets and accurate regulation of the printing velocity. However, these two parameters are governed by several physical factors [16]. The droplet size is influenced by various factors, incorporating factors like nozzle size, printing mechanism, material thickness, and the material's capacity to be wetted [17].

Inkjet printing technology provides numerous benefits compared to alternative patterning methods, including high precision, high throughput, and cost-effectiveness. However, various bio-based and physical limitations restrict the capabilities and structure of inkjet bioprinters. Nonbiological constraints of bioprinter design include ink viscosity, surface tension, and resolution. On the other hand, biological restrictions encompass biocompatibility, biodegradability, mechanical strength, release profile, and cell viability. Additionally, approval might be necessary for bioprinting technologies from governing bodies, like the US FDA, for business applications within the medical and pharmaceutical sectors. Numerous research has been undertaken to overcome these limitations in addition to challenges associated with inkjet bioprinting technology. Further research is required to create inkjet printing technologies that are more effective and dependable and can cater to the specific demands of the medical and pharmaceutical sectors. In order to commercialize this bioprinting technology, certain limitations of conforming inkjet dispensing technology must be dealt with. Proper ink formulation can be challenging to satisfy both engineering and biological requirements. For instance, inks with low viscosity tend to radiate more on medium, leading to lower-resolution features, whereas elevating viscosity may result in nozzle clogging. The incorporation of cells in the inks also affects their viscosity. The successful creation and production of scaffolds or drug carriers require collaboration across multiple disciplines involving chemistry, material science, pharmacology, and engineering. The properties of inks, substrates, and their interactions need to undergo modifications to enhance printing clarity and effectiveness. The effects of non-Newtonian behavior, such as the coffee drop effect and Marangoni flow, should be minimized to maintain print quality. Customizing inkjet printer designs for each application, developing polymer inkjet printers, and advancing automation and microscale devices are some of the upcoming trends in inkjet bioprinting technology. Cell patterning using inkjet printers is beneficial for studying diseases that cannot be recreated in animal models, necessitating optimized inks for commercial translation. Inkjet printing enables precise drug concentration deposition, adjustment of drug release profiles, and the development of homogenous API drug carriers. Complex in vitro models can be fabricated to analyze delivery systems for drugs and strategies for administering newly developed medications [11].

5.3.3 Stereolithography-Based Bioprinting (SL)

This is a hasty pilot model using a visible laser beam that solidifies the photo-crosslinked substances by pint-point curing technique to fabricate 3D complex products having greater fidelity. From this approach, we can synthesize micro-scale bionic structures [18, 19]. SL, being a nozzle-less approach, yields remarkable cell viability exceeding 95%, along with a print repetition accuracy of 10 μm [20]. Moreover, the major advantages of SL are biocompatibility and manufacturing accuracy. Furthermore, photosensitive resin print is marketable SL equipment and is inadequate for soft photocrosslinkable hydrogels. There are certain SL devices for printing photocrosslinkable hydrogels. Besides, there is a high cost and inconvenience related to heavy machinery operators which dwindle the biological applications of SL. Consequently, a novel SL bioprinting system has been created, comprising effective printing methods and a compact, affordable device that demonstrates excellent biocompatibility [21]. Figure 5.4 A visual depiction of the stereolithography bioprinting system showcasing its components and operational processes [1].

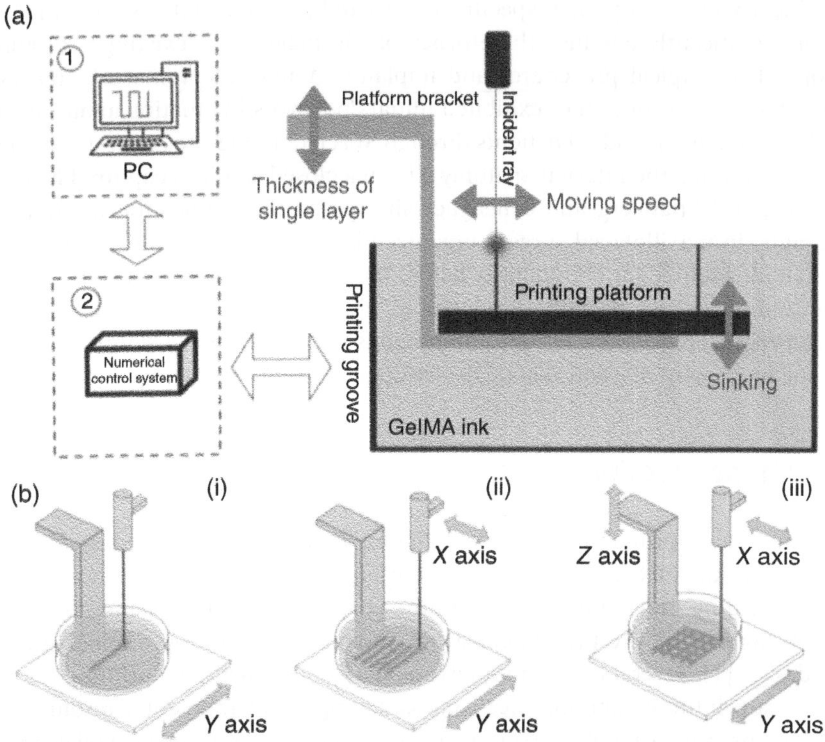

Figure 5.4 Schematic representation of stereolithography bioprinting system. (a) graphical representation of the SL system. (b) modeling of SL bioprinting system. *Source:* Schwab et al. [1] / with permission of American Chemical Society.

Peng et al. put forward a compact and cost-effective lightweight system, with greater cell-viable SL bioprinting. This user-friendly and convenient equipment, equipped with a visible laser transmitter and a Petri dish, demonstrates the rapid printing of GelMA hydrogel as its initial proof. Furthermore, it aims to tackle the challenges of limited printability and extend the duration for crosslinking of photo-crosslinkable hydrogels. Peng et al. concentrated on studying the incrementally curing-based 3D bioprinting approach to precisely fabricate intricate frameworks. For this purpose, they utilized the GelMA hydrogel, selected as an exemplary light-curable hydrogel for the examination of printability, considering its versatile applications and exceptional biological properties. In conclusion, the research consistently explored the sequential process of solidification in 3D bioprinting, evaluated printing outcomes using defined metrics, and confirmed its biocompatibility. These findings serve as a valuable resource for drug screening, understanding pathological mechanisms, and constructing biological disease models [21].

The incorporation of patient-specific models and personalized design parameters can significantly enhance the efficacy of the majority of existing pre- and post-operative surgical procedures and implants. A few research works are as follows: Stoker and colleagues exhibited initial endeavors in creating anatomical models tailored to individual patients through SLA technology, to generate plastic models illustrating the internal anatomy within a closed skull, as captured by CT imaging [22]; Wittenberg and colleagues showcased the application of stereolithography in maxillofacial surgery organized by assessing such suitability of a human donor source for cranioplasty before the actual operative procedure [23]; Bashir and colleagues recently developed a "biobot," a biological machine that relies on engineered cardiac tissue's autonomous and synchronized contraction to drive the movement of a soft robotic device created using SLA fabrication [24].

5.4 Shape Fidelity

The evaluation of printability and shape fidelity represents a pivotal stage in the process of evaluation of a bioink. Shape fidelity, often referred to as print accuracy, is determined by the ability to maintain the shape of individual filaments after extrusion and the entire printed construct, in comparison to the original computer design [1]. Several factors influence shape fidelity, including printing resolution, post-printing combining effectiveness, and rigidity. The potential resolution for lithographic and vat polymerization techniques is primarily determined by the printing hardware, ranging from 5 nm for SLA [25], VBP [26], and DLP processes to 10 nm for MPP [27]. The smallest capacity unit, or voxel, determined during the irradiation-merging process, dictates the surface's configuration and roughness characteristics in the printed structure's item. Yet, accomplishing thoroughly encapsulated photopolymerization within the exposed area can be difficult.

Interestingly, the term "printability" lacks a standardized and clear definition in the field of bioprinting. For the purpose of this study, printability is defined as the ability to extrude a hydrogel and dispense it in a pattern with a satisfactory level of shape fidelity, which indicates how closely the printed structure matches the original CAD design. Form fidelity, therefore, plays a critical role in printability, and the development of novel bioinks could greatly benefit from a standardized and quantitative definition of form fidelity. Currently, form fidelity is commonly assessed visually using macro- and microscopic photographs of extruded filaments or printed objects [28].

The ability to produce objects with exceptional resolution and shape fidelity is a pivotal aspect of 3D printing. Consequently, the desire for quick phase transitions, shifting from a liquid-like solution to a solid state, aims to improve the printability of hydrogels [29]. Furthermore, an ideal ink should exhibit shear-thinning behavior during printing, possess adequate mechanical stability to maintain its form post-printing and exhibit high structural integrity under physiological conditions. Additionally, if cells are enclosed and printed alongside the hydrogel, they must be cytocompatible [30].

Hydrogel-based ink printability remains a concern due to its stiffness and tendency to shrink, which poses limitations on form integrity. To enhance the precision of hydrogel printing, several methods have been explored. One approach involves utilizing highly viscous polymeric solutions, which aid in creating solid filaments, enabling the deposition of robust hydrogel scaffolds with self-supporting capability. Increasing the overall polymer concentration, employing high molecular weight polymers, or combining two different polymers to form polyelectrolyte complexes are all viable strategies for creating viscous inks [31, 32].

5.4.1 Shape Fidelity in Planar Structures

In addition to strand development, uniformity as well as regularity of the filament are crucial factors in the production of planar structures. Planar structures are made primarily along two dimensions (the *x* and *y* planes), which are far broader than the structure's height. In more qualitative and quantitative examinations of shape fidelity documented within the literature, this often equates to structures made up of one to two layers. For instance, filament diameter can be indirectly determined by measuring the distance between filaments or directly assessed on photographs of the filament taken at various points (Figure 5.1) [31, 33–35]. The spreading ratio is produced by normalizing the measured diameter of the printed filament in relation to the needle diameter [1].

Additionally, it is possible to determine how effectively two filaments are stacked and how surface tension could cause the filaments to bend by looking at the diameter of the filament at their junction [36]. In image-based analysis, the height of the manufactured filament is also looked at in addition to filament width [37, 38], Geunseon et al. 2017. Geometric accuracy is the measurement of the fiber diameter (*x*–*y* plane) and breadth (*z*-axis) at various locations along the longitudinal axis of a specific thread or planar framework, and the comparison of those measurements to the initial dimensions of the computer-aided design (CAD) file at identical locations (Figure 5.5c(ii)) [39].

5.4.2 Shape Fidelity in Multilayered Structures

The structure consistency of structures grounded on many layers should be assessed after a reliable rule for the deposition of strands in a solo layer has been established. Geometric precision, layer stacking, and structural integrity are significant factors that have been suggested for this evaluation. A multilayered setup's quality of layer stacking, also known as critical height, is indicated by the estimation of the highest peak obtained while imprinting a certain configuration

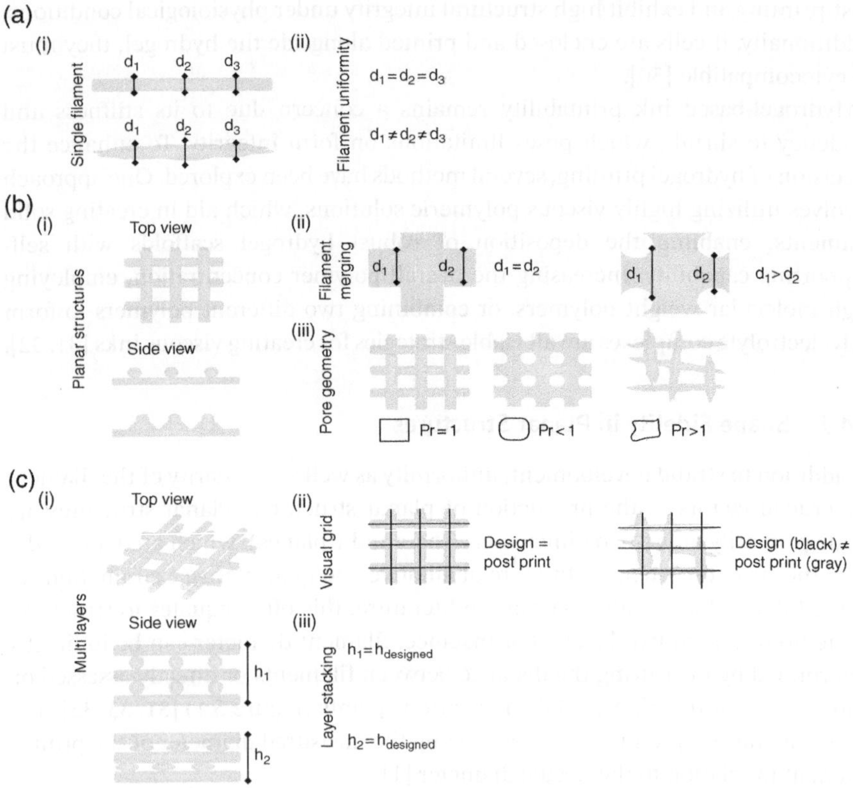

Figure 5.5 Quantitative examinations are conducted to evaluate the accuracy of a bioink's shape both during the printing process and after fabrication. For single filaments (Ai), their uniformity is assessed based on their fiber diameter (d1, d2, and d3), where consistent diameters indicate a homogeneous filament. In 2D planar structures (Bi), which primarily extend in two directions and are typically composed of 1–2 layers, the evaluation focuses on filament diameter and fusion, particularly at the intersection/overlay of two filaments, as well as the transversal pore geometry aiming for an optimal rectangular shape to facilitate filament stacking (printability index Pr = 1). Multilayered constructs (Ci) are analyzed for filament circularity, and a visual grid is used post-printing to compare the printed structure (gray lines) with the computer-designed shape (black lines). Additionally, the retention of circular filaments' shape in multilayered constructs is examined by comparing the height of the computer-designed sample to the actual height (h1 and h2) after layer stacking. *Source:* Schwab et al. [1] / with permission of American Chemical Society.

in addition to the spatial arrangement correctness discussed in the preceding part [40]. The determined percentage of dimensions of the construct after the printing process in relation to the theoretically planned ones is a key component in both shape fidelity analysis and integrity index calculations [4, 41].

3D computed tomography (CT) imaging has been suggested to evaluate shape fidelity and repeatability, as well as to spot flaws or artifacts inside a build post-printing. CT may be used to visualize structures, despite the fact that inks consisting exclusively of hydrogels have a low CT contrast due to their high water content [42]. However, morphometric analysis of micro-CT scans of hydrogel-based structures is possible by utilizing reduced radiant power and focusing on the bottom limit of the grey levels (Figure 5.6) [41].

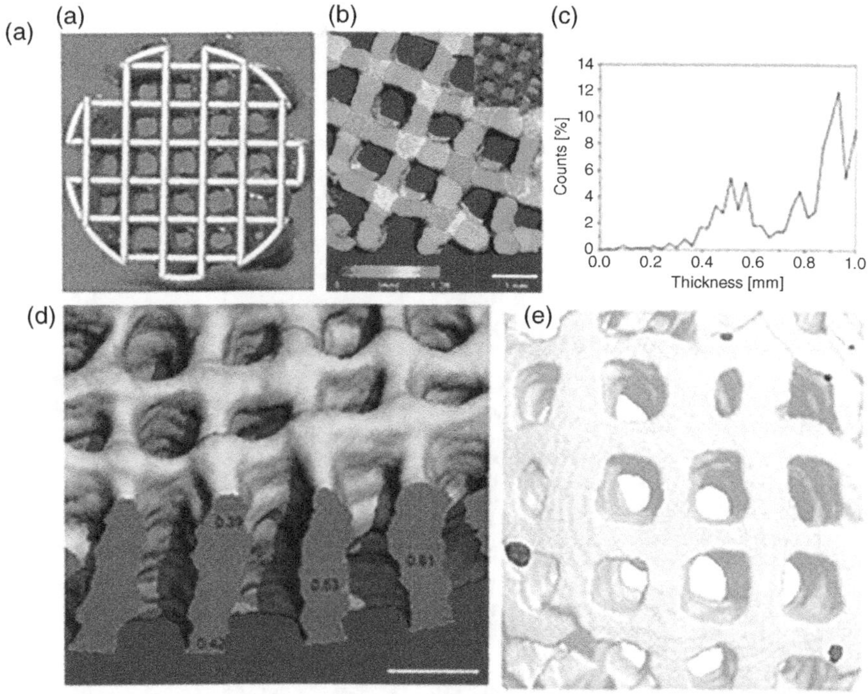

Figure 5.6 CT and OCT are utilized to visualize filaments and pore structures in printed hydrogels, aiding in the assessment of shape fidelity. Micro-CT is employed to evaluate the shape fidelity of 3D printed HA-based hydrogels:- Optical imagery (a) displays a 3D printed lattice grid superimposed with its CAD model.- Micro-CT's 3D reconstruction (b) of the printed construct uses color to depict thickness.- Strut thickness distribution (c) is analyzed within the 3D reconstruction.- A micro-CT cross-section (d) of a multi-layered 3D printed construct showcases overlaying accuracy.- The 3D reconstruction image (e) highlights air pockets in red, with scale bars set at 1 mm. In addition, optical coherence tomography (OCT) is used to image gelatin/alginate hydrogels with varying architectures, showing:- Cross-section (A1, 3) and surface (B1, 3) views.- Hydrogel structure at a 1 mm depth (C1, 3).- A 3D observation of the hydrogel matrix (D1, 3), with scale bars of 500 μm and UMP indicating undefined micropores. *Source:* Petta et al. [41] /with permission of American Chemical Society.

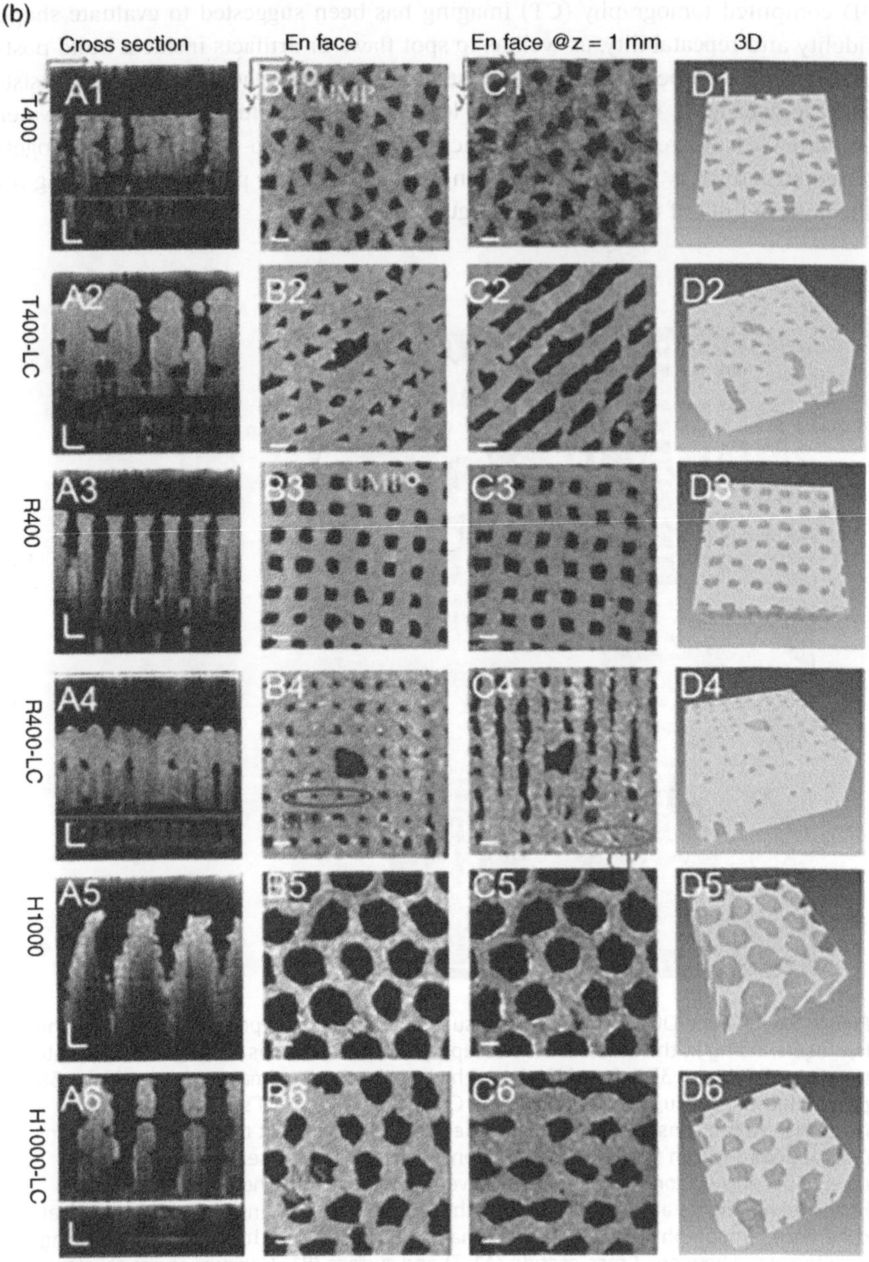

Figure 5.6 (Continued)

Analyzing shape fidelity and printability is a crucial stage in creating a new bioink. Printability, also known as shape fidelity, is the capacity of elements to be applied in a premade design while keeping its original shape (i.e., akin to CAD design) [43, 44]. To maximize the utility of bioprinting, it is crucial to evaluate the form consistency of lone fiber, which is necessary to calculate impression correctness. Bioinks employed in EBB should exhibit shear-reducing behavior, which is characterized by a reduction in viscosity when the shear rate is increased [45]. Cells could face potential damage during the printing process, either through shear or heat stressors, or subsequent to printing, due to filaments with limited biocompatibility. Cell viability is another crucial factor to take into account. High-viscosity bioinks encounter larger shear stress during printing, which reduces cell viability [46], but they can preserve the 3D bioprinted designs over the incubation phase. Due to their resemblance to natural extracellular matrix (ECM), natural hydrogels such as alginate, gelatin [47, 48], collagen, and chitosan have frequently been employed for bioprinting. However, the printability of these natural bioink components is often low due to their weak mechanical qualities. By modifying the characteristics of the materials using additive materials, this restriction can be overcome [46].

There have been attempts to statistically assess the printability and shape fidelity of bioinks. In this context, the degree of pore circularity in grid arrangements is utilized as a gauge for good printability (i.e., less circularity after replication indicates a closer match between the print result and the computer design) [49]. However, this approach ignores the rheology of bioink and the deformation of hanging filaments in more intricate patterns, focusing primarily on the *X*–*Y* resolution of patterns. Different bioink compositions and concentrations were printed on support structures with variable spacings to test if a filament could maintain its form after printing [46, 50].

One of the key factors affecting a patch's functionality is its fidelity. After serving its purpose, a patch must keep its shape and framework on the substantial tissue or organ for a while before biodegrading. Mechanical loading is a crucial design factor; for instance, patches have been suggested or created to act as pressure sensors on organs and veins as well as repair materials for cardiac reconstruction in individuals with complicated congenital heart abnormalities [46].

5.4.3 Characterization Approaches

5.4.3.1 Rheological Characterization

Utilizing an AR-G2 rheometer from TA instruments, located in New Castle, DE, USA, equipped with a rotatable cone palate measuring 40 mm in diameter and set at a 150 m gap, the rheological characteristics of bioinks and their constituents were examined. All tests were conducted at a temperature of 25 °C, and samples were given a minute to level off before being added to the test platform. To test viscosity, a range of shear rates between 0.01 and 100 s-1 was used. Using the same cone plate, a dynamic strain sweep was conducted to determine the region where viscoelastic behavior is linear (LVR). The test was conducted using the 1% strain sweep. Then, with angular frequencies ranging from 0 to 100 rad/s, the storage modulus (G′) and loss modulus (G″) were measured, and estimated using the frequency sweep [46, 51].

5.4.3.2 Mechanical Characterization

Utilizing a dynamic mechanical analyzer (DMA Q800, TA Instrument, New Castle, DE, USA), samples were mechanically tested in order to determine Young's modulus in the slope in the early straight area of the stress-strain curve. Specimens with dimensions of 20 mm × 10 mm × 0.5 mm were subjected to ramping force (stress) at a momentum of 0.1 N/min. Until the sample failed, concurrent deformation (strain) monitoring was done. Additionally, the maximum deformation before failure ("elongation at break") was established. In order to simulate movement in the skin, cyclic loading experiments were used to measure hysteresis (deformation in reaction to cycling) over time. To prevent drying, the cycle tensile was carried out in DI water. A total of 100 cycles were completed. In order to simulate the maximal pressure applied to the skin on the human wrist, the strain was ramped up and down at a rate of 10%/min between 0% and 20% [52]. The stress-strain curve of the interior region of applying and removing the load was used to compute hysteresis. At 25 °C, this experiment was conducted [46].

5.4.3.3 Swelling Test

Samples from 3D printing were weighed after the samples were left to dry overnight in the incubator at a temperature of 37 °C. Subsequently, the dried samples were immersed for up to 96 hours in deionized water at ambient temperature. At defined intervals (1, 2, 4, 6, 8, 24, and 96 hours), the measurement of the extent of swelling, and extra surface water on the sample was blotted out prior to weighing. The samples were put back in the DI water bath when the measurement process was finished. The following equation [53] was used to compute the swelling ratio for samples:

$$S = (ws - wd/wd) \times 100$$

Here, S represents the swelling percentage, wd, and ws are the dry weight and swollen weight of the sample, respectively. The experiment was performed with three distinct samples for each bioink, and the error bars indicated the standard deviation.

5.4.3.4 Viability Characterization

The cell viability characterization is done as follows: Ethidium homodimer-1 (EthD), which spots dead cells red, and calcein AM, which imparts a green color to living cells, were both utilized in this study. After staining, fluorescent microscopy was used to see the cells. After the samples had been cleaned with PBS, a solution of calcein at a concentration of 1 part in 2000 and ethidium homodimer at a concentration of 1 part in 500 was added to the PBS. For 15 minutes, the staining solution was incubated with each sample. In the *z*-axis, images were taken across a 100 m distance in six separate focus planes. To calculate the cell viability, the six pictures were integrated into a *z*-stack, and an image using *z*-projection was created for each channel independently using the highest value among all six planes for each (x, y) pixel. This process was carried out using ImageJ from the National Institutes of Health (NIH) in Maryland, USA, and a noise tolerance setting of 20 to identify local maxima in each picture was found using the "find maxima" tool. Each local maximum in the red-channel (EthD) picture served as a dead cell, whereas the local maximum in the green-channel (calcein) rendering served like a living cell. Each image's cell integrity was determined by applying:

$$\text{Cell viability} = \text{live cells} / \left(\text{live cells} + \text{dead cells}\right)$$

5.4.3.5 Bioprinting Procedure

Utilizing mouse fibroblast cells of the NIH/3T3 strain from the American Type Culture Collection in Manassas, Virginia, USA, the biocompatibility of hydrogel was assessed. To achieve the required cell density, fourth-stage cells were grown for a week in accordance with the ATCC methodology. Hydrogels 2A4CNC and 4A1CNF were used to make syringes, which were then sterilized by UV for 20 minutes. Before printing, $3 \times 10\hat{}6$ cells were combined with the hydrogels while still in a sterile environment. Air pressures for 2A4CNC and 4A1CNF were established at 12 PSI and 20 PSI, respectively, and extrusion velocity was adjusted at 7 mm/s. One layer of the bioink mixes was applied. After the printing process, the samples underwent two washes with cell media immediately after replacing the calcium solution with warm PBS. For up to 10 days, the bioink samples were stored in cell medium at 37 °C and were moved into a well plate, and the CaCI2 solution was used to crosslink the bioink for two minutes.

The continuum-mechanics models rely on the idea that granular materials may be thought of as a continuum, for which the mechanical flow behavior can be described by the constitutive laws. Varied fundamental regulations or varied values associated with the inherent material parameters were used to realize the difference in species of biomass, size of particles, and particle shape. For instance, the authors have used the altered Drucker–Prager/Cap model, the NorSand model, and the hypoplastic model to forecast the flow characteristics of milled loblolly pine with different particle size distributions. To simulate milled corn stover, Douglas Yi and colleagues (2020a, b) utilized the Drucker-Prager/Cap model and the modified Cam-Clay representation.

Individual particles are tracked using discrete particle models, which also forecast particle trajectories based on collisions with nearby particles. Discrete particle modeling explicitly describes and simulates the variance in particle size, morphology, and mechanical characteristics brought on by the species and pre-processing of the biomass. For instance, flexible clumped spheres were employed by [54] to mimic various particle morphologies for loblolly pine and switchgrass that have been processed. Polyhedral discrete element modeling was used by [55] to precisely represent the intricately formed pine chips. An experimentally guided, semiempirical elastoplastic bonding model was created by [56] for modeling woody biomass particles using discrete element methods. A set of sophisticated particle contact rules was recently presented to explain particle interactions with monospheres.

Individual particles are tracked using discrete particle models, which also forecast the paths of particles as they collide with neighboring particles. Discrete particle modeling explicitly accounts for the variance in particle size, morphology, and mechanical characteristics resulting from biomass species and pre-processing. For example, flexible clumped spheres were employed to mimic various particle morphologies for milled loblolly pine and switch grass [55]. Xia et al. [54] utilized polyhedral discrete element modeling to accurately represent the complex-shaped pine chips, while [56] created A semiempirical elastoplastic bond model, guided by experimental data, for the discrete element modeling of particles in woody biomass. Recently introduced a set of sophisticated particle contact rules to explain particle collisions with individual spheres.

Both the continuum mechanics and discrete particle models have been effectively utilized to replicate the flow of granular biomass in diverse characterization tests, including linear compression, axial shear, Schulz ring shear, and pilot-scale hopper simulations [57]. Continuum mechanics models are effective in predicting large flow systems with reduced order precision, while discrete particle models, despite being more resource-intensive in terms of computation, offer higher accuracy and are better suited for studying granular movement on a miniature scale. However, a quantitative comparison of the processing costs and their ability to capture intricate flow mechanics at various sizes remains unexplored [58].

In the field of intelligent biofabrication, product modeling, and design play a critical role. The ultimate objective of TERM biofabrication is to fulfill a patient's clinical requirement for regeneration, necessitating the creation of biomimetic, biofabricated products that accurately reproduce biological structures. Information retrieval is vital to achieving biomimetic fidelity in TERM biofabrication products, and image-based biofabrication of patient-specific tissues relies on processed pictures from MRI or CT scans, which are converted into CAD models for printing. Computer-assisted design (CAD), computer-aided manufacturing (CAM), and computer-aided engineering (CAE) are applied to biological processes and sub-procedures to enable the creation of tissue and organ blueprints, the production of biological products, and the validation and optimization of biological products and biomanufacturing tools. Computer-controlled rapid prototyping (RP) using CAD is widely used for scaffold creation and enabling physiologically relevant structures in multiple dimensions.

For biomechanical research, finite element modeling (FEM) is a powerful technique used for modeling and optimization, aiding in the design and improvement of scaffolds for TERM. FEM simulates biomechanical and geometrical properties, as well as dynamic characteristics and cell responses during construct maturation. The computer modeling of 3D bone tissue regeneration, including new bone production and scaffold deterioration, relies on the voxel finite element approach as a foundational strategy [59].

5.5 Case Studies and Applications

Tissue engineering has arisen as an interdisciplinary area involving doctors and biologists, together with engineers, to create tissue constructs that accurately replicate anatomy, addressing the shortage of donor tissues or organs. Basic cell sowing on pre-made polymeric frameworks is insufficient to properly reproduce the complex cell-matrix interactions found in native tissues, even though there are significant advancements in the field of tissue engineering. The diversity in the composition of the ECM, in both the epidermal and dermal regions of the skin, performs various functions, from controlling cellular proliferation to influencing the destiny of stem cells. Bioprinting, a relatively new technology, uses 3D printing technology to build sophisticated tissues or organs by incorporating viable living cells with biomaterials. This technology has a lot of potential for fabricating complex 3D multicellular tissue constructs, hence it allows the concurrent placement of various biomaterials and cell types, along with the creation and assembly of personalized tissue-engineered structures tailored to individual patients.

Bioprinting has demonstrated considerable potential in fabricating skin, cartilage, and liver constructs with multiple layers. The synthesis of basic human tissues or organs, such as skin, was widely expected to become a reality soon. Fabrication of hydrogel frameworks consisting of diverse skin cells (keratinocytes and fibroblasts), and in situ printing of skin cells and biomaterials right above the wound site, are two recent works on bioprinting of skin constructions. The thin layer of human skin has a distinctive pattern generated by the natural compartmentalization of various skin cell types that are arranged relative to one other at a high degree of precision, contrary to popular belief that skin is a relatively basic 2D structure. This unique arrangement of skin cells is required for cell–cell interactions that trigger autocrine and paracrine signaling [60, 61].

The utilization of 3D printers for the fabrication of scaffolds is gaining increased popularity in the field of tissue engineering and regenerative medicine. 3D printers are capable of producing complicated geometric shapes with great resolution and reproducibility using a variety of biocompatible polymeric materials. The BioScaffolder 3.1 (GeSiM, Germany) 3D printer used in the study allows for the creation of scaffolds with desired properties (e.g., roughness, topography, functional groups, porosity, pore size, and surface area/volume ratio). Controlling all of these properties is critical for creating 3D structures that can mimic the skin's natural role in cell proliferation, and blood circulation, and thus supply nutrients to healing tissues. The primary structural proteins of the skin, such as elastin and collagen, exist biologically in the form of fibers with sizes ranging from 10 to a few hundred nanometers. The skin's adaptive qualities and integrity (e.g., elasticity and mechanical support) are provided by a mesh-like matrix, i.e., ECM made up of these fibers. Structures with a critical resemblance to the skin ECM can be generated using electrospinning and a formulation based on biodegradable polymers, which can imitate the ECM's morphological features. As a result, electrospun nanofibers have almost reached "state-of-the-art" status in the creation of wound care materials. Electrospun matrices have also been found to have a large surface area and microporosity, making them excellent for drug or biomolecule loading [62].

Day-to-day improvements in numerous domains of tissue engineering are documented in research reports. Several innovative ways for 3D scaffold construction have been investigated for easy, rapid, and precise processing in regenerative treatment. 3D printing is a cutting-edge tissue engineering technique that has the potential to outperform traditional solid scaffold-based tissue engineering. Scaffolds with specified geometries and regulated and linked porous structures are created using this method. These scaffolds are made of a variety of polymers that are divided into natural and synthetic categories. There are several obstacles to be solved in the creation of 3D scaffolds in tissue engineering applications, notwithstanding the promise of technological improvement [63]. To tackle the numerous challenges in the fabrication of 3D scaffolds for organ tissue engineering, future progress in the field of tissue engineering will encompass the use of diverse biomaterials, refinement of scaffold designs, and a thorough understanding of cells and developmental biology. The material processability, degradation pattern, scaffold deterioration, and mechanical performance are all determined by the kind of material used. Second, the design of the scaffolds is a critical factor in determining the pore size, morphology, and topography of the scaffolds. Finally, the majority of the obstacles in printing living cells are restricted to cell survival and vascularization. A handheld bioprinting device that helps transport cells into diverse tissues like skin or cartilage might be a viable future option for regenerative medicine. Moreover, the advancement of materials that are both biodegradable and biocompatible will exert a substantial impact on future tissue engineering applications.

5.6 Conclusion

Bioprinting technology is advancing through diverse methods, leading to the emergence of novel platforms on a global scale. The chapter delves into the remarkable progress and obstacles encountered in the cutting-edge domain of printability and shape fidelity in diverse bioprinting processes. The exploration within this chapter encompasses a range of bioprinting methods, each showcasing their capacity to fabricate intricate structures with exceptional precision and accuracy. From extrusion-based bioprinting methods to inkjet and stereolithographic techniques, each approach has its distinct strengths and drawbacks.

The importance of printability and shape fidelity cannot be underestimated, as they significantly influence the quality and functionality of printed structures, whether applied in tissue engineering, restorative medicine, or drug testing. Attaining the best printability and shape fidelity necessitates a detailed comprehension of material characteristics, printer settings, and bioink compositions.

In conclusion, the quest to improve printability and shape fidelity in bioprinting techniques remains a continuous endeavor, holding tremendous promise for transforming healthcare and regenerative treatments. By persistently dedicating ourselves to innovation and collaborating across disciplines, we can create a pathway toward a future in which bioprinting plays a central role in personalized medicine, bringing about a positive influence on innumerable lives across the globe.

References

1 Schwab, A., Levato, R., D'Este, M. et al. (2020). Printability and shape fidelity of bioinks in 3D bioprinting. *Chem. Rev.* 120 (19): 11028–11055.

2 Habib, A. and Khoda, B. (2019). Development of clay based novel hybrid bio-ink for 3D bio-printing process. *J. Manuf. Process.* 1 (38): 76–87.

3 Ahn, G., Min, K.H., Kim, C. et al. (2017). Precise stacking of decellularized extracellular matrix based 3D cell-laden constructs by a 3D cell printing system equipped with heating modules. *Sci. Rep.* 7 (1): 8624.

4 Deo, K.A., Singh, K.A., Peak, C.W. et al. (2020). Bioprinting 101: design, fabrication, and evaluation of cell-laden 3D bioprinted scaffolds. *Tissue Eng. A* 26 (5–6): 318–338.
5 Hong, S., Kim, J.S., Jung, B. et al. (2019). Coaxial bioprinting of cell-laden vascular constructs using a gelatin–tyramine bioink. *Biomater. Sci.* 7 (11): 4578–4587.
6 Huang, J., Fu, H., Wang, Z. et al. (2016). BMSCs-laden gelatin/sodium alginate/carboxymethyl chitosan hydrogel for 3D bioprinting. *RSC Adv.* 6 (110): 108423–108430.
7 Mir, T.A., Iwanaga, S., Kurooka, T. et al. (2019). Biofabrication offers future hope for tackling various obstacles and challenges in tissue engineering and regenerative medicine: A Perspective. *Int. J. Bioprinting* 5 (1): 153.
8 Zhang, Z., Jin, Y., Yin, J. et al. (2018). Evaluation of bioink printability for bioprinting applications. *Appl. Phys. Rev.* 5 (4): 041304.
9 Mishra, R.N., Singh, M.K., and Kumar, V. (2022). Biomechanical analysis of human femur using finite element method: A review study. *Mater. Today Proc.* 1 (56): 384–389.
10 Mora Boza, A., Wlodarczyk-Biegun, M.K., Del Campo, A. et al. (2019). Chitosan-based inks: 3D printing and bioprinting strategies to improve shape fidelity, mechanical properties, and biocompatibility of 3D scaffolds. *Biomecánica* https://doi.org/10.5821/sibb.27.1.9199.
11 Azizi Machekposhti, S., Mohaved, S., and Narayan, R.J. (2019). Inkjet dispensing technologies: recent advances for novel drug discovery. *Expert Opin. Drug Discov.* 14 (2): 101–113.
12 Jia, J., Richards, D.J., Pollard, S. et al. (2014). Engineering alginate as bioink for bioprinting. *Acta Biomater.* 10 (10): 4323–4331.
13 Panwar, A. and Tan, L.P. (2016). Current status of bioinks for micro-extrusion-based 3D bioprinting. *Molecules* 21 (6): 685.
14 Jin, W., Lu, Y., Chen, F. et al. (2022). On the fidelity of computational models for the flow of milled loblolly pine: A benchmark study on continuum-mechanics models and discrete-particle models. *Front. Energy Res.* 28 (10): 855848.
15 Saunders, R.E. and Derby, B. (2014). Inkjet printing biomaterials for tissue engineering: bioprinting. *Int. Mater. Rev.* 59 (8): 430–448.
16 Lee, V.K., Dias, A., Ozturk, M.S. et al. (2015). 3D bioprinting and 3D imaging for stem cell engineering. *Bioprinting in regenerative medicine* 33–66.
17 Caviezel, D., Narayanan, C., and Lakehal, D. (2008). Adherence and bouncing of liquid droplets impacting on dry surfaces. *Microfluid. Nanofluid.* 5: 469–478.
18 Zhou, X., Cui, H., Nowicki, M. et al. (2018). Three-dimensional-bioprinted dopamine-based matrix for promoting neural regeneration. *ACS Appl. Mater. Interfaces* 10 (10): 8993–9001.
19 Zhu, W., Cui, H., Boualam, B. et al. (2018). 3D bioprinting mesenchymal stem cell-laden construct with core–shell nanospheres for cartilage tissue engineering. *Nanotechnology* 29 (18): 185101.

20 Herschdorfer, L., Negreiros, W.M., Gallucci, G.O., and Hamilton, A. (2021). Comparison of the accuracy of implants placed with CAD-CAM surgical templates manufactured with various 3D printers: an in vitro study. *J. Prosthet. Dent.* 125 (6): 905–910.

21 Zhang, P., Wang, H., Wang, P. et al. (2021). Lightweight 3D bioprinting with point by point photocuring. *Bioact. Mater.* 6 (5): 1402–1412.

22 Mankovich, N.J., Cheeseman, A.M., and Stoker, N.G. (1990). The display of three-dimensional anatomy with stereolithographic models. *J. Digit. Imaging* 3: 200–203.

23 Bill, J.S., Reuther, J.F., Dittmann, W. et al. (1995). Stereolithography in oral and maxillofacial operation planning. *Int. J. Oral Maxillofac. Surg.* 24 (1): 98–103.

24 Chan, V., Park, K., Collens, M.B. et al. (2012). Development of miniaturized walking biological machines. *Sci. Rep.* 2 (1): 857.

25 Raman, R., Bhaduri, B., Mir, M. et al. (2016). Bioprinting: high-resolution projection microstereolithography for patterning of neovasculature (adv. healthcare mater. 5/2016). *Adv. Healthcare Mater.* 5 (5): 622.

26 Bernal, P.N., Delrot, P., Loterie, D. et al. (2019). Biofabrication: volumetric bioprinting of complex living-tissue constructs within seconds (adv. mater. 42/2019). *Adv. Mater.* 31 (42): 1970302.

27 Vaezi, M., Seitz, H., and Yang, S. (2013). A review on 3D micro-additive manufacturing technologies. *Int. J. Adv. Manuf. Technol.* 67: 1721–1754.

28 Ribeiro, A., Blokzijl, M.M., Levato, R. et al. (2017). Assessing bioink shape fidelity to aid material development in 3D bioprinting. *Biofabrication* 10 (1): 014102.

29 Gopinathan, J. and Noh, I. (2018). Recent trends in bioinks for 3D printing. *Biomater. Res.* 22: 1–5.

30 Mora-Boza, A., Włodarczyk-Biegun, M.K., Del Campo, A. et al. (2020). Glycerylphytate as an ionic crosslinker for 3D printing of multi-layered scaffolds with improved shape fidelity and biological features. *Biomater. Sci.* 8 (1): 506–516.

31 Negrini, N.C., Celikkin, N., Tarsini, P. et al. (2020). Three-dimensional printing of chemically crosslinked gelatin hydrogels for adipose tissue engineering. *Biofabrication* 12 (2): 025001.

32 Park, S.H., Kang, B.K., Lee, J.E. et al. (2017). Design and fabrication of a thin-walled free-form scaffold on the basis of medical image data and a 3D printed template: its potential use in bile duct regeneration. *ACS Appl. Mater. Interfaces* 9 (14): 12290–12298.

33 Kim, M.K., Jeong, W., Lee, S.M. et al. (2020). Decellularized extracellular matrix-based bio-ink with enhanced 3D printability and mechanical properties. *Biofabrication* 12 (2): 025003.

34 Müller, M., Fisch, P., Molnar, M. et al. (2020). Development and thorough characterization of the processing steps of an ink for 3D printing for bone tissue engineering. *Mater. Sci. Eng. C* 1 (108): 110510.

35 Soltan, N., Ning, L., Mohabatpour, F. et al. (2019). Printability and cell viability in bioprinting alginate dialdehyde-gelatin scaffolds. *ACS Biomater. Sci. Eng.* 5 (6): 2976–2987.

36 Shin, J.Y., Yeo, Y.H., Jeong, J.E. et al. (2020). Dual-crosslinked methylcellulose hydrogels for 3D bioprinting applications. *Carbohydr. Polym.* 15 (238): 116192.

37 Göhl, J., Markstedt, K., Mark, A. et al. (2018). Simulations of 3D bioprinting: predicting bioprintability of nanofibrillar inks. *Biofabrication* 10 (3): 034105.

38 Wilson, S.A., Cross, L.M., Peak, C.W., and Gaharwar, A.K. (2017). Shear-thinning and thermo-reversible nanoengineered inks for 3D bioprinting. *ACS Appl. Mater. Interfaces* 9 (50): 43449–43458.

39 Athirasala, A., Tahayeri, A., Thrivikraman, G. et al. (2018). A dentin-derived hydrogel bioink for 3D bioprinting of cell laden scaffolds for regenerative dentistry. *Biofabrication* 10 (2): 024101.

40 Duarte Campos, D.F., Blaeser, A., Korsten, A. et al. (2015). The stiffness and structure of three-dimensional printed hydrogels direct the differentiation of mesenchymal stromal cells toward adipogenic and osteogenic lineages. *Tissue Eng. A* 21 (3–4): 740–756.

41 Petta, D., Grijpma, D.W., Alini, M. et al. (2018). Three-dimensional printing of a tyramine hyaluronan derivative with double gelation mechanism for independent tuning of shear thinning and postprinting curing. *ACS Biomater. Sci. Eng.* 4 (8): 3088–3098.

42 Wang, L., Xu, M., Zhang, L. et al. (2016). Automated quantitative assessment of three-dimensional bioprinted hydrogel scaffolds using optical coherence tomography. *Biomed. Opt. Express* 7 (3): 894–910.

43 D'Amore, A., Yoshizumi, T., Luketich, S.K. et al. (2016). Bi-layered polyurethane–ECM cardiac patch improves ischemic ventricular wall remodeling in a rat model. *Biomaterials* 1 (107): 1–4.

44 Mancha Sánchez, E., Gómez-Blanco, J.C., López Nieto, E. et al. (2020). Hydrogels for bioprinting: a systematic review of hydrogels synthesis, bioprinting parameters, and bioprinted structures behavior. *Front. Bioeng. Biotechnol.* 6 (8): 776.

45 Unagolla, J.M. and Jayasuriya, A.C. (2020). Hydrogel-based 3D bioprinting: a comprehensive review on cell-laden hydrogels, bioink formulations, and future perspectives. *Appl. Mater. Today* 1 (18): 100479.

46 Temirel, M., Hawxhurst, C., and Tasoglu, S. (2021). Shape fidelity of 3D-bioprinted biodegradable patches. *Micromachines* 12 (2): 195.

47 Ramu, M., Ananthasubramanian, M., Kumaresan, T. et al. (2018). Optimization of the configuration of porous bone scaffolds made of Polyamide/Hydroxyapatite composites using Selective Laser Sintering for tissue engineering applications. *Biomed. Mater. Eng.* 29 (6): 739–755.

48 Rastogi, P. and Kandasubramanian, B. (2019). Review of alginate-based hydrogel bioprinting for application in tissue engineering. *Biofabrication* 11 (4): 042001.

49 Heid, S. and Boccaccini, A.R. (2020). Advancing bioinks for 3D bioprinting using reactive fillers: a review. *Acta Biomater.* 1 (113): 1–22.

50 Top, N., Şahin, İ., and Gökçe, H. (2021). Computer-aided design and additive manufacturing of bone scaffolds for tissue engineering: state of the art. *J. Mater. Res.* 15: 1–21.

51 Markstedt, K., Mantas, A., Tournier, I. et al. (2015). 3D bioprinting human chondrocytes with nanocellulose–alginate bioink for cartilage tissue engineering applications. *Biomacromolecules* 16 (5): 1489–1496.

52 Lu, N., Lu, C., Yang, S., and Rogers, J. (2012). Highly sensitive skin-mountable strain gauges based entirely on elastomers. *Adv. Funct. Mater.* 22 (19): 4044–4050.

53 Huq, T., Salmieri, S., Khan, A. et al. (2012). Nanocrystalline cellulose (NCC) reinforced alginate based biodegradable nanocomposite films. *Carbohydr. Polym.* 90 (4): 1757–1763.

54 Xia, Y., Chen, F., Klinger, J.L. et al. (2021). Assessment of a tomography-informed polyhedral discrete element modelling approach for complex-shaped granular woody biomass in stress consolidation. *Biosystems Eng.* 205: 187–211.

55 Guo, Y., Chen, Q., Xia, Y. et al. (2020). Discrete element modeling of switchgrass particles under compression and rotational shear. *Biomass Bioenergy.* 141: 105649.

56 Guo, Y., Chen, Q., Xia, Y. et al. (2021). A nonlinear elasto-plastic bond model for the discrete element modeling of woody biomass particles. *Powder Technol.* 385: 557–571.

57 Gao, T., Gillispie, G.J., Copus, J.S. et al. (2018). Optimization of gelatin–alginate composite bioink printability using rheological parameters: A systematic approach. *Biofabrication* 10 (3): 034106.

58 Bardini, R. and Di Carlo, S. (2023). Computational modeling and optimization of biofabrication in Tissue Engineering and Regenerative Medicine-a literature review. *bioRxiv* 23: 2023.

59 Dellaquila, A., Campodoni, E., Tampieri, A., and Sandri, M. (2020). Overcoming the design challenge in 3d biomimetic hybrid scaffolds for bone and osteochondral regeneration by factorial design. *Front. Bioeng. Biotechnol.* 7 (8): 743.

60 Wu, Q., Therriault, D., and Heuzey, M.C. (2018). Processing and properties of chitosan inks for 3D printing of hydrogel microstructures. *ACS Biomater. Sci. Eng.* 4 (7): 2643–2652.

61 Yan, Q., Dong, H., Su, J. et al. (2018). A review of 3D printing technology for medical applications. *Engineering* 4 (5): 729–742.

62 Maver, T., Smrke, D.M., Kurečič, M. et al. (2018). Combining 3D printing and electrospinning for preparation of pain-relieving wound-dressing materials. *J. Sol-Gel Sci. Technol.* 88: 33–48.

63 Yu, N., Nguyen, T., Cho, Y.D. et al. (2019). Personalized scaffolding technologies for alveolar bone regenerative medicine. *Orthod. Craniofacial Res.* 22: 69–75.

6

Advancements in Bioprinting for Medical Applications

Kevin Y. Wu[1], *Maxine Joly-Chevrier*[2], *Laura K. Gorwill*[3], *Michael Marchand*[1], *and Simon D. Tran*[4]

[1] *Department of Surgery, Division of Ophthalmology, University of Sherbrooke, Sherbrooke, Quebec, Canada*
[2] *Faculty of Medicine, University of Montreal, Montreal, Quebec, Canada*
[3] *Department of Biology, University of Toronto Mississauga, Mississauga, Ontario, Canada*
[4] *Faculty of Dental Medicine and Oral Health Sciences, McGill University, Montreal, Quebec, Canada*

6.1 Introduction

The innovative introduction of three-dimensional (3D) printing technology has led to extraordinary transformations in various areas, with healthcare at the forefront. The use of 3D printing in biomedicine, also known as bioprinting, is a game-changing paradigm with the potential to reshape the landscape of medical science in the coming years. This chapter discusses the importance of bioprinting in numerous aspects of medical applications, focusing on tissue engineering, regenerative medicine, organ transplantation, pharmaceutical development, and testing.

Bioprinting applies 3D printing principles and combines them with living cells, biomaterials, and chemicals to create biomedical parts with utmost precision and clinical functionality. These biofabricated constructs are designed to mimic the functional and anatomical properties of living tissue, paving the way for significant clinical applications. Bioprinting's possibilities range from skin grafts to sophisticated organ tissues, making it one of the most exciting areas of modern medical research.

The applicability of bioprinting extends beyond clinical procedures. It has enormous potential to streamline drug development and testing processes. Bioprinting provides more precise and reliable preclinical drug testing by producing 3D bioprinted tissues that closely replicate in vivo settings. Additionally, organ-on-a-chip

3D Bioprinting from Lab to Industry, First Edition. Edited by Prosenjit Saha, Sabu Thomas, Jinku Kim, and Manojit Ghosh.

(OoC) technology employs bioprinting to miniaturize human organs on microchips, allowing for high-throughput drug screening and disease modeling.

While the potential of bioprinting is undeniably great, it is crucial to recognize the limitations this new sector faces. Obstacles range from technical limits in bioprinting to ethical, regulatory, and translational issues. Nonetheless, with ongoing research, stakeholder collaboration, and innovation, these barriers are expected to decrease, fostering a more favorable environment for the maturation of this groundbreaking technology.

This chapter's objective extends beyond an investigation of bioprinting's possibilities in medical applications; it also offers a comprehensive review of the field's current state, barriers, and future prospects. By understanding bioprinting's capabilities, we can unlock innovative therapeutic approaches and progress toward personalized medicine, potentially revolutionizing the way we study, diagnose, treat, and prevent diseases.

Our journey of discovery starts with a thorough examination of bioprinting's significance in tissue engineering and regenerative medicine, covering diverse tissue types, their potential therapeutic uses, and specific challenges to overcome. The focus then shifts to the intriguing possibilities of bioprinting in pharmaceutical development and testing. Finally, we explore the challenges bioprinting faces in its quest for clinical translation, discussing potential solutions and future directions.

6.2 Bioprinting for Drug Development and Testing

6.2.1 Overview

The development of a single pharmaceutical drug is exorbitantly costly, time-consuming, and subject to multiple points of failure. At present, a decade or more is typically required for a novel drug to progress through the stages of discovery, preclinical and clinical trials, and market launch [1]. The financial costs of this process, which are inflated by the high candidate attrition rate, can reach 2.6 billion USD [1]. During this lengthy window, the clinical need prompting the drug's development remains unmet, and the associated physical, emotional, and economic burden is significant.

In the discovery phase, a vast pool of compounds known to interact with the biochemical target is narrowed to a few lead drug candidates with the strongest binding affinities [2]. These candidates are then subjected to rigorous preclinical trials that evaluate their pharmacodynamics, pharmacokinetics, and toxicity. Despite the robustness of the preclinical screening process, the failure rate at the subsequent clinical stage is approximately 90% [3]. That is, most compounds that are deemed safe and efficacious in preclinical studies fail in one or both respects when trialed in humans. The extreme attrition rate of the drug development process, with its associated financial losses, warrants the need for more effective screening platforms.

Failure at the clinical stage can largely be attributed to the fact that conventional isolated systems and animal disease models incompletely recapitulate in vivo human pathophysiology. Two-dimensional (2D) monolayer cell cultures, which involve the cultivation of cells on a planar glass or polystyrene substrate, are widely used to screen drug candidates in vitro [4]. While efficient and cost-effective, the use of 2D cultures for this purpose presents significant limitations. Changes in cell polarity, intracellular and extracellular signaling, morphology, and differentiation render 2D cell cultures poor simulators of the native cell microenvironment. In vivo diffusion kinetics are generally inaccurate when modeled using 2D systems, such that effective drug dosages cease to be so when administered to patients after scaling [5]. Finally, 2D cultures fail to mimic essential cell–cell and cell–extracellular matrix (ECM) interactions of the native microenvironment [6]. As such, findings obtained using 2D cultures are often poorly predictive of those of animal and human trials, increasing the attrition rate and financial costs of the developmental process [7]. Animal disease models, while superior to 2D cultures in most respects, are inherently limited in their ability to emulate human pharmacokinetics and pharmacodynamics. Significant anatomical, immunological, genetic, and hormonal discrepancies exist between humans and traditional animal models, such as mice [8]. As such, animal models are restricted in their capacity to predict clinical findings. Animal tests of human pharmaceuticals can falsely ascribe toxicity to safe therapeutic agents and, conversely, ascribe safety to those that are toxic in humans [8]. Furthermore, the provision of care, housing, and food contributes to their high financial costs, and their use is fraught with ethical concerns.

The considerable limitations of conventional screening modalities have spurred the search for those that more accurately mimic the human cell microenvironment. 3D bioprinted models, produced by the automated layer-by-layer deposition of cell-laden bioinks, exhibit considerable promise in this regard [7]. Because they can precisely deposit various cell types, growth factors, proteins, and pharmaceuticals in three dimensions, they achieve complex spatial architecture that closely resembles that of the native cell microenvironment. As an automated process, it is also highly reproducible, meaning that it can be scaled for the mass production of models required for high-throughput drug screening. While the applications of 3D bioprinting are vast and ever-growing, this chapter will focus specifically on the applicability of 3D bioprinting to drug discovery and development.

6.2.2 3D Bioprinted Organoids

Broadly speaking, an organoid is a miniature, in vitro version of an organ of interest that consists of one or a few organ-specific cell types [9]. Because organoids exhibit microanatomy that closely matches that of the native organ, they are useful for pharmacokinetic and toxicological assessments of drug candidates. Bioprinting has been used to improve the reproducibility and scalability of organoid production, while also producing constructs that are more biomimetic than their traditionally cultured counterparts [10].

Human stem cells and/or progenitor cells are seeded to produce 3D organoids that emulate key functional and anatomical characteristics of their corresponding organ [11]. In a manner akin to in vivo organogenesis, traditionally cultured organoids self-organize through cell sorting and spatially restricted lineage differentiation. A suitable ECM, such as Matrigel or collagen, and appropriate organ-specific cues are required to direct the proliferation, migration, and differentiation of isolated stem cells [10]. Growth factors and/or small-molecule drugs that modulate signaling pathways recapitulate the functions of soluble cues that guide in vivo organogenesis [11]. Because the presentation of such cues occurs in a manner that is spatiotemporally delimited, their application to in vitro organoid cultures must faithfully replicate this precision [12]. In addition to biochemical signals, many physical parameters, such as pH, temperature, oxygen, and nutrient levels, must be continually monitored during organoid formation.

However, the limitations of traditional, non-printed organoid culturing methods greatly constrain their utility. Because traditional systems rely on the propensity of seeded cells to self-organize, control over the final spatial architecture is limited [11]. However, 3D bioprinters can precisely deposit cells and other biomaterials in *X*, *Y*, and *Z* orientations, thereby achieving a level of architectural complexity that more closely mimics that of the native organ [2]. Additionally, traditionally cultured systems lack perfusable vasculature, meaning that the exchange of nutrients and wastes is constrained by diffusion kinetics [10, 13]. For example, the maximum diffusion distance of oxygen in an avascular tissue is 100–200 μm [13]. As such, while cells located near the periphery of the construct receive an optimized oxygen and nutrient supply, diffusion alone cannot sustain the cells of the core [14]. The lack of nutrients, combined with the accumulation of metabolic waste products, generally renders the innermost cells quiescent, apoptotic, and/or necrotic, especially in the later stages of the organoid culture [15]. Additionally, upon reaching a threshold size, organoids transition from a proliferative, stem-like state to one that is terminally differentiated and no longer proliferative [14]. Therefore, traditionally cultured organoids are restricted in the ultimate size that they can attain and the length of time that they remain viable. However, 3D bioprinting can overcome such limitations via the creation of porous structures and/or complex vasculature that permits the influx of oxygen and nutrients, and the efflux of metabolic wastes [7, 16]. The controlled spatiotemporal deposition of growth factors and other biological signals necessary for specifying cell fate and organization is also readily achieved by bioprinting [15]. Additionally, significant batch-to-batch variations of non-printed organoid cultures with respect to size, degree of cell heterogeneity, and spatial arrangement hinder reproducibility [9]. Because 3D bioprinting is a tightly controlled, inherently reproducible process, this method overcomes challenges in standardization and scalability associated with traditional organoid culturing methods. For this reason, 3D bioprinting is optimized for the mass-production of tissue models required for high-throughput drug screening [14].

Bioprinted organoids mimicking various native organs have been produced as platforms for drug screening. For instance, Lawlor et al. in 2021 used automated extrusion-based 3D bioprinting to produce kidney organoids, markedly improving throughput, scale, structure, and quality control as compared to labor-intensive manual production approaches [17]. As proof of concept for drug testing, they assessed the relative nephrotoxicity of various aminoglycosides (a class of broad-spectrum antibiotics), ultimately concluding that their bioprinted model can be practically applied in toxicological studies. Recognizing that drug development for dry eye disease is constrained by current in vitro models, Rodboon et al. in 2022 bioprinted lacrimal gland organoids using primary cells derived from murine and porcine lacrimal glands [18]. They used a magnetic bioprinting system, in which cells are tagged with biocompatible NanoShuttle™ magnetic nanoparticles prior to their deposition with the aid of a magnetic field. Like other 3D-printed organoids, the reproducibility, scalability, and biomimicry of this model renders it useful for high-throughput drug screening.

Although in its infancy, the bioprinting of organoids for drug development shows considerable promise. As outlined above, printed organoids are superior to their traditionally cultured counterparts with respect to size, complexity, and anatomical and functional similarity to their native tissues. As such, they can generate in vitro preclinical data that are more predictive of clinical outcomes than standard 2D, 3D, and animal models. Ultimately, bioprinted organoids can reduce attrition rates and the financial costs associated with clinical failures.

6.2.3 Organ-on-a-Chip/Microfluidic Systems

Broadly speaking, organ-on-a-chip (OoC) technologies model human organs in vitro using a 3D microfluidic cell culture [19]. Like organoids, they recapitulate many of the structural and functional characteristics of their respective simulated organs. Therefore, they have immense potential to streamline the processes of drug discovery and development. To date, a variety of OoC systems, including heart-on-a-chip (HoC), liver-on-a-chip, lung-on-a-chip, liver-on-a-chip, and tumor-on-a-chip (ToC), have been successfully engineered [20].

In OoC systems, a microfluidic device integrates perfused microchannels (which have dimensions ranging from tens to hundreds of micrometers) with structured cellular components that mimic those of the native organ [20]. Additionally, OoCs can be equipped with optical, electrochemical, and/or mass spectrometry biosensors [21]. These sensors permit the automated, real-time monitoring of both culture conditions and responses of the culture to applied stimuli, such as perfused pharmaceuticals. Because microscale channels require only minute sample volumes, they minimize reagent waste and financial costs. This, combined with their excellent mimicry of native tissues, renders them an increasingly attractive option for efficacy and toxicological evaluations of drug candidates in preclinical trials [20].

3D bioprinting of OoCs offers numerous advantages over traditional OoC production approaches. First and foremost, bioprinted OoCs can achieve precise spatial arrangement of seeded cells, closely mimicking the microarchitectures of the tissues they are designed to emulate [21]. In contrast, traditional preparation of OoCs requires that cells be manually pipetted onto microfluidic devices by laboratory personnel [20]. In addition to the inherent inefficiency and limited scalability of this approach, it carries an elevated risk of exposure to contaminants. By largely eliminating the need for human intervention during the seeding process, bioprinters better maintain the sterility of OoCs. Additionally, bioprinted OoCs permit the rapid immobilization of cells, eliminating the six-to-eight-hour time delays typically required for manually seeded cells to express attachment proteins and achieve necessary adherence to the substrate. Bioprinting can also be used to achieve highly specific channel flow patterning within the microfluidic device that cannot be achieved via traditional methods [22]. Finally, bioprinted technologies facilitate the replication of native tissue cell–cell communication more effectively. Finely tuned control of cell ratios, necessary to prompt biomimetic intercellular signaling, is made possible by the capacity of bioprinters to deposit a wide range of cells per droplet [20]. In fact, bioprinters have been used to achieve resolutions as high as a single cell per droplet, providing unparalleled cell ratio control that is infeasible by traditional modalities [23]. Evidently, bioprinting technologies for OoCs greatly surpass traditional approaches in terms of quality control, biomimicry, reproducibility, and scalability, and thus provide promising new avenues for drug discovery and screening.

Various bioprinting methods have been studied for their applicability to OoC production and can be broadly divided into nozzle-based and optical-based approaches [22]. As the name implies, nozzle-based methods involve the transfer of bioink through a nozzle that is in direct contact with the construct. This method is exemplified by sacrificial inkjet and extrusion-based techniques, in which a support material necessary to prevent the collapse of the uppermost layers into the channel lumen is removed post-printing. Optical-based methods, which include laser-induced forward transfer (LIFT), two-photon-lithography (TFL), and stereolithography (SLA), among others, are dependent on interactions between printed materials and light.

To date, a variety of OoC systems have been engineered and exhibit many of the key structural and functional attributes of their respective in vivo tissues.

Liver-on-a-chip models are particularly valuable in preclinical drug screening, as they are more highly predictive of clinical hepatotoxicity than 2D cultures and animal models [24]. Drug-induced liver injury (DILI), defined as acute or chronic injury to the liver secondary to drug usage, is the leading cause of attrition at the clinical stage. Additionally, DILI is the most common reason for the withdrawal of FDA-approved drugs, at which time substantial harm is likely to have occurred. Therefore, it is advantageous, both from a financial and human health standpoint, that hepatotoxic compounds be identified early in the developmental process. Various liver-on-a-chip models have been produced for the purpose of drug screening, including that of Knowlton and Tasoglu in 2016. Using HepG2/C3A human hepatocarcinoma cells, hepatic spheroids were bioprinted into a microfluidic platform and an acute toxic dose of acetaminophen, a well-known hepatotoxin, was administered. The resultant decreases in cell density, metabolic activity, and biomarker production demonstrate the utility of their model for in vitro hepatotoxicity analyses. It should also be noted that OoCs can also represent natural population variability in preclinical drug screening by using induced pluripotent stem cells from multiple patients, rather than a single cell line. This approach helps to predict rare hepatotoxic events resulting from genetic anomalies.

In addition to hepatic injury, drug-induced cardiotoxicity is an exceedingly common cause for attrition during the developmental process and withdrawals post-market launch [25]. Bioprinted HoC systems, like other OoCs, are more predictive of clinical outcomes than traditional 2D cultures and animal models because of their ability to recapitulate the precise microarchitecture and physiological functions of the human heart. Human-induced pluripotent stem cell-derived cardiomyocytes (hiPSC-CMs), which are both expensive and difficult to manufacture, are well-suited for use in bioprinted OoCs. The small size of an OoC (approximately that of a USB stick) and the highly precise deposition of cells by bioprinters ensure that comparatively few cells are required for production. Faulker-Jones et al. in 2022 took advantage of this property in their fabrication of a HoC using microvalve-based bioprinting. When compared to monolayer cultures plated by manual pipetting, their model exhibited no significant differences in voltage, intracellular calcium, and contractility, demonstrating that the hiPSC-CMs were unharmed by the bioprinting process. To validate the use of their bioprinted HoC for drug assays, they administered three well-characterized cardiac drugs to both the model and manually seeded control. They found no significant differences in drug sensitivity between controls and the model, both of which were exposed to the antiarrhythmic E4031, the L-type Ca^{2+} channel blocker nifedipine, and flecainide, a Na^{+} channel blocker. By showing that the physiological functions and drug sensitivity of hiPSC-CMs are preserved in their bioprinted HoC, they provide a promising, reproducible platform for streamlining the drug screening process.

Integrated multiorgans-on-a-chip (MOoC) can be used to better predict the systemic effects of novel compounds, such as the effects of the metabolic by-products of a drug on off-target organs, or the effects of a drug on both target and secondary tissues simultaneously [26]. For example, liver-heart-on-a-chip platforms fluidically connect hepatic and cardiac constructs in a single device, thereby modeling pharmacodynamic processes in vitro in a highly integrated manner [27]. Because unexpected hepatotoxicity and cardiotoxicity are the primary causes of safety-related market withdrawal, the use of such models in drug development is both financially and socially advantageous. As an example, Ferrari et al. in 2023 bioprinted a liver-heart-on-a-chip to predict the off-target cardiotoxicity of terfenadine. Terfenadine, an H1 antagonist known to cause arrhythmia, was first administered to the hepatic construct of the model, and the viability and functionality of the associated cardiac counterpart were then observed. By reproducing terfenadine's well-characterized cardiotoxicity in vitro, they demonstrated the utility of their model in preclinical tests of drug safety. However, MOoCs are not limited to the in vitro representation of only two organs. Skardal et al. in 2017 developed a three-tissue (liver, heart, and lung) MOoC by bioprinting cardiac and hepatic organoids and connecting them to a lung organoid at the air–liquid interface [5]. The three constructs were then serially connected to a circulatory perfusion system, and bleomycin was used to assess the utility of the integrated system for toxicology screening. Models such as these, which combine the advantages of bioprinting and microfluidic technology in a single USB-sized device, have the potential to revolutionize drug development.

6.2.4 Bioprinted Models for Cancer Research

Globally, cancer is a leading cause of death, and the associated physical, emotional, and financial burdens are immense [28]. Although early detection strategies have improved the prognoses for many cancers, treatment resistance (both intrinsic and acquired) remains a major concern [29]. Several factors, including tumor mass, heterogeneity, complexity, and dynamism contribute to treatment resistance, particularly to "gold standard" chemotherapeutic agents [30]. In many countries, an aging population will contribute to the increased overall incidence of cancer [31]. This, combined with the difficulty of eradicating cancer and preventing relapse, calls for the development of better pharmaceutical treatments.

At present, a mere 5.3% of anticancer drugs that enter phase I clinical trials ultimately obtain FDA approval, highlighting the need for technologies that refine the drug development process [32]. In this respect, 3D bioprinted in vitro models are used to reconstitute the complex cancer microenvironment, an endeavor that has proven difficult using 2D models [33]. Monolayer cancer cultures, while cost-effective and simple to use, fail to recapitulate various determinants of tumor development, including biochemical signaling, ECM stiffness, topography, interstitial flow, and shear stresses. They also lack the array of stromal and immune cell types present in solid tumors, as well as perfusable vasculature. In a typical tumor, aberrant vascularization and the restricted perfusion of oxygen, nutrients, and metabolic wastes about the tumor core often induce compensatory mechanisms that contribute to treatment inefficacy and tumor recurrence [34]. For instance, hypoxic tumors harbor larger populations of pluripotent cancer stem cells (CSCs) and exhibit increased resistance to chemotherapeutic agents. 2D monolayer cultures fail to emulate the hypoxic core of the in vivo tumor microenvironment (TME), further underscoring the need for better model systems [33]. Although traditionally cultured 3D models have been successfully employed in oncological studies, they lack the microscale biomimicry, reproducibility, and scalability afforded by bioprinted models. For instance, 3D cancer organoids produced by standard culturing methods also generally lack vasculature, comprise only a few cell types, and fail to exhibit biomimetic, tumor-induced stromal cell epigenetic modifications [32]. Altogether, traditional models often yield findings that are inconsistent with those of clinical studies, further increasing the exorbitant costs and protracted timelines of anticancer drug development.

On the other hand, 3D bioprinters precisely control the layer-by-layer deposition of various cell types found in solid tumors, including cancer, immune, and stromal cells [32]. Bioprinting also facilitates the precise spatiotemporal distribution of proteins and growth factors (such as VEGFs, FGFs, and TGFβ) implicated in tumor signaling, proliferation, and metastasis [35]. In this way, they yield models with the complex and biomimetic spatial architecture necessary to study the etiology and/or pathophysiology of various cancers. In addition to providing a platform for studying cancer biology, 3D bioprinted models are invaluable in the development of safe and efficacious anticancer therapies. To date, a variety of 3D cancer models, including breast, pancreatic, brain, skin, and liver have been bioprinted for these purposes. Additionally, 3D bioprinting provides promising new avenues for personalized anticancer treatment. Patient-derived cells can be incorporated into in vitro cancer models, which can be used to screen for anticancer drugs (or combinations thereof) that are tailored to the patient's tumor type, cancer stage, and genotype.

As mentioned previously, non-printed 3D tumor models are limited by their lack of perfusable vasculature. The tumor microvasculature is abnormal in both architecture and function, promoting hypoxia in the tumor core that promotes tumor progression, metastasis, and treatment resistance [13, 36]. The incorporation of a biomimetic vascular network into in vitro cancer models is integral to the study of tumor cell intravasation, drug extravasation, and interactions of circulating cancer cells with stromal and immune cells, among other phenomena [37].

A 3D bioprinted glioblastoma model, developed by Neufeld et al. in 2021, exemplifies the ways by which cancer constructs outfitted with biomimetic vasculature faithfully mimic the TME [38]. Their model, which features a peristaltic pump and capillaries lined with endothelial cells and pericytes, allows for perfusion with circulating tumor cells, peripheral blood mononuclear cells, and various anticancer agents. Beyond simply incorporating vasculature into tumor constructs, bioprinters can fine-tune the anatomical and functional features of that vasculature. For instance, the abnormal permeability of tumor vessels ("vessel leakiness"), in addition to promoting tumor cell intravasation, significantly affects the delivery of anticancer agents to target tissues [30]. In this respect, 3D-printed models are extremely valuable. Because the integrity of the vascular–endothelial barrier can be precisely controlled in bioprinted constructs, they more accurately model drug delivery and efficacy in vitro.

However, when printing vasculature, specific steps must be taken to prevent low-viscosity, mechanically weak bioinks from collapsing into the channel lumen immediately after their deposition [39]. Sacrificial bioinks, which can be removed after printing, can be used to establish prespecified tubular architecture (Figure 6.1) [40]. Thermo-reversible Pluronic F-127 or water-soluble polyvinyl alcohol (PVA), among other materials, can be used for this purpose [30, 40]. Following their removal, the cellularization of the remaining porous template creates biomimetic vasculature that can supply the construct with nutrients, dissolved gasses, and pharmaceuticals [40]. Alternatively, suspension-bath-based bioprinting, which involves the direct deposition of materials into semi-solid support media, can be used to create hollow vessels [39, 41]. The suspension bath, which encapsulates the printed bioink as it solidifies, remains solid in the absence

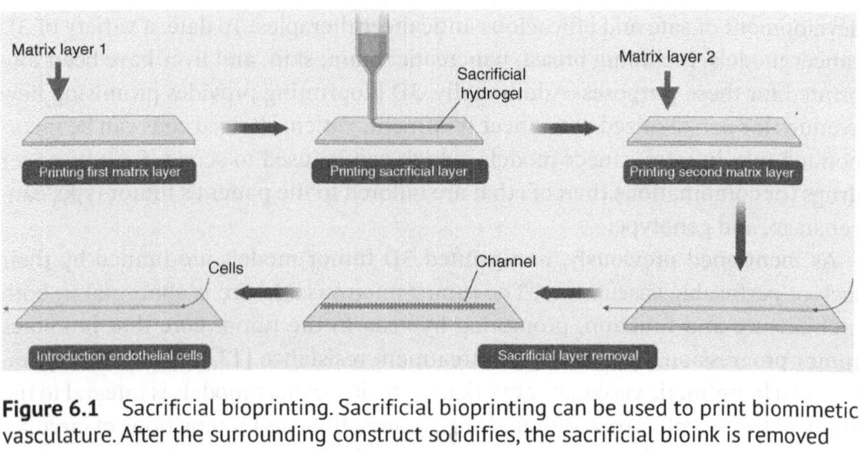

Figure 6.1 Sacrificial bioprinting. Sacrificial bioprinting can be used to print biomimetic vasculature. After the surrounding construct solidifies, the sacrificial bioink is removed and the hollow tubes are seeded with endothelial cells.

of applied stress but is temporarily fluidized by disturbance (i.e., the displacement of the medium by newly deposited bioink) [39]. The embedded vascular construct, once solidified, can then be extracted from the suspension bath and seeded with endothelial cells.

3D bioprinted models can also be used as an in vitro platform for studying metastasis, an incompletely understood phenomenon that causes approximately 90% of cancer-related deaths [42]. The term describes the development of secondary tumors at sites distant from the primary tumor. Traditionally, Boyden chamber assays have been used to study the migration and invasion of cancer cells in metastasis [43]. Though simple and cost-efficient, these assays are laborious and limited in their physiological relevance. On the other hand, bioprinted platforms, owing to their reproducibility, scalability, and biomimetic potential, are invaluable for studying the mechanisms of metastasis. Additionally, they facilitate the development of anti-metastatic drugs when used as a platform for high-throughput drug screening. For instance, Jung et al. in 2022 used a drop-on-demand (DOD) approach to deposit cancer cell-laden bioink into 96-well plates [44]. Their model, which allows for real-time monitoring of cell movement within the construct, provides a novel screening platform for anti-metastatic drugs. To determine the feasibility of their model for this purpose, they administered two known inhibitors of cell migration, blebbistatin (a global myosin inhibitor) and Y-27632 (a Rho-associated protein kinase inhibitor), and found that cell migration was significantly impeded.

Furthermore, 3D bioprinted models are useful for studying the properties of CSCs and can thus be used as a platform for developing CSC-targeting anticancer therapies [45]. In addition to various differentiated cancer cells, tumors harbor a small population of these cells, which are characterized by altered gene expression and symmetrical division [46]. Also called tumor-initiating cells, they are pluripotent, highly motile, self-renewing, apoptosis-resistant, and above all, exceptionally tumorigenic. Their integration into in vitro cancer models, made possible by 3D bioprinting, is necessary to develop treatments that address the stubborn resistance of CSCs to current approaches [45].

Alongside factors such as tumor mass and complexity, extensive patient-to-patient pharmacodynamic variation further limits the efficacy and tolerability of standard treatments [35]. Microenvironmental and cellular heterogeneity within a single tumor compounds this inefficacy, often necessitating the use of multi-drug treatment regimens [35]. These issues call for the development of highly personalized approaches to treatment, in which pharmaceutical regimens are tailored to the patient's unique genetic profile (including age, sex, and ethnicity), tumor subtype, cancer stage, and comorbidities. Patient-derived xenograft models are one such approach to personalized oncology and involve the implantation of a patient's resected tumor and stromal cells into an immunodeficient mouse [47].

However, the patient-derived stroma is quickly replaced by murine stroma, such that model fidelity declines steeply with time. Carcinoma-associated fibroblasts (CAFs), the predominant component of tumor stroma, are replaced by murine equivalents the most rapidly, diminishing the relevance of the model. Drug screening using PDX models is further limited by high costs, ethical concerns, and the six-to-nine-month time window required for establishment. During this lengthy window, the outcome may be rendered irrelevant, either because another treatment has been found or the patient has succumbed to their illness. 3D bioprinted models, however, represent a promising new approach to personalized oncology that circumvents many of these limitations [35]. 3D bioprinting facilitates the high-throughput production of physiologically relevant, personalized cancer models using patient-sourced cancer and stromal cells. Such cells, obtained via tissue biopsy or debulking surgery, are bioprinted along with ECM protein and relevant growth factors. These models can then be used to rapidly screen for patient-specific, efficacious oncological agents that minimize adverse reactions. Moreover, they can be used to optimize therapeutic dosages and the timing of administration, as well as to identify the most effective multi-drug combinations [30]. It should be noted that the relevance of personalized cancer models is contingent upon the appropriate integration of various cell types and proteins. Patient-specific primary cancer cells, circulating tumor cells, and CSCs can be integrated into the construct alongside stromal cells (such as fibroblasts and endothelial cells) that are representative of the tumor's tissue of origin. Angiogenic factors, such as epidermal growth factor and vascular endothelial growth factor, should be applied in a spatiotemporally delimited manner to maximize the biomimetic potential of the construct.

The OoC technology described in the previous section can also be adapted to recapitulate the TME [48]. This technology, called ToC, uses tumor and stromal cells printed onto a perfusable microfluidic device to study cancer biology and develop novel anticancer therapies. As with OoCs, perfusable microchannels permit the influx of oxygen, nutrients, and pharmaceuticals and the efflux of metabolic wastes. Rather than relying on the propensity of seeded cells to self-organize, printing ToCs achieves a precise spatial arrangement of materials and results in greater reproducibility and scalability. Like OoCs, biosensors can be integrated directly into the chip for real-time monitoring of various parameters.

Moreover, ToCs can incorporate patient-derived cancer cells obtained via biopsy or debulking surgery [49]. This approach provides an avenue for highly personalized treatment that maximizes efficacy and tolerability and minimizes recurrence. As an example, Yi et al. in 2019 used extrusion-based bioprinting with patient-derived tumor cells, vascular endothelial cells, and decellularized ECM from brain tissue to create a personalized glioblastoma-on-a-chip. The device, which featured a radial oxygen gradient mimicking that of the native tumor, reproduced the resistance to concurrent treatment with chemoradiotherapy and temozolomide that was previously observed in the patient. In doing so, the model demonstrated its value as a tool for identifying effective patient-specific treatments, especially for cancers refractory to traditional approaches.

3D bioprinting, therefore, has enormous potential to streamline the discovery and development of novel anticancer treatments. 2D monolayer, PDX, and traditionally cultured 3D models fail to emulate the spatial and physiological complexity of the TME to the degree that printed models can. As the incidence of cancer, particularly in aging populations, continues to grow, the value of biomimetic printed models as screening platforms does as well [31].

6.2.5 3D Bioprinting for Immunotherapy and Cell Therapy

In addition to cancer and stromal cells, tumors contain a heterogeneous array of immune cells representing both the innate and adaptive immune systems [33]. The inflammatory cascade directed by these cells can enhance tumor growth by increasing rates of angiogenesis and cell proliferation. For example, neutrophils, macrophages (M2), and myeloid-derived suppressor cells of the innate immune system, and T-helper 2 CD4+ T cells, CD4+ T regulatory cells, and B lymphocytes of the adaptive immune system generally facilitate tumor progression and metastasis. Alternatively, the immune system may mount a tissue-destructive response following tumor antigen recognition by cytotoxic T cells [50]. Notably, the efficacy of treatment with radiotherapy and/or chemotherapy is partially contingent upon the innate and adaptive immune responses [51]. As such, physiologically accurate models are needed to better understand the complex tumor-immune interplay and to develop safe and efficacious immunotherapies in accordance with these findings.

Broadly speaking, immunotherapeutic approaches selectively target cancer cells using the host's adaptive and/or innate immune system [52]. Various immunological agents, including monoclonal antibodies, cancer vaccines, cytokines, and immune checkpoint blockers, have demonstrated efficacy in the treatment of a range of malignancies [53]. Because these therapies target malignant cells with high specificity, they spare patients from many of the adverse systemic effects associated with the attack of healthy tissue (as is common with radiotherapy and chemotherapy). The success of immunological therapy varies widely between cancer types, with renal cell carcinoma, hematologic malignancies, and melanomas responding particularly well, and pancreatic cancer, prostate cancer, and glioblastoma tending to respond more poorly [54]. Moreover, acquired resistance to immunotherapy, in which efficacy declines with repeated treatments, further limits the applicability of this approach [55]. Further research is needed to address these shortcomings and broaden the scope of immunotherapy.

Several bioprinted models have been fabricated to examine the effect of immunological agents on various simulated cancer types. For example, Kim et al. in 2021 used a bioprinted bladder cancer-on-a-chip model to assess responses to varying dosages of a known immunotherapeutic agent, Bacillus Calmette–Guérin (BCG) [56]. The construct was prepared using bladder cancer cells (5637/T24), stromal fibroblasts (MRC-5), and human umbilical vein endothelial cell (HUVEC). By observing cytokine (including IL-6, IL-12, TNF-α, and IFN-γ) expression, THP-1 monocyte migration, and the proliferation and viability of cancerous cells, they experimentally validated the utility of their model for assessing dose-dependent BCG response. Because the efficacy of immunotherapy for bladder cancer varies widely between patients, bioprinted models are useful for studying the underlying mechanisms.

Chimeric antigen receptor-modified CD8+ T (CAR-T) cell therapy, a type of adoptive T-cell therapy, has also been studied using 3D bioprinted platforms [38]. For instance, Dey et al. in 2022 bioprinted a vascularized breast cancer model that was responsive to immunotherapy with CAR-T cells recognizing human epidermal growth factor receptor 2 (HER2) [57]. Specifically, CAR-T cell perfusion significantly suppressed tumor growth, demonstrating the utility of the model for understanding the complex interplay between solid tumors and the immune system. Generally speaking, CAR-T therapy is less effective against solid tumors than hematological malignancies. The heterogeneity of solid tumor antigens, restricted CAR-T penetration of the tumor, and poor CAR-T viability in the TME have all been implicated as causes of this discrepancy. Biomimetic printed models such as this are needed to further explore these phenomena and to address them to expand the scope of CAR-T therapy. Grunewald et al. in 2021 also sought to evaluate CAR-T cell efficacy using a bioprinted model; neuroblastoma was the in vitro simulated malignancy [58]. SLA was used to create several models from SK-N-BE [2] cells suspended in gelatin methacrylate (GelMA). Seventy-six percent of neuroblastoma cells (averaged across five models) remained viable at 11 days post-printing, a time window that far exceeds the 3–5-day window typically required to test CAR-T efficacy. Therefore, their model is well-suited for use in lengthy experiments, including those that test encounters with multiple antigens.

The study of T-cell migration patterns is yet another application of 3D bioprinted tumor models. For instance, Morley et al. in 2022 bioprinted a glioma model using KR158B murine glioma cells, hematopoietic stem cells, and T cells to characterize spatiotemporal T-cell migration [59]. Additionally, they used their model to examine mechanisms governing the effectiveness of adoptive T-cell therapy. They concluded, via time-lapse imaging, that models with HSCs recruited T cells more quickly than those without. They suggest that their methods can be generalized to create various 3D-printed models for the study of immune–tumor interactions.

Immunotherapy is an attractive alternative to standard anticancer treatments, which are often characterized by poor tolerability and intrinsic and/or acquired treatment resistance. However, the scope of cancer immunotherapy has been bottlenecked, to a degree, by the limitations of traditional model systems. Because bioprinted models reproduce the complex interplay between cancer, stromal, and immune cells, they are valuable platforms for the development of effective immunotherapies that are applicable to a broader range of cancers.

6.3 Bioprinting in Tissue Engineering, Regenerative Medicine, and Organ Transplantation

Although tissue bioengineering has been present for several decades, the development and manufacture of functional vital organs still face numerous challenges. Common obstacles encountered in vital organ bioprinting include replicating the complex vascular network and the complex interactions between multiple cell types within a tissue while recreating its function. Moreover, many vital organs display extreme sensitivity upon exposure to changes through the printing process and external environment. In this section, an overview of the application area specific to tissues and organs is discussed.

6.3.1 Ocular Tissue Engineering

6.3.1.1 Retina

Studies on bioprinting functional retinal tissue present numerous opportunities in the development of alternative clinical approaches to address debilitating retinal pathologies, such as Stargardt disease and age-related macular degeneration. Though there is great interest in this field, recent studies remain limited due to the challenge of replicating the retina's advanced biological complexity.

In one study, Kim et al. in 2022 used extrusion-based bioprinting to recreate a component of the retina, known as Bruch's membrane (BM) [60]. This component is an important part for retina tissue structure and engineering due to its scaffolding properties that can support cell growth. The researchers developed a hybrid membrane printing approach to create an outer blood–retinal barrier model composed of a retinal pigment epithelium (RPE) monolayer on top of an engineered BM. Compared to traditional collagen-based bioinks, their RPE- and BM-based bioink allows for the preservation of RPE functions in vitro, such as barrier clearance and maintenance functions. This model can serve as a foundational platform for future retina bioprinting research and drug development projects.

Another important structural and functional component of the retina are Müller cells, which span the entire layer of the retina, regulating the strength of blood–retinal barrier. The death of Müller cells is highly detrimental to the blood–retinal barrier, which protects neural and vascular cells. Jung et al. in 2022 were the first to develop a 3D biomimetic culture platform for Müller cells [61]. They used Müller cells extracted from Sprague-Dawley rats combined with a fibrinogen-based bioink, composed of gelatin, fibrinogen, and hyaluronic acid, and a sacrificial-based bioink, composed of various polymers. Using these two bioink formulas, they created a construct that successfully enhanced the aligned morphology of Müller cells in a 3D environment. Subsequent analyses showed that within the 3D-printed structure, the morphology of Müller cells was preserved, such as the endfeet, soma, and microvilli. Additionally, the expression of its specific markers, such as GS, Kir4.1, and AQP4, remained comparable to its native counterpart, thus confirming the functionality of the cells. A technical limitation of this study, however, was that the cells were only able to be printed vertically, so future studies should investigate horizontal printing of these cells in addition to different printing methods and printing patterns.

Despite immense progress in retinal tissue regeneration and engineering, there are many challenges with implementing this technology in a clinical setting. Since the retina is directly connected to specific areas of the brain, this generates discussion on whether a 3D bioprinted retina would be functionally compatible with the native retina-connected areas of the brain or even if such areas of the brain itself could be bioprinted and implanted into humans [62].

6.3.1.2 Cornea

3D bioprinting of corneal tissue presents a promising treatment option for severe corneal disorders, which are the main cause of blindness worldwide [63]. The applications of cornea bioprinting allow for more flexibility and optimization of the current gold standard treatment for severe cornea-related disease, which is corneal transplantation.

Selecting the perfect bioink poses as an intriguing task, since it necessitates consideration of tissue characteristics, such as its transparency, mechanical and biological stability, complex geometry, and potential for transplantation. Many recent studies therefore center around developing the most optimal bioink for corneal tissue bioprinting. In a recent study, Mörö et al. [64] in 2023 utilized an extrusion-based bioprinter to fabricate corneal stroma structures. Their hyaluronic acid-based bioink formula included human adipose stem cells (hASCs), hASC-derived corneal stromal keratocytes, and human pluripotent stem cell (hPSC)-derived neurons. After implanting the stromal structures into ex vivo porcine corneas, the bioink showed excellent printability, shape fidelity, and regenerative properties. In another extrusion-based bioprinting study, Boix-Lemonche et al. in 2023 used different types of porcine mesenchymal stromal cell, adipose derived-mesenchymal stromal cells, bone marrow-derived mesenchymal stromal cells, and corneal stroma-derived mesenchymal stromal cells suspended in a bioink containing 80% nanocellulose and 20% alginate or 70% nanocellulose, 20% alginate and 10% Col I to construct a corneal stroma substitute [65]. Not only did these cells survive, grow, and proliferate after being 3D bioprinted and implanted, but they also produced and remodeled their own ECM in vitro.

While this research has great potential for future applications in corneal tissue bioprinting and engineering research, the bioink materials used, such as nanocellulose and alginate, may not be ideal choices for corneal surgeries and transplantation, due to their weak biocompatibility with biomaterials used in eye surgery [66]. For this reason, in a different study, Kostenko et al. in 2022 aimed to reduce alginate bioink concentration, while improving its printability through the introduction of calcium sulfate ($CaSO_4$) crosslinking [67]. This bioink was combined with human corneal stromal fibroblasts (CSFs) and used to print human corneal ring constructs. After testing different concentrations of alginate and the presence or absence of $CaSO_4$ crosslinking in the bioinks, analysis showed that crosslinking alginate with $CaSO_4$ increased the viscosity of the bioink, while still maintaining cell viability and fidelity. In another study looking at ways to alter the viscosity of corneal tissue bioink, Li et al. in 2022 used extrusion-based bioprinting to develop a novel GelMA/alginate-based hydrogel formula with additives polyethylene glycol dimethacrylate (PEGDMA) and xanthan gum [68]. After combining these hydrogel components with human mesenchymal stem cells (hMSCs), they found that the PEGDMA adjusted the hydrogel's mechanical properties and the xanthan gum improved the viscosity of the hydrogel. Future research on 3D bioprinting corneal tissue for surgery and transplantation should first investigate how the bioink's viscosity, cell viability, and cell fidelity change with varying concentrations of $CaSO_4$ or xanthan gum.

Although more studies are still needed to establish a gold standard bioink formula and method for printing corneal stroma cells, many recent studies have started looking into the applications of transplanting 3D bioprinted cornea stromal tissues. He et al. in 2022 proposed a digital light processing (DLP)-based 3D printing approach to create a biomimetic epithelium/stroma bilayer implant for corneal regeneration [69]. Their bioink included a blend of GelMA and long-chain poly(ethylene glycol) diacrylate (PEGDA) along with rCECs and rabbit adipose-derived mesenchymal stem cells (rASCs). To mimic the native corneal structure, they printed the hydrogel into two layers: an epithelial layer and an orthogonal aligned fibrous stroma layer. Subsequent analyses of the constructs using slit-lamp biomicroscopy, histological assessment, and PCR showed that the hydrogel also met corneal transplant requirements with respect to cell adhesion, cell proliferation, cell migration, high light transmittance, nutrient permeation, and cytocompatibility. Though this hydrogel mixture met corneal transplant surgical requirements, and thus the scaffolds could be used as a variable treatment approach for corneal diseases, suture resistance of the hydrogel may need to be thoroughly evaluated before it is used in surgery.

In another study, Jia et al. in 2023 proposed the use of pre-designed refractive lenticules using poly-NAGA-gelatin methacryloyl (GelMA) (PNG) bioinks to address refractive error disorder [70]. They demonstrated this application using DLP-based bioprinting and used human corneal epithelial cells (hCE-Ts), human keratocytes (hKs), human umbilical cord mesenchymal stem cells (hUCMSCs), human periphery blood mononuclear cells (PBMCs), and rabbit corneal endothelial cells (prCECs) to print the PNG lenticules. The PNG lenticules developed by this study demonstrated optical stability and nutritional functionality compared to their native counterpart. In vitro immune responses analyzed by Illumina RNA sequencing in the PBMCs showed that the PNG lenticules activated type-2 immunity, encouraging tissue regeneration and inflammation suppression. Additionally, in vivo performance analysis showed that intrastromal implementation of the PNG lenticules avoided common complications caused by surgeries to correct refractive error syndrome, such as calcification, stromal melt, appearance of scars, and harm to the host. This method is not only convenient and safe but is also a reversible, individualized, and customizable alternative to the currently available refractive glasses, lenses, and surgeries.

Additional studies place emphasis on the development of platforms and models as starting points for clinical implementation of the fabricated corneal tissue. In one study, Zheng et al. in 2021 demonstrated the opportunities for using GelMA-based hydrogel along with a hyaluronic acid glycidyl methacrylate (HAGM)-based bioink combined with rabbit limbal stem cells (LSCs) [71]. Using DLP-based bioprinting, they generated hydrogel scaffolds that could support rabbit LSCs in culture. This platform presents a future opportunity for therapy development in the field of regenerative medicine and corneal reconstruction to address LSC deficiency and various corneal diseases. Lastly, Song et al. in 2022 demonstrated the potential patient-customizable aspects of engineered corneal tissue for transplantation [72]. They used an extrusion-based bioprinter along with collagen type I (Col-I) bioink combined with human corneal epithelial cells or human corneal stromal cells to print a round structure that mimicked the corneal interior. They found that the Col-I bioink possesses many advantageous features, including shorter crosslinking time, high transparency, and high cell compatibility. Additionally, using a Col-I-only bioink allowed for the adjustment of cornea size and curvature, which could be used for patient-customized cornea implants. A limitation, however, to the clinical implantation of this technology is that it may be difficult to evaluate the safety of such a treatment since human variability would prevent extrapolation of the results during clinical trials.

6.3.2 Neural Tissue

Bioprinting 3D neural tissue offers great promise for advancements in the fields of neural tissue transplantation, disease modeling, and drug testing and development. Neural tissue is a challenging tissue to replicate due to its cellular complexity, vascularization, and plasticity. At this stage, many researchers have primarily been focused on investigating and optimizing bioink formulas that are biologically compatible with electrical conductivity, an important characteristic of neural tissue.

In a recent study, Song et al. in 2022 developed a novel bioink formula, consisting of GelMA, PEGDA, poly(3,4-ethylenedioxythiophene) (PEDOT:CSMA), and neural stem cells (NSCs), to construct a neuronal stem cell-laden 3D electroconductive hydrogel (ECH) scaffold [73]. By utilizing extrusion-based bioprinting, Song and collaborators were able to create an ECH scaffold that promoted electron propagation for enhanced nerve regeneration and allowed for the differentiation of NSCs to peripheral neurons. Their scaffold was able to maintain soft-tissue consistency throughout the culture period, and it exhibited high cell viability, cell proliferation, and significant recovery from cold shock at −80 °C (Figure 6.2).

In another study, Wang et al. in 2021 utilized extrusion-based bioprinting to develop an oxidized konjac glucomannan (OKGM)-branched polyethyleneimine (PEI)-based hydrogel [74]. After testing the electroactive characteristics of this hydrogel, Wang and collaborators added carbon nanotubes, which further enhanced the electrical conductivities of the OKGM/PEI composite hydrogel. Subsequent analyses revealed that the hydrogels were sensitive to pH and possessed self-regenerative characteristics. Gao et al. in 2023 also investigated the electrical conductivity of hydrogels by creating a GelMA, hyaluronic acid methacrylate (HAMA), and PEDOT:sulfonated lignin (LS) bioink that displayed mechanical and conductivity properties similar to native spinal cord tissues [75]. They used this bioink combined with NSCs to fabricate NSC-laden scaffolds that would be placed at the spinal cord injury site in rats. After implantation and subsequent analyses, the cells of the scaffold exhibited good survival rate (higher than 90%), regeneration of neurons at the injury site, prevention of glial scarring, and nerve axon regeneration and myelination. As electrical conductivity can only be obtained through the careful selection of specific electrical-conducting hydrogels, future research should investigate combining the use of traditional scaffolds, which have been extensively studied, utilized, and established in the clinical setting due to their superior mechanical integrity, well-defined tissue architecture and ability to be mass produced, with these hydrogel formulas.

Many other studies, on the other hand, have focused on the structural and cell proliferation-promoting characteristics of various bioink combinations. In one study, Cheng et al. in 2023 tested a novel ternary soft segment-based biodegradable polyurethane-based bioink, consisting of PCL, poly(D,L-lactide), and

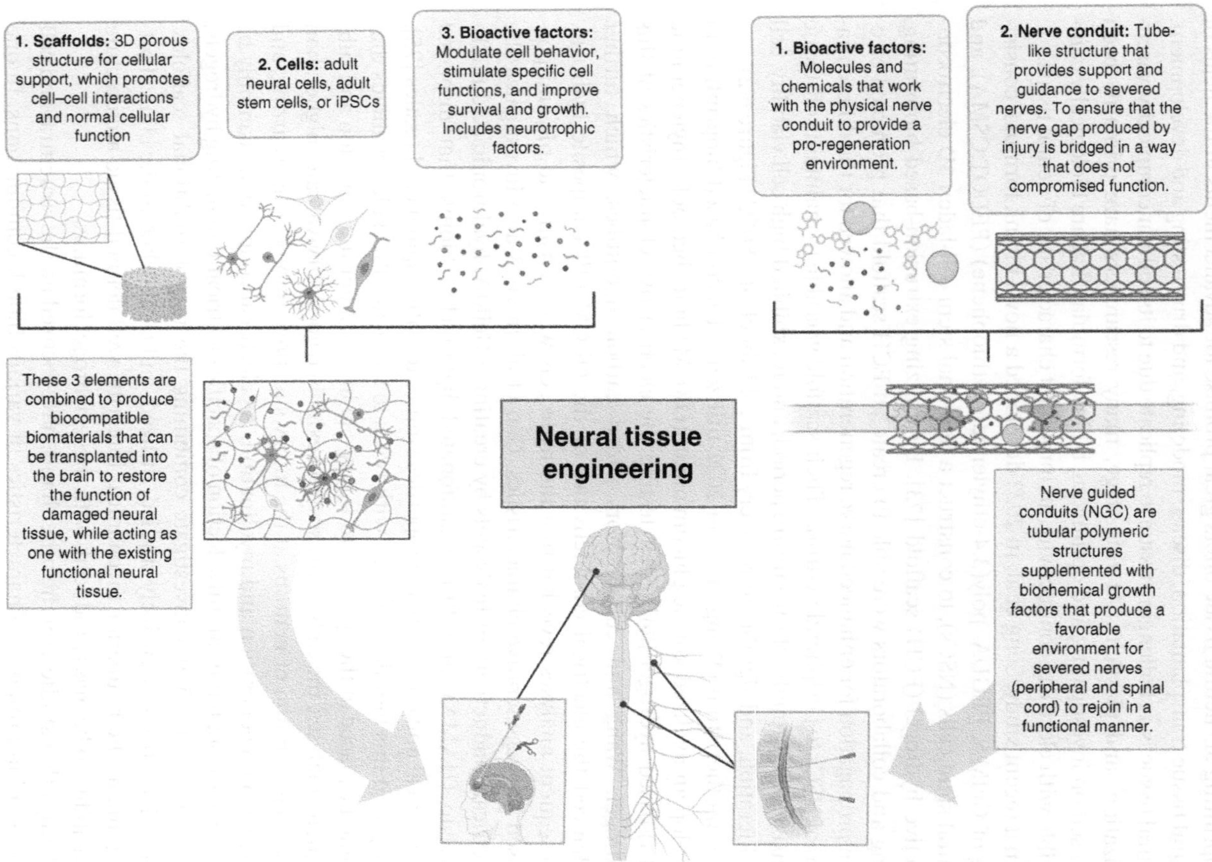

Figure 6.2 Overview of 3D bioprinting process of 3D scaffolds (right) and conductive spinal cord biomimetic scaffolds (left) that can be used to restore function of damaged neural tissue and spinal cord injuries, respectively.

poly(3-hydroxybutyrate) (PHB), combined with mouse NSCs and printed neural soft tissue [76]. They found that the fabricated tissue had good cytocompatibility, stacking ability, resolution, cell proliferation, and high printing resolution (~105 μm). In another study, Cruz et al. in 2023 tested different GelMA-based bioinks for neuronal differentiation using extrusion-based bioprinting combined with human induced pluripotent stem cells (hiPSCs) or human neural progenitor cells (hNPCs) [77]. After fabricating murine cortical astrocytes and hNPCs, they found that a mixture of both Geltrex™ and GelMA was able to support neural cell survival and differentiation. This finding corroborated those of previous studies that showed that combining bioinks with ECM materials such as Matrigel® or Geltrex™ improves neural cell adhesion and proliferation. Lastly, Peng et al. in 2022 tested a combination of alginate and different combinations of gelatin in bioink using extrusion-based bioprinting [78]. They found that a specific bioink, consisting of 2% gelatin and 1 g of sodium alginate (SA), was able to create a cell scaffold for soft tissues, such as nerve tissues, that retained its shape 24 hours post-printing, allowing for cell growth and proliferation.

While many studies have focused on developing bioink formulations that optimize the printing of neuronal tissue and that still maintain all of the unique and complex qualities of neuronal tissue, there have also been many studies demonstrating the applications of bioprinting for directly addressing injuries to the neurological system. In a recent study, Liu et al. in 2021 developed a novel, biocompatible bioink consisting of hydroxypropyl chitosan (HBC), thiolated hyaluronic acid (HA-SH), vinyl sulfonated hyaluronic acid (HA-VS), and Matrigel® (MA) crosslinked with an elastic modulus for mechanical support, and combined it with neuronal stem cells from the telencephalons of postnatal SD rats [79]. By using extrusion-based bioprinting, they created a living neuro scaffold that was able to mimic the native spinal cord, providing an optimal environment for neuron regeneration. This study was the first to demonstrate the feasibility of NSC-laden scaffolds for spinal cord injury repair in vivo. In another study, researchers investigated the applications of 3D bioprinting for peripheral nerve injury repair. Min et al. in 2022 developed a microgel printing bath to 3D print a peripheral nervous system decellularized extracellular matrix (PNSdECM) nerve graph for peripheral nerve regeneration [80]. They used PNSdECM bioink combined with porcine sciatic nerve and an extrusion-based bioprinting technique to create a 3D-printed dECM bath with a PCL conduit. Histological analysis of the regenerated nerves showed that the 3D-printed dECM with PCL conduit graph at the nerve defect site showed a higher total axon count and distribution of the axon diameter. This research lays the foundation for future studies on replacing autological nerve grafts.

While current studies have made great strides in the fabrication of effective neural tissue scaffolds, there is a great challenge with its clinical implementation, since the nervous system is highly influenced by genetic and environmental factors, something that has not yet been addressed by bioprinting. It is crucial first to develop a solid model that can replicate natural neurological development and diseased states before implanting fabricated constructs in humans.

6.3.3 Skin

Bioprinting in skin tissue engineering and regenerative medicine is one of the most developed fields across all tissues and organs. While the skin is a complex organ with different layers and cell type distribution, it can be simplified by printing simple skin layers that can be used in preclinical drug testing, in studying disease pathophysiology, and potentially in clinical practice with skin grafts. Mimicking all skin functionalities remains a challenge. Similarly, vascularization is difficulty, near non-existent, in current skin constructs given the printing complexity. The lack of vascularization prevents the long-term survival of skin tissues and full skin functionality similar to native tissues. Different types of bioinks are assessed to increase cell viability post-printing. In order to print skin constructs, a sufficient quantity of cells needs to be sourced, which can be time-consuming with the current methods. Novel cell sourcing protocols are being developed through decellularization of the dermis. Two main clinical areas of research emerge with burn treatment and chronic wound healing, followed by skin grafts for reconstructive surgery.

6.3.3.1 Disease and Pharmaceutical Studies

An interesting application of skin bioprinting is the possibility to study cosmetic and drug characteristics on different skin constructs. However, this application requires pigmented skin constructs, which few studies previously worked on. Printing uniformly pigmented skin is a major challenge. The DOD bioprinting technique could represent a novel approach given the advantage of its precise cell deposition. This was demonstrated by Long Ng et al. in 2018, who developed a two-step technique with a mixture of keratinocytes, melanocytes, and fibroblasts from human donors [81]. This study was a proof of concept to show the potential of drug testing on uniformized pigmented skin constructs. Similarly, Min et al. in 2018 developed a novel bioprinting approach to produce biomimetic pigmented skin structures referred as ephelides [82]. The biomimetic skin showed different layers of skin including the dermis, epidermis, and differentiated stratum corneum. While this approach remains limited to simple skin structures, it can be further optimized to produce pigmented skin grafts. These constructs can be used in disease studying.

Derr et al. in 2019 developed a novel approach to create skin equivalent constructs based on hydrogels [83]. After analysis, the skin constructs showed skin resemblance in terms of morphology and function. The applications of such constructs would be helpful in research to better study skin diseases and interactions of drugs and cosmetics with the skin.

Yim et al. in 2022 developed phantom-based human skin prototypes from gelatin methacryloyl (GelMA) bioink [84]. The prototypes can be used to assess the properties of different skin tones using optical devices. The social impacts of this work are tremendous, as they facilitate the development of standardized imaging modalities that are equally effective across all skin tones.

Masri et al. in 2022 designed single dermal layers with gelatin–PVA–genipin (GNP) hydrogels for compound testing. Different bioink formulations were assessed for cell viability [85]. The gelatin–PVA crosslinked with GNP (GPVA-GNP) bioink was the most successful in preserving human dermal fibroblast viability compared to the gelatin crosslinked (GE-GNP). However, results remain preliminary given the lack of immunohistochemical analyses.

6.3.3.2 Wound Healing

Issues related to wound healing constitute a worldwide healthcare burden due to the paucity of treatments that can accelerate healing processes. Bioprinting in this field could offer novel treatments by printing constructs that provide an optimized environment for tissue regeneration or accelerate slow healing processes in chronic skin conditions.

Modulating Bioink Formulation to Enhance Tissue Viability

Preserving cell viability is a universal challenge in bioprinting. While hydrogel-based bioink has shown the best biocompatibility profile, combining it with different biological and synthetic components has enabled it to gain new properties. Therefore, different combinations of hybrid bioinks have been assessed to optimize cell viability while preserving structural integrity after skin printing. Jorgensen et al. in 2020 developed and assessed a decellularized human skin-derived extracellular matrix (dsECM) fibrinogen hydrogel-based bioink [86]. The hybrid hydrogel showed improved rheological properties that suggest potential improvement in cell viability.

Cell viability is impeded during the printing process given the cells' exposure to high shear stresses. It is estimated that the viability of cells post-printing with extrusion-based techniques ranges from 40% to 80%. Lameirinhas et al. in 2023 designed a nanofibrillated cellulose gellan gum hydrogel-based bioink that showed 70% cell viability at three days post-printing [87]. Hafezi et al. in 2020 modified the degree of crosslinking between chitosan and GNP, as well as the proportions used [88]. The bioink viscosity was improved, thus allowing a lower printing pressure during the extrusion-based printing process. The bioink exhibited over 85% keratinocyte and human dermal fibroblast cell viability at seven days post-printing. However, cell proliferation could not be achieved, which suggests that additional biocomponents in the bioink may be required. Pereira et al. in 2018 developed a novel hydrogel bioink composed solely of methacrylate-modified pectin with modulable rheological properties [89]. Dermal fibroblasts were printed and were able to secrete ECM. Fibroblast viability was assessed throughout 14 days of culture and showed preserved structural integrity. The authors discuss the applicability of their novel hydrogel-based bioink to other tissues given the application of fibroblasts beyond skin tissue engineering. Liu et al. in 2019 used human amniotic epithelial cells and Wharton's jelly-derived mesenchymal stem cells to create bilayered membranous skin constructs [90]. The bioink was hydrogel-based and composed of alginate/gelatin. A high cell viability of over 95% was achieved. However, this was only measured over six days.

Cell sourcing is a challenge in bioprinting due to the difficulty of extracting enough cells for the bioink formulation. Bera et al. in 2022 developed a novel protocol to decellularize goat dermal tissue [91]. Their detergent-free approach yielded a higher quantity of cells than current methods that use detergent. A hydrogel-based bioink was formulated from these decellularized dermal extracellular matrices (ddECM). After conducting rheological studies, the bioink exhibited shear recovery above 90%, which preserved cell structure.

Li et al. in 2023 assessed a new bioink composed of different GelMA nanocellulose formulations [92]. This strategy enabled the creation of an 80 μm thick dermis over the course of 14 days in vitro. Granulation tissue formation was reported and was significantly more present in the skin construct than in the control group. This bioink formulation shows promise for use in chronic wound healing. The tissue-engineered skin was also assessed in vivo in a mouse model. In vivo hair follicles and preliminary stage rete ridges from the skin construct were reported and mimicked those of the mouse. Yang et al. in 2022 used gelatin methacryloyl-recombinant human type III collagen (rhCol3) to print a full-thickness human skin equivalent [93]. They investigated the use of recombinant collagen, which was previously studied by Huang et al. in 2022 [94]. Previous work by Huang et al. had shown that recombinant collagen improved rat skin regeneration. After conducting multiple rheological studies, they found that adding rhCol3 to GelMA increased the printability of the skin construct. In a mouse model, it is suggested that the rhCol3 would have an important role in promoting cell spreading.

Different Printing Techniques in Skin Regeneration

Miguel et al. in 2019 employed a combined electrospinning and extrusion-based technique for the first time to create asymmetric skin constructs [95]. The epidermis was produced through electrospinning, while an extrusion-based technique was used to print the dermis. Their approach demonstrates that different skin layers can be produced by combining techniques, which might enable better skin function mimicry. In fact, a dense epidermis was obtained, thus offering a more protective skin barrier. On the other hand, the chitosan/SA hydrogel facilitated the creation of a porous dermis that promoted cell spreading.

Applications of 3D-Printed Models

The use of scaffolds has shown promise in recreating the native skin environment and for providing additional skin regeneration functions to printed constructs (Figure 6.3). This was shown by Wang et al. in 2019 [96], who designed a bilayer membrane scaffold from PLGA and alginate hydrogel layer [95]. In vitro and in vivo studies in a rat model showed improved wound healing properties of the scaffold by promoting neovascularization and production of Col-I and Col-III. In the rat model, compared to control groups, the scaffold was the most effective in closing wounds. Wound measures were taken over the course of 12 days. In a recent study, Ma et al. in 2023 developed a gelatin–alginate hydrogel with adipose-derived stem cells (ADSCs) [97]. The hydrogel was used to create a gradient stiffness scaffold to emulate the skin's mechanical microenvironment.

The novel gradient stiffness scaffold was tested both in vitro and in vivo as a skin graft on a mouse model. Promising results were achieved with the mouse model within 7 and 14 days, as angiogenesis and cell proliferation were accelerated. This study demonstrated the possibility of gaining skin regeneration properties with gradient bioprinted scaffolds. Shi et al. in 2019 designed a scaffold derived from decellularized small intestinal submucosa (dSIS) slurry for skin regeneration [98]. By using cryogenic and extrusion-based methods, the authors report the novel possibility of using dSIS, which commonly has poor formability. However, the use of lower temperatures via a cryogenic approach could circumvent this issue, as demonstrated in this study. As dSIS is biocompatible and could promote cell spreading, applications of this scaffold in skin regeneration should be explored.

The use of microfat in skin grafts for improving chronic wound healing has been investigated and has shown the potential to provide a supportive regenerative environment. However, poor cell viability and graft retention impede the healing process. Schmitt et al. in 2021 developed a novel system to bioprint through extrusion-based technique microfat grafts from methacrylated collagen (CMA) bioink [99]. The system enables the production of microfat clusters with high viable properties. These clusters were bioprinted with CMA bioink as microfat-laden collagen to increase cell retention. This construct enabled the cells to remain active and viable for up to 10 days in vitro. The construct was also able to release in a controlled manner proinflammatory cytokines, which are important in the wound healing process. Bioprinting technique played a crucial role in successfully positioning microfat in the collagen construct to ensure adequate distribution of regenerative cells and controlled release of bioactive molecules such as the proinflammatory cytokines.

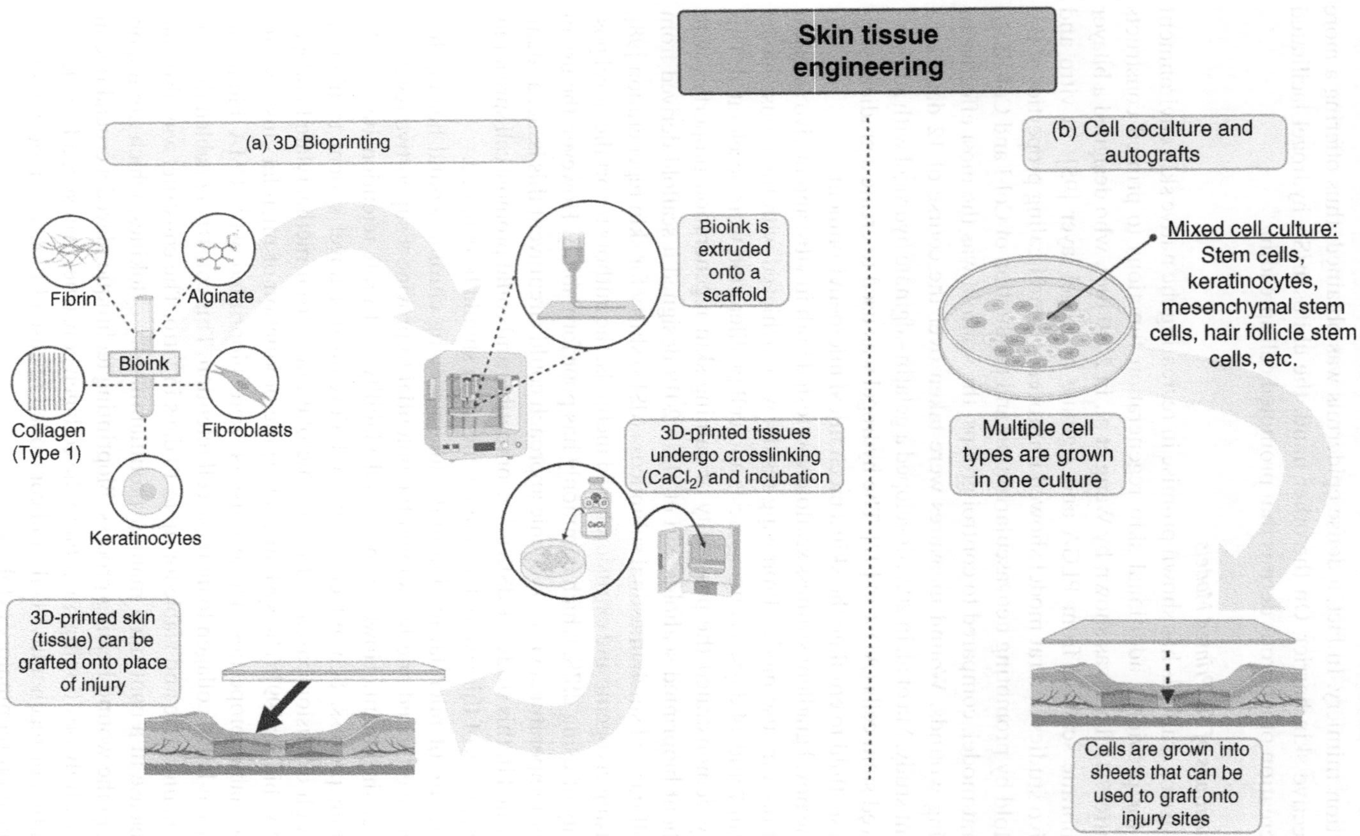

Figure 6.3 (a) Applications of 3D bioprinting in skin tissue engineering (b) The process of cell coculture and autografts onto skin injury sites.

Tissue-engineered skin grafts are used as treatments in nonhealing cutaneous ulcers. However, given the absence of vascular networks, these skin grafts cannot be used for a long period of time. Bioprinting vascular networks in a skin graft construct could thus prolong their therapeutic use in chronic wound applications. Kim et al. in 2018 initially demonstrated the feasibility of developing a vascularized skin patch [100]. Their construct was 1 cm in diameter and 1 mm thick and was derived from porcine decellularized extracellular matrix (dECM). The skin patch was tested in an immunodeficient mouse model. Re-epithelization at 90% was reported after 14 days. However, no specific evidence with respect to the functional vascularization and perfusion of the dermis was reported. Baltazar et al. in 2020 later developed a functional multilayered vascularized bioengineered skin graft derived from different human cell lines [101]. The incorporation of human pericytes in the dermal bioink showed cell maturation into a stratified epidermal layer. The skin grafts showed evidence of vascularization and perfusion in an immunodeficient mouse model at four weeks post-engraftment. From these findings, Baltazar et al. in 2022 presented a xeno-free skin graft methodology to avoid immunological complications [102]. Although animal-derived skin grafts are currently used in clinical practice for persistent ulcers and in burn victims, allergic reactions and the development of antibodies specific to xenografts have been reported [103]. There is thus a need to develop xeno-free skin grafts to avoid such complications. Baltazar et al. in 2022 bioprinted a vascularized bilayer xeno-free skin graft. After being grafted on an immunodeficient mouse, the skin construct created novel blood vessels and supported cell differentiation in a stratified epidermis within two weeks [102]. This study shows the possibility of using a xeno-free approach and its potential biocompatibility in a mouse model.

Albanna et al. in 2019 developed a mobile skin bioprinting system able to personalize cell deposition and print according to wound characteristics [104]. The advantage of using bioprinting is to create organized, distinct skin layers. The bioprinting system showed accelerated wound closure compared to control groups. Further long-term studies are required to assess additional cell types and to better understand bioprinting cell clearance over time. Nonetheless, this study opens the door to personalized wound healing treatment.

6.3.3.3 Reconstructive Surgery

As hydrogels alone present limitations, the use of composite hydrogels might avoid skin collapsing as suggested by Zou et al. in 2021 [105]. They reported using three different polysaccharides (agarose, SA, and nanocellulose) as a composite hydrogel to print a customized, vascularized human face. A sacrificial scaffold composed of PVA was used to provide structure to the face. This study represents a first start at bioprinted facial reconstructive constructs. Encouraging results show vascularization potential through coculture of HUVECs and human fibroblasts.

6.3.4 Cartilage and Bone

Studies in this field are focused on developing engineering cartilage tissue that can be used in the case of osteochondral defects repair. As cartilage has limited regenerative properties, bioprinting represents a promising avenue for cartilage grafts.

6.3.4.1 Cartilage Printing Modalities

Wu et al. in 2020 developed a new extrusion-based bioprinting method to print stratified cartilage from human and predifferenciated ADSCs [106]. The use of cartilage tissue strands with predifferenciated ADSCs was the most successful combination for obtaining stratified cartilage similar to native cartilage. The mechanical properties of this articular cartilage were also comparable to those of native human cartilage.

Adult stratified cartilage is particularly challenging to recreate because of its distinct stratified zones. Daly et al. in 2019 designed cartilage that can be used in joint resurfacing [107]. They developed a novel framework that features inkjet bioprinting and a microchamber system involving cellular spheroids with self-assembly and fusion properties. Their model showed similarities to native cartilage, thus suggesting future applications as cartilage implants for the treatment of osteoarthritis. Similarly, Lam et al. in 2019 created varying density cartilage constructs for possible applications in articular cartilage defects (Figure 6.4) [108]. Gelatin and hyaluronic acid were chosen in the bioink because of their ability to support ECM formation. Interestingly, gelatin was more useful for promoting chondrogenesis compared to hyaluronic acid, the latter of which improved cell viability. These findings suggest that bioink combination can be considered as a method for enhancing cartilage constructs. The authors highlight the utility of the stereolithographic approach for creating different cartilage concentration models because of the high resolution and gentle printing process. SLA could enable the large-scale deployment of such constructs, as it can produce multiple constructs simultaneously.

6.3.4.2 Cartilage Regeneration

A common challenge is the development of a suitable bioink. The latter should provide mechanical, viable, cell differentiation and tissue regeneration support while being biocompatible and easy to print. Different bioinks are being tested to optimize and create one that is biocompatible, promotes high cell viability, and supports cartilage stiffness and stability. In these respects, Haupstein et al. in 2020 assessed hyaluronic acid (HA) crosslinked with allyl-modified poly(glycidol) [109]. They found that by combining HA to gels with a low percentage density polymer, cartilage stiffness and distribution were increased. These findings highlight the

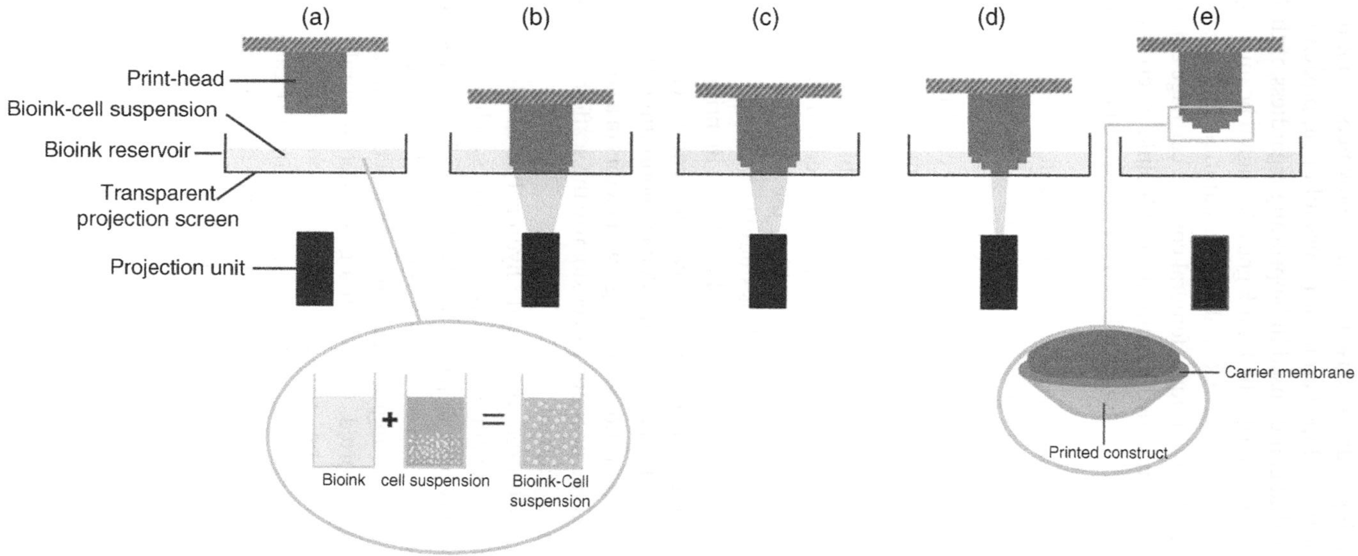

Figure 6.4 Printing of cartilage constructs with various densities using a stereolithographic technique. (a) Identification of printing components involved in the stereolithographic technique, (b–d) this process involves superposing cells in layers, and (e) after the construct is printed, the cells are incubated for cultivation in vitro.

advantages of combining hydrogel-based bioink to polymers to optimize bioinks. Galarraga et al. in 2019 developed an in situ crosslinking approach that expands the variety of bioinks that can be used in bioprinting regardless of their intrinsic printability characteristics [110]. The bioink goes through a beam of light immediately prior to cell deposition, thus enabling its curation and the use of non-viscous bioinks. Using this technique, femoral condyle cartilage constructs were printed from norbornene-modified hyaluronic acid bioink. Afterward, they were cultured for 56 days in vitro. Histological analyses confirmed that cell maturation was achieved in the construct.

Sun et al. in 2020 used mesenchymal stem cells to create an anisotropic gradient-structured knee articular cartilage (Figure 6.5) [105]. The construct was tested in vivo in a rabbit model over the course of six months. The mesenchymal stem cells were able to differentiate as chondrocytes and exhibit human cartilage phenotypes. Cell integrity was maintained after a six month follow up. As seen with other organs, hydrogel-based bioink alone is insufficient for providing adequate physical properties to the cartilage construct. To this end, the authors found that adding a poly-ε-caprolactone (PCL) scaffold created a more stable environment. The fabrication of constructs mimicking the structural and functional aspect of native cartilage was achieved. This study shows the possibilities of using human-scale cartilage constructs as surgical implants. Bioprinting is the central key to these results. The gradient-structured cartilage construct was made possible through the precise cell deposition provided by bioprinting. This also allowed the successful encapsulation and release of growth factors in a controlled manner. Bioprinting also offered the possibility to customize the scaffold's pores and fiber spacing, thus enabling a gradient.

Critchey et al. in 2020 used an alginate hydrogel to develop a bi-construct cartilage implant for regeneration of cartilage and subchondral bone [112]. They assessed fiber networks to pair with the hydrogel and proceeded to test in vivo in a mouse model. The construct was implanted in a nude mouse model and successfully promoted angiogenesis in the endochondral bone. Cartilage phenotypes were also reported. In vivo in a goat osteochondral defect model, both subchondral and hyaline cartilage repair were reported in six months. This study demonstrated the improved regeneration of hyaline-like cartilage with the engineering bi-phasic construct compared to lesions treated with scaffolds as controls. These findings suggest that the bi-phasic construct could become an alternative surgical treatment in osteochondral defect repairs with improved cartilage regeneration.

Like cartilage, bones have limited regenerative properties. Therefore, successful bone printing would revolutionize treatment for patients in need of bone grafts.

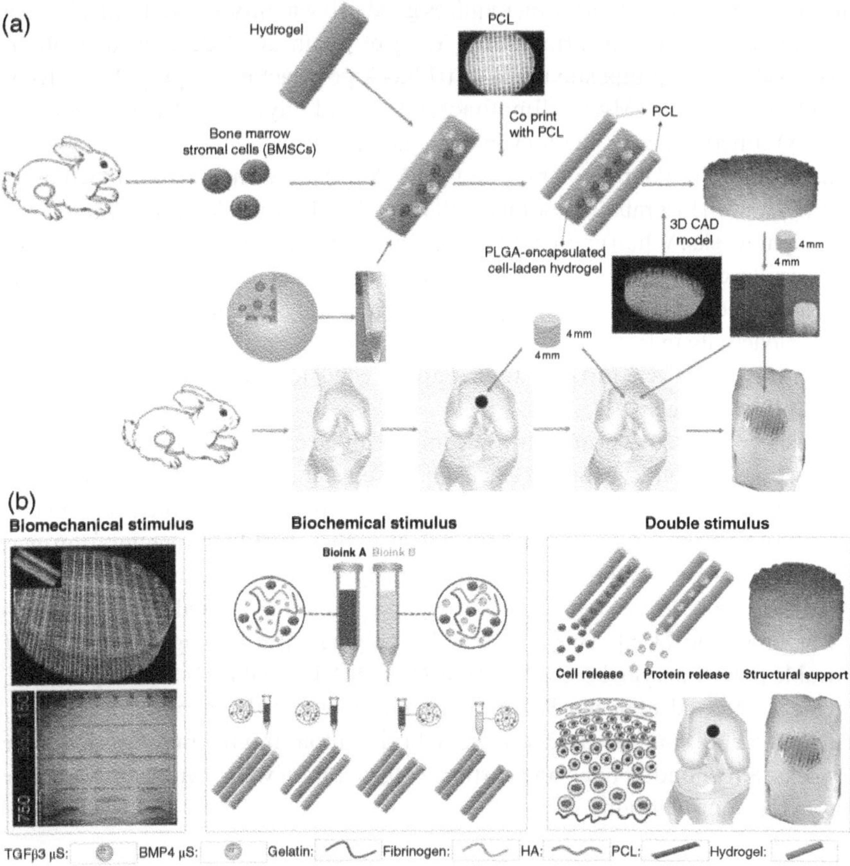

Figure 6.5 Schematic presentation of the study design and scaffold construction. Development of a Gradient-spiepr DHYPAnisotropic Knee Articular Cartilage: (a) (i) The study employs 3D bioprinting to create mesenchymal stem cell (MSC)-spiepr DHYPladen constructs with dual-spiepr DHYPfactor release and gradient structuring for articular cartilage regeneration in rabbits. The schematic outlines the construction of the anisotropic cartilage scaffold and overall study design. (ii) Utilizing a computer-spiepr DHYPaided design (CAD) model, a four-spiepr DHYPlayer gradient polycaprolactone (PCL) scaffold was designed to facilitate anisotropic chondrogenic differentiation and deep-spiepr DHYPlayer nutrient supply. This scaffold was fabricated through one-spiepr DHYPstep spiepr NONBREAK3D bioprinting, incorporating dual protein-spiepr DHYPreleasing composite hydrogels and bioinks loaded with bone marrow stromal stem cells (BMSCs) and either BMP4 or TGFβ3 spiepr NONBREAKmicrospheres.
(b) This construct offers both structural support and sustained release of differentiative factors, enabling biomimetic regeneration of anisotropic articular cartilage in animal models. *Source:* [110]. Sun et al. 2020 / American Association for the Advancement of Science/CC BY 4.0.

The ideal bioink in bone printing requires good cell viability, resistant mechanical characteristics, and easy printability. Yang et al. in 2020 developed a photo-crosslinkable nanocomposite made of tri-block poly(lactide-co-propylene glycol-co-lactide) (PmLnDMA), dimethacrylate, and hydroxyethylmethacrylate (HEMA)-functionalized hydroxyapatite nanoparticles [113]. The interaction between the PmLnDMA polymer and HEMA imparted the desired mechanical stability. In vivo in a rabbit femoral condyle model, the scaffold partially remained up to eight weeks. When compared to the control group, the scaffold enabled osteogenesis. Based on these preliminary results, this ink could potentially be used to overcome mechanical (e.g., insufficient strength) and biological (e.g., biocompatibility, bioactivity in fostering osteogenesis) challenges encountered in bone tissue bioprinting. However, further studies are needed to assess generalizability in other animal models.

Idaszek et al. in 2019 printed hyaline and calcified cartilage with two cell-laden hydrogel-based bioinks using an extrusion-based technique [114]. Stratified cartilage started to form after 21 days in culture, demonstrating the potential for stem cell differentiation. In a rat trochlea defect model, cartilage regeneration was significantly greater in the scaffold bioprinted group compared to the control.

Gao et al. in 2019 developed a hydrogel-based scaffold to facilitate osteochondral regeneration [115]. They combined a polymer in a hydrogel composed of cleavable poly(*N*-acryloyl 2-glycine) (PACG) and GelMA (PACG–GelMA). Adding polymer conferred high tensile strength and increased the stability of the construct. Gao et al. in 2019 assessed the hybrid scaffold in vivo in a rat model for 12 weeks. Subchondral bone and cartilage regeneration were significantly greater than in the control group.

6.3.5 Vascular Tissue

Cardiovascular diseases represent the primary cause of death worldwide. Creating vascular grafts and blood vessels is an interesting avenue to avoid common autologous graft failure. Vascularization is a common challenge in all organs because of the complexity of the vessel networks. The current bioprinting equipment is also not adapted to print such small vessel constructs. Some studies have focused on solely printing blood vessels. Gold et al. in 2021 engineered a nano-hydrogel that is easily printable with good resistance to shear forces. When tested in vitro, the hydrogel preserves its phenotype for one month [116]. Blood vessels were printed with endothelial and vascular smooth muscle cells. Impressively, the thromboinflammatory phenomenon, which is commonly induced by a cytokine cascade following injury to the endothelium, lesion to the endothelium, was reproduced. This indicates potential applications of this model in pathophysiological study and compound testing. The use of bioprinting in this context is more advantageous than using regular scaffolds given the superiority of bioprinting in spatial and temporal cell precision. The high precision achieved through cell deposition enables the creation of complex vascular structures, which can be difficult to achieve with regular scaffolds.

OoC models have been widely employed in medical research to better study organ function and physiology. Abudupataer et al. in 2019 combined OoC technology with an extrusion-based bioprinting technique to create vascular tissues [117]. By utilizing both techniques, the OoC approach adds hemodynamic properties to the vascular construct, which cannot be solely achieved by bioprinting. Cell-laden hydrogel-based GelMA bioink concentrated at 5% was the best concentration to obtain the highest printability properties. The hydrogel bioink with endothelial cells and smooth muscle cells was printed on the microchip. Smooth muscle cell protein expression similar to the native microvascular environment was reported. Therefore, this study shows the possibility of printing vessels on a chip through bioprinting to study blood vessels, particularly the ECM. The vessel-on-a-chip model mimics cell behavior under dynamic conditions through the microfluidic channels. This study shows that bioprinting enables the creation of complex tissue models onto microfluidic chips given its high printing precision as demonstrated with the cell-laden hydrogel construct. Bioprinting can replicate in vivo natural tissue environments and cell arrangements, which improves the accuracy of tissue constructs.

Freeman et al. in 2019 used a bioink blend of fibrinogen and gelatin to create small-diameter vascular constructs [118]. By adjusting printing parameters such as heat and gelation concentration, printability was improved. Fibrinogen stands out as a component in the bioink that is particularly useful in blood vessel formation. Using a rotary system, as suggested in this approach, might facilitate the printing of tubular constructs.

6.3.6 Cardiac Tissue Engineering

While advancements in cardiac tissue engineering have been achieved in the last five years, the heart is a particularly challenging organ to recreate given its highly complex cell organization and compartments (e.g., valves and chambers). Heart components with limited functions were successfully created. Lee et al. in 2019 printed collagen constructs that can be used in various heart components [119]. A contractible ventricle construct was printed from human stem cell–derived cardiomyocytes by using a previous approach published in 2018 with freeform reversible embedding of suspended hydrogels (FRESH). Different heart components were printed as a proof-of-concept, including collagen vessels, tri-leaflet heart valve, and even a neonatal-scale human heart.

Ensuring cardiac tissue structural and functional integration is a major challenge. Not only do cardiomyocytes need to be deposited in a precise and controlled manner, but they need to integrate and function properly in synergy. Building upon the previous work, Mirdamadi et al. in 2020 were the first to develop a 3D bioprinted alginate heart with contractile properties [120]. The cardiac construct was similar to a ventricle and derived from human cardiomyocytes. A novel FRESH approach was designed, which can potentially be applied to broader cardiac tissues. The eminent application of this groundbreaking research is to create hearts usable in clinical practice to prepare for surgical procedures. Similar groundbreaking research was conducted by Tracy et al. in 2020, who successfully printed cardiac Purkinje cell networks [121]. The cell networks exhibited electric stimulation activity, which shows promising applications for studying cardiac physiology and compound testing. The team published a protocol to create printable Purkinje cell networks that were derived from reprogrammed human adipogenic mesenchymal stem cells. Despite needing refinements in terms of mass production and durability, this important work can be used for future studies involving electric stimulation activity.

Biomaterial selection such as the bioink is crucial to replicate the cardiac ECM. In general, the ideal bioink should be able to ensure proper cell distribution and growth, while being biocompatible and biodegradable. However, it is even more challenging in cardiac tissue engineering due to the propagation of electrical signal propagation between cells. Therefore, the ideal bioink needs to support cell contractions and electrical signal propagation. Jung Shin et al. in 2021 assessed a novel bioink formulation incorporating porcine cardiac decellularized extracellular matrix (cdECM), laponite-XLG nanoclay, and poly(ethylene glycol)-diacrylate (PEG-DA) [122]. This formulation enabled 94% cell survival after seven days and exhibited promising mechanical and biocompatible properties. This composite bioink could be further considered to print fibrotic cardiac disease constructs. Koti et al. in 2019 found that combining different cell populations with an enriched hydrogel-based bioink improved cell survival, spreading, and interactions [123]. Three major parameters were identified to support cell survival: (i) adding adhesive cells to the bioink, (ii) combining cardiac myocytes and fibroblasts bioinks, and (iii) lowering GelMA concentration. This study found that cardiac fibroblasts can release ECM proteins that foster attachment and adhesion between cells within the hydrogel scaffold, thus replicating the structure of native tissues. The combination of cardiac myocytes and fibroblasts enabled to reproduce the functional tissue-like environment. The combination of cardiac myocytes and fibroblasts exhibited synergistic effects, which resulted in improved tissue function and structure. Cardiac myocytes ensure the heart's functional contractions as opposed to cardiac fibroblasts, which provide structural integrity and preservation of the ECM. Lowering GelMA concentration was beneficial to making the hydrogel scaffold more porous, thus enhancing the transit of nutrients and facilitating cell interactions.

Similarly, Basara et al. in 2021 assessed different hydrogel-based bioinks composed of human heart dECM and GelMA alone or accompanied by methacrylated hyaluronic acid [124]. They also found that the composite GelMA-MeHA-dhECM (GME) hydrogel-based bioink had the highest mechanical properties (e.g., structural stability). All hydrogel-based bioinks were usable with hiPSC-derived cardiomyocytes and human cardiac fibroblasts. The findings from this study showed that hydrogels can be readily tunable to different physiological contexts, thus enabling the study of cardiac diseases. Hydrogels can be used to replicate healthy and pathological cardiac tissues. The authors suggest that the pathophysiology of cardiac diseases such as myocardial infraction, which is among the leading causes of death worldwide, could be better understood and studied through hydrogel-based bioinks. While myocardial infraction has been studied on some animal models, the underlying disease mechanisms remain uncovered. Therefore, 3D bioprinting represents a novel avenue to effectively replicate disease models through mimetic cardiac tissues.

Liu et al. in 2019 developed humanized cardiac constructs from human embryogenic stem cell-derived cardiomyocytes, gelatin hydrogel, and GelMA through light-based micro-continuous optical (μCOP) printing [125]. The μCOP approach enables the assessment of mechanical forces and calcium signaling within the cardiac construct, which can be applied in compound testing and drug development.

Bejleri et al. in 2018 printed a cardiac patch releasing regenerative human cardiac progenitor cells (hCPCs) [126]. This cardiac patch was developed in the hope of repairing heart tissues. A decellularized cardiac ECM hydrogel was used with hCPCs and GelMA. When tested in vivo in a rat model, the patches remained stable and vascularized over 14 days. Patches can also be customized to pediatric patients for possible applications in pediatric cardiac surgery.

As discussed in the vascularization section, vascularization is a major challenge in cardiac tissue bioprinting given the high complexity to replicate the complex miniatured vascular networks. Given that most printed cardiac component lack vascularization, their viability and function are limited compared to native tissues. Maiullari et al. in 2018 developed vascularized cardiac tissues printed from induced pluripotent stem cell-derived cardiomyocytes (iPSC-CM) [127]. The cardiac constructs were implanted in a mouse model for two weeks with evidence of vascularization. The advantage of using iPSC-CM is their high proliferative and differentiative potential. iPSC can also be easily extracted from adult donors. By using a microfluidic printing head, a higher resolution of the cardiac tissue model was achieved. Blood-vessel-like structures were also printed with endothelial cells and induced pluripotent stem cells.

6.3.7 Pancreas

6.3.7.1 Modulating Bioink Formulation to Enhance Tissue Viability

Increasing the viability of artificial pancreatic models has been a topic of interest for many researchers in the field of pancreatic bioengineering. In particular, many studies have focused on modulating the bioink formulation so that the fabricated construct bears the greatest resemblance to the native tissue.

Wang et al. in 2023 developed a new bioink composition containing pancreatic extracellular matrix (pECM) and HAMA to construct a 3D-printed islet organoid [128]. The HAMA/pECM hydrogel was used in a DLP-based printer, and the resulting model displayed high levels of viability. Furthermore, when implanted in diabetic mice, the islet organoid produced from the HAMA/pECM hydrogel elevated insulin levels, sustained a healthy blood glucose level for 90 days, and released insulin following glucose stimulation. Additionally, because the HAMA/pECM hydrogel promoted vascularization of the islet organoid, nutrients were able to reach the transplanted islets, resulting in the long-term stabilization of blood glucose levels in diabetic mice. Although the new bioink composition represents a promising approach to diabetic treatment, a major area of concern is the lack of mechanical stability in HAMA/pECM hydrogel. This means that the HAMA/pECM hydrogel must be reinforced with additional materials in order to achieve lasting mechanical stability.

Hu et al. in 2021 created a cell-laden construct by using a new tri-component bioink consisting of pectin, alginate, and pluronic for use in a Biobots 1 desktop bioprinter [129]. Results from the study showed that the new bioink enabled the islet organoid to experience less tissue rejection and that it enhanced the survival of β-cells when cells were subjected to stress conditions. Moreover, in vivo implantation showed that pectin-containing constructs have various advantages, including less fibrotic overgrowth and the ease of removal when a replacement is needed for the implanted tissue. Although the findings present an innovative approach to creating durable pancreatic islet grafts for transplantation, picking a biomaterial that enables the survival of cells while still ensuring excellent printing outcomes remains difficult. Compared to synthetic polymers, natural polymers such as alginate or pectin have relatively fleeting stability and are much harder to produce on a mass scale.

In another study, Duin et al. in 2019 employed alginate/methylcellulose (Alg/MC) as the bioink to generate macroporous constructs containing pancreatic islets [130]. Specifically, the MC was not crosslinked, while the alginate was crosslinked with either $CaCl_2$ to represent "clinical-grade" alginate or $SrCl_2$ to mimic "research-grade" alginate. Although no variation in speed of insulin uptake was recorded for hydrogels with and without MC, the Alg/MC blend generally had higher levels of insulin uptake. Additionally, both clinical-grade and research-grade scaffolds greatly maintained the viability, structure, and functions of the embedded islets. The researchers also observed higher levels of glucose secretion in the clinical-grade alginate. Meanwhile, no major differences in the insulin uptake level were detected for both types of Alg/MC blend. Compared to the clinical-grade alginate, the research-grade exhibited greater stability and crosslinking density. Because human transplantation requires the use of clinical-grade alginate, the fleeting mechanical stability of the clinical-grade alginate remains a major concern.

6.3.7.2 Controlling Other Printing Parameters to Enhance Tissue Viability

Klak et al. in 2021 employed an extrusion-based printing technique using a BioX 3D printer to measure the ability of rat and porcine pancreatic islets to withstand printing pressures ranging from 15 to 100 kPa [131]. The islets demonstrated great sensitivity to the applied pressure, with the optimal pressure being less than 30 kPa and the carrier containing 3% alginate. Moreover, the mechanical stress induced during printing was noted to play a significant role in reducing the size of the islets following the printing process. Ultimately, pancreatic bioengineering remains a challenge because the production of fully viable islets is greatly compromised by the reduction in print resolution. Additionally, very little is understood about how different printing pressures can damage the pancreatic islets on a molecular level.

Hwang et al. in 2022 proposed the encapsulation of insulin-producing β-cells by creating a hybrid system consisting of a macroporous polymer capsule with stagger-type membrane and tissue-derived decellularized ECM (dECM) hydrogel [132]. The printing technique used was micro extrusion-based. Overall, the embedded β-cells exhibited various cellular activities, including maturation and insulin secretion. In addition, the biocompatibility of the encapsulation system both in vitro and in vivo was established based on the polarization of macrophages following pro-inflammatory conditions. Viability of the artificial islet was observed for up to 14 days post-printing, with signs of systemic proliferation and functional improvement. Intercellular interactions were also detected, highlighting the high resemblance between the islet-like aggregates and the native tissue. However, further studies on protecting cell aggregates from allogenic response are required before this encapsulation system could become a widespread treatment for type 1 diabetes.

6.3.7.3 Using Printed Models to Study Pancreatic Cancer

In a recent study, Huang et al. in 2023 demonstrated the use of dot extrusion-based printing (DEP) to construct a 3D pancreatic ductal adenocarcinoma (PDAC) microtissue [133]. Specifically, the printed model consisted of GelMA hydrogel beads that contained both pancreatic cancer cells and stromal fibroblasts.

The study highlighted the ability of the hydrogel beads to foster cell proliferation. In addition, a dense pancreatic tumor–stroma network was observed after one week of coculturing, accurately mimicking the in vivo conditions. When compared to mono-cultured cancer, the bioprinted PDAC microtissue exhibited higher levels of resistance to the anticancer drug gemcitabine. Although the DEP system is a viable method for developing cancer drug screening platforms, the detailed molecular mechanisms of tumor–stroma interactions remain unclear. Moreover, a coculture of endothelial cells and tumor cells should be developed to closely examine the progression of PDAC.

Hakobyan et al. in 2020 investigated the onset of PDAC by creating artificial pancreatic cell spheroids through laser-assisted bioprinting via a solid Nd:YAG crystal laser [134]. The resulting spheroids consisted of both acinar and ductal cells and accurately represented the phenotypic features manifested in the early stages of PDAC. Although no instances of a necrotic spheroid core were noted, the researchers did observe a dismantled core that was a possible indication of the larger spheroids experiencing a hypoxic effect. Indeed, ensuring a sufficient supply of oxygen and nutrients to spheroids with larger dimensions remains a challenge in the field of pancreatic bioengineering. Future studies are also needed to establish the mechanisms by which pre-cancer and acinar cells communicate, as well as the impact of agricultural xenobiotics in accelerating PDAC lesions and cancer formation.

Langer and colleagues in 2019 manufactured pancreatic tumor tissues by using Organovo's Novogen MMX Bioprinter Platform and creating a bioink that consisted of both cancer and stromal cells [135]. Overall, the bioprinted model demonstrated high resemblance to in vivo tumors, as a dense network of stromal cells encompassed the cancer cells within the fabricated tissues. Moreover, proliferation, migration, and drug resistance patterns of the engineered tissues were also found to closely replicate those observed in in vivo tumors. Although the multicellular bioprinted tissues provide an excellent platform for understanding complex TMEs, further research on additional stromal cell types in metastatic conditions is required. Moreover, further examination of the impact of anti-tumor drugs on patient-derived cancer cells in bioprinted tissues is needed to develop individualized treatments.

6.3.8 Liver

6.3.8.1 Developing Suitable In Vitro Models

Advancements in the field of liver bioengineering are limited, as preserving the viability of 3D-printed hepatic tissues is difficult. Current objectives in this field include developing viable tissues that can be used as in vitro models. Because prior studies had only focused on bioprinting at low temperatures, Khati et al. in

2002 utilized extrusion-based printing to produce a crosslinked 3D structure at 37 °C [136]. More specifically, the researchers combined multi-material decellularized liver matrix (dLM) with gelatin and polyethylene glycol to create a mechanically durable bioink. Maturation of the HepG2 cells embedded in the artificial 3D structure was observed, and hepatic functions were detected for up to seven days. Although this study serves as a critical foundation for the future of cytotoxicity research in HepG2 cells, construction of hepatic structures that recapitulate the human drug response is needed. Moreover, high-throughput printing is still limited as current 3D printing systems are not able to maintain consistent heating conditions that can support temperature-sensitive bioinks. This deficiency greatly constrains the possibility of generating complex, high-resolution models that can replicate native hepatic lobules.

The fabrication of 3D multicellular constructs has also been an area of great interest. Kang and colleagues in 2020 demonstrated the successful creation of hepatic lobules that contain a lumen, hepatic cells, and endothelial cells using an extrusion-based technique with alginate as the bioink [137]. The engineered lobules generated consistent levels of albumin, urea, proteins, and enzymatic activity. Despite the optimistic outcomes, the cultivation of primary human hepatocytes remains a challenge, as immortalized cell lines are greatly limited in quantity. In another study, Zein et al. in 2013 used CT scans and MRI imaging to fabricate liver models that highly resembled the native tissue of six patients [138]. Although this research highlights the potential to use engineered liver tissues in humans, the production of 3D-printed models is greatly dependent upon imaging, which is susceptible to imaging errors that can occur from low image resolution. Additionally, in order for liver bioprinting to become widespread, a wide array of printing parameters (such as the quality or properties of materials) must be tightly regulated to ensure the survival of the resulting construct. Presently, it has been proposed by Zein et al. in 2013 that a soft material can be used to closely mimic the mechanical properties of the native liver tissue [138].

6.3.9 Lungs

6.3.9.1 Developing Suitable In Vitro Models

In recent years, there has been an increased interest in the study of lung tissue bioengineering, mainly because it provides opportunities to closely study pulmonary diseases. Specifically, various approaches have been proposed to enhance the development of in vitro lung models that are suitable for investigating disease pathogenesis. In a recent study, Kim et al. in 2023 presented the potential of combining lung-on-a-chip platform and inkjet bioprinting (Figure 6.6) [139]. Such a combination not only enhances product consistency during mass production but also allows the printed tissue to highly resemble the

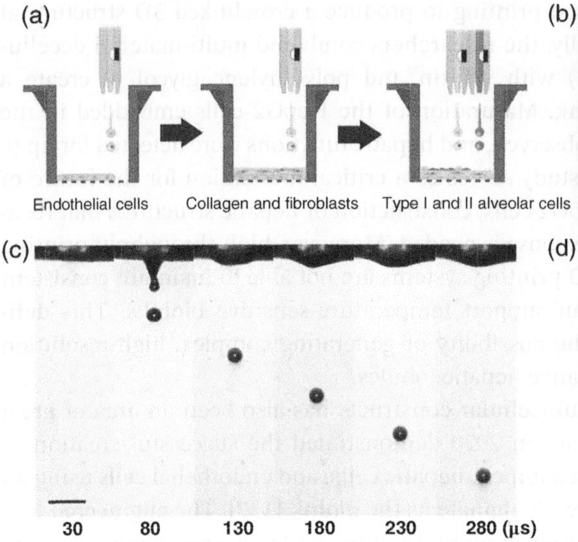

Figure 6.6 Using inkjet bioprinting to manufacture alveolar barrier models. (a) Printing protocol for alveolar barrier model, (b) using a single nozzle of the piezo-inkjet printer to automate the printing process, (c) cell-laden drop stroboscopic images, and (d) printed cell-laden droplets on a substrate. *Source:* Kim et al. [139] / Adapted from American Chemical Society.

native model. The researchers implanted inkjet-printed lung tissue on a culture insert that can subsequently be used in a microfluidic system. Although this study highlights the utility of this approach for the mass production of lung models, further research on constructing more physiologically relevant environments is required. In another study, da Rosa and colleagues in 2023 reported the use of alginate and gelatin as bioink in extrusion-based printing to fabricate lung cells that were appropriate for in vitro drug tests [140]. The study showed that the majority of the printed lung cells were viable, with high levels of proteins and secretory cell expression. Despite the promising findings, current limitations include constructing a 3D structure that contains both lung and endothelial cells while still replicating the native environment. Ng et al. in 2021 proposed the use of polyvinylpyrrolidone-based bioink to print alveolar models via DOD bioprinting [141]. The printed models were able to survive for at least 14 days while displaying high viability with proliferation patterns similar to those of non-printed cells. Although the DOD bioprinting technique is a promising method for developing sustainable in vitro lung models, the application of primary alveolar cells in pathogen translocation studies and cytotoxicity testing is still very limited.

6.3.9.2 Application of 3D Construct

Numerous studies have examined the utility of bioprinted constructs as models to examine the pathology of pulmonary diseases, as well as evaluate drug efficacy. In a recent study, Dong et al. in 2023 used SA, hyaluronic acid, and arginine-glycine-aspartic acid peptide as the bioink in droplet-based bioprinting to produce lung cancer organoids for anticancer drug examination [142]. Results demonstrated that the A549 cells in the organoids exhibited higher levels of functionality, cell activity, and anticancer drug resistance when compared to the 2D model. In another study that studied lung cancer metastasis, Das and colleagues in 2022 used extrusion-based printing to build a 3D-lung-cancer-on-a-chip model that closely reproduces in vitro pathogenesis [143]. Specifically, the bioink consisted of 3D hydrogels that contained cultures of lung cancer cells (A549) and human lung fibroblast. The high viability of both cell types at 14 days post-printing showed that the printed model can serve as a powerful platform for studying metastasis and assessing the efficacy of respiratory drugs. However, in order to construct a microenvironment that better mimics the in vivo environment, future studies on co-printing macrophages and human lung fibroblasts are needed. Additionally, the application of patient-derived lung primary cells in investigating the impact of external stimuli on lung cancer progression remains a challenge.

In another study, Kang et al. in 2022 illustrated the use of an inkjet-printed 3D alveolar barrier construct for exploring drugs for pulmonary fibrosis [144]. Using a piezoelectric inkjet printing system, the researchers created a model that mimicked the fibrotic physiological conditions. Subsequently, the printed model was treated with two existing drugs for fibrosis, nintedanib and pirfenidone. The results confirmed the capability of the fabricated tissue to replicate the disease pathology as well as the therapeutic impacts observed in the human body.

Berg et al. in 2018 presented the use of an extrusion-based printed lung model in investigating influenza A virus strain H3N2, the human immune response, and the effects of oseltamivir on stopping infection [145]. Two bioinks were developed: the first consisted of alginate, gelatin, Col-I, $CaSO_4$, and A549 cells, while the second contained alginate, gelatin, Col-I, $CaSO_4$, fibroblasts, THP-1 cells, and A549 cells. The bioprinted models released cytokines when exposed to bacterial toxins, demonstrating the capability of the artificial tissues to develop an immune response. Furthermore, when the lung constructs were infected with influenza A virus and treated with oseltamivir, the researchers observed that inhibition of viral infection was dependent upon the drug dosage. Although the findings confirm the effective application of bioprinted models in studying pathogenesis and developing new drugs, the use of primary cells remains a challenge. Precisely, type II alveolar epithelial cells are available in limited quantities in the human lungs and are difficult to preserve as they are highly susceptible to mechanical stress during the printing process.

6.3.10 Renal/Kidney

6.3.10.1 Printing Parameters Affecting the Viability of Printed Model

Currently, maintaining long-term viability in bioprinted kidney constructs remains a challenge as current printing technology is unable to fully replicate the complicated vascular network of the kidney. Several approaches addressing different printing parameters have been examined to extend the survivability rates of artificial kidney models. The parameters that have been investigated thus far include the bioink and printing technique being used.

Bioink

Sobreiro-Almeida et al. in 2021 developed decellularized kidney ECM (dKECM) as a bioink for the BioX, CELLINK printers to produce renal progenitor cells [146]. The dKECM-based bioink maintained the precise microenvironment that was critical to the growth and development of progenitor cells. The progenitor cells also expressed high viability, which highlighted the great potential of dKECM-based bioink in constructing an artificial kidney tissue. However, it remains unknown whether the bioprinted cells can perform optimally when implanted in an animal that has kidney disease.

Carreno-Galeano et al. in 2020 synthesized photocrosslinkable kidney ECM-derived bioink (KdECMMA) for use in a NovoGen MMX bioprinter [147]. Results from this research indicated that the KdECMMA-based bioink was able to replicate the renal microenvironment that could foster cell maturation. The engineered tissue displayed high viability with structures and functional elements that greatly resembled those of normal tissue. Moreover, in one to two months following implantation of the bioprinted constructs in rats with renal defects, maturation of tubular and glomerular-like structures was seen in the implanted region. Despite the promising results seen in rat implantation, it is still unclear whether this bioink will produce optimal constructs for use in human transplantation. Because humans' renal structures are much more complex than those of rats, there may be additional, undiscovered aspects of the human renal microenvironment when using KdECMMA-based bioink for printing.

Printing Techniques

Trondle et al. in 2021 combined bioprinting with cellular self-assembly to generate renal spheroids and nephron-like tubules [148]. Specifically, the researchers utilized Matrigel, Collagen I, and Fibrin (or hydrogel ECM) as a bioink for the DOD bioprinting technique. Although the researchers were able to produce complex tissues that had more minute features, current difficulties include the fleeting stability of the bioink itself and the inability of the hydrogel to adhere well to the chip interface. Altogether, these two limitations greatly impact the overall sturdiness of the engineered model.

Lawlor et al. in 2021 developed automated extrusion-based printing using NovoGen MMX printer to fabricate hPSC-derived kidney structures [17]. The researchers utilized cells alone, or cells combined with hydrogels as the bioink for the printing process. Because the printing process was automated, the production of the renal structure was increased ninefold compared to traditional culturing methods, with the resulting constructs meeting the standards for cell number, viability, and diameter. Compared to the manual organoids, the bioprinted models had higher survivability rates, with bigger and more complex structures. Most importantly, given the great consistency in the differentiation patterns of the bioprinted structures, automated extrusion-based printing exemplified the possibility of applying this printing technique to mass produce kidney models that could truly recapitulate the structure and functions of the native tissue.

6.3.10.2 Applications of 3D-Printed Model

Due to various challenges in the field of kidney bioengineering, the application of printed kidney constructs in drug screening is rather limited. However, in one recent study, Trondle et al. in 2022 utilized artificial spheroids to conduct an automated toxicity test with cisplatin [149]. In particular, the researchers used the DOD printing technique to fabricate the spheroids on Matrigel bioink. Subsequently, the researchers employed various microscopic pictures of the printed spheroids to train the algorithm to classify the images based on three categories of no, mild, or acute therapeutic effects. Although the deep learning system was able to label the images with high accuracy, more images of cell morphologies produced from different assays and biomarkers could be used to enhance the current dataset. When the dataset is reinforced with more images, improvement in performance level may be observed in the algorithm, which could result in greater reliability in automated image-based readout systems.

6.3.11 Composite Tissues

With the rapid advancements in single-tissue bioprinting, many researchers have been interested in bioprinting applications to recreate the complex anatomy and functionality of composite tissues. Composite tissues are ideal candidates for 3D bioprinting due to their multiple cell type composition, specific spatial patterns, microarchitecture, and complex vascularization. While the realm of bioprinting composite tissues highlights the exceptional abilities of 3D bioprinting compared to traditional printing approaches, there are still many technical challenges left to tackle, such as bioprinting the structure in a way that satisfies the metabolic requirements of each individual tissue and creating tissue complexes that maintain the vascular fidelity of their native biological counterparts. However, there are a few recent studies that have investigated simple tissue complexes (i.e., those that consist of two tissue types) that would later be used to synthesize composite tissue structures.

In comparison to nonbioprinting hydrogels, bioprinting hydrogels enable the replication of complex tissue architecture and tissue-specific microenvironments, such as in synthesizing cartilage-related complexes. Chen et al. in 2022 recently used extrusion-based bioprinting and SA-Col-I hybrid bioink to fabricate arterial-cartilage tissue mimicries [150]. The concentration of these two components impacted the filament diameter and merging; however, Yan et al. found that a higher concentration of their bioink solution led to better printing fidelity and less deformation [150]. There was also a lower shrinkage ratio after crosslinking. In another study, Chakraborty et al. in 2023 investigated different types of bioinks for fabrication of a cartilage–collagen network [151]. They used two types of bioinks, GelMa and silk fibroin-gelatin (SF-G), combined with human bone marrow MSCs (BM-MSCs) to recreate the articular cartilage constructs. The SF-G constructs inside of the human BM-MSCs demonstrated increased proliferation compared to the GelMA constructs. Further analyses, including biochemical analyses, gene and protein analyses, and expression assays revealed that SF-G played an important role in forming the fibrous collagen network and chondrogenesis. Chakraborty and collaborators thus investigated protein–protein interactions and demonstrated the important role of SF-G in regulating the Wnt/β-catenin and TGF-β signaling pathways. Both arterial-cartilage tissue and cartilage–collagen networks consist of a highly organized and precise spatial arrangement of cells that can only be faithfully replicated using bioprinting hydrogels.

A few recent studies have investigated bone-tissue fabrication, since there is a growing demand for bone graft substitutes that support bone stem cell proliferation. In one study, Liu et al. in 2023 used extrusion-based bioprinting to create a GelMA/nano-attapulgite (nano-ATP)-composite hydrogel combined with mouse bone mesenchymal stem cells (BMSCs) and mouse umbilical vein endothelial cells (MUVECs) [152]. Because GelMA has poor mechanical properties, they tested the potential of nano-ATP to improve bioink stability, bioactivity, and mechanical properties. They printed scaffolds layer-by-layer and tested different concentrations of nano-ATP (0, 1%, 2%, and 4% w/v) in GelMA (5%, 10%, 15%, and 20% w/v) to evaluate the printability of the nano-ATP/GElMA bioink. They found that the nano-ATP-containing scaffolds had greater mechanical strength and pore regularity in only the GelMA-containing scaffolds, thus reinforcing the versatile function of GelMA in tissue engineering, regeneration, and repair. In another study, Allen et al. in 2022 demonstrated the use of a bioink composed of GelMA, gelatin, hydroxyapatite (HA), and osteoblasts for bone tissue engineering [153]. They printed porous hydrogel scaffolds using bioinks with varying HA additive concentrations (0, 5, 10, and 20 mg/ml). They found that the addition of HA to the scaffolds significantly decreased swelling from baseline GElMA-gelatin hydrogels, decreased hydrogel degradation in a dose-dependent manner, supported osteogenic differentiation and mineralization, and increased osteogenic gene expression. However, the HA additive had no effect on cell proliferation. In another study, Gharacheh et al. in 2022 developed a novel composite bone-mimetic bioink composed of methacrylate alginate (MeALG) hydrogel filled with human bone allograft particles and combined with hMSCs [154]. Using an extrusion-based bioprinting technique, the hydrogel showed low shear viscosity, but enhanced shear thinning after the addition of the bone particles. After printing, various cell viability analyses of in vitro cultures showed up to 28 days of high cell viability (~90%) for hMSCs. Additionally, differentiation studies confirmed that there was high alkaline phosphatase activity, calcium deposition, and osteocalcin expression, which further confirms the applications of their composite bioink for scaffolds for bone tissue regeneration.

There have only been a few studies investigating bone vascularization. In one recent study, Yang et al. in 2022 used extrusion-based bioprinting to fabricate bone marrow-derived mesenchymal stem cells-laden scaffold using a bioink consisting of GelMA/methacrylated silk fibroin (SFMA) along with the release of the small drug fingolimod (FTY-720) to mimic the synergistic interactions of vascularization and osteogenesis during bone repair [155]. They tested different concentrations of GelMA and SFMA bioinks when fabricating the MSCs scaffolds and found that an increase in the GelMA and SFMA concentration reduced the pore diameter of all hydrogels, thus encouraging the diffusion of nutrients and oxygen and leading to better cell adhesion, proliferation, and differentiation.

Additionally, they found that the addition of SFMA to the GelMA bioink slowed biodegradation of the hydrogel, thus confirming that GelMA/SFMA may present an ideal environment for tissue regeneration. The hydrogel release of FTY-720 behavior was analyzed using UV–vis spectroscopy at 220 nm, which showed that the GelMA/SFMA could serve as a promising candidate for localized drug delivery.

Being able to compartmentalize different components and establish tissue-specific microenvironments through 3D bioprinting offers new research avenues for drug development and testing, personalized therapies, and models for understanding composite tissue-related disorders, such as Marfan syndrome or cleft lip and palate.

6.3.12 Other Tissues

In addition to the aforementioned tissue types, there have been a few recent studies looking at bioprinting urological and reproductive tissues. As there is currently a lack of models that can capture the human bladder's biological complexity and function, many researchers have turned to 3D bioprinting to help achieve this. Chae et al. in 2022 used bladder tissue-derived dECM bioink combined with

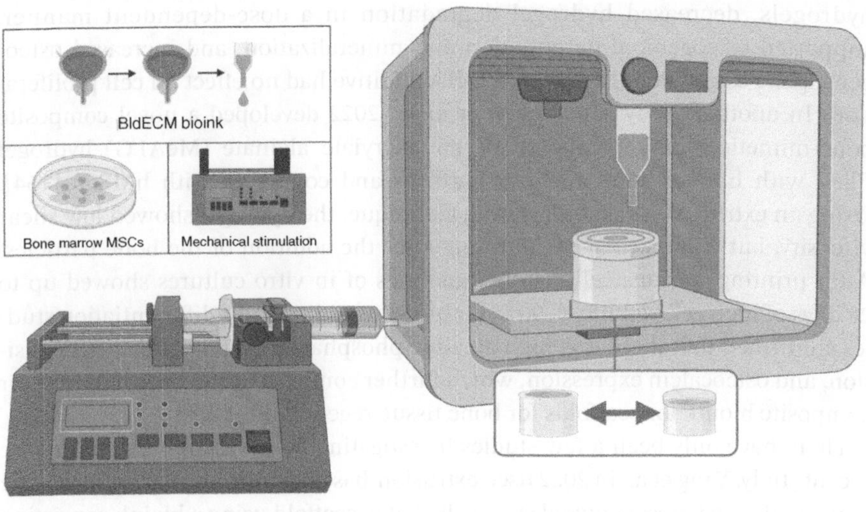

Figure 6.7 Development of an in vitro bladder model system. A bladder-specific dECM bioink was used to create an in vitro bladder model. The model included a contract-release system (CRS) that mirrored the smooth muscle functions of a native human bladder. In the future, this platform could be used for drug and/or treatment studies on bladder-related diseases.

human bone marrow-derived MSCs and a custom-made 3D printer to develop a human bladder model that could accurately replicate the bladder's functionality (Figure 6.7) [156].

Their model included a contract-release system (CRS) that mirrored the smooth muscle functions of a native human bladder and was thus presented as a proof-of-concept for the development of a bladder platform that could respond to mechanical and physiological stimuli.

Like the bladder, current models of the vagina lack the ability to effectively capture its functionality, complexity, and biological mechanisms. In addition, vaginal congenital diseases and injuries are very painful for patients and can significantly decrease their quality of life, so there is a great need for vaginal reconstruction options [157]. In one recent study, Hou et al. in 2021 used a bioink mixture of 15% gelatin, 3% SA, and a pig acellular vaginal matrix (AVM) combined with BM-MSCs [158]. They used an extrusion-based bioprinting technique to fabricate 3D vaginal scaffold tissues and then transplanted them subcutaneously into rats. Results from various staining techniques revealed that the fabricated vaginal tissue had good epithelization, vascularization, and function, since the BM-MSCs were able to successfully acquire the phenotypic characteristics of native vaginal epithelial cells, which was in line with past studies [159, 160]. Though this work presents a promising and more natural approach for vaginal reconstruction, there are a few limitations to this study as naturally derived bioinks, such as one made from pig AVM, may not be as mechanically stable as synthetic materials. Nevertheless, they possess high biocompatibility and accurate tissue-mimicking properties, so future work in this field should focus on testing and strengthening the varying mechanical abilities of naturally derived bioinks.

While research within these two fields is still very limited, they serve as good foundational platforms for future research studies not only on drug development and disease modeling but also on less commonly printed tissue types, such as adipose, lymphatic, esophageal, and thyroid tissue.

6.4 Bioprinting in Tissue: Challenges, Barriers to Clinical Translation, and Future Directions

6.4.1 Introduction

Since the mid-1950s, organ transplantation has been introduced as a field at the intersection of surgery, immunology, and regenerative medicine [161]. Organ transplantation enables patients to increase life expectancy and enhance their quality of life [162].

6.4.1.1 Current Challenges in Organ Transplantation

Modern challenges in organ transplantation hinder the number of patients that can benefit from it. The scarce availability of donor organs remains a predominant obstacle in face of the increasing populational demand. Additionally, given that preservation of organs is limited, organs cannot stay viable for a long period of time, which impedes access to transplantation. Due to different immunological profiles between the donor and the receiver, organ rejection represents a medical and social pressing issue. Immunosuppressive drugs help to counteract the host's immune system. However, numerous complications emerge as side effects from these drugs. The entire organ transplantation process relies upon a high-cost system ranging from organ harvesting, storage, and preparation to host transplantation and post-treatments needed.

6.4.1.2 Potential of Bioprinted Organs for Transplantation

Considering these obstacles, the field of regenerative medicine is introducing new sources of tissues and organs through bioengineering. A promising avenue is bioprinting, which aims to create viable tissues and organs made through assisted printing techniques. The latter use cells, synthetic and biocomponents to form a bioink that can be deposited in a determined sequence to form complex biological structures.

Bioprinting presents unique advantages to overcome the current challenges faced with traditional organ transplantation. The cells used in bioprinting can be of any source, including the patient's own cells. A potential scenario would be to harvest the own's patient cells, which would be multiplied in laboratory, and used to create specific tissue or organs tailored to the patient. The advantage of using one's own cells is to potentially bypass histocompatibility issues between donors and receivers. Thus, shortage of donor organs due to histocompatibility and rejection after organ transplantation could potentially be avoided.

6.4.1.3 Challenges and Limitations in Bioprinting Tissues and Organs

The field of bioprinting remains difficult to develop given multiple obstacles related to its feasibility, access, regulation, and development. The complexity of organs is particularly challenging to replicate. Functional organs contain an ecosystem of supporting cells, vascular and neural networks, which are extremely complicated to recreate in terms of functions and viability [163]. Bioprinting small vessels is difficult given the limited printer resolution [119]. As demonstrated in the previous section on "Bioprinting in Tissue Engineering, Regenerative Medicine, and Organ Transplantation," few studies successfully printed organs and when it occurred, organs were proof of concepts with very limited functions. However, potential bioprinting applications can already be foreseen in simple tissue constructs. Skin studies have successfully printed simple two-layer tissues that have direct applications in assessing drug pharmacokinetic characteristics and profiles.

6.4.2 Insight on Barriers to Clinical Translation of Bioprinting Technology

Technical limitations: A central issue to bioprinting is to recreate complex tissue structures and to emulate tissue functions. This is largely due to the limitations of the current equipment, including the bioprinter itself and the employed techniques, that can compromise cell viability and difficulty to reproduce the proper mechanical environment for tissue function. The field is currently researching novel and optimized bioink formulations, biocomponents and biocompatible synthetic polymers to preserve and replicate tissue functioning. Organ bioprinting is still very far from reaching clinical bench side.

Lack of regulatory and production guidelines: The development of bioprinting remains a challenge due to a lack of product and standardized techniques offered by manufacturers. This impedes on research productivity and development. As the field is relatively new, important ethical considerations are raised and currently unaddressed by different stakeholders worldwide. Establishing regulations and limitations remains crucial in effectively and safely guiding the future of this growing field. Ethical issues are currently country-dependent. Most countries do not have specific regulations or policies for bioprinting, which makes it difficult for researchers to understand the scope and limits of their projects. Given the high costs of bioprinting, developing this field is difficult for laboratories. There is limited funding by the industry due to multifactorial causes like the uncertainty related to regulations and the high costs related to the entire process.

High cost: Researching this field remains currently difficult due to the cost of materials and equipment required [164]. Additionally, bioprinting requires cell cultivation prior to in vivo grafting. Therefore, the entire process from printing to host grafting requires expensive human and material resources. The lack of standardized high-throughput manufacturing processes contributes to keeping high in-house production costs. Bioprinting could be more expensive than expected. Given that every patient is unique, the variation in tissues and organs can require printing personalization, which makes it patient-dependent [165]. There is also variation in batches of cells used. Certain cell batches have a higher differential potential than others, which creates discrepancies. Therefore, regulations or verification protocols would be required in the future for mass production [166].

Limited availability of donor tissues: Cell sourcing is a major limitation given the limited number of cells that can be harvested from patients and reintroduced through bioprinting. Additionally, given limited regulations on tissue extraction and storage, it remains unclear what are the limitations and processes that will be legally allowed in a specific country. Optimized pathways to cell sourcing need to be assessed to ensure easy access to cells with an acceptable quantity. Otherwise, it is quite inconvenient for patients to regularly donate cells. Preserving cell viability through the different bioprinting techniques remains another technical challenge that currently hinders research studies. Cells are not able to survive beyond months, which makes it impossible to currently treat patients.

6.4.3 Future Directions

Current research aims to optimize combinations of biocompatible materials that support cell growth and function. Novel bioink formulations are developed and assessed to increase tissue viability. Hydrogel-based bioinks are the most promising formulations as they initially have a high biocompatible profile. However, when used alone they can be limited in their properties. However, combining and merging them with synthetic materials has been proven to be the most successful. New physical properties are obtained, which enable cell nutrition and protection [167].

Bioprinting in tissue and organ transplantation remains limited, near non-existent, given the high complexity of replicating viable organ structures. However, viable simple structured tissue can successfully be printed and thus represents a novel avenue in the near future for drug testing and wound regeneration. Simple layered tissues can be used in laboratory to further test tissue behaviors across pharmacological studies.

Studies currently remain at the preclinical phase with in vitro and in vivo testing in rat and mouse models. Eminent challenges are to increase tissue viability in the span of days to months and to better assess the long-term impact of tissue grafting on animal models. While bioprinting is still at its beginning, its promising applications in tissue regeneration and organ transplantation could offer advantageous patient outcomes and facilitate access to care of patients in need of tissues and organs. Further research is needed to assess the long-term viability and safety of simple structured tissues in animals and humans.

6.5 Conclusions

As we conclude this chapter, it is evident that bioprinting has begun to reshape the medical landscape, providing substantial contributions in areas such as tissue engineering, organ transplantation, and drug testing. The promise of bioprinting lies in its potential to address some of the most formidable medical challenges, leading to potentially transformative changes in healthcare delivery.

However, the field of bioprinting is still in its infancy, grappling with numerous technical, ethical, and regulatory obstacles that need resolution. Despite these challenges, continuous scientific breakthroughs indicate steady progress toward overcoming these hurdles, thereby paving the way toward the clinical application of bioprinting.

Looking toward the future, bioprinting holds extensive possibilities in medical applications. It paves a pathway toward an era of personalized medicine, where bioprinted tissues and organs could become commonplace in clinical practice. Although the journey to this future entails challenges, the transformative potential of bioprinting suggests that every stride taken is worthwhile. As we continue to navigate this emerging field, we can anticipate a wave of scientific breakthroughs, positioning bioprinting as a foundational pillar in the next generation of medical advancements.

Acknowledgments

Danielle Brister and Doanh K. Truong – contributing to section 3; Ashley Jong and Nisa Notta – contributing to figures; Gerard K. Kim – referencing and formatting.

References

1 Berdigaliyev, N. and Aljofan, M. (2020). An overview of drug discovery and development. *Future Med. Chem.* 12 (10): 939–947.

2 Satpathy, A. (2018). Developments with 3D bioprinting for novel drug discovery. *Expert Opin. Drug Discov.* 13 (12): 1115–1129.

3 Mullard, A. (2016). Parsing clinical success rates. *Nat. Rev. Drug Discov.* 15 (7): 447–447.

4 Fontoura, J.C., Viezzer, C., dos Santos, F.G. et al. (2020). Comparison of 2D and 3D cell culture models for cell growth, gene expression and drug resistance. *Mater. Sci. Eng. C Mater. Biol. Appl.* 107: 110264–110264.

5 Skardal, A., Murphy, S.V., Devarasetty, M. et al. (2017). Multi-tissue interactions in an integrated three-tissue organ-on-a-chip platform. *Sci. Rep.* 7 (1): 8837–8816.

6 Kapałczyńska, M., Kolenda, T., Przybyła, W. et al. (2018). 2D and 3D cell cultures – a comparison of different types of cancer cell cultures. *Arch. Med. Sci.* 14 (4): 910–919.

7 Peng, W., Datta, P., Ayan, B. et al. (2017). 3D bioprinting for drug discovery and development in pharmaceutics. *Acta Biomater.* 57: 26–46.

8 Van Norman, G.A. (2019). Limitations of animal studies for predicting toxicity in clinical trials. *JACC Basic Transl. Sci.* 4 (7): 845–854.

9 Ren, Y., Yang, X., Ma, Z. et al. (2021). Developments and opportunities for 3D bioprinted organoids. *Int. J. Bioprinting* 7 (3): 18–36.

10 Badr-Eldin, S.M., Aldawsari, H.M., Kotta, S. et al. (2022). Three-dimensional in vitro cell culture models for efficient drug discovery: progress so far and future prospects. *Pharmaceuticals* 15 (8): 926.

11 Gao, G., Ahn, M., Cho, W.W. et al. (2021). 3D printing of pharmaceutical application: drug screening and drug delivery. *Pharmaceutics* 13 (9): 1373.

12 Zhao, Z., Chen, X., Dowbaj, A.M. et al. (2022). Organoids. *Nat. Rev. Methods Primers* 2 (1): 1–21.

13 Cai, Y., Zhang, J., Wu, J. et al. (2015). Oxygen transport in a three-dimensional microvascular network incorporated with early tumour growth and preexisting vessel cooption: numerical simulation study. *Biomed. Res. Int.* 2015: 476964–476910.

14 Grebenyuk, S. and Ranga, A. (2019). Engineering organoid vascularization. *Front. Bioeng. Biotechnol.* 7: 39–39.

15 Rawal, P., Tripathi, D.M., Ramakrishna, S., and Kaur, S. (2021). Prospects for 3D bioprinting of organoids. *Bio-Des. Manuf.* 4 (3): 627–640.

16 Szűcs, D., Fekete, Z., Guba, M. et al. (2023). Toward better drug development: three-dimensional bioprinting in toxicological research. *Int. J. Bioprinting* 9 (2): 663–663.

17 Lawlor, K.T., Vanslambrouck, J.M., Higgins, J.W. et al. (2021). Cellular extrusion bioprinting improves kidney organoid reproducibility and conformation. *Nat. Mater.* 20 (2): 260–271.

18 Rodboon, T., Yodmuang, S., Chaisuparat, R., and Ferreira, J.N. (2022). Development of high-throughput lacrimal gland organoid platforms for drug discovery in dry eye disease. *SLAS Discovery* 27 (3): 151–158.

19 Zhang, B., Korolj, A., Lai, B.F.L., and Radisic, M. (2018). Advances in organ-on-a-chip engineering. *Nat. Rev. Mater.* 3 (8): 257–278.

20 Chliara, M.A., Elezoglou, S., and Zergioti, I. (2022). Bioprinting on organ-on-chip: development and applications. *Biosensors (Basel)* 12 (12): 1135.

21 Rahmani Dabbagh, S., Rezapour Sarabi, M., Birtek, M.T. et al. (2023). 3D bioprinted organ-on-chips. *Aggregate (Hoboken)* 4 (1): e197.

22 Miri, A.K., Mostafavi, E., Khorsandi, D. et al. (2019). Bioprinters for organs-on-chips. *Biofabrication* 11 (4): 042002–042002.

23 Yamaguchi, S., Ueno, A., Akiyama, Y., and Morishima, K. (2012). Cell patterning through inkjet printing of one cell per droplet. *Biofabrication* 4 (4): 045005–045005.

24 Knowlton, S. and Tasoglu, S. (2016). A bioprinted liver-on-a-chip for drug screening applications. *Trends Biosci. (Regular ed)* 34 (9): 681–682.

25 Faulkner-Jones, A., Zamora, V., Hortigon-Vinagre, M.P. et al. (2022). A bioprinted heart-on-a-chip with human pluripotent stem cell-derived cardiomyocytes for drug evaluation. *Bioengineering (Basel)* 9 (1): 32.

26 Ferrari, E. and Rasponi, M. (2021). Liver–Heart on chip models for drug safety. *APL Bioeng.* 5 (3): 31505–031505.

27 Ferrari, E., Visone, R., Monti, E. et al. (2023). LivHeart: a multi organ-on-chip platform to study off-target cardiotoxicity of drugs upon liver metabolism. *Adv. Mater. Technol.* 8 (8): 2201435.

28 Mattiuzzi, C. and Lippi, G. (2019). Current cancer epidemiology. *J. Epidemiol. Glob. Health* 9 (4): 217–222.

29 Loud, J.T. (2017). Cancer screening and early detection in the 21st century. *Semin. Oncol. Nurs.* 33 (2): 121–128.

30 Germain, N., Dhayer, M., Dekiouk, S., and Marchetti, P. (2022). Current advances in 3D bioprinting for cancer modeling and personalized medicine. *Int. J. Mol. Sci.* 23 (7): 3432.

31 Pedersen, J.K., Engholm, G., Skytthe, A. et al. (2016). Cancer and aging: Epidemiology and methodological challenges. *Acta Oncol.* 55 (Suppl 1): 7–12. https://doi.org/10.3109/0284186X.2015.1114670. Epub 2016 Jan 12. PMID: 26825001; PMCID: PMC4957549.

32 Neufeld, L. (2022). 3D bioprinted cancer models: from basic biology to drug development. *Nat. Rev. Cancer* 22 (12): 679–692.

33 Datta, P. (2020). 3D bioprinting for reconstituting the cancer microenvironment. *npj Precis. Oncol.* 4 (1): 18–18.

34 Walsh, J.C., Lebedev, A., Aten, E. et al. (2014). The clinical importance of assessing tumor hypoxia: relationship of tumor hypoxia to prognosis and therapeutic opportunities. *Antioxid. Redox Signal.* 21 (10): 1516–1554.

35 Augustine, R., Kalva, S.N., Ahmad, R. et al. (2021). 3D Bioprinted cancer models: revolutionizing personalized cancer therapy. *Transl. Oncol.* 14 (4): 101015.

36 Li, Y., Zhao, L., and Li, X.F. (2021). Hypoxia and the tumor microenvironment. *Technol. Cancer Res. Treat.* 20: 15330338211036304–15330338211036304.

37 Meng, F., Meyer, C.M., Joung, D. et al. (2019). 3D bioprinted in vitro metastatic models via reconstruction of tumor microenvironments. *Adv. Mater. (Weinheim)* 31 (10): e1806899.

38 Neufeld, L. (2021). Microengineered perfusable 3D-bioprinted glioblastoma model for in vivo mimicry of tumor microenvironment. *Sci. Adv.* 7 (34).

39 McCormack, A., Highley, C.B., Leslie, N.R., and Melchels, F.P. (2020). 3D printing in suspension baths: keeping the promises of bioprinting afloat. *Trends Biotechnol. (Regular ed)* 38 (6): 584–593.

40 Liu, S., Wang, T., Li, S., and Wang, X. (2022). Application status of sacrificial biomaterials in 3D bioprinting. *Polymers* 14 (11): 2182.

41 Li, S., Jin, J., Zhang, C. et al. (2023). 3D bioprinting vascular networks in suspension baths. *Appl. Mater. Today* 30: 101729.

42 Fares, J. (2020). Molecular principles of metastasis: a hallmark of cancer revisited. *Signal Transduct. Target. Ther.* 5 (1): 28.

43 Guy, J.B., Espenel, S., Vallard, A. et al. (2017). Evaluation of the cell invasion and migration process: a comparison of the video microscope-based scratch wound assay and the boyden chamber assay. *J. Vis. Exp.* 2017 (129): e56337.

44 Jung, M., Skhinas, J.N., Du, E.Y. et al. (2022). A high-throughput 3D bioprinted cancer cell migration and invasion model with versatile and broad biological applicability. *Biomater. Sci.* 1 (2): 5876–5887.

45 Ruiz-Garcia, H., Alvarado-Estrada, K., Schiapparelli, P. et al. (2020). Engineering three-dimensional tumor models to study glioma cancer stem cells and tumor microenvironment. *Front. Cell. Neurosci.* 14: 558381–558381.

46 Moharil, R. (2017). Cancer stem cells: an insight. *J. Oral Maxillofac. Pathol.: JOMFP* 21 (3): 463–463.

47 Liu, Y., Wu, W., Cai, C. et al. (2023). Patient-derived xenograft models in cancer therapy: technologies and applications. *Signal Transduct. Target. Ther.* 8 (1): 160–160.

48 Fang, L. (2022). Bioprinting and its use in tumor-on-a-chip technology for cancer drug screening: a review. *Int. J. Bioprinting* 8 (4): 603–603.

49 Yi, H.G., Jeong, Y.H., Kim, Y. et al. (2019). A bioprinted human-glioblastoma-on-a-chip for the identification of patient-specific responses to chemoradiotherapy. *Nat. Biomed. Eng.* 3 (7): 509–519.

50 Gonzalez, H., Hagerling, C., and Werb, Z. (2018). Roles of the immune system in cancer: from tumor initiation to metastatic progression. *Genes Dev.* 32 (19–20): 1267–1284.

51 Ghiringhelli, F. and Apetoh, L. (2014). The interplay between the immune system and chemotherapy: emerging methods for optimizing therapy. *Expert Rev. Clin. Immunol.* 10 (1): 19–30.

52 Naran, K. (2018). Principles of immunotherapy: implications for treatment strategies in cancer and infectious diseases. *Front. Microbiol.* 21 (9): 3158–3158.

53 Esfahani, K. (2020). A review of cancer immunotherapy: from the past, to the present, to the future. *Curr. Oncol. (Toronto)* 27 (2): 87–S97.

54 Lee, V.C. (2017). Cancer immunotherapy, part 2: efficacy, safety, and other clinical considerations. *P&T (Lawrenceville, NJ)* 42 (7): 452–463.

55 Bai, R. (2020). Mechanisms of cancer resistance to immunotherapy. *Front. Oncol.* 6 (10): 1290–1290.

56 Kim, J.H., Lee, S., Kang, S.J. et al. (2021). Establishment of three-dimensional bioprinted bladder cancer-on-a-chip with a microfluidic system using bacillus calmette–guérin. *Int. J. Mol. Sci.* 22 (16): 8887.

57 Dey, M., Kim, M.H., Dogan, M. et al. (2022). Chemotherapeutics and CAR-T cell-based immunotherapeutics screening on a 3D bioprinted vascularized breast tumor model. *Adv. Funct. Mater.* 32 (52): 2203966.

58 Grunewald, L. (2021). A reproducible bioprinted 3D tumor model serves as a preselection tool for CAR T cell therapy optimization. *Front. Immunol.* 29 (12): 689697–689697.

59 Morley, C.D., Flores, C.T., Drake, J.A. et al. (2022). Spatiotemporal T cell dynamics in a 3D bioprinted immunotherapy model. *Bioprinting* 28: e00231.

60 Kim, J., Kong, J.S., Kim, H. et al. (2022). A bioprinted bruch's membrane for modeling smoke-induced retinal pigment epithelium degeneration via hybrid membrane printing technology. *Adv. Healthc. Mater.* 11 (24): e2200728.

61 Jung, S.S., Son, J., Yi, S.J. et al. (2023). Development of Müller cell-based 3D biomimetic model using bioprinting technology. *Biomed. Mater. (Bristol)* 18 (1): 15009.

62 Samanipour, R., Tahmooressi, H., Rezaei Nejad, H. et al. (2022). A review on 3D printing functional brain model. *Biomicrofluidics* 3 (16): 011501.

63 Tidke, S.C. and Tidake, P. (2022). A review of corneal blindness: causes and management. *Curēus (Palo Alto, CA)* 14 (10): e30097–e30097.

64 Mörö, A., Samanta, S., Honkamäki, L. et al. (2022). Hyaluronic acid based next generation bioink for 3D bioprinting of human stem cell derived corneal stromal model with innervation. *Biofabrication* 13: 15.

65 Boix-Lemonche, G. (2023). Intracorneal implantation of 3D bioprinted scaffolds containing mesenchymal stromal cells using femtosecond-laser-assisted intrastromal keratoplasty. *Macromol. Biosci.* 26: e2200422–e2200422.

66 Ruiz-Alonso, S. (2021). Tendon tissue engineering: cells, growth factors, scaffolds and production techniques. *J. Control. Release* 10 (333): 448–486.

67 Kostenko, A., Swioklo, S., and Connon, C.J. (2022). Effect of calcium sulphate pre-crosslinking on rheological parameters of alginate based bio-inks and on human corneal stromal fibroblast survival in 3D bio-printed constructs. *Front. Mech. Eng.* 8: 867685.

68 Li, J. (2023). Development and systematic characterization of GelMA/alginate/PEGDMA/xanthan gum hydrogel bioink system for extrusion bioprinting. *Biomaterials* 293: 121969–121969.

69 He, B., Wang, J., Xie, M. et al. (2022). 3D printed biomimetic epithelium/stroma bilayer hydrogel implant for corneal regeneration. *Biomed. Mater.* 17: 234–247.

70 Jia, S., Yang, J., Lau, A.D.S. et al. (2023). Digital light processing-bioprinted poly-NAGA-GelMA-based hydrogel lenticule for precise refractive errors correction. *Biofabrication* 15 (3): 035011.

71 Zhong, Z. (2021). Bioprinting of dual ECM scaffolds encapsulating limbal stem/progenitor cells in active and quiescent statuses. *Biofabrication* 13 (4): 44101.

72 Song, Y., Hua, S., Sayyar, S. et al. (2022). Corneal bioprinting using a high concentration pure collagen I transparent bioink. *Bioprinting* 28: e00235.

73 Song, S., Liu, X., Huang, J., and Zhang, Z. (2022). Neural stem cell-laden 3D bioprinting of polyphenol-doped electroconductive hydrogel scaffolds for enhanced neuronal differentiation. *Biotechnol. Adv.* 133: 112639–112639.

74 Wang, Y.L., Han, L., Zhang, X.L. et al. (2022). 3D bioprinting of an electroactive and self-healing polysaccharide hydrogels. *J. Tissue Eng. Regen. Med.* 16 (1): 76–85.

75 Gao, C., Li, Y., Liu, X. et al. (2023). 3D bioprinted conductive spinal cord biomimetic scaffolds for promoting neuronal differentiation of neural stem cells and repairing of spinal cord injury. *Chem. Eng. J.* 1 (451): 138788.

76 Cheng, K.C., Sun, Y.M., and Hsu, S.H. (2023). Development of double network polyurethane-chitosan composite bioinks for soft neural tissue engineering. *J. Mater. Chem. B* 11 (16): 3592–3606.

77 Cruz, E.M., Machado, L.S., Zamproni, L.N. et al. (2023). A gelatin methacrylate-based hydrogel as a potential bioink for 3D bioprinting and neuronal differentiation. *Pharmaceutics* 15 (2): 627.

78 Peng, B.Y., Ou, K.L., Liu, C.M. et al. (2022). A three-dimensional bioprinted copolymer scaffold with biocompatibility and structural integrity for potential tissue regeneration applications. *Polymers (Basel)* 14 (16): 3415.

79 Liu, X., Hao, M., Chen, Z. et al. (2021). 3D bioprinted neural tissue constructs for spinal cord injury repair. *Biomaterials* 272: 120771.

80 Min, K., Kong, J.S., Kim, J. et al. (2022). Three-dimensional microfilament printing of a decellularized extracellular matrix (dECM) bioink using a microgel printing bath for nerve graft fabrication and the effectiveness of dECM graft combined with a polycaprolactone conduit. *ACS Appl. Bio. Mater.* 5 (4): 1591–1603.

81 Ng, W.L. (2018). Proof-of-concept: 3D bioprinting of pigmented human skin constructs. *Biofabrication* 10 (2): 025005.

82 Min, D., Lee, W., Bae, I.H. et al. (2018). Bioprinting of biomimetic skin containing melanocytes. *Exp. Dermatol.* 27 (5): 453–459.

83 Derr, K., Zou, J., Luo, K. et al. (2019). Fully three-dimensional bioprinted skin equivalent constructs with validated morphology and barrier function. *Tissue Eng. Part C Methods* 25 (6): 334–343.

84 Yim, W., Zhou, J., Sasi, L. et al. (2023). 3D-bioprinted phantom with human skin phototypes for biomedical optics. *Adv. Mater.* 35 (3): e2206385.

85 Masri, S., Fauzi, F.A.M., Hasnizam, S.B. et al. (2022). Engineered-skin of single dermal layer containing printed hybrid gelatin-polyvinyl alcohol bioink via 3D-bioprinting: in vitro assessment under submerged vs. air-lifting models. *Pharmaceuticals (Basel)* 15 (11): 1328.

86 Jorgensen, A.M., Chou, Z., Gillispie, G. et al. (2020). Decellularized skin extracellular matrix (dsECM) improves the physical and biological properties of fibrinogen hydrogel for skin bioprinting applications. *Nanomaterials (Basel)* 10 (8): 1484.

87 Lameirinhas, N.S., Teixeira, M.C., Carvalho, J.P.F. et al. (2023). Nanofibrillated cellulose/gellan gum hydrogel-based bioinks for 3D bioprinting of skin cells. *Int. J. Biol. Macromol.* 28 (229): 849–860.

88 Hafezi, F., Shorter, S., Tabriz, A.G. et al. (2020). Bioprinting and preliminary testing of highly reproducible novel bioink for potential skin regeneration. *Pharmaceutics* 12 (6): 550.

89 Pereira, R.F., Sousa, A., Barrias, C.C. et al. (2018). A single-component hydrogel bioink for bioprinting of bioengineered 3D constructs for dermal tissue engineering. *Mater. Horiz.* 5 (6): 1100–1111.

90 Liu, P., Shen, H., Zhi, Y. et al. (2019). 3D bioprinting and in vitro study of bilayered membranous construct with human cells-laden alginate/gelatin composite hydrogels. *Colloids Surf. B. Biointerfaces* 1 (181): 1026–1034.

91 Bera, A.K., Sriya, Y., and Pati, F. (2022). Formulation of dermal tissue matrix bioink by a facile decellularization method and process optimization for 3D bioprinting toward translation research. *Macromol. Biosci.* 22 (8): e2200109.

92 Li, M., Sun, L., Liu, Z. et al. (2023). 3D bioprinting of heterogeneous tissue-engineered skin containing human dermal fibroblasts and keratinocytes. *Biomater. Sci.* 11 (7): 2461–2477.

93 Yang, Y., Xu, R., Wang, C. et al. (2022). Recombinant human collagen-based bioinks for the 3D bioprinting of full-thickness human skin equivalent. *Int. J. Bioprinting* 8 (4): 611.

94 Huang, J., Lei, X., Huang, Z. et al. (2022). Bioprinted gelatin-recombinant type III collagen hydrogel promotes wound healing. *Int. J. Bioprinting* 8 (2): 517.

95 Miguel, S.P., Cabral, C.S.D., Moreira, A.F., and Correia, I.J. (2019). Production and characterization of a novel asymmetric 3D printed construct aimed for skin tissue regeneration. *Colloids Surf. B Biointerfaces* 1 (181): 994–1003.

96 Wang, S., Xiong, Y., Chen, J. et al. (2019). Three dimensional printing bilayer membrane scaffold promotes wound healing. *Front. Bioeng. Biotechnol.*

19 (7): 348. https://doi.org/10.3389/fbioe.2019.00348. PMID: 31803738; PMCID: PMC6877717.

97 Ma, Y., Wang, Y., Chen, D. et al. (2023). 3D bioprinting of a gradient stiffened gelatin-alginate hydrogel with adipose-derived stem cells for full-thickness skin regeneration. *J. Mater. Chem. B* 11 (13): 2989–3000.

98 Shi, L., Hu, Y., Ullah, M.W. et al. (2019). Cryogenic free-form extrusion bioprinting of decellularized small intestinal submucosa for potential applications in skin tissue engineering. *Biofabrication* 11 (3): 035023.

99 Schmitt, T., Katz, N., and Kishore, V. (2021). A feasibility study on 3D bioprinting of microfat constructs towards wound healing applications. *Front. Bioeng. Biotechnol.* 9: 707098.

100 Kim, B.S., Kwon, Y.W., Kong, J.S. et al. (2018). 3D cell printing of in vitro stabilized skin model and in vivo pre-vascularized skin patch using tissue-specific extracellular matrix bioink: a step towards advanced skin tissue engineering. *Biomaterials* 168: 38–53.

101 Baltazar, T., Merola, J., Catarino, C. et al. (2020). Three dimensional bioprinting of a vascularized and perfusable skin graft using human keratinocytes, fibroblasts, pericytes, and endothelial cells. *Tissue Eng. Part A* 26 (5–6): 227–238.

102 Baltazar, T., Jiang, B., Moncayo, A. et al. (2023). 3D bioprinting of an implantable xeno-free vascularized human skin graft. *Bioeng. Transl. Med.* 8 (1): e10324.

103 Meyer, A.A., Manktelow, A., Johnson, M. et al. (1988). Antibody response to xenogeneic proteins in burned patients receiving cultured keratinocyte grafts. *J. Trauma* 28 (7): 1054–1059.

104 Albanna, M., Binder, K.W., Murphy, S. et al. (2019). In situ bioprinting of autologous skin cells accelerates wound healing of extensive excisional full-thickness wounds. *Sci. Rep.* 9 (1): 1856–1856.

105 Zou, Q., Tian, X., Luo, S. et al. (2021). Agarose composite hydrogel and PVA sacrificial materials for bioprinting large-scale, personalized face-like with nutrient networks. *Carbohydr. Polym.* 269: 118222–118222.

106 Wu, Y., Ayan, B., Moncal, K.K. et al. (2020). Hybrid bioprinting of zonally stratified human articular cartilage using scaffold-free tissue strands as building blocks. *Adv. Healthcare Mater.* 9 (22): e2001657.

107 Daly, A.C. and Kelly, D.J. (2019). Biofabrication of spatially organised tissues by directing the growth of cellular spheroids within 3D printed polymeric microchambers. *Biomaterials* 1 (197): 194–206.

108 Lam, T., Dehne, T., Krüger, J.P. et al. (2019). Photopolymerizable gelatin and hyaluronic acid for stereolithographic 3D bioprinting of tissue-engineered cartilage. *J. Biomed. Mater. Res. B Appl. Biomater.* 107 (8): 2649–2657.

109 Hauptstein, J., Böck, T., Bartolf-Kopp, M. et al. (2020). Hyaluronic acid-based bioink composition enabling 3D bioprinting and improving quality of deposited cartilaginous extracellular matrix. *Adv. Healthcare Mater.* 9 (15): e2000737.

110 Galarraga, J.H., Kwon, M.Y., and Burdick, J.A. (2019). 3D bioprinting via an in situ crosslinking technique towards engineering cartilage tissue. *Sci. Rep.* 9: 19987. https://www.ncbi.nlm.nih.gov/pmc/articles/PMC6934815/.

111 Sun, Y., You, Y., Jiang, W. et al. (2020). 3D bioprinting dual-factor releasing and gradient-structured constructs ready to implant for anisotropic cartilage regeneration. *Sci. Adv.* 6 (37): eaay1422.

112 Critchley, S., Sheehy, E.J., Cunniffe, G. et al. (2020). 3D printing of fibre-reinforced cartilaginous templates for the regeneration of osteochondral defects. *Acta Biomater.* 1 (113): 130–143.

113 Yang, Y., Zhang, Q., Xu, T. et al. (2020). Photocrosslinkable nanocomposite ink for printing strong, biodegradable and bioactive bone graft. *Biomaterials* 1(263): 120378.

114 Idaszek, J., Costantini, M., Karlsen, T.A. et al. (2019). 3D bioprinting of hydrogel constructs with cell and material gradients for the regeneration of full-thickness chondral defect using a microfluidic printing head. *Biofabrication* 11 (4): 044101.

115 Gao, F., Xu, Z., Liang, Q. et al. (2019). Osteochondral regeneration with 3D-printed biodegradable high-strength supramolecular polymer reinforced-gelatin hydrogel scaffolds. *Adv. Sci. (Weinh)* 6 (15): 1900867.

116 Gold, K.A., Saha, B., Rajeeva Pandian, N.K. et al. (2021). 3D bioprinted multicellular vascular models. *Adv. Healthcare Mater.* 10 (21): e2101141.

117 Abudupataer, M., Chen, N., Yan, S. et al. (2019). Bioprinting a 3D vascular construct for engineering a vessel-on-a-chip. *Biomed. Microdevices* 22 (1): 10.

118 Freeman, S., Ramos, R., Alexis Chando, P. et al. (2019). A bioink blend for rotary 3D bioprinting tissue engineered small-diameter vascular constructs. *Acta Biomater.* 1 (95): 152–164.

119 Lee, A., Hudson, A.R., Shiwarski, D.J. et al. (2019). 3D bioprinting of collagen to rebuild components of the human heart. *Science* 365 (6452): 482–487.

120 Mirdamadi, E., Tashman, J.W., Shiwarski, D.J. et al. (2020). FRESH 3D bioprinting a full-size model of the human heart. *ACS Biomater. Sci. Eng.* 6 (11): 6453–6459.

121 Tracy, E.P., Gettler, B.C., Zakhari, J.S. et al. (2020). 3D bioprinting the cardiac purkinje system using human adipogenic mesenchymal stem cell derived purkinje cells. *Cardiovasc. Eng. Technol.* 11 (5): 587–604.

122 Shin, Y.J., Shafranek, R.T., Tsui, J.H. et al. (2021). 3D bioprinting of mechanically tuned bioinks derived from cardiac decellularized extracellular matrix. *Acta Biomater.* 1 (119): 75–88.

123 Koti, P., Muselimyan, N., Mirdamadi, E. et al. (2019). Use of GelMA for 3D printing of cardiac myocytes and fibroblasts. *J. 3D Print. Med.* 3 (1): 11–22.

124 Basara, G., Ozcebe, S.G., Ellis, B.W., and Zorlutuna, P. (2021). Tunable human myocardium derived decellularized extracellular matrix for 3D bioprinting and cardiac tissue engineering. *Gels* 7 (2): 70.

125 Liu, J., He, J., Liu, J. et al. (2019). Rapid 3D bioprinting of in vitro cardiac tissue models using human embryonic stem cell-derived cardiomyocytes. *Bioprinting* 13: e00040.

126 Bejleri, D., Streeter, B.W., Nachlas, A.L.Y. et al. (2018). A bioprinted cardiac patch composed of cardiac-specific extracellular matrix and progenitor cells for heart repair. *Adv. Healthcare Mater.* 7 (23): 1800672.

127 Maiullari, F., Costantini, M., Milan, M. et al. (2018). A multi-cellular 3D bioprinting approach for vascularized heart tissue engineering based on HUVECs and iPSC-derived cardiomyocytes. *Sci. Rep.* 8 (1): 13532.

128 Wang, D., Guo, Y., Zhu, J. et al. (2023). Hyaluronic acid methacrylate/pancreatic extracellular matrix as a potential 3D printing bioink for constructing islet organoids. *Acta Biomater.* 165: 86–101.

129 Hu, S., Martinez-Garcia, F.D., Moeun, B.N. et al. (2021). An immune regulatory 3D-printed alginate-pectin construct for immunoisolation of insulin producing β-cells. *Mater. Sci. Eng. C* 123: 112009–112009.

130 Duin, S., Schütz, K., Ahlfeld, T. et al. (2019). 3D bioprinting of functional islets of langerhans in an alginate/methylcellulose hydrogel blend. *Adv. Healthcare Mater.* 8 (7): e1801631.

131 Klak, M., Kowalska, P., Dobrzanski, T. et al. (2021). Bionic organs: shear forces reduce pancreatic islet and mammalian cell viability during the process of 3D bioprinting. *Micromachines (Basel)* 12 (3): 304.

132 Hwang, D.G., Jo, Y., Kim, M. et al. (2022). A 3D bioprinted hybrid encapsulation system for delivery of human pluripotent stem cell-derived pancreatic islet-like aggregates. *Biofabrication* 14 (1): 14101.

133 Huang, B., Wei, X., Chen, K. et al. (2023). Bioprinting of hydrogel beads to engineer pancreatic tumor-stroma microtissues for drug screening. *Int. J. Bioprinting* 9 (3): 676.

134 Hakobyan, D., Medina, C., Dusserre, N. et al. (2020). Laser-assisted 3D bioprinting of exocrine pancreas spheroid models for cancer initiation study. *Biofabrication* 4: 12.

135 Langer, E.M., Allen-Petersen, B.L., King, S.M. et al. (2019). Modeling tumor phenotypes in vitro with three-dimensional bioprinting. *Cell Rep.* 26 (3): 608–623.e6.

136 Khati, V., Ramachandraiah, H., Pati, F. et al. (2022). 3D bioprinting of multi-material decellularized liver matrix hydrogel at physiological temperatures. *Biosensors* 12 (7): 521.

137 Kang, D., Hong, G., An, S. et al. (2020). Bioprinting of multiscaled hepatic lobules within a highly vascularized construct. *Small* 16 (13): 1905505.

138 Zein, N.N., Hanouneh, I.A., Bishop, P.D. et al. (2013). Three-dimensional print of a liver for preoperative planning in living donor liver transplantation. *Liver Transpl.* 19 (12): 1304–1310.

139 Kim, W., Lee, Y., Kang, D. et al. (2023). 3D inkjet-bioprinted lung-on-a-chip. *ACS Biomater. Sci. Eng.* 9 (5): 2806–2815.

140 da Rosa, N.N., Appel, J.M., Irioda, A.C. et al. (2023). Three-dimensional bioprinting of an in vitro lung model. *Int. J. Mol. Sci.* 24 (6): 5852.

141 Ng, W.L., Ayi, T., Liu, Y.C. et al. (2021). Fabrication and characterization of 3D bioprinted triple-layered human alveolar lung models. *Int. J. Bioprinting* 9: 7.

142 Dong, Q., Su, X., Li, X. et al. (2023). In vitro construction of lung cancer organoids by 3D bioprinting for drug evaluation. *Colloids Surf. A Physicochem. Eng. Asp.* 666: 131288.

143 Das, P., Najafikhoshnoo, S., Tavares-Negrete, J.A. et al. (2022). An in-vivo-mimicking 3D lung cancer-on-a-chip model to study the effect of external stimulus on the progress and inhibition of cancer metastasis. *Bioprinting* 28: e00243.

144 Kang, D., Lee, Y., Kim, W. et al. (2022). 3D pulmonary fibrosis model for anti-fibrotic drug discovery by inkjet-bioprinting. *Biomed. Mater.* 18 (1): 015024.

145 Berg, J., Hiller, T., Kissner, M.S. et al. (2018). Optimization of cell-laden bioinks for 3D bioprinting and efficient infection with influenza A virus. *Sci. Rep.* 8 (1): 13877–13813.

146 Sobreiro-Almeida, R., Gómez-Florit, M., Quinteira, R. et al. (2021). Decellularized kidney extracellular matrix bioinks recapitulate renal 3D microenvironment in vitro. *Biofabrication* 13 (4): 45006.

147 Carreno-Galeano, G., Ali, M., Jackson, J. et al. (2020). MP52-08 3D bioprinted renal tissue constructs using a novel kidney ECM-derived bioink. *J. Urol.* 203 (Supplement 4): e775.

148 Tröndle, K., Rizzo, L., Pichler, R. et al. (2021). Scalable fabrication of renal spheroids and nephron-like tubules by bioprinting and controlled self-assembly of epithelial cells. *Biofabrication* 13 (3): 35019.

149 Tröndle, K., Miotto, G., Rizzo, L. et al. (2022). Deep learning-assisted nephrotoxicity testing with bioprinted renal spheroids. *Int. J. Bioprinting* 8 (2): 164–173.

150 Chen, Y., Zhou, Y., and Wang, C. (2022). Investigation of collagen-incorporated sodium alginate bioprinting hydrogel for tissue engineering. *J. Comp. Sci.* 6 (8): 227.

151 Chakraborty, J., Fernández-Pérez, J., van Kampen, K.A. et al. (2023). Development of a biomimetic arch-like 3D bioprinted construct for cartilage regeneration using gelatin methacryloyl and silk fibroin-gelatin bioinks. *Biofabrication* 15 (3): 035009.

152 Liu, C., Dai, T., Wu, X. et al. (2023). 3D bioprinting of cell-laden nano-attapulgite/gelatin methacrylate composite hydrogel scaffolds for bone tissue repair. *J. Mater. Sci. Technol.* 1 (135): 111–125.

153 Allen, N.B., Abar, B., Johnson, L. et al. (2022). 3D-bioprinted GelMA-gelatin-hydroxyapatite osteoblast-laden composite hydrogels for bone tissue engineering. *Bioprinting* 1 (26): e00196.

154 Gharacheh, H. and Guvendiren, M. (2022). Cell-laden composite hydrogel bioinks with human bone allograft particles to enhance stem cell osteogenesis. *Polymers* 14 (18): 3788. https://pubmed.ncbi.nlm.nih.gov/36145933/.

155 Yang, J., Deng, C., Shafiq, M. et al. (2022). Localized delivery of FTY-720 from 3D printed cell-laden gelatin/silk fibroin composite scaffolds for enhanced vascularized bone regeneration. *Smart Mater. Med.* 1 (3): 217–229.

156 Chae, S., Kim, J., Yi, H.G., and Cho, D.W. (2022). 3D bioprinting of an in vitro model of a biomimetic urinary bladder with a contract-release system. *Micromachines (Basel)* 13 (2): 277.

157 McQuillan, S.K. and Grover, S.R. (2014). Systematic review of sexual function and satisfaction following the management of vaginal agenesis. *Int. Urogynecol. J.* 25 (10): 1313–1320.

158 Hou, C., Zheng, J., Li, Z. et al. (2021). Printing 3D vagina tissue analogues with vagina decellularized extracellular matrix bioink. *Int. J. Biol. Macromol.* 180: 177–186.

159 Li, Y., Liu, F., Zhang, Z. et al. (2015). Bone marrow mesenchymal stem cells could acquire the phenotypes of epithelial cells and accelerate vaginal reconstruction combined with small intestinal submucosa. *Cell Biol. Int.* 39 (11): 1225–1233.

160 Zhang, H., Zhang, J., Huang, X., and Li, Y. (2018). The methods and mechanisms to differentiate endothelial-like cells and smooth muscle cells from mesenchymal stem cells for vascularization in vaginal reconstruction. *Mol. Biotechnol.* 60 (6): 396–411.

161 Edgar, L., Pu, T., Porter, B. et al. (2020). Regenerative medicine, organ bioengineering and transplantation. *Br. J. Surg.* 107 (7): 793–800.

162 Kupiec-Weglinski, J.W. (2022). Grand challenges in organ transplantation. *Front. Transplant.* 1: 897679. https://www.frontiersin.org/articles/10.3389/frtra.2022.897679.

163 Fa, A., Gibot, L., and Lacroix, D. (2023). The pivotal role of vascularization in tissue engineering. *Annu. Rev. Biomed. Eng.* 15: 177. https://pubmed.ncbi.nlm.nih.gov/23642245/.

164 Hunsberger, J.G., Goel, S., Allickson, J., and Atala, A. (2017). Five critical areas that combat high costs and prolonged development times for regenerative medicine manufacturing. *Curr. Stem Cell Rep.* 3 (2): 77–82.

165 Kyttälä, A., Moraghebi, R., Valensisi, C. et al. (2016). Genetic variability overrides the impact of parental cell type and determines iPSC differentiation potential. *Stem Cell Rep.* 6 (2): 200–212.

166 Nishizawa, M., Chonabayashi, K., Nomura, M. et al. (2016). Epigenetic variation between human induced pluripotent stem cell lines is an indicator of differentiation capacity. *Cell Stem Cell* 19 (3): 341–354.

167 Murphy, S.V., De Coppi, P., and Atala, A. (2020). Opportunities and challenges of translational 3D bioprinting. *Nat. Biomed. Eng.* 4 (4): 370–380.

7

4D-Printed, Smart, Multiresponsive Structures and Their Applications

Jinku Kim[1], D.A. Gouripriya[2], and Prosenjit Saha[2]

[1] *Department of Biological and Chemical Engineering, Hongik University, Sejong, Republic of Korea*
[2] *Centre for Interdisciplinary Sciences, JIS Institute of Advance Studies and Research, Kolkata, West Bengal, India*

7.1 Introduction

In recent years, the convergence of cutting-edge technologies has given rise to remarkable advancements in the field of additive manufacturing, popularly known as 3D printing, which becomes a critical tool to fabricate complex 3D structures for various biomedical applications, including tissue engineering and regenerative medicine (TERM) [1]. One significant evolution within this domain has been the emergence of 4D bioprinting: It is an advanced technique in the field of bioprinting that goes beyond traditional 3D bioprinting and also a paradigm-shifting approach that transcends the boundaries of traditional 3D fabrication [2]. Whereas 3D bioprinting involves the precise layer-by-layer deposition of bioinks (materials containing living cells) to create 3D structures resembling tissues or organs, 4D bioprinting adds an additional dimension of time, allowing the printed structures to change shape, function, or properties over time in response to specific stimuli. Therefore, by introducing the dimension of time into printing process, 4D bioprinting has opened up new avenues for the creation of smart multiresponsive structures with transformative potential across various biomedical applications, such as TERM [3, 4].

This chapter describes a comprehensive exploration of the realm of "4D-printed, smart, multiresponsive structures and their applications." We delve into the fundamental principles that underlie 4D bioprinting, examining how it builds upon the foundations of 3D printing and extends them into a dynamic realm. Through a detailed analysis of materials, biomedical applications, and sign

3D Bioprinting from Lab to Industry, First Edition. Edited by Prosenjit Saha, Sabu Thomas, Jinku Kim, and Manojit Ghosh.

strategies involved, we aim to provide a holistic understanding of the field's current state and its immense potential. The multiresponsiveness inherent in 4D-printed structures is a hallmark of their ingenuity. These structures have the capability to sense and respond to a diverse range of stimuli, which may include changes in temperature, humidity, light, pH, or the presence of specific chemicals [5]. This responsiveness is a reflection of the careful integration of advanced materials and precise fabrication techniques, resulting in smart structures that can adapt and transform in ways that have previously been reserved for the natural world. Moreover, biomedical applications of 4D-printed smart structures span a wide spectrum of biomedical applications, such as health-care applications and regenerative medicine. These structures hold the promise of revolutionizing fields that require adaptable and intelligent materials [6, 7]. Their potential to create self-assembling medical implants, dynamically changing architectural elements, and shape-shifting devices will place 4D bioprinting as a game-changer in the pursuit of innovation.

As we begin this journey through the world of 4D-printed smart, multiresponsive structures, we invite readers to better understand the intricacies of this rapidly evolving field. Herein, we will excavate deeper understanding into technologies, materials, and the plethora of applications that showcase the remarkable potential of 4D bioprinting. By the end of this exploration, we hope that readers will gain a profound appreciation for the transformative power of engineering structures that come alive in the fourth dimension.

7.2 4D-Printing Technologies

The concept of 4D printing draws inspiration from nature's own dynamic processes. Just as living organisms respond and adapt to their environments, 4D-printed structures can autonomously undergo shape changes, self-assembly, or functional alterations in response to external stimuli. This ability to harness the fourth dimension, time, enables a new realm of possibilities for engineering complex structures that mirror the intricate behaviors of natural systems.

There are several available 4D-printing technologies, each of which has its own advantages and limitations, which have been published elsewhere [8]. The most widely used printing is extrusion-based methods, also known as fused filament fabrication (FFF) or fused deposition modeling (FDM) (Figure 7.1a). This process melts a thermoplastic filament in a heated nozzle to form a 3D object in a layer-by-layer manner, and the printer precisely deposits the material with computer-controlled movements. The extrusion-based methods are widely available for 4D printing, relatively easy to use, and can print a wide range of thermoplastic materials [11]. Due to layer-by-layer nature of this technology, the

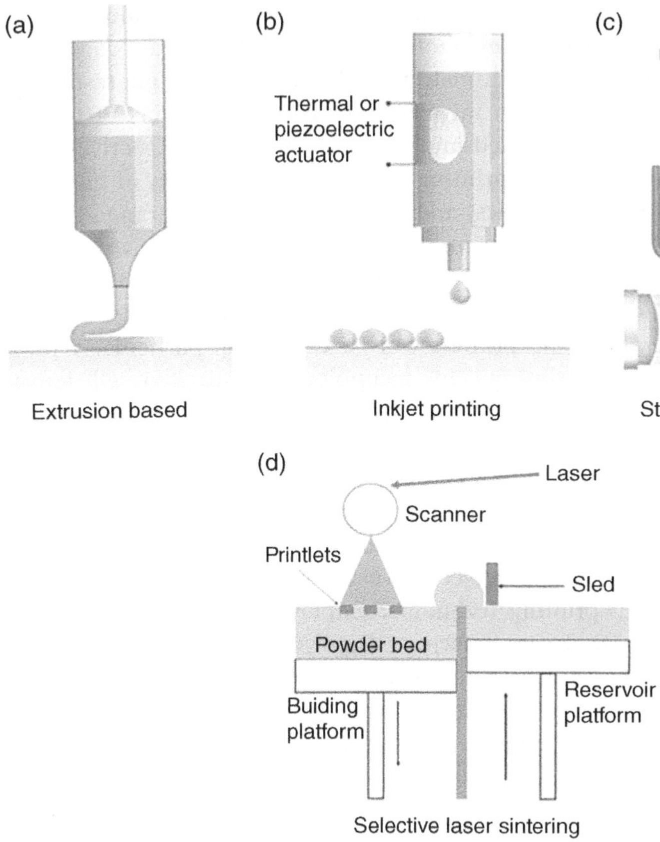

Figure 7.1 Schematics of common 4D-printing technologies. (a) Extrusion-based, (b) inkjet printing, (c) stereolithography, *Source:* Bedell et al. [9] / with permission of American Chemical Society and (d) selective laser sintering. *Source:* Gueche et al. [10] / MDPI / Licensed Under CC BY 4.0.

resolution is limited by the nozzle diameter, typically in the range of 100–200 μm, thus, achieving high resolution of resulting structures can be challenging [12]. Another limitation of the extrusion-based method is the incapability to incorporate living cells during the process, especially when high processing temperature is needed to create 3D structures [13]. In addition, it is difficult to print multiple materials using the FFF method because different thermoplastics may have different printing temperatures.

Inkjet printing, on the contrary, can be used for multimaterial printing if two polymers are compatible with each other [14] (Figure 7.1b). Inkjet 3D printing operates on principles similar to traditional 2D inkjet printing and uses multiple printheads to simultaneously spray different photocurable polymers to form a 3D structure, followed by photocuring. This enables the convenient 3D fabrication composed of multimaterials at a relatively high resolution. Typical in-plane resolution of an inkjet printer is approximately 30–40 and 200~400 μm for single-material and multimaterial printing, respectively [15]. Depending on the printer, inkjet printing can work with a variety of materials, including photopolymers, ceramics, and metals, although the range is more limited compared with other 3D-printing methods, such as selective laser sintering (SLS), especially for metals and ceramics [9]. However, inkjet 3D-printing process can be relatively slow, especially for large or complex objects, resulting in product efficiency, and the printers can be expensive, particularly those capable of multimaterial printing. This technology also limited cell density (10^6~10^7), compared with other 3D-printing methods although it yields higher cell viability (>90%) [16].

Another fast-growing method for 3D printing of polymers is the so-called vat photopolymerization (or light-based printing). This technology can be called stereolithography (SLA) or digital light processing (DLP), depending on light sources (laser beam or digital light projector is used for photocuring) (Figure 7.1c). Stereolithography, a 3D-printing technique, employs photopolymerization to build successive polymeric layers to form a 3D polymeric network. Along with FDM, stereolithography is becoming increasingly popular for 3D-printing method in recent years due to its high resolution, accuracy, and smooth surface finish, making it particularly suitable for applications requiring intricate details and precise dimensions. However, it comes with trade-offs in terms of equipment cost and material options, which may influence its suitability for specific application [17].

The last major group of polymer 3D-printing methods is the droplet-based (or jetting-based) methods, including SLS, which is a versatile 3D-printing technology known for its ability to work with various materials and produce intricate, functional parts (Figure 7.1d). Although cost and material considerations can be limiting factors, this technology is widely used in industries, such as aerospace, automotive, and biomedical applications for rapid prototyping and end use part production [18].

7.3 Biomaterials for 4D Bioprinting

4D printing relies on materials that undergo changes in response to external stimuli over time, allowing the printed 3D structure to transform or adapt. The choice of materials is crucial to achieving the desired architecture and functionality of 4D-printed structures. A biomaterial can be defined as any natural or synthetic materials used in a medical device that is intended to interact with a biological system [19]. Several types of biomaterials can be used for 4D printing, each with its unique properties and stimuli-responsiveness.

7.3.1 Water-Responsive Polymers

The ability to respond to changes in humidity is a common feature seen in nature. Many biological systems release or absorb moisture in response to variations in the surrounding humidity. These natural occurrences have, in turn, served as inspiration for the creation of humidity-responsive materials. As water is one of the common stimuli for altering the shape of bioprinted structures, this shape transformation can occur temporally as well as spatially due to varying degrees of water (moisture) absorption (i.e., swelling) in different sections of the printed structures [4].

Water-responsive polymers play a crucial role in 4D printing by enabling dynamic and responsive behavior in printed structures. These polymers respond to changes in moisture or humidity levels by altering their shape, mechanical properties, or function. Water-responsive polymers can undergo changes in shape when exposed to moisture. They may swell or contract in a controlled and predictable manner, allowing for the creation of structures that can self-fold, self-assemble, or change shape in response to changes in humidity. This is particularly valuable for creating complex, dynamic geometries. For example, self-folding 3D-printed structures consisting of photopatterned poly(ethylene glycol) (PEG)-based hydrogel bilayers with two different molecular weights (MWs) was printed (Figure 7.2a). The self-folding is driven by differential swelling of the two PEG bilayers in aqueous solutions. Authors encapsulated insulin-secreting cells within bilayers, and cell-laden structures subsequently self-fold them into cylinders with different radii. The calcein AM (acetoxymethyl) staining and ELISA (enzyme-linked immunosorbent assay) data revealed that the encapsulated cells were viable and showed robust insulin production within the structures [22]. With a proper design, such hydroresponsive scaffolds can be fully customizable to fabricate various anatomically relevant microscale geometries.

Similarly, Ionov and colleagues developed shape-morphing biopolymer hydrogels based on alginate and hyaluronic acid [21]. In the first step, methacrylated alginate or hyaluronic acid polymer solutions with or without live cells were printed onto glass or polystyrene (PS) surfaces in 2D rectangular shapes (Figure 7.2b (i)). The printed films were photocross-linked with green light and mildly dried (Figure 7.2b (ii)). After immersion of the cross-linked films in water, PBS, or cell culture media in the third step, they instantly fold into tubes (Figure 7.2b (iii)). In the case of cell-laden tubes, cells were viable and evenly distributed throughout the self-folding tubes. Authors demonstrated that

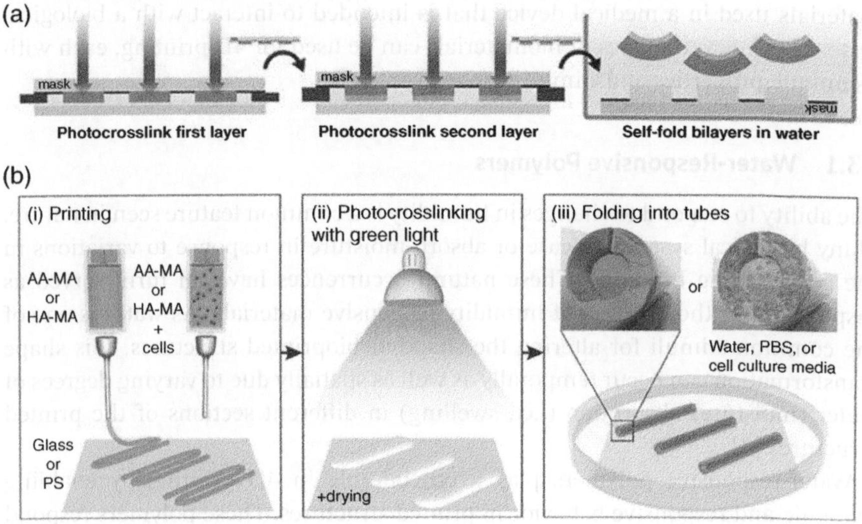

Figure 7.2 Schematics of the 4D-printed structure fabricated by water-responsive hydrogels. (a) PEG-based hydrogel bilayers. *Source:* Groll et al. [20] / IOP Publishing/ Public Domain and (b) methacrylated alginate or hyaluronic acid-based hydrogels. *Source:* Kirillova et al. [21] / John Wiley & Sons.

self-folding of the printed film can be formed into tubes with the cross-linking gradient in the films as top layer of the film absorbs more light that the bottom, leading to different volume expansions. In addition, Thiele and colleagues investigated the potential of a moisture-responsive, cellulose-based material [23]. They produced a two-layer film using cellulose stearoyl esters with varying substitution degrees on each side. One ester had a high hydrophilicity with a substitution degree of 0.3, whereas the other had a high hydrophobicity with a substitution degree of 3. As the sample was cooled from 50 °C to 22 °C, and the relative humidity was increased from 5.9% to 35%, the hydrophobic side contracted and the hydrophilic side expanded, leading to the sample to tightly roll up. Importantly, this process is reversible; and reverting the temperature and humidity changes caused the sample to unroll once again.

7.3.2 Temperature-Responsive Polymers (Hydrogels)

Temperature is another most frequently used stimulus to transform bioprinted constructs. Temperature-responsive polymers have a wide range of applications in 4D printing, offering opportunities to create dynamic, adaptive, and functional structures. Their ability to transform shape, properties, or function in response to temperature variations contributes to the versatility and potential of 4D-printing technology [24]. In particular, the thermoresponsive hydrogels can transform their shape based on the wettability and solubility change following the temperature variations. Poly(*N*-isopropylacrylamide), or pNIPAM, is a widely used thermoresponsive material. It has a unique property, which becomes hydrophilic and gets swollen up to its low critical solution temperature (32 °C). Temperatures above that start to dehydrate the hydrogel and cause it to shrink, thus, achieving shape transformation. For example, Bakarich et al. developed a new type of 4D-printed smart valve based on thermoresponsive pNIPAM, ionic–covalent entanglement (ICE) hydrogels for actuator, which was designed to control water flow by closing and opening the valve, when touching hot water (60 °C, contracted) and touching cold water (20 °C, swollen), respectively [25].

Breger and colleagues explored photocross-linked pNIPAM-co-acrylic acid (pNIPAM-AAc) to create reversible self-folding, soft, photopatterned microstructures in response to temperature, which composed of stiff PPF atop a pNIPAM-AAc layer [26]. As the temperature increased to above 36 °C, the pNIPAM-AAc component transformed to a hydrophobic state to expose the PPF segments, which causes the shape transformation of microgrippers to open and close such that the PPF segments face outward. Below 36 °C, the pNIPAM-AAc layer swells, which causes the pNIPAM-AAc layer face outward (Figure 7.3a). Furthermore, a bilayered starlike patterned polycaprolactone (PCL)–pNIPAM thermoresponsive self-folding capsules, which reversibly encapsulate/release yeast cells in response to a temperature variation, were printed [28]. When the temperature increased, the bilayered capsule shrunk, resulting in unfolding of the capsule. In contrast, a decrease in the temperature led to the swelling of the structure, thus, resulting in folding of the polymer capsules and encapsulation of the cells (Figure 7.3b). These 4D-printed structures are commonly used in various applications, including self-assembling structures, medical technology, soft robotics, and sensor technology.

In addition, another thermoresponsive system based on poly(acrylic acid)–PNIPAAm and ceramic powder (Al_2O_3), can undergo a controllable sol–gel transformation when subjected to a thermal trigger [29]. Typically, this process involves combining the thermosensitive polymer with cells, nutrients, or growth factors, and then introducing the mixture into the body. Upon injection, as the temperature increases, the polymer engages in a phase transition to form a gel, which subsequently releases its contents from the 3D-printed structure [30].

7.3.3 Electrical/Magnetic-Responsive Polymers

Electrically responsive polymers play a role in 4D printing by enabling structures to respond to electrical stimuli (e.g., electrically conductive characteristic). These polymers exhibit changes in size and/or shape, mechanical properties, or functionality in response to electrical stimulation.

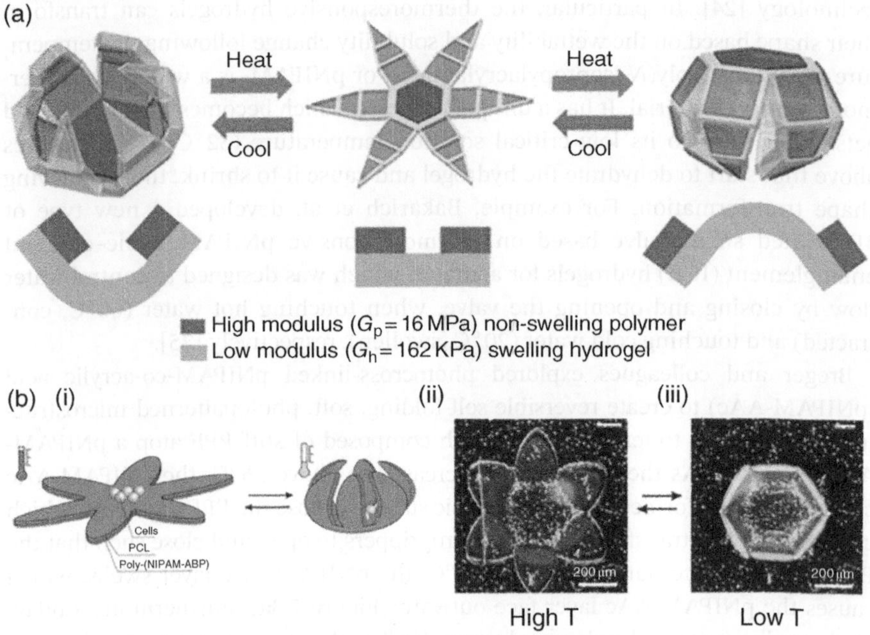

Figure 7.3 Schematics of the 4D-printed structure fabricated by temperature-responsive hydrogels. (a) The transformation behaviors of the thermoresponsive microgripper. *Source:* Breger et al. [26] / American Chemical Society, (b) (i) the folding star-shaped polymer bilayer, (ii) a thermoresponsive hydrogel swells and folds into a capsule-like structure when the temperature decreases, (iii) when the temperature increases, the capsule unfolds and releases the encapsulated cells. *Source:* Xu et al. [27] / John Wiley & Sons.

Polypyrrole (PPy) has been explored as an electrically conductive material for shape transformation. For instance, Okuzaki and colleagues created the accordion-shaped origami actuator by folding a conductive PPy film, which was doped with tetrafluoroborate in an L-shaped copper attached to the corner with a silver paste (Figure 7.4a) [31]. After turning the film over, one side was folded to the opposite side and then the other side was folded in the same manner. By alternately folding each side of the film six times, the film was folded into an accordion-shaped structure, and then a copper wire is attached on its top. When the electric potential was applied to the structure, it expands up to 147% at 2 V due to the water vapor desorption by Joule heating. Using this technique, they created a biomorphic origami robot by connecting two accordion-shaped structures in series.

Magnetic stimulation has also been used to fabricate 4D-printed structures. In this case, magnetic-responsive hydrogels can be used for the controlled release of drug and/or cells [33]. For example, Mooney and colleagues developed porous

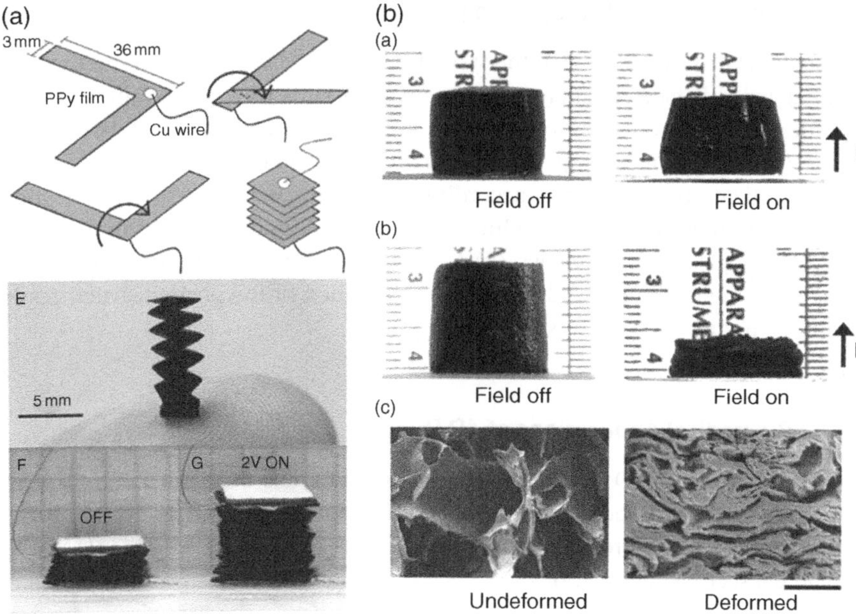

Figure 7.4 (a) Schematics of the fabrication of the accordion-shaped actuator based on PPy film and the images of the actuator (E) before and after annealing at (F) 0 V and (G) 2 V. *Source:* Reproduced with permission [31]. IOP Publishing, (b) images of nanoporous ferrogel reduced its height by ~5% under moderate magnetic field and microporous ferrogel reduced its height of ~70% under the same magnetic field. SEM images of microporous ferrogel at undeformed and deformed states. *Source:* [33]. Zhao et al. 2011 / Proceedings of the National Academy of Sciences of the United States of America [PNAS].

magnetic-responsive alginate-based hydrogels, which can be used as scaffolds in tissue engineering or cell-based therapies [32]. The active porous ferrogel gives strong and rapid compression and volume change of more than 70% under a moderate magnetic field. The deformation and volume variation drove the outward movement of water from the internal pores, allowing the release of various drugs, including mitoxantrone, plasmid DNA, and a chemokine from the scaffold (Figure 7.4b).

7.4 Biomedical Applications for 4D Bioprinting

The need and significance of 4D printing get its space when 3D printing fails to give the ideal results and attractive features. So, it is the need of the hour to check limitations of 3D bioprinting, and how 4D printing helps to overcome bottlenecks.

7.4.1 Limitations of 3D Bioprinting

Although 3D printing has taken the standard of biomedical treatment methods to a higher level, it faces challenges as follows: (i) Difficulty in retaining printed structure due to the effect of temperature, humidity, pH, etc., (ii) installation and maintenance costs of high-quality, automated, and high-resolution bioprinters, (iii) customization of bioprinter according to the nature of bioinks, (iv) selection of proper candidates as bioinks, which are compatible and support cells and have proper rheological characteristics, (v) designing complicated structures and replicating it through printing, and (vi) clinical validation of the printed structures in producing tissues and organs successfully [3].

7.4.2 Biomedical Applications of 4D Printing

The introduction of 4D printing using smart materials enabled the control of properties of the printed structure by external variables, such as temperature, heat, pH, etc. This offered numerous possibilities to make developments in the biomedical sector (Figure 7.5).

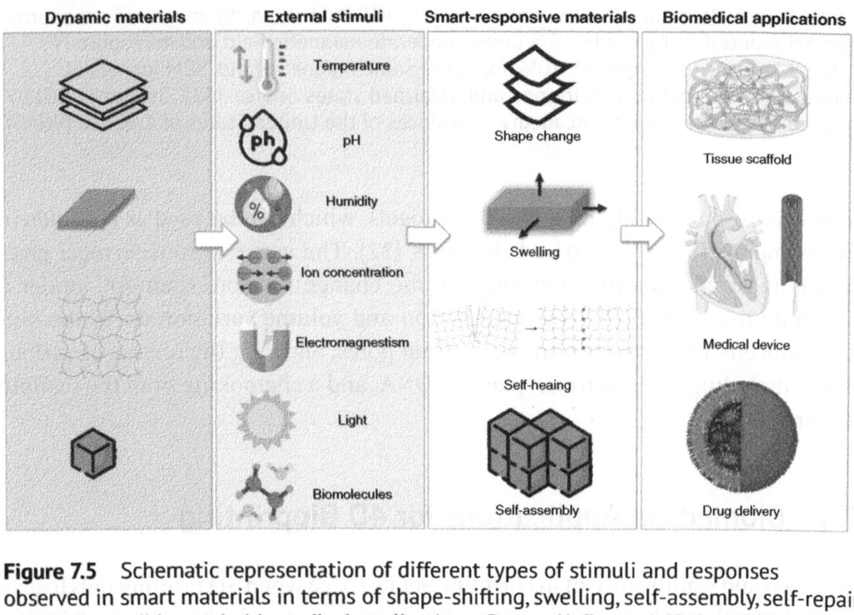

Figure 7.5 Schematic representation of different types of stimuli and responses observed in smart materials in terms of shape-shifting, swelling, self-assembly, self-repair, and their possible use in biomedical applications. *Source:* Kafle et al. [34] / MDPI / Licensed Under CC BY 4.0.

7.4.3 Scaffold Preparation

Printing technology can be used to produce structural constructs that can undergo shape-shifting and functional flexibility, including mechanical and electrical behaviors with effect to the external stimuli. This property has been used in scaffold preparation for organ and tissue development. Studies have been carried out using plant-based biopolymers to have temperature sensitivity and support to cell attachment. Miao et al. used Soybean oil epoxidized acrylate to make a shape memory scaffold using SLA, which supports human bone marrow mesenchymal stem cells (hMSCs). The scaffolds which were kept at −18 °C returned to their initial shape at 37 °C [35]. Similarly, porous complex scaffolds were prepared through castor oil-based polymers with PCL as a control and checked for hMSC adhesion and proliferation [36]. Polyurethane-based shape memory polymers with porous structures are studied for their time-dependent shape changes and cell modulation. Wei Zhao et al. developed a bio-designed tracheal scaffold using 4D printing with magnetic stimulated shape memory composites. 4D printing of shape memory polymer tracheal scaffolds with glass sponge structure exhibited better mechanical performance and strength [37]. Another set of polymers that are extensively used for stimuli-based scaffold preparation is lignin, cellulose, chitosan, hyaluronic acid, alginate, etc. [38, 39]

7.4.4 Drug Delivery

The stimuli-responsive behavior of materials used for 4D printing can be used in specific and controlled drug delivery. Wang et al. loaded an anticancer drug, methotrexate, into a printed shape memory hydrogels (SMHs) and analyzed its drug delivery properties. The in vitro studies showed that drug release was higher in printed structures than in traditional (SMH) structures because of the increased surface area. They combined sodium alginate and pluronic F127 diacrylate macromer (F127DA) to make SMHs. These SMHs have been shown to have a higher recovery ratio and drug-releasing rate [40]. Melocchi et al. used shape memory polymers based on polyvinyl alcohol, which can respond to the presence of water for fabricating devices for intravesical drug delivery. The device is administered in a temporary shape into the body and it recovers its shape in the presence of water. The device is then eliminated through urine [41]. As mentioned earlier, Breger et al. developed thermoresponsive gripper structures from poly(*N*-isopropylacrylamide-co-acrylic acid) hinges using a photolithographic technique. The grippers, if applied from a cold state to the body, will close above 32 °C and get attached to the tissue, which facilitates the delivery of therapeutic agents [26]. Despite these advancements in fabricating drug delivery systems with the help of 4D printing, the optimization, sensitivity, and spontaneity in drug release should be given importance to achieve better treatment standards.

7.4.5 Sensors

4D printing becomes relevant because of its sensitivity to external stimuli and shape-shifting. Researchers are trying to make use of shape memory polymers to treat hyperglycemia. Temperature, electromagnetic waves, humidity levels, pH, and electrolyte intensity of the external conditions can act as external stimuli the composites [42]. The real-time monitoring of wounds is necessary to give proper treatment and protection from the risk of infection. In this regard, Mostafalu et al. developed smart wound dressing bandages with temperature and pH sensors for monitoring wounds [43]. In addition, Wang et al. introduced a strategy to manufacture a reversible actuator that can self-sense the strain via structural design and a combination of bilayer composites. The actuator has reversible deformation ability, which can be controlled by voltage facilitated using conductive graphene polylactic acid (GPLA), and the wiggle structure can control the direction of deformation [44]. Lv et al. fabricated a hydrogel film through photopolymerization of poly(ethylene glycol) diacrylate (PEG-DA) monomer. The hydrogel films are sensitive to changes in humidity levels and undergo spontaneous deformation and motion. In different relative humidity, it shows different shades of pink color and fluorescence under 365 nm illumination. So, these hydrogel films can be used as humidity test strips [45]. Chen et al. reported a high-performance integrated sensor–actuator capable of simultaneous actuation and sensation. Nanocarbon/PLA composites with gradient microgap structures enabled the decoupled thermal stimulation and strain sensation. [46]. Chen et al. developed a novel multifunctional preceramic polymer composite with a shape memory effect for sensing and actuation applications. By tuning material compositions, the precursor inks are pyrolyzed into lightweight self-shaping ceramics to generate complex structures [47]. Lee developed a UV exposure sensor to address the problem of skin cancer due to overexposure to UV radiation. The sensor is a flexible trilayer structure with a UV-responsive polymer sandwiched between two protective transparent layers. With the increase in concern for diabetes, glucose sensors are also in the line of research. Song et al. developed a chemo-resistive glucose sensor by incorporating glucose oxidase enzyme with platinum nanoparticles. An inkjet printer is used for printing carbon nanotubes and polyaniline nanowires to make low sheet-resistant electrodes and a glucoresponsive area, respectively [48].

7.4.6 Medical Devices

Medical devices can diagnose, monitor, and forecast diseases and injuries. Shape memory polymers have wide applications in medical devices. For example, these materials are used in ingestible electronic devices, percutaneous feeding tubes, and intragastric balloons. However, one of the challenges faced while using these polymers is their nondegradable nature. Zhang et al. developed a pH-responsive supramolecular elastomer that consists of two synthetic polymers, poly(acryloyl-6-aminocaproic acid) and poly(methacrylic acid-co-ethyl acrylate) to overcome the issue of degradation. Cell experiments proved that it is noncytotoxic and compatible with the intestine environment [49]. 4D printing and soft robotics can function coherently to carry out tasks with minimum human intervention. Zarek et al. suggested that 4D printing may be used to make customized medical devices, such as bifurcated stents designed as per the patient's anatomy using shape memory polymers. These stents were designed in such a way that shape-shifting takes place with changes in body temperature [50]. Kim et al. jointly used CAD and 4D printing to fabricate bifurcated stents, which can be deployed in the targeted site without surgical implantation. The stents are folded and deployed in a Kirigami-inspired manner [50]. Waseem et al. created a T-shaped vascular bifurcation by 3D bioprinting with shape transformation ability based on a mathematical model. The tubular constructs acquire a T junction shape with a diameter in the millimeters range upon immersion in water. It showed excellent cell viability and can serve as a self-actuated vascular graft [51].

7.4.7 Tissue Engineering and Organ Regeneration

3D printing has marked a revolutionary step in organ regeneration and improved the quality of treatment by overcoming organ shortages. 4D printing adds a time-dependent dynamic process during the fabrication of the design. In this regard, Cui et al. developed a near-infrared (NIR)-responsive 4D-printed nanomaterials by combining photothermal materials, polycaprolactone–polyethylene glycol copolymer, and black phosphorus quantum dots using two-photon polymerization [52]. Su et al. fabricated an NIR-responsive tumoral injectable hydrogel with photothermal therapy (PTT) for in situ tongue tumors. The prepared soft materials show a strong photothermal effect due to the network formed by anionic proteins, chitosan, and Ag_3AuS_2 nanoparticles (NPs) [53]. Hwangbo et al. prepared a biomimetic collagen/hydroxyapatite scaffold via 3D printing combined with one-way shape morphing, thereby, falling under the category of 4D printing. The microchannel scaffold showed great potential for osteogeneration and growth of blood vessels [54].

While talking about bone regeneration, one of the important categories is the use of piezoelectric materials in repairing damaged bone by mimicking the tissue's electrical environment. Chen et al. have mentioned that 4D-printed flexible bio-piezoelectric scaffolds can provide time-dependent electrical stimuli responses for bone regeneration [55]. Wang et al. composed photothermal responsive shape memory composite scaffolds with β-tricalcium phosphate/poly(lactic acid-co-trimethylene carbonate) (TCP/P(DLLA-TMC) and incorporated black phosphorus nanosheets and osteogenic peptides in it. The scaffold temperature increases to 45 °C on applying NIR radiation and undergoes shape shifting to fit into irregular bone defects. Then the temperature is decreased to 37 °C and mimics the properties of natural bone [56]. Wang et al. made use of the photothermal sensitivity of MXene nanosheets and thermoresponsive behavior of pNIPAM hydrogels in the MXene–hollow fibers (MX-HF) scaffolds for vascularization and skin flap regeneration. The scaffolds showed controlled delivery of vascular endothelial growth factor (VEGF) and promoted proliferation and angiogenesis [56].

Another great leap toward 4D printing in cardiac tissue engineering was by Cui et al. by 4D-printing cardiac patch for the treatment of myocardial infarction. The patch has self-morphing capacity and improved vascularization under suitable mechanical stimuli [57]. 4D-printed structures with microgrooves are also experimented with for neural tissue engineering. Miao et al. developed a stereolithography-based, 4D-printed, multiresponsive construct that has stress-induced shape transformation properties. The printed structures experience light-induced internal stress followed by solvent-induced relaxation that gives reversible shape-changing properties after printing [58]. Shin-Da Wu et al. developed a self-healing injectable hydrogel with polyurethane and gelatin considering time as the fourth dimension. The healing and shape recovery were satisfying, and it also supported neural stem cell proliferation [30]. Kim et al. developed trachea implants with photocurable silk fibroin hydrogel through the DLP technique. The implants were integrated into the host and growth of both epithelium and cartilage could be seen in the implanted area [59]. Betsch et al. also have taken the assistance of 4D printing for the development of an artificial fibrocartilage. Here, they have used a magnetic-based mechanism for the directional alignment of collagen fibers in agarose hydrogel. Steptavidin-coated iron nanoparticles are embedded in the hydrogels to impart magnetic sensitivity to the collagen fibers. These aligned collagen fibers have high compression moduli, which makes them ideal to be used in cartilage tissue engineering applications [60].

7.5 Future Perspectives

Despite all these impressive biomaterials and applications that are already established and currently under concern, 4D-printed structures face obvious limitations in its design and successful fabrication. Challenges still exist in replicating complex tissue structures and matching the response time of shape memory polymers with biological processes. The coherent degradation of 4D-printed moieties with the regeneration of tissues is to be optimized in a narrow time frame. The transformation of constructs, duration of stimuli, stability in the tissue

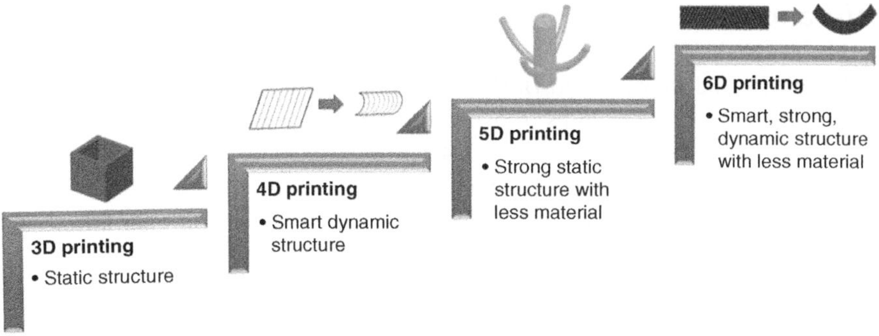

Figure 7.6 High-dimensional printing. *Source:* Nida et al. [61] / MDPI / Licensed Under CC BY 4.0.

environment, and biocompatibility should be carefully monitored to ensure safe and effective application and results. 4D printing is now being modified to 5D and 6D printing to improve efficiency and avoid the usage of supporting material. In 5D printing, printing will be done with three translational axes and two rotational axes, and in 6D printing, an additional factor will be present. Thus, the printing techniques can revolutionize the field of medicine and biomedical applications [61] (Figure 7.6).

The future of 4D printing is likely to be marked by significant advancements in a wide range of fields. As materials, techniques, and applications continue to develop, 4D printing has the potential to transform industries, enhance customization, and contribute to more sustainable and responsive solutions to a variety of challenges. However, it is also important to consider and address ethical and regulatory aspects of this technology to ensure its responsibilities and safe use. In particular, the manufacturers of bioprinted structures have to consider corporate regulatory goals and compliance risks in planning a biocompatibility testing of devices for 4D bioprinted structures. Inevitably, the determination of the biocompatibility of a device is a risk assessment exercise. There is no risk-free device or device material. The goal of device designers is to minimize risks, while maximizing benefits to patients. Ultimately, the establishment of standardized protocols for both in vitro and in vivo testing of smart 4D-printed structures is imperative, especially in biomedical applications regarding their safety and efficacy in clinical realization. While aforementioned innovative biomaterials may not see immediate widespread adoption to various biomedical applications in standardized manners, they hold great promises to produce more relevant outcomes to realize smart 4D-printed structures in a clinical setting.

References

1 Murphy, S. and Atala, A. (2014). 3D bioprinting of tissues and organs. *Nat. Biotechnol.* 32: 773–785. https://doi.org/10.1038/nbt.2958.

2 Choi, J., Kwon, O., Jo, W. et al. (2015). 4D printing technology: a review. *3D Print. Addit. Manuf.* 2: 159–167. https://doi.org/10.1089/3dp.2015.0039.

3 Agarwal, K., Srinivasan, V., Lather, V. et al. (2023). Insights of 3D bioprinting and focusing the paradigm shift towards 4D printing for biomedical applications. *J. Mater. Res.* 38: 112–141. https://doi.org/10.1557/s43578-022-00524-2.

4 Gao, B., Yang, Q., Zhao, X. et al. (2016). 4D bioprinting for biomedical applications. *Trends Biotechnol.* 34: 746–756. https://doi.org/10.1016/j.tibtech.2016.03.004.

5 Ge, Q., Dunn, C., Qi, H., and Dunn, M. (2014). Active origami by 4D printing. *Smart Mater. Struct.* 23. https://doi.org/10.1088/0964-1726/23/9/094007.

6 Gladman, A., Matsumoto, E., Nuzzo, R. et al. (2016). Biomimetic 4D printing. *Nat. Mater.* 15: 413. https://doi.org/10.1038/NMAT4544.

7 Li, G., Yang, S., Wu, W. et al. (2023). Biomimetic 4D printing catapult: from biological prototype to practical implementation. *Adv. Funct. Mater.* https://doi.org/10.1002/adfm.202301286.

8 Momeni, F., Hassani, N., Liu, X., and Ni, J. (2017). A review of 4D printing. *Mater. Des.* 122: 42–79. https://doi.org/10.1016/j.matdes.2017.02.068.

9 Mueller, J., Raney, J., Shea, K., and Lewis, J. (2018). Architected lattices with high stiffness and toughness via multicore-shell 3D printing. *Adv. Mater.* 30: https://doi.org/10.1002/adma.201705001.

10 Panda, S., Hajra, S., Mistewicz, K. et al. (2022). A focused review on three-dimensional bioprinting technology for artificial organ fabrication. *Biomater. Sci.* 10: 5054–5080. https://doi.org/10.1039/d2bm00797e.

11 Bedell, M.L., Navara, A.M., Du, Y. et al. (2020). Polymeric systems for bioprinting. *Chem. Rev.* 120: 10744–10792. https://doi.org/10.1021/acs.chemrev.9b00834.

12 Gueche, Y., Sanchez-Ballester, N., Cailleaux, S. et al. (2021). Selective laser sintering (SLS), a new chapter in the production of solid oral forms (SOFs) by 3D printing. *Pharmaceutics* 13: https://doi.org/10.3390/pharmaceutics13081212.

13 Kuang, X., Roach, D., Wu, J. et al. (2019). Advances in 4D printing: materials and applications. *Adv. Funct. Mater.* 29: https://doi.org/10.1002/adfm.201805290.

14 Sun, K., Wei, T., Ahn, B. et al. (2013). 3D printing of interdigitated Li-ion microbattery architectures. *Adv. Mater.* 25: 4539–4543. https://doi.org/10.1002/adma.201301036.

15 Chia, H. and Wu, B. (2015). Recent advances in 3D printing of biomaterials. *J. Biol. Eng.* 9: https://doi.org/10.1186/s13036-015-0001-4.

16 Hölzl, K., Lin, S., Tytgat, L. et al. (2016). Bioink properties before, during and after 3D bioprinting. *Biofabrication* 8: https://doi.org/10.1088/1758-5090/8/3/032002.

17 Cui, X., Gao, G., Yonezawa, T., and Dai, G. (2014). Human cartilage tissue fabrication using three-dimensional inkjet printing technology. *Jove-J. Vis. Exp.* https://doi.org/10.3791/51294.

18 Ng, W.L., Lee, J.M., Zhou, M. et al. (2020). Vat polymerization-based bioprinting-process, materials, applications and regulatory challenges. *Biofabrication* 12: 022001. https://doi.org/10.1088/1758-5090/ab6034.

19 von Recum, A.F. and Laberge, M. (1995). Educational-goals for biomaterials science and engineering – prospective view. *J. Appl. Biomater.* 6: 137–144.

20 Kirillova, A., Maxson, R., Stoychev, G. et al. (2017). 4D biofabrication using shape-morphing hydrogels. *Adv. Mater.* 29: https://doi.org/10.1002/adma.201703443.

21 Jamal, M., Kadam, S., Xiao, R. et al. (2013). Bio-origami hydrogel scaffolds composed of photocrosslinked PEG bilayers. *Adv. Healthcare Mater.* 2: 1142–1150. https://doi.org/10.1002/adhm.201200458.

22 Groll, J., Burdick, J., Cho, D. et al. (2019). A definition of bioinks and their distinction from biomaterial inks. *Biofabrication* 11: https://doi.org/10.1088/1758-5090/aaec52.

23 Zhang, K., Geissler, A., Standhardt, M. et al. (2015). Moisture-responsive films of cellulose stearoyl esters showing reversible shape transitions. *Sci. Rep.* 5: https://doi.org/10.1038/srep11011.

24 Li, Y., Zhang, Y., Akpek, A. et al. (2017). 4D bioprinting: the next-generation technology for biofabrication enabled by stimuli-responsive materials. *Biofabrication* 9: https://doi.org/10.1088/1758-5090/9/1/012001.

25 Bakarich, S., Gorkin, R., Panhuis, M., and Spinks, G. (2015). 4D printing with mechanically robust, thermally actuating hydrogels. *Macromol. Rapid Commun.* 36: 1211–1217. https://doi.org/10.1002/marc.201500079.

26 Breger, J., Yoon, C., Xiao, R. et al. (2015). Self-folding thermo-magnetically responsive soft microgrippers. *ACS Appl. Mater. Interfaces* 7: 3398–3405. https://doi.org/10.1021/am508621s.

27 Stoychev, G., Puretskiy, N., and Ionov, L. (2011). Self-folding all-polymer thermoresponsive microcapsules. *Soft Matter.* 7: 3277–3279.

28 Xu, F., Celli, J., Rizvi, I. et al. (2011). A three-dimensional in vitro ovarian cancer coculture model using a high-throughput cell patterning platform. *Biotechnol. J.* 6: 204–212. https://doi.org/10.1002/biot.201000340.

29 Wang, X., Sun, Y., Peng, C. et al. (2015). Transitional suspensions containing thermosensitive dispersant for three-dimensional printing. *ACS Appl. Mater. Interfaces* 7: 26131–26136. https://doi.org/10.1021/acsami.5b07913.

30 Gandhi, A., Paul, A., Sen, S., and Sen, K. (2015). Studies on thermoresponsive polymers: phase behaviour, drug delivery and biomedical applications. *Asian J. Pharm. Sci.* 10: 99–107. https://doi.org/10.1016/j.ajps.2014.08.010.

31 Okuzaki, H., Saido, T., Suzuki, T. et al. (2008). A biomorphic origami actuator fabricated by folding a conducting paper. *J. Phys.* 127: 012001.

32 Kokkinis, D., Schaffner, M., and Studart, A. (2015). Multimaterial magnetically assisted 3D printing of composite materials. *Nature. Communications* 6: https://doi.org/10.1038/ncomms9643.

33 Zhao, X., Kim, J., Cezar, C. et al. (2011). Active scaffolds for on-demand drug and cell delivery. *Proc. Natl. Acad. Sci. USA* 108: 67–72. https://doi.org/10.1073/pnas.1007862108.

34 Kafle, A., Luis, E., Silwal, R. et al. (2021). 3D/4D printing of polymers: fused deposition modelling (FDM), selective laser sintering (SLS), and stereolithography (SLA). *Polymers* 13: https://doi.org/10.3390/polym13183101.

35 Miao, S., Zhu, W., Castro, N. et al. (2016). 14D printing smart biomedical scaffolds with novel soybean oil epoxidized acrylate. *Sci. Rep.* 6: https://doi.org/10.1038/srep27226.

36 Miao, S., Zhu, W., Castro, N. et al. (2016). Four-dimensional printing hierarchy scaffolds with highly biocompatible smart polymers for tissue engineering applications. *Tissue Eng. Part C Methods* 22: 952–963. https://doi.org/10.1089/ten.tec.2015.0542.

37 Zhao, W., Zhang, F., Leng, J., and Liu, Y. (2019). Personalized 4D printing of bioinspired tracheal scaffold concept based on magnetic stimulated shape memory composites. *Compos. Sci. Technol.* 184: https://doi.org/10.1016/j.compscitech.2019.107866.

38 Mulakkal, M., Trask, R., Ting, V., and Seddon, A. (2018). Responsive cellulose-hydrogel composite ink for 4D printing. *Mater. Des.* 160: 108–118. https://doi.org/10.1016/j.matdes.2018.09.009.

39 Arif, Z., Khalid, M., Zolfagharian, A., and Bodaghi, M. (2022). 4D bioprinting of smart polymers for biomedical applications: recent progress, challenges, and future perspectives. *React. Funct. Polym.* 179: https://doi.org/10.1016/j.reactfunctpolym.2022.105374.

40 Wang, Y., Miao, Y., Zhang, J. et al. (2018). Three-dimensional printing of shape memory hydrogels with internal structure for drug delivery. *Mater. Sci. Eng. C-Mater. Biol. Appl.* 84: 44–51. https://doi.org/10.1016/j.msec.2017.11.025.

41 Melocchi, A., Inverardi, N., Uboldi, M. et al. (2019). Retentive device for intravesical drug delivery based on water-induced shape memory response of poly(vinyl alcohol): design concept and 4D printing feasibility. *Int. J. Pharm.* 559 299–311. https://doi.org/10.1016/j.ijpharm.2019.01.045.

42 Sonatkar, J., Kandasubramanian, B., and Ismail, S. (2022). 4D printing: pragmatic progression in biofabrication. *Eur. Polym. J.* 169: https://doi.org/10.1016/j.eurpolymj.2022.111128.

43 Mostafalu, P., Tamayol, A., Rahimi, R. et al. (2018). Smart bandage for monitoring and treatment of chronic wounds. *Small* 14: https://doi.org/10.1002/smll.201703509.

44 Wang, Y. and Li, X. (2021). 4D printing reversible actuator with strain self-sensing function via structural design. *Composites, Part B-Eng.* 211: https://doi.org/10.1016/j.compositesb.2021.108644.

45 Lv, C., Xia, H., Shi, Q. et al. (2017). Sensitively humidity-driven actuator based on photopolymerizable PEG-DA films. *Adv. Mater. Interfaces* 4: https://doi.org/10.1002/admi.201601002.

46 Chen, D., Liu, Q., Han, Z. et al. (2020). 4D printing strain self-sensing and temperature self-sensing integrated sensor-actuator with bioinspired gradient gaps. *Adv. Sci.* 7: https://doi.org/10.1002/advs.202000584.

47 Chen, S., Li, J., Shi, H. et al. (2023). Lightweight and geometrically complex ceramics derived from 4D printed shape memory precursor with reconfigurability and programmability for sensing and actuation applications. *Chem. Eng. J.* 455: https://doi.org/10.1016/j.cej.2022.140655.

48 Song, E., da Costa, T., and Choi, J. (2017). A chemiresistive glucose sensor fabricated by inkjet printing. *Microsyst. Technol. Micro Nanosyst. -Inf. Storage Process. Syst.* 23: 3505–3511. https://doi.org/10.1007/s00542-016-3160-4.

49 Zhou, W., Qiao, Z., Zare, E. et al. (2020). 4D-printed dynamic materials in biomedical applications: chemistry, challenges, and their future perspectives in the clinical sector. *J. Med. Chem.* 63: 8003–8024. https://doi.org/10.1021/acs.jmedchem.9b02115.

50 Kim, D., Kim, T., and Lee, Y. (2019). 4D printed bifurcated stents with kirigami-inspired structures. *Jove-J. Vis. Exp.* https://doi.org/10.3791/59746.

51 Kitana, W., Apsite, I., Hazur, J. et al. (2023). 4D biofabrication of T-shaped vascular bifurcation. *Adv. Mater. Technol.* 8: https://doi.org/10.1002/admt.202200429.

52 Cui, H., Miao, S., Esworthy, T. et al. (2019). A novel near-infrared light responsive 4D printed nanoarchitecture with dynamically and remotely controllable transformation. *Nano Res.* 12: 1381–1388. https://doi.org/10.1007/s12274-019-2340-9.

53 Su, J., Lu, S., Jiang, S. et al. (2021). Engineered protein photo-thermal hydrogels for outstanding in situ tongue cancer therapy. *Adv. Mater.* 33: https://doi.org/10.1002/adma.202100619.

54 Hwangbo, H., Lee, H., Roh, E. et al. (2021). Bone tissue engineering via application of a collagen/hydroxyapatite 4D-printed biomimetic scaffold for spinal fusion. *Applied. Phys. Rev.* 8: https://doi.org/10.1063/5.0035601.

55 Chen, A., Su, J., Li, Y. et al. (2023). 3D/4D printed bio-piezoelectric smart scaffolds for next-generation bone tissue engineering. *Int. J. Extreme Manuf.* 5: https://doi.org/10.1088/2631-7990/acd88f.

56 Wang, C., Yue, H., Liu, J. et al. (2020). Advanced reconfigurable scaffolds fabricated by 4D printing for treating critical-size bone defects of irregular shapes. *Biofabrication* 12: https://doi.org/10.1088/1758-5090/abab5b.

57 Cui, H., Liu, C., Esworthy, T. et al. (2020). 4D physiologically adaptable cardiac patch: A 4-month in vivo study for the treatment of myocardial infarction. *Sci. Adv.* 6: https://doi.org/10.1126/sciadv.abb5067.

58 Miao, S., Cui, H., Nowicki, M. et al. (2018). Stereolithographic 4D bioprinting of multiresponsive architectures for neural engineering. *Adv. Biosyst.* 2: https://doi.org/10.1002/adbi.201800101.

59 Kim, S., Seo, Y., Yeon, Y. et al. (2020). 4D-bioprinted silk hydrogels for tissue engineering. *Biomaterials* 260: https://doi.org/10.1016/j.biomaterials.2020.120281.

60 Betsch, M., Cristian, C., Lin, Y. et al. (2018). Incorporating 4D into bioprinting: real-time magnetically directed collagen fiber alignment for generating complex multilayered tissues. *Adv. Healthcare Mater.* 7: https://doi.org/10.1002/adhm.201800894.

61 Nida, S., Moses, J.A., and Anandharamakrishnan, C. (2022). Emerging applications o 5D/6D printing in the food industry. *J. Agric. Food Res.* 24: 100392.

8

Toxicity Aspects and Ethical Issues of Bioprinting

Noura Al Hashimi[1] *and Sanjairaj Vijayavenkataraman*[1,2]

[1] *The Vijay Lab, Division of Engineering, New York University Abu Dhabi, Abu Dhabi, UAE*
[2] *Department of Mechanical and Aerospace Engineering, Tandon School of Engineering, New York University, Brooklyn, NY, USA*

8.1 Introduction

Tissue engineering is an interdisciplinary field that combines the principles of science and engineering to replace or restore damaged tissue. This is done by creating scaffolds that support and enhance cell proliferation, adhesion, and growth into functionalized tissues or biofabrication of cell-laden constructs that would mature into functionalized tissues. The long-term goal is to provide full organ replacements and transplants. Tissue engineering has been successful in formulating simple tissues such as retina and cartilage, while more complex tissues and organs still pose a major challenge, particularly in terms of spatial and functional integration and viability of cells [1]. There are many techniques that are used to engineer tissues in the labs, most notable of which is 3D bioprinting. Much like the name suggests, 3D bioprinting is just 3D printing but with cells; it is a bottom-up manufacturing approach that involves using computer-aided manufacturing to design tool paths that are read and executed by the printer in a precise manner, allowing for the creation of accurate, complex tissue structures [2, 3]. Several additional factors are kept in mind, such as autonomous self-assembly and biomimicry, which aim to fabricate functional tissues with biomechanical properties similar to those found in nature. Not only does this allow for large-scale organ manufacturing, but also patient-specific customization [4, 5]. To ensure proper 3D bioprinting, specific requirements need to be met in terms of bioprinting

3D Bioprinting from Lab to Industry, First Edition. Edited by Prosenjit Saha, Sabu Thomas, Jinku Kim, and Manojit Ghosh.

hardware and biomaterials. The bioprinter needs to be easy to use, sterile, and nontoxic to cells, allowing for simultaneous material deposition, and having high throughput and resolution [6]. The biomaterial needs to be biocompatible, biodegradable, and tunable to have optimal biomechanical properties with suitable printability, porosity, and morphology to support different cell niches [7, 8]. Given the rapid growth in research and commercial bioprinting, visibly seen in the increasing number of publications, journals, and rapid development of new biomaterials and commercial bioprinters [9, 10], the technical requirements are being met, yet the path to clinical application is lagging behind, mainly due to ethical and regulatory issues. Some guidelines already exist for specific parts of the technology, such as the approval for a biomaterial being used, however, regulations are still lacking in terms of assessing this technology as a whole.

This chapter reviews and analyzes 3D bioprinting from an ethical, legal, and social (ELSA) viewpoint, showcased in Figure 8.1, raising the most pressing concerns with regard to 3D bioprinting advancement and future commercialization.

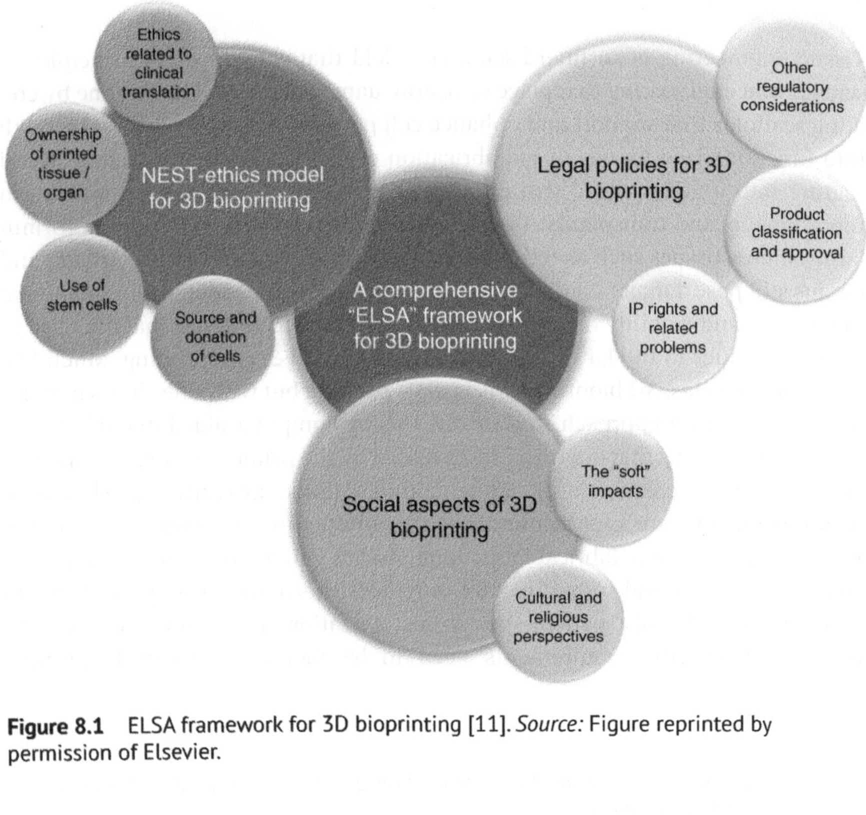

Figure 8.1 ELSA framework for 3D bioprinting [11]. *Source:* Figure reprinted by permission of Elsevier.

8.2 Toxicity Issues in Bioprinting

8.2.1 Cell Harvesting and Culture

In 3D bioprinting, the cells of most interest are undifferentiated human embryonic stem cells (hESCs) and induced pluripotent stem cells (iPSCs) due to their ability to grow and give way to different cell lines and ultimately whole organs [12]. This brings about ethical issues with regard to cell harvesting which will be discussed in Section 8.3.2. This section will focus on the protocols used for hESC and iPSC harvesting and culture.

The first hESC lines were successfully isolated in 1998 from human embryos [13]. In order to obtain hESCs, donated eggs and sperm are fertilized and cultured for four to five days until they achieve the blastocyst stage. At this stage, the embryo contains 64 to a few hundred cells organized in an outer shell and an inner clump of polarized cells called inner cell mass (ICM). The ICM is the focal point of all pluripotent cells that yield all the tissues in an adult organism. To reach the ICM, the outer shell is removed by surgery which results in the "death" of the embryo, and the ICM is separated and plated for culture. The cultured cell colonies are then screened for stem cells, which can divide symmetrically unlimited times or differentiate into diverse cell types, depending on the growth factors given. Once harvested, the hESC lines can be cultured steadily for years [14]. While this procedure of obtaining hESCs by *in vitro* fertilization (IVF) of donated sperm and eggs is recommended, the other source of hESCs is the embryos obtained via abortion. The main concern with stem cells lies in the mainstream debate of the embryo's status upon cell harvesting which is mostly done via abortion [15, 16], which is discussed in Section 8.3.2.

iPSCs, on the other hand, were first obtained in 2006 by Yamanaka and Takahashi when they first reprogrammed adult fibroblasts [17]. Fibroblasts are the main source of somatic cells used to generate iPSCs; this is due to the ease of isolation using skin biopsies as well as simple, easy cell culturing requirements [18]. However, due to their invasive, painful nature, some alternatives are being explored, such as keratinocytes from hair and hematopoietic cells from blood samples [19, 20]. The reprogramming protocol for all the somatic cell types is mainly focused on expressing OCT3/4, SOX2, KLF4, and C-MYC using retroviral transduction; therefore, here we will use fibroblasts as an example [21].

Human dermal fibroblasts are harvested using either skin biopsies or from circumcisions of neonatal foreskin. In the case of 3D bioprinting and where autogenous cells are required, adult human patients will need to get skin biopsies with 3.5–4 mm diameter punch size, most commonly from the arm [22]. The fibroblasts are then reprogrammed to pluripotency by retroviral transfection using Oct3/4, SOX2, KLF4, and C-MYC and start expanding in culture within weeks, at

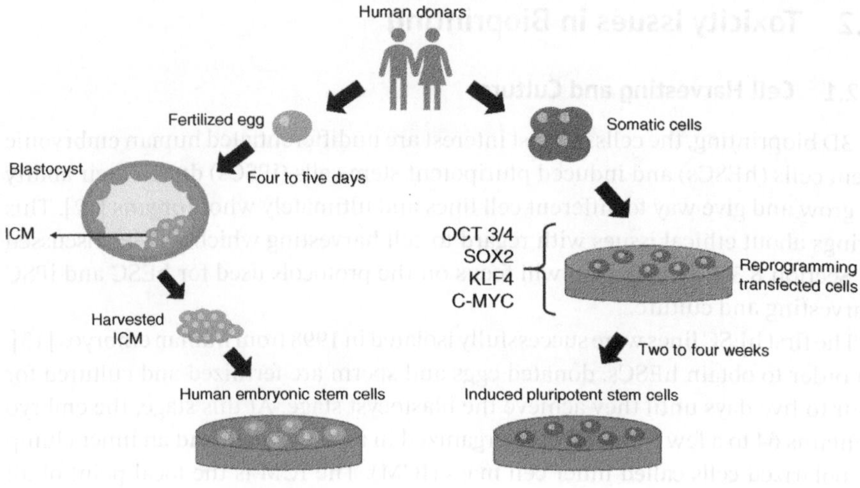

Figure 8.2 Schematic diagram showing the pathways and protocols followed to harvest and culture human stem cells, either hMSCs from fertilized embryos or iPSCs from somatic cells.

very high proliferation rates. This is not optimal since there is a possibility that the unprogrammed cells will outgrow the reprogrammed ones and use up all the culture media. Therefore, when fibroblasts are being used, the reprogramming efficiency decreases with each passage, and it is better to use less than passage five to lower the chances of genetic changes [23]. Figure 8.2 provides a schematic diagram summarizing the key cell harvesting and culture protocols followed to achieve human stem cells, either from a fertilized egg into hMSCs or from somatic cells into iPSCs.

Each of these steps requires careful handling and any contamination will render the cells unusable (waste of time, money, effort, and at the expense of the donors/patients).

8.2.2 Aseptic Techniques in Bioprinting

Due to the use of cells in bioprinting, it is vital to have sterile equipment and ensure proper aseptic conditions at all times. Aseptic and sterile are nowadays used interchangeably. However, strictly speaking, "asepsis" means being free of pathogens whereas "sterile" means being free of all forms of life, pathogenic or otherwise [24, 25]. Even with aseptic techniques, the introduction of microbes cannot be completely eliminated but can only be reduced below a certain threshold to prevent infection [26]. This section will discuss the aseptic techniques used throughout the bioprinting process.

The first stage involves the preparation of the bioink and cell suspension. The cell culturing *in vitro* must be done in a sterile incubator, following strict aseptic conditions. The bioink cell suspension is prepared in a laminar flow biosafety cabinet which is disinfected with 70% ethanol and kept on UV overnight. Similarly, anything the cells come or will come in contact with, such as the bioprinter and its components (syringes, needles), are sprayed with 70% ethanol and sterilized under UV overnight. The easiest yet most important aseptic technique is maintaining proper hygiene and following basic lab protocols such as wearing gloves, clean lab coat, and tying hair. Once the bioink cell suspension is ready to be printed, it is poured into a sterile, single-use plastic syringe and placed into the bioprinter (in case of extrusion bioprinting). Other bioprinting methods might require pouring the bioink into vats or containers instead of plastic syringes. The bioprinter itself is sprayed with ethanol and located in a sterile biosafety cabinet. Ideally, the printing should be completed quickly to minimize cell death in external conditions. After this, the bioprinted constructs are carefully collected, placed in sterile well plates, and placed in the incubator or bioreactor for further studies.

8.3 Ethical Issues in Bioprinting

The speed with which 3D bioprinting is progressing brings about not only significant innovations and scientific improvements but also raises a myriad of unaddressed ethical issues. This section will delve into the main ethical problems related to this new technology and how they could possibly be addressed.

8.3.1 Purpose

The leading concern when it comes to bioprinting is "Why?" What purpose does this product serve and what need does it meet? More importantly, is it a need or a want? Is there anything that should not be printed? Where do we draw the line on legal and ethical restrictions on 3D bioprinting?

While the purpose and need for 3D bioprinted organs seem obvious in the need to replace or regenerate damaged tissue, especially with shortage of organ donations, the difficulty in answering those questions remains with regard to where to draw the line between "human enhancement" and "therapy." On one hand, "enhancement" is seen positively as it leads to individual and social well-being, but, on the other, it is seen negatively since adding or implanting "machines" is seen to decrease the "naturalness" and thus the "morality" of humans [27–29]. This brings about a controversial debate and begs the question of whether society will accept or shun people with bioprinted organs. Will they be seen as less or more or just treated equally? Looking at an example of a revolutionary medical breakthrough from the past, IVF was seen by some as an abnormal, "ungodly" procedure, yet here we are now with major IVF commercialization and acceptance with little to no societal resistance. Ultimately, society could go one of two ways: they can be swayed into acceptance or total rejection; consequently, it is very important to have informed and educated people. Philosophical debates aside, most governmental organizations accept and justify 3D bioprinting due to its therapeutic effect and unmatched potential [30]. Therefore, it is important to study the ethical problems and place proper regulations and legislation before the technology is ready for commercialization and widespread use [11, 31].

8.3.2 Cell Source

Another ethical concern is related to the components of the bioink used in bioprinting. The cell is the most vital element in bioprinting which brings about several issues with regard to cell source and harvesting, donor confidentiality, and consent. One of the cell types most commonly used in bioprinting is hESCs, which raises critical ethical issues [32]. The main concern with stem cells lies in the mainstream debate of the embryo's status upon cell harvesting which is mostly done via abortion [15, 16]. There are multiple religious viewpoints on this topic. Orthodox and Catholic Christians as well as Buddhists completely prohibit the use of hESCs because they believe "life" begins at fertilization and the human embryo is a person who must be protected [33]. Muslims and Jews accept the use of hESCs because they believe "life" begins after a few months of fertilization and the embryo has no moral status before that time. Protestant Christians, on the other hand, accept hESC research as long as it is done ethically [34–36]. This debate is not only religious or philosophical, but it also extends into the scientific community as the definition of "embryo" ranges from egg activation, penetration, fertilization, or loss of totipotency [37].

To avoid this debate, another viable cell source is xenogeneic cells which are taken from other animals. However, similar to hESCs, this brings up religious issues such as Muslim and Jewish patients refusing to use porcine cells due to religious beliefs and vegans who refuse to use any animal-derived cells out of moral beliefs [38]. Moreover, xenotransplantation may lead to problems with personal identity and pose a risk of zoonosis [39]. Current advances in reprogramming mature cells into iPSCs eliminate the ethical and religious issues of using hESCs or xenogeneic cells but come with issues of their own. In the case of iPSCs, cells are removed from the patient's body and reprogrammed into stem cells; however, this reprogramming technology is still new and unmastered [40, 41], with many concerns still unknown, such as how the cells will behave after printing? Will they migrate? Or possibly mutate? Safety issues with stem cells have been reported in cases like the formation of bone tissue around a patient's eye after a stem cell therapy facelift and development of kidney lesions after stem cell treatment, showing the risk of unpredictable iPSC behavior and tumor formation [42, 43]. To make the use of iPSCs safer, clinical genetic testing of the cell lines is recommended [44–46], which brings about another ethical issue with regard to the use of data and information consent.

8.3.3 Data and Consent

Every major biotechnological breakthrough brings about critical issues with regard to the accuracy and ethical correctness of the information at the public level. A key aspect is the balance between transparency and public awareness of the benefits of the technology without causing alarm with misleading information. A lot of the time, mass media builds up unrealistic expectations to have catchy headlines or provide hope of a new "solution" but this just creates a misinformed and often alarmed society, which may hinder the progress of the research. It is important that researchers get involved in communicating the science, purpose, and risks of their research accurately and truthfully. This will help foster a more symbiotic relationship between society and science based on fact and mutual benefit [47]. By tackling the fear and ignorance of the new technology, human subjects can willingly be involved in experimental research and clinical trials which can help progress the science. Participants and donors in the clinical trial can express their consent to participate with full knowledge and awareness of the benefits, complications, and risks associated [48, 49]. Currently, there is still a lot to explore and discover in terms of the possibilities and risks of 3D bioprinting, this uncertainty must clearly be reported to the patients and the public [50, 51]. For instance, when it comes to using nonautologous cells derived from donor stem cells, issues arise in terms of donor consent and coercion, where obtaining viable informed consent may be difficult due to the patient's mental deterioration or fatigue and lack of decision-making capacity. Therefore, some form of alternative consent process needs to take place such as obtaining consent from next of kin.

Another important aspect of data and donor consent is data ownership. 3D bioprinting technology is set in a digital model, where tissues and organs are "digitized" and the line between the real physical world and the digital world seems to fade [52]. Printed organs made based on digital models will eventually replace natural ones, thus ultimately models will replace nature. Typically, models are owned by their designers or programmers but in this case when the model is built from personalized patient data, who owns the model? And can it be used without patient consent even for studies beyond the original scope? If the patient owns the model, that brings about a popular debate about altruistic donation, can the donor choose the recipient, or should it be on a first come first serve basis? [53]. If the donor owns the organ, they will favor their loved ones and move them to the top of the list. Questions such as those regarding privacy and patient confidentiality arise when it comes to human digitization, and special regulations are required for the handling of this information.

8.3.4 Safety

The rate at which 3D bioprinting is gaining traction and being investigated brings about unpredictable safety risks and long-term consequences of using this technology and uncertain product outcomes [54, 55]. It is important to assess this risk using the risk to benefit ratio. For instance, there is an urgent need for organ supply, which means the benefit of this technology would be huge and this may entail accepting a greater level of risk. But by whom? Who ensures the quality of the organs? Who controls organ production? Given the wide availability and commercialization of 3D printers, almost anyone at home could find the digital model and bioprint organs and tissues given the proper components [56]. However, the printed organ would be of no use without a surgeon and a proper medical team. Therefore, the risks associated with 3D bioprinting lie in the hands of professional researchers in the design, manufacturing, and medical areas. The best approach to deal with the safety concern is a multidisciplinary one involving experts from engineering, biomaterials, and medicine, with specific regulations and standards imposed by legal bodies.

8.3.5 Cost and Equity

Money has and always will play a major role in stigmatizing and dividing society into levels. Does that mean the less fortunate are not allowed the same luxury or access to healthcare? Ethically speaking, the exploitation of the human body is banned as it is seen as "sacred" and the commodification of the human body is a complex ethical topic [57]. However, with the rise of organ 3D bioprinting, this could change, and given the costs necessary to bioprint and produce the organ, the price will definitely be too high. This will deny people of all social classes access to the bioprinted organs and will be a violation of justice and equity. Yet, this is seen in all types of technologies and, in a free market, prohibiting something because it is too expensive or unjust will eventually imply the prohibition of any and all technologies. In addition, once the data and technology are commercialized, given the huge market for organs, no form of prohibition would suffice. With a simple ban and no proper legislation on the market, we will likely witness the rise of a 3D bioprinted organ black market, where anyone can buy and sell organs, especially to the rich [58]. Therefore, prohibition will only create more problems, safety concerns, and injustice in the availability of bioprinted organs. Alternatively, by allowing the well-off people to benefit from this technology would encourage research development and potentially allow for lower, more sustainable costs in the future [33].

8.3.6 Reproductive Organs

Under the umbrella of 3D bioprinting of organs, reproductive organs stand slightly apart as there is no direct threat to life, but rather an improvement to the quality of life. Patients with ovarian cancer or other reproductive complications may, in some cases, require removal of the reproductive organ [59, 60]. After the organ removal, unfortunately, there are no current replacement options. Development of artificial ovaries could provide hope to this group of patients to have their own biological children and restore endocrine functions. Animal experiments on artificial ovaries have been successful but human clinical trials and subsequent clinical translation are likely to take a long time, mainly due to ethical and regulatory issues [61].

Similar to other 3D bioprinted organs, the major concern is with the risk to benefit ratio when it comes to artificial ovary implantation. In this case, the removal of the naturally damaged or tumor-ridden ovary is a necessary procedure to save the patient's life and is seen as good when successful. However, the loss of reproductive functionality is a harm and the introduction of bioprinted artificial ovaries could restore the natural functions and right the wrong. This is a strong argument for the development of artificial ovaries, but there are many things that are to be considered to make it a reality. The main opposing concern with ovary bioprinting is that the risks faced could harm not only the patient herself but also all future offspring. Because it is a reproductive organ, any manipulation could possibly lead to genetic changes in the egg cells and affect all future generations. Thus, the artificial ovary could transfer genetic errors and variants across generations, thereby increasing their likelihood of developing cancer [62, 63]. Due to the oncological treatment and its effects on the body, several years need to pass from the time of ovary removal to artificial ovary transplantation [64]. With the passage of time, the patient might change their mind and refuse transplantation when the ovary is printed, but due to the expense and cost, the patient might feel obliged and pressured to proceed with the procedure. Additionally, many questions arise in terms of the inclusion criteria for trials of artificial ovaries. If a patient already has children, can she still get the procedure? Is there an upper or lower age limit? All these questions require proper, specific legal regulations to be put in place before the technology is implemented.

8.4 Issues in Clinical Trials

3D bioprinting is an emerging tool in the fields of tissue engineering and stem cell therapy, both of which are complex fields on their own. This makes the technology and its use for 3D organ bioprinting highly risky, therefore, before its use, the technology and its products have to be safety tested and approved to be clinically used in humans. This section delves into the issues faced in clinical trial designs.

8.4.1 Personalized Treatment

3D bioprinting aids in the development of personalized treatments, with unique and precise solutions to patients' needs, where personalized data from CT scans can be used to design implants on an ad hoc basis per patient. Clinical trials involving 3D bioprinting require safety and efficacy techniques specially designed for each individual patient, which goes against the norm of performing randomized, double-blind, placebo clinical trials [65]. The question then is how proper safety procedures can be implemented, given the uncertainties and risks of the personalized treatment regime. Theoretically, the use of autologous patient-specific iPSCs does not require safety and efficacy trials on a randomized group of people, because the treatment is specific to the patient and the findings will never be the same. Taking into account the risk to benefit ratio, one possibility is to restrict eligibility in initial trials and pace the studies, starting with simple initial tests such as external, easily explantable tissue patches like skin, before moving on to more complex tissues and whole organs. Furthermore, although the personalized biomaterials themselves cannot be properly evaluated, the protocols and criteria for making the biomaterials could instead be standardized and thus subject to evaluation. Most of the materials, media, and production processes are commonplace and have been used before, this provides insightful information relevant to the treatment.

Since 3D bioprinting treatments are customized using the patient's stem cells and specifically target the condition of that one patient, it is pointless and unethical to use a randomized trial of nonspecific participants who do not and cannot, share the exact same condition. Therefore, in this regard, each patient undergoing or waiting for a bioprinted tissue replacement would serve as their own test subject and no one else. Taking this scenario into consideration, having an assessment of efficacy alongside safety during the clinical trial may be more ethically acceptable, given the limited replicability. It may seem unethical to include therapeutic efficacy in an early-stage clinical trial; however, given the specificity of treatment and lack of repeatability, a trial focusing on both efficacy and safety is justifiable, especially when the efficacy tests could lead to immediate benefit. When the untested efficacy component is necessary for patient survival, such as in life-threatening conditions, then it is justifiable that the safety and efficacy requirements are indistinguishable [66]. For instance, a terminally ill patient could undergo efficacy treatments due to the life-preserving aspect of the case where these patients have the most to gain and little to lose in case of failure or complications. On the other hand, in non-life-threatening cases, the potential for complications and harm outweighs the improvement in quality of life and could deny them access to future therapies. The necessity of efficacy testing is proportional to the improvement in the patient's quality of life and well-being.

8.4.2 Inability to Withdraw or Access Alternate Treatments

Unlike patients involved in regular clinical trials for drugs, patients in trials assessing 3D bioprinting will have difficulty withdrawing from the trial if things go south or if they change their minds. Since the bioprinted organ is implanted in the body, it is challenging to remove all of it without causing even more harm, i.e., 3D bioprinting trials have limited reversibility and any attempt to reverse the procedure could potentially lead to more harm, leaving patients with limited to no chances of withdrawal. In most cases of bioprinted interventions, the patient is in a life-threatening position and the implant is their last-case scenario, which means they are unlikely to walk out or reverse the procedure, and they cannot do so either. The fact that they cannot withdraw also eliminates their hopes and chances of accessing alternative future treatments which could be better given the rapid technological and biological advances. The issue is that not only does the patient lose a shot at another treatment option, but also access to a better quality of life, only due to an experimental trial [67]. These risks are exacerbated in cases where the treatment targets a patient with a psychiatric condition. The implanted biomaterials could cause harmful neurological effects as seen with other types of brain implants, and even worse, the removal of the implant could worsen the psychiatric problems, leading to more harm and no gain for the patient [68, 69]. Given such high risks and long-term irreversible damage, it is important to determine whether it is morally just to undergo the implant procedure for that patient. The answer ultimately lies in whether the benefits outweigh the harms in that particular case. Implanting 3D bioprinted tissues just to test the technology with no promise of benefit for the patient is arguably unethical because, in this case, the harm is far greater than the good [70, 71].

Furthermore, given that 3D bioprinting is a novel, experimental technology, it is difficult to foresee all the major risks and complications in advance, which makes it almost impossible for surgeons to provide full risk disclosure and, therefore, obtain informed consent. With 3D bioprinted tissues, there is a range of possible irreversible risks and side effects, such as tumor formation and differentiation of cells into unintended, different cell lines post-operation. However, 3D bioprinting of removable, explantable tissue structures has much lower risks and complications compared to biomaterials injected into circulation or directly onto functioning organs. In the former, because the bioprinted treatment is maintained within a scaffold construct as opposed to injected into the body, it is easier to extract and remove the implant, thus withdrawing from the treatment and having the opportunity to pursue alternative treatments in the future.

Another issue with treatment options that use autologous iPSCs is replicability in terms of safety and efficacy. Because the treatment options are so personalized, the results obtained with one patient are not necessarily replicable in other patients [72, 73]. Furthermore, the information gathered from these personalized trials is much more limited than in traditional trials. In the personalized trials, the bioprinted implant is made from the patient's own cells, making it a completely unique product; redoing the exact same process using someone else's cells may yield completely different results. With traditional trials, it is much easier where a specific agreed-upon design and material is tested and studied on a cohort of patients and, if effective, recreated and used for future clinical applications. So, how can a personalized treatment be commercialized, when positive clinical results may be unique to the patient's genetics or other unknown, unaccounted personal factors, and where having one successful trial does not guarantee any future results? This makes a case for why it is increasingly important to develop a new regulatory framework to deal with the issues faced with 3D bioprinting in clinical evaluations.

8.5 Legal Issues in Bioprinting

Having looked at the ethical issues and gray areas in clinical trials with respect to the field of bioprinting in the sections above, it is clear that a robust, comprehensive, legal framework is required. This would include right to patents, specific regulatory policies, and governmental interventions, which would require a specialized, qualified team who understands this technology to be at the forefront of lawmaking, to avoid complications that could occur once bioprinting is actually tested and commercialized.

8.5.1 Intellectual Property Rights and Product Classification

Any novel, fast-growing technology with high societal impact and potential for commercialization gets a rush of patents and intellectual property (IP) applications. In the case of 3D bioprinting, the first question is what can and should be patentable. This question divides bioprinting into a process and a product and the answer depends on whether either one passes the eligibility criteria of being novel and non-naturally occurring [74]. Several products and processes passed these criteria, namely the choice of hydrogel and biomaterial used, the process of cell isolation, growth, and differentiation, any bioreactor or method of culturing the printed organ, the printing technique used, and finally any novel 3D biofabrication techniques [75]. However, the final bioprinted product is nonpatentable as it is naturally occurring and does not meet the second eligibility criteria. This brings about another issue previously touched upon, in regard to ownership of the bioprinted organ and the data associated with it [76].

The data obtained for stem cell and 3D bioprinting research is very important and classified as it contains personal genomic data, this opens the possibility of it being leaked and used maliciously. In 2012, a report titled "Privacy and Progress in Whole Genome Sequencing" was published by the Presidential Commission for the Study of Bioethical Issues in the United States [77]. The report entails some recommendations on maintaining individual privacy and protection even while allowing the exchange of genetic data. However, there is no existing legislation on obtaining donor consent for participation and use of their genetic information. Current legislation also does not protect the rights of donors or regulate the use of biological materials [78], which leaves many questions unanswered such as who owns the final bioprinted product? The cell donor? The engineer who printed the cells? Is the ownership shared? And if so, then how?

The next issue with this technology is IP theft. If the designs and models of the bioprinted scaffold were open source, there would be significant concerns with regard to patient data safety, product quality control, and the potential to have organ or tissue bioprinting by unauthorized individuals who merely have access to a 3D printer and the internet. However, if patents are granted for the models and processes of the bioprinting technology, this will cause major delays in the technology development and research growth. Given how all technologies can be a double-edged sword, this is an important point to address to make sure the sword is pointed in the right direction.

8.5.2 Lack of Regulatory Guidelines

Given the recent jump in interest in bioprinting technologies, no legislation yet exists that entails the whole process; some regulations address some parts that overlap with other technologies but given the newness of this technique, new and major legislations need to be formed to keep up with its rapid advancement.

The main problem with legislation is that the personalized medical treatments achieved by 3D bioprinting do not fit under any of the current regulatory categories. While it does fall under the regenerative medicine umbrella, 3D bioprinting is technically considered a combination product under the Australian Regulatory Guidelines for Medical Devices (ARGMD), as well as the Food and Drug Administration (FDA) in the US, which means it is classified as part device, and part "biologic" [31]. The issue with regulatory classification is that it determines the future approval processes down the line. Each product has different agencies that review the product based on the classification and provide approval pathways. Under the FDA, there is the Center for Biologics Evaluation and Research (CBER), Center for Drug Evaluation and Research (CDER), and Center for Devices and Radiological Health (CDRH), with each responsible for its different product classification of biologics, drugs, and devices, respectively. When it comes to bioprinting and bioprinted products, the review process is not straightforward as it is a combination of device and biologic. Faced with such complex regulatory pathways, there will surely be major arguments when bioprinted products reach the market. Ideally, those issues can and should be addressed by a global, international regulatory agency, so that the market is ready for 3D bioprinting when it takes off.

Another potential legislative issue is that these personalized, one-off treatments are exempt from regulation by most regulatory bodies since they only provide single, subject-specific treatments [79, 80]. Hence, it is difficult to evaluate this technology through regular clinical trials or to check whether it meets regulatory requirements. Across the globe, regulatory agencies are struggling to place proper legislation to address the benefits and risks associated with bioprinting [81, 82]. In the US, the FDA has placed regulations on additive manufacturing technology but excluded biological or tissue-based manufacturing as it may require additional or different regulatory pathways and considerations. Due to the lack of more specific, robust guidelines, the FDA follows the regulations set by the CBER. CBER's regulatory scope includes cellular therapies, however, they have no mention whatsoever of 3D bioprinting [83]. While the FDA does recognize the need for specific legislation targeting 3D bioprinting, industry specialists speculate that the technological advancements will far outpace the legislation at this rate and this would cause a ban on the use of this technology on humans.

Globally, only South Korea and Japan have addressed the legal implications of 3D bioprinting and provided loose guidelines specifically targeting this technology. The regulations are still broad and address either additive manufacturing without bioprinting or only address bioprinting from an academic research perspective [84]. The rest of the world is still scrambling to fill the gap and gray areas in legislation. In Australia, the Therapeutic Goods Administration (TGA) is evaluating whether the existing act on 3D printing can be modified and applied to 3D bioprinted products as well without the need for major new legislation. The TGA also states that human cells extracted and used by a medical practitioner on the same person in a single treatment are not to be regulated by the TGA, which means personalized stem cell bioprinting may be exempt from legislation, especially if the scaffold materials are already approved for implantation [85, 86].

In Europe, the European Commission (EC) and European Medicine Agency apply the same principles developed for gene and cell therapy to different stages of the 3D bioprinting process [87, 88]. An important issue to address with regulation is product quality and who is responsible for the quality of the 3D bioprinted tissues or organs. In Russia, legal issues with the use of bioprinting human organs are increasingly being deliberated [89]. The absence of current legislation is seen as a deterrent factor in the development and research of 3D bioprinting [90]. Presently, like elsewhere in the world, the technology of 3D bioprinting falls in a gap in Russian legislation, where it does not lie under the human organ transplant clause as it is artificial, or under any umbrella completely [91]. Therefore, the current, temporary solution is that the patients, 3D bioprinting providers, and medical facilities can agree and sign a written contract. This is used in cases of personalized treatment and fabrication of autologous artificial organs; otherwise, in cases where it is depersonalized, a sales agreement can be applied. The school of thought followed is that artificial organs have features that distinguish them from human organs and have been created under specific, controlled conditions. The bioprinted organs are formed outside the human body and impose no risk on the donor during fabrication. Moreover, the donor consent provided is limited to the cells and not the organ fabrication as a whole [92]. These differences in artificial vs natural organs may allow them to be commercialized at a faster pace. However, after implantation, they become an integral part of the human body and cannot be removed after death unless there is a termination of title of ownership.

Given all the different governmental regulations imposed by different countries, it is essential that they collaborate and form a global task force aiming at addressing regulations on bioprinting. It is vital that this is done at the early stages of technological development so that the legislation is established by the time 3D bioprinting is ready to hit the market and transform the medical sphere.

8.6 Conclusion

This chapter explored issues related to the use of 3D bioprinting, starting from cell harvesting and various ethical concerns to the importance of placing proper legislation and patent controls. The major ethical concerns are centered around the risk to benefit ratio. Is bioprinting necessary? When is it okay? Will it do more harm than good? Other ethical and moral issues exist with regard to the digitization of humans, clarity on donor ownership rights, and the importance of providing informed consent, knowing all the risks and benefits associated with the technology. The use of hESCs and xenogeneic cell lines also brings about major religious and moral concerns. The bioprinting of artificial reproductive organs poses ethical challenges of its own, especially with the higher risk of transmitting genetic errors to all future generations.

It is evident from the numerous concerns expressed above that there are more ethical issues at play than simply our freedom to print any biological object. It is also necessary to consider the short- and long-term consequences on patients and society at large. The ability to print new, customized organs seems to be a promising medical development with broad implications for both public and individual health. However, given the novelty of this technique and its uncertain benefits and unique risk profile, it is necessary to evaluate the ethical obstacles that need to be addressed, particularly in the early stages of exploratory clinical trials. Because 3D bioprinting trials require a single-person personalized treatment to establish their safety and efficacy for the person for whom the intervention has been specially designed, a patient waiting for an artificial organ would likely serve as their own guinea pig. In that context, it is vital to have clear, honest communication with the patient in order to safeguard both them and society from any potential risks related to this technology. The shift of 3D bioprinting from research laboratories to the clinic brings about well-established ethical concerns regarding the inability to replicate, reverse, or withdraw from the study, with the potentially added harm of not being able to seek alternative treatment in the future. The development of more specific, robust regulations for 3D bioprinting, rather than broad generalizations, is an area of global interest. While it may be challenging to establish at first, there is an urgent need to provide a consistent framework that addresses questions such as IP rights, evaluation of safety, and quality control of personalized organs and to fill this regulatory gap soon! Having specific regulations and guidelines with proper enforcement could decrease the negative impacts of 3D bioprinting raised in this chapter and help the technology revolutionize the world.

References

1 Lorber, B., Hsiao, W.K., and Martin, K.R. (2016). Three-dimensional printing of the retina. *Curr. Opin. Ophthalmol.* 27 (3): 262–267.

2 Agarwal, S., Saha, S., Balla, V.K. et al. (2020). Current developments in 3d bioprinting for tissue and organ regeneration–a review. *Front. Mech. Eng.* 6: 589171.

3 Xiongfa, J., Hao, Z., Liming, Z., and Jun, X. (2018). Recent advances in 3D bioprinting for the regeneration of functional cartilage. *Regen. Med.* 13 (1): 73–87.

4 Jakab, K., Norotte, C., Marga, F. et al. (2010). Tissue engineering by self-assembly and bio-printing of living cells. *Biofabrication* 2 (2): 022001.

5 Murphy, S.V. and Atala, A. (2014). 3D bioprinting of tissues and organs. *Nat. Biotechnol.* 32 (8): 773–785.

6 Ozbolat, I.T., Moncal, K.K., and Gudapati, H. (2017). Evaluation of bioprinter technologies. *Addit. Manuf.* 13: 179–200.

7 Agrawal, P., Soni, S., Mittal, G., and Bhatnagar, A. (2014). Role of polymeric biomaterials as wound healing agents. *Int. J. Low Extrem Wounds* 13 (3): 180–190.

8 Maaz Arif, M., Khan, S.M., Gull, N. et al. (2021). Polymer-based biomaterials for chronic wound management: promises and challenges. *Int. J. Pharm.* 598: 120270.

9 Cui, H., Nowicki, M., Fisher, J.P., and Zhang, L.G. (2017). 3D bioprinting for organ regeneration. *Adv. Healthcare Mater.* 6 (1): 1601118.

10 Monti, M. (2016). Bioprinting in regenerative medicine. *Eur. J. Histochem.* 60 (1): 2627. https://doi.org/10.4081/ejh.2016.2627.

11 Vijayavenkataraman, S., Lu, W.F., and Fuh, J.Y.H. (2016). 3D bioprinting – an ethical, legal and social aspects (ELSA) framework. *Bioprinting* 1–2: 11–21.

12 Matai, I., Kaur, G., Seyedsalehi, A. et al. (2020). Progress in 3D bioprinting technology for tissue/organ regenerative engineering. *Biomaterials* 226: 119536.

13 Thomson, J.A., Itskovitz-Eldor, J., Shapiro, S.S. et al. (1998). Embryonic stem cell lines derived from human blastocysts. *Science* 282 (5391): 1145–1147.

14 Landry, D.W. and Zucker, H.A. (2004). Embryonic death and the creation of human embryonic stem cells. *J. Clin. Invest.* 114 (9): 1184–1186.

15 Daar, A.S., Bhatt, A., Court, E., and Singer, P.A. (2004). Stem cell research and transplantation: science leading ethics. *Transplant. Proc.* 36 (8): 2504–2506.

16 Abouna, G.M. (2003). Ethical issues in organ transplantation. *Med. Princ. Pract.* 12 (1): 54–69.

17 Takahashi, K. and Yamanaka, S. (2006). Induction of pluripotent stem cells from mouse embryonic and adult fibroblast cultures by defined factors. *Cell* 126 (4): 663–676.

18 Raab, S., Klingenstein, M., Liebau, S., and Linta, L. (2014). A comparative view on human somatic cell sources for iPSC generation. *Stem Cells Int.* 2014: 768391.

19 Loh, Y.H., Agarwal, S., Park, I.H. et al. (2009). Generation of induced pluripotent stem cells from human blood. *Blood* 113 (22): 5476–5479.

20 Novak, A., Shtrichman, R., Germanguz, I. et al. (2010). Enhanced reprogramming and cardiac differentiation of human keratinocytes derived from plucked hair follicles, using a single excisable lentivirus. *Cell Rep.* 12 (6): 665–678.

21 Warlich, E., Kuehle, J., Cantz, T. et al. (2011). Lentiviral vector design and imaging approaches to visualize the early stages of cellular reprogramming. *Mol. Ther.* 19 (4): 782–789.

22 Alguire, P.C. and Mathes, B.M. (1998). Skin biopsy techniques for the internist. *J. Gen. Intern. Med.* 13 (1): 46–54.

23 Takahashi, K., Tanabe, K., Ohnuki, M. et al. (2007). Induction of pluripotent stem cells from adult human fibroblasts by defined factors. *Cell* 131 (5): 861–872.

24 Block, S.S. (2001). *Disinfection, Sterilization, and Preservation.* Lippincott Williams & Wilkins.

25 Penn, R.G. (2007). *Antisepsis, Disinfection, and Sterilization: Types, Action, and Resistance.* LWW.

26 Mellish, J., Tuomi, P., Hindle, A. et al. (2010). Skin microbial flora and effectiveness of aseptic technique for deep muscle biopsies in Weddell seals (Leptonychotes weddellii) in McMurdo Sound, Antarctica. *J. Wildl. Dis.* 46 (2): 655–658.

27 Harris, J. (2020). Moral enhancement and freedom. In: *The Ethics of Sports Technologies and Human Enhancement*, 399–408. Routledge.

28 Chan, S. and Harris, J. (2011). Moral enhancement and pro-social behaviour. *J. Med. Ethics* 37 (3): 130–131.

29 J. SM. (2007). *The Case Against Perfection: Ethics in the Age of Genetic Engineering*. Cambrige: Harvard University Press.

30 Jovic, T.H., Combellack, E.J., Jessop, Z.M., and Whitaker, I.S. (2020). 3D bioprinting and the future of surgery. *Front. Surg.* 7: 609836.

31 Gilbert, F., O'Connell, C.D., Mladenovska, T., and Dodds, S. (2018). Print me an organ? Ethical and regulatory issues emerging from 3D bioprinting in medicine. *Sci. Eng. Ethics* 24 (1): 73–91.

32 Gilbert, F., Viaña, J.N.M., O'Connell, C.D., and Dodds, S. (2018). Enthusiastic portrayal of 3D bioprinting in the media: ethical side effects. *Bioethics* 32 (2): 94–102.

33 Patuzzo, S., Goracci, G., Gasperini, L., and Ciliberti, R. (2018). 3D bioprinting technology: scientific aspects and ethical issues. *Sci. Eng. Ethics* 24 (2): 335–348.

34 Henon, P.R. (2003). Human embryonic or adult stem cells: an overview on ethics and perspectives for tissue engineering. *Adv. Exp. Med. Biol.* 534: 27–45.

35 Volarevic, V., Markovic, B.S., Gazdic, M. et al. (2018). Ethical and safety issues of stem cell-based therapy. *Int. J. Med. Sci.* 15 (1): 36–45.

36 Lo, B. and Parham, L. (2009). Ethical issues in stem cell research. *Endocr. Rev.* 30 (3): 204–213.

37 Condic, M.L. and Policy, P. (2013). When does human life begin? The scientific evidence and terminology revisited. *U. St. Thomas JL & Pub.* 8: 44.

38 Enoch, S., Shaaban, H., and Dunn, K.W. (2005). Informed consent should be obtained from patients to use products (skin substitutes) and dressings containing biological material. *J. Med. Ethics* 31 (1): 2–6.

39 Cooper, D.K.C., Gaston, R., Eckhoff, D. et al. (2018). Xenotransplantation-the current status and prospects. *Br. Med. Bull.* 125 (1): 5–14.

40 Nori, S., Okada, Y., Nishimura, S. et al. (2015). Long-term safety issues of iPSC-based cell therapy in a spinal cord injury model: oncogenic transformation with epithelial-mesenchymal transition. *Stem Cell Rep.* 4 (3): 360–373.

41 Blum, B. and Benvenisty, N. (2009). The tumorigenicity of diploid and aneuploid human pluripotent stem cells. *Cell Cycle* 8 (23): 3822–3830.

42 Thirabanjasak, D., Tantiwongse, K., and Thorner, P.S. (2010). Angiomyeloproliferative lesions following autologous stem cell therapy. *J. Am. Soc. Nephrol.* 21 (7): 1218–1222.

43 Jabr, F.J.S.A. (2012). In the flesh: the embedded dangers of untested stem cell cosmetics. *Sci. Am.* 17: 2012.

44 Andrews, P.W., Ben-David, U., Benvenisty, N. et al. (2017). Assessing the safety of human pluripotent stem cells and their derivatives for clinical applications. *Stem Cell Rep.* 9 (1): 1–4.

45 Jo, H.-Y., Han, H.-W., Jung, I. et al. (2020). Development of genetic quality tests for good manufacturing practice-compliant induced pluripotent stem cells and their derivatives. *Sci. Rep.* 10 (1): 1–15.
46 Sullivan, S., Stacey, G.N., Akazawa, C. et al. (2018). Quality control guidelines for clinical-grade human induced pluripotent stem cell lines. *Regen. Med.* 13 (7): 859–866.
47 Isa, N.M. and Safian Shuri, M.F.J.S. (2018). ethics e, Ethical concerns about human genetic enhancement in the Malay science fiction novels. *Sci. Eng. Ethics* 24 (1): 109–127.
48 Cornwall JJTAmj. (2016). The ethics of 3D printing copies of bodies donated for medical education and research: what is there to worry about? 9 (1): 8.
49 Povarov, Y.S. (2019). Requirements for consent to conduct research, treatment or diagnostics in the area of human. *Genome* 5 (2): 23–28.
50 Patuzzo, S. and Manganotti, P. (2014). Deep brain stimulation in persistent vegetative states: ethical issues governing decision making. *Behav. Neurol.* 2014: 641213.
51 Molinelli, A., Bonsignore, A., Rocca, G., and Ciliberti, R. (2009). Medical treatment and patient decisional power: the Italian state of the art. *Minerva Med.* 100 (5): 429–434.
52 Osborn, L.S. (2014). Regulating three-dimensional printing: the converging worlds of bits and atoms. *San Diego L. Rev.* 51: 553.
53 Abouna, G.M. (2003). Ethical issues in organ transplantation. *Med. Princ. Pract.* 12 (1): 54–69.
54 Neely, E.L. and Ethics, E. (2016). The risks of revolution: ethical dilemmas in 3D printing from a US perspective. *Sci. Eng. Ethics* 22 (5): 1285–1297.
55 Walther, G. (2015). Printing insecurity? The security implications of 3d-printing of weapons. *Sci. Eng. Ethics* 21 (6): 1435–1445.
56 Wolf, M.J. and Fresco, N. (2016). My liver is broken, can you print me a new one? In: *Computing and Philosophy Selected Papers from IACAP 2014* (ed. V.C. Muller), 259–269. Springer.
57 Greasley, K. (2014). A legal market in organs: the problem of exploitation. *J. Med. Ethics* 40 (1): 51–56.
58 Vermeulen, N., Haddow, G., Seymour, T. et al. (2017). 3D bioprint me: a socioethical view of bioprinting human organs and tissues. *J. Med. Ethics* 43 (9): 618–624.
59 Medrano, J.V., del Mar, A.M., García, S. et al. (2018). Basic and clinical approaches for fertility preservation and restoration in cancer patients. *Trends Biotechnol.* 36 (2): 199–215.
60 Amorim, C.A. and Shikanov, A. (2016). The artificial ovary: current status and future perspectives. *Future Oncol.* 12 (19): 2323–2332.
61 Laronda, M.M., Rutz, A.L., Xiao, S. et al. (2017). A bioprosthetic ovary created using 3D printed microporous scaffolds restores ovarian function in sterilized mice. *Nat. Commun.* 8 (1): 1–10.

62 Amorim, C.A. (2016). *Artificial Ovary. Gonadal Tissue Cryopreservation in Fertility Preservation*, 175–192. Springer.

63 Krotz, S.P., Robins, J.C., Ferruccio, T.-M. et al. (2010). In vitro maturation of oocytes via the pre-fabricated self-assembled artificial human ovary. *J. Assist. Reprod. Genet.* 27 (12): 743–750.

64 Cho, E., Kim, Y.Y., Noh, K., and Ku, S.Y. (2019). A new possibility in fertility preservation: the artificial ovary. *J. Tissue Eng. Regen. Med.* 13 (8): 1294–1315.

65 Radenkovic, D., Solouk, A., and Seifalian, A. (2016). Personalized development of human organs using 3D printing technology. *Med. Hypotheses* 87: 30–33.

66 Gilbert, F., Harris, A.R., and Kapsa, R.M. (2012). Efficacy testing as a primary purpose of phase 1 clinical trials: is it applicable to first-in-human bionics and optogenetics trials? *AJOB Neurosci.* 3 (2): 20–22.

67 Gilbert, F. (2015). A threat to autonomy? The intrusion of predictive brain implants. *AJOB Neurosci.* 6 (4): 4–11.

68 Gilbert, F. and Cook, M. (ed.) (2015). Are predictive brain implants an indispensable feature of autonomy. *Bioethica Forum* 8: 121.

69 Gilbert, F. (2013). Deep brain stimulation and postoperative suicidality among treatment resistant depression patients: should eligibility protocols exclude patients with a history of suicide attempts and Anger/Impulsivity? *AJOB Neurosci.* 4 (1): 28–35.

70 Bretzner, F., Gilbert, F., Baylis, F., and Brownstone, R.M. (2011). Target populations for first-in-human embryonic stem cell research in spinal cord injury. *Cell Stem Cell* 8 (5): 468–475.

71 Hess, P. (2012). Intracranial stem cell-based transplantation: reconsidering the ethics of phase 1 clinical trials in light of irreversible interventions in the brain. *AJOB Neurosci.* 3 (2): 3–13.

72 Kimbrel, E.A. and Lanza, R. (2015). Current status of pluripotent stem cells: moving the first therapies to the clinic. *Nat. Rev. Drug Discov.* 14 (10): 681–692.

73 Nakano-Okuno, M., Borah, B.R., and Nakano, I. (2014). Ethics of iPSC-based clinical research for age-related macular degeneration: patient-centered risk-benefit analysis. *Stem Cell Rev. Rep.* 10 (6): 743–752.

74 Tran, J.L. (2015). Patenting bioprinting. *Harvard Journal of Law and Technology Digest* 29.

75 Yoo, S.-S. (2015). *3D-printed Biological Organs: Medical Potential and Patenting Opportunity*, 507–511. Taylor & Francis.

76 Salvaterra, E., Lecchi, L., Gobbi, S., and Rebulla, P. (2006). The ethics of cord blood banking in light of ownership, informed consent, and solidarity. *Cell Preserv. Technol.* 4 (2): 91–96.

77 Capron, A.M. (2017). Building the next bioethics commission. *Hastings Cent. Rep.* 47: S4–S9.

78 Stambolsky, D., Bryzgalina, E., and Efimenko, A.Y. (2018). Informed consent to the receipt and use of human cellular material: juristic and ethical regulation. *Russ. J. Cardiol.* 12: 84–90.

79 Morrison, R.J., Hollister, S.J., Niedner, M.F. et al. (2015). Mitigation of tracheobronchomalacia with 3D-printed personalized medical devices in pediatric patients. *Sci. Transl. Med.* 7 (285): 285ra64–ra64.

80 Morrison, R.J., Kashlan, K.N., Flanangan, C.L. et al. (2015). Regulatory considerations in the design and manufacturing of implantable 3D-printed medical devices. *Clin. Transl. Sci.* 8 (5): 594–600.

81 Van Schothorst, M, Weeda, J, Schiffers, K, et al. (2016). Study on the Regulation of Advanced Therapies in Selected Jurisdictions: 20147306 RfS Chafea/2014/ Health/24-Final Report. European Union.

82 Heinonen, M., Oila, O., and Nordström, K. (2005). Current issues in the regulation of human tissue-engineering products in the European Union. *Tissue Eng.* 11 (11–12): 1905–1911.

83 Food and Drug Administration (2016). Technical considerations for additive manufactured devices.

84 Willemsen, K., Nizak, R., Noordmans, H.J. et al. (2019). Challenges in the design and regulatory approval of 3D-printed surgical implants: a two-case series. *Lancet Digital Health* 1 (4): e163–e171.

85 Australian Regulatory Guidelines for Medical Devices (ARGMD). 2011.

86 *Australian Regulatory Guidelines for Biologicals (ARGB).* 2014.

87 Mourby, M. and Morrison, M. (2020). Gene therapy regulation: could in-body editing fall through the net? *Eur. J. Hum. Genet.* 28 (7): 979–981.

88 Meij, P., Canals, J.M., Lowery, M., and Scott, M. (2019). Advanced therapy medicinal products. *Brief Pap.* 3: 2020.

89 Bogdanov, D. (2019). Bioprinting technology as a legal challenge: determining the model of legal regulation. *Lex Russica* 6: 80–91.

90 Grin, O., Grin, E., and Solovyov, A. (2019). The legal design of the smart contract: the legal nature and scope of application. *Lex Russica* 8: 51–62.

91 Kelly, E. (2017). FDA regulation of 3D-printed organs and associated ethical challenges. *U. Pa. L. Rev.* 166: 515.

92 Kirillova, A., Bushev, S., Abubakirov, A., and Sukikh, G. (2020). Bioethical and legal issues in 3D bioprinting. *Int. J. Bioprinting* 6 (3): 2020.

9

Planning Bioprinting Project

Anish Deb[1], *Prosenjit Saha*[1], *and Debashis Sarkar*[2]

[1] *Centre for Interdisciplinary Sciences, JIS Institute of Advance Studies and Research, Kolkata, West Bengal, India*
[2] *ME Department, Asansol Engineering College, Asansol, West Bengal, India*

9.1 Introduction

People from diverse industries are currently accepting bioprinting more and more, comparable to engineering, healthcare, bioengineering, biomedical research, and so forth. Bioprinting is advancing in a similar manner to how additive manufacturing is doing so quickly. Any organ cannot live for a longer period of time since organ preservation is exceedingly tough. Additionally, the procedure for replacing an organ is quite expensive because there are so many steps required, such as taking the organ from the donor, transporting it via air or land, and then performing the transplant at a hospital. Additionally, all these processes are time-consuming. In numerous instances, patients pass away while waiting for an organ transplant [1, 2]. Since a long time ago, scientists have been searching for alternatives to organ replacement [3]. Computed tomography (CT) scans and magnetic resonance imaging (MRI) scans are two types of medical imaging that are available [4]. Various image processing programs can be used to process these images. MIMICS is a program that can be used to process images. This software processes medical picture raw data in accordance with the patient's needs. This software exports the data in STL file format. Graphics programs like AutoCAD, SOLIDWORKS, CATIA, and ANSYS can read the STL file format. These programs allow for the import of files in the STL file format. In the

3D Bioprinting from Lab to Industry, First Edition. Edited by Prosenjit Saha, Sabu Thomas, Jinku Kim, and Manojit Ghosh.

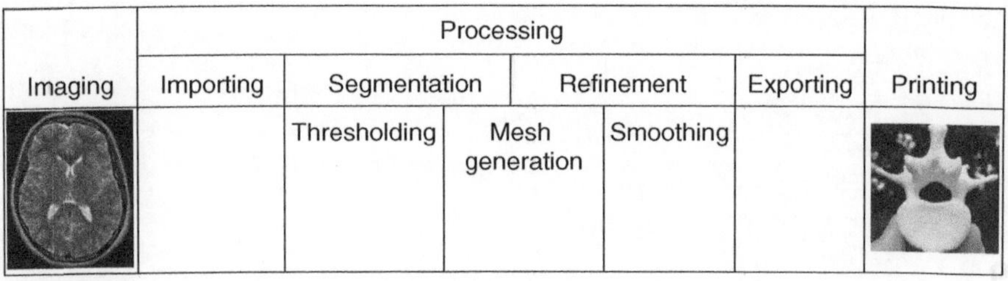

Figure 9.1 Basic workflow for dealing out the patient image volume into 3D printable models.

graphics software, a three-dimensional (3D) model was created (Figure 9.1). Not only models of regular-size components are developed, but critical geometry shapes can also be generated without much effort. Regular additive manufacturing works on acellular type of structures. The 3D shape is achieved by layer-by-layer deposition of fused filament material in case of fused deposition modeling (FDM) of 3D printing. In bioprinting, cellular material is deposited through the nozzle. Cells are placed in the hydrogel medium called as bioink. In this environment, cell differentiation is possible. Liquid hydrogel is called as bioink. Bioink is deposited on the substrate as layer on layer [5, 6].

Currently, clinicians are using 3D printing for surgical planning, patient education, and preparing prostheses. The common cases used include customized implants, surgical cutting guides, and anatomic models [7]. Anatomical models help in planning for surgery with a better surgical outcome. Even though anatomic models reduce the risks relatively with the cutting guides and implants, these are treated as medical tools that require meeting clinical requirements for the safety of the patients. A more systematic approach and experience are collected from critical attributes in the digital design and fabrication process. Several steps are required to convert digital images into 3D printable format. Software algorithms isolate the anatomic regions of interest (ROI) (indicated while segmentation). They transform ROI volume into the surface mesh in addition to removing the edges in turn errors, which are known in the character of smoothing. Several algorithms are developed that assist users in identifying the boundaries joining the complex anatomical regions within the medically captured images by segmentation along with further increasing accuracy by refining processes (Figure 9.2). When users have knowledge about the key algorithm and parameters, they can create anatomic models with better accuracy, which finally leads to patient safety. Studies by many researchers identified differences in accuracy between specific software programs.

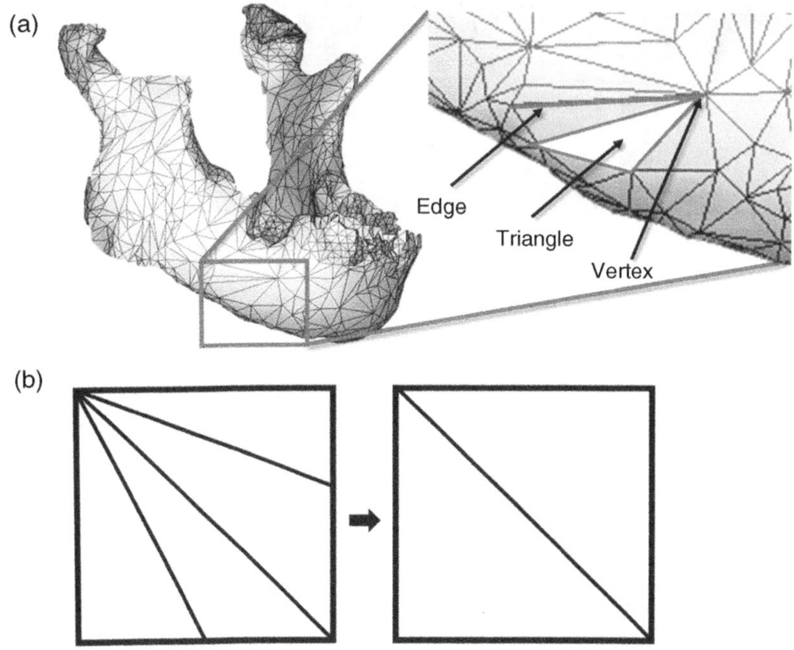

Figure 9.2 Meshing fundamentals: (a) solid mesh of mandible through key mesh features recognized. (b) Decimation of features as maintaining fidelity of features. *Source:* [8] / Springer Nature / Licensed Under CC BY 4.0.

9.2 Background: Image Capturing and Solid Model Preparation of Virtual Anatomical Model for 3D Printing

Volumetric medical imaging data such as CT with MRI are commonly the basis data for 3D printed models. They consist of volumetric pixels, indicated as voxels. The size uniformity (isotropy) of voxels depend upon the slice thickness (z) and pixels (xy). These are again dependent on scanned hardware parameters, assigned field of view, and matrix size. For example, for the same exposure of the images, larger voxels do carry a higher value of signal-to-noise ratio, but decrease spatial resolution and increase in volume averaging from an adjacent structure with the voxel. Segmentation is the process of transferring voxels that contain an anatomical ROI, by creating a new follow-on volume, referred to as a mask. The mask, a stir-shaped appearance, is for the segmentation of similar voxel size. It does not represent the true contour of tissue or bony structures. It contains a portion of adjacent tissue that

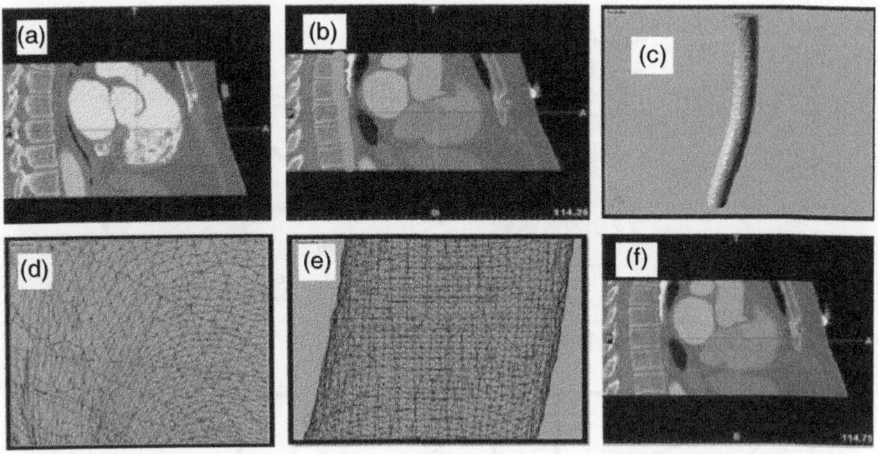

Figure 9.3 (a–c) Segmentation of an aorta from a spine process based on region growing, (d–f) STL file of the aorta, mesh structure. *Source:* Can et al. [1] / with permission from Elsevier.

is considered within a given voxel size. Mesh generation is the next step after the segmentation and preparation of the mask (Figure 9.3). The outer contour of the mask is considered for mesh generation. In mesh generation, the outer contour of the volumetric data is considered. Surface meshes are made from triangles of different sizes. In general, the surfaces begin with uniform size triangles. The area between any three vertices is called the face. The intersection of the vertices is connected by edges. The more the triangles in the mesh, the more accuracy of contour can be achieved. Curved features can be approximated. However, more dense mesh will make computation difficult.

So a balance between complexity and dimensional fidelity needs to be maintained. Largest influence on the accuracy of the model depends on the quality of the scan, segmentation done manually for complex soft tissue [10]. Availability of automated methods increased with the improvement of algorithms for segmentation and refinement [11]. Two image processing steps have a significant impact on model correctness be mask editing with mesh smoothening. As the first mask has been decided, the next process is region growing. It examines the voxels in the neighboring region for refinement of the threshold of the mask. With recent developments in algorithms, programs may semiautomatically refine the mask by several methods to find the ROI contour along with the generation of 3D surface mesh. Segmentation creates a volumetric ROI and then changes into a surface mesh. By the smoothing process, the formed mask is smoothened by refining sharp contours, overlapping between two surfaces, and flattening attributes as per the user's definite setting [12]. Decimation reduces the number of triangles created in the mesh, which reduces the file sizes and also the complexity of maintaining topology. Geometrically irregular shapes images are created from the patient images; it creates a very complex model and decimation reduces the computation load.

Commonly, the mesh smoothening algorithm works on the Laplacian smoothing technique, a vertex-based nonlinear method that converges to a point. Another modification such as Taubin smoothing (Table 9.1) includes a compensating inflation step after each mesh shrinkage takes place. Methods like angle based, bilateral, and curvature are considered for the preservation of details like sharp corners, very small radius curvature, and thin walls. Most of the programs are based on user selection options to maintain small features or boundaries. New methods are investigated for mesh smoothening, and they are introduced for new intricate shapes (Figure 9.4) [13].

Tissue engineering, an interdisciplinary field in biology, biomaterials, and engineering, is trying to obtain restoration of tissue function by designing soft tissue construct 3D models. It also provides the required environment for cell growth and attachment, regeneration of tissue, fluid flow, and structural integrity. The tissue 3D model tries to mimic the external geometry (anatomical shape) and also generates the internal geometry of the replaced tissue. Both the geometries are very crucial for exactly representing the correct shape of the replaced tissue. Noninvasive medical imaging systems like CT and MRI techniques are used on a large scale to collect accurate patient information. On completion of the 3D volumetric outer layer of the tissue construct based on the patient-specific information collected from CT or MRI images, the next step is internal geometry. Internal geometry consists of internal channels and interconnected pores, enabling cell attachment, nutrient flow, tissue regeneration, and cell proliferation. This internal architecture has a large influence on the proliferation and growth of the cell body. It is directly related to accelerated tissue regeneration. In case a designed tissue requires transplantation, it must have anatomically correct size of organ and tissue of the host patient to engraft. Additionally, the acceptance of mechanical forces, biochemical and mechanical excitement, and stretchability to house the cell development in the patient's body are all factors that need to be considered while designing tissue constructs in order to facilitate cellular activity following bioprinting. For instance, the bulk of the bone tissues must endure mechanical strain. The bone structures were therefore created to withstand mechanical loads. Similar to how articular cartilage tissue regenerates, dynamic loads must be taken into account when tissue is repaired. Therefore, until the entire regeneration process has taken place, the intended tissue must tolerate the dynamic loads. The biological requirements for cell attachment, proliferation, signaling system, transportation of nutrients, and transfer of metabolism waste product must also be addressed by tissue constructs in a 3D environment. The mechanical, biochemical, and biological requirements should therefore be met for regenerating tissue. The bioprinting of planned tissue constructs must take into account all these properties. A 3D model is created using the medical pictures from CT or MRI scans using CAD-based software such as AutoCAD, SOLIDWORKS, CATIA, etc.

Table 9.1 Summary of nearly all common smoothing algorithms use in segmentation software. Images in the Example column show implementation of the algorithm before (left) with after (right) execution [28].

Smoothening Type	Definition	General Equation	Restrictions	Example
Laplacian	Vertex based Vertex is relocated to the centre of neighboring vertices	Closed Curve Convergence to point $p_i \leftarrow p_i + \lambda(p_i)$ *for* $0 < \lambda < 1$ $\lambda =$ *"dampening coefficient"* $p_i^{(t+1)} = p_i^{(t)} + \lambda \Delta p_i^{(t)}$ *for meshes*	Triangle size effects Without weighting shrinks mesh volume with each iteration may invert edges/triangles	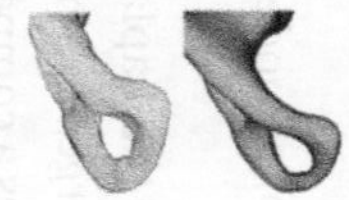
Recursive Gaussian	Vertex based Iteratively set each vertex to be linear combination of all neighbors; variant set's vertex to centroid of weighted sum of centroids of adjacent faces	"Zero-Pass" Filtering with adjustments $g(t;\sigma) = \frac{1}{\sqrt{2\pi\sigma}} e^{-t^2/2\sigma^2}$ $\sigma = Gaussian \frac{1}{2} width$	Too many iterations results in loss of surface details. Choosing too high σ results in no smoothening because of noise/jitter	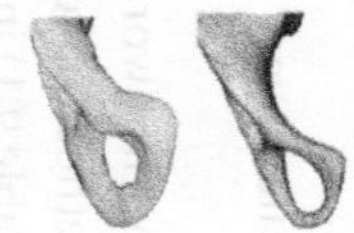
Taubin	Vertex based Laplacian with correction for shrinkage	$p_i \leftarrow p_i + \lambda\Delta(p_i)$ (*shrink*) $p_i \leftarrow p_i + \mu\Delta(p_i)$ (*inflate*); *for* $\lambda > 0$ *and* $\mu < 0$ *OR* $\mu < 0$ & $\lvert\mu\rvert > \lambda$	Requires vertex interpolation to minimize shrinkage	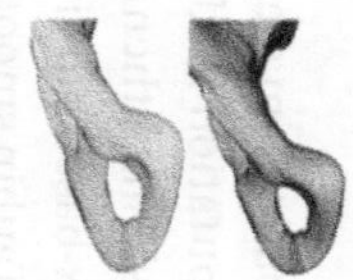
Angle	Angle based Assigns minimal angle, iterates through mesh and retriangulates until all of the angles are less than minimal angle.	Node acts like torsion spring $E = \sum_{i=0}^{2(n-1)} \frac{1}{2} k\theta^2$ $n =$ # node connected to internal node $k =$ constant $\theta =$ angle between internal and i-th node	Works best on concave domains	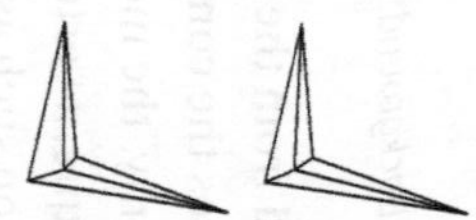

Bilateral	Approximate surface of points via tangent plane and down weight points with poor fit. Based on similarity and closeness.	Low pass filter Domain $h(x) = k_d^{-1}(x)\int_{-\infty}^{\infty}\int_{-\infty}^{\infty} f(\xi)c(\xi,x)d\xi$ Range $h(x) = k_r^{-1}(x)\int_{-\infty}^{\infty}\int_{-\infty}^{\infty} f(\xi)s(f(\xi),f(x))d\xi$ x = *neighborhood center* ξ = *nearby point*	Parameters can be selected for edge preservation but are dependent on mesh quality	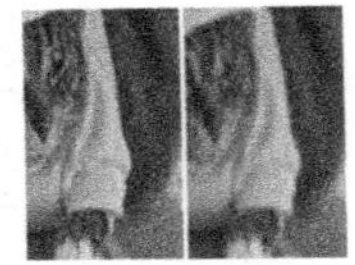
Curvature	Mean curvature based Adjust local curvature Anisotropic smoothening	$P_{new} \leftarrow P_{old} + \lambda H(P_{old})n(P_{old})$ $n(P)$ = *vector of all inward pointing normals* $H(P)$ = *amount to move each point inward*	Considering the curvature continuity and maximum curvature constraints can be addressed by using cubic Bezier curves	

Source: [28]. Magdalene et al. 2022 / Springer Nature / CC BY 4.0.

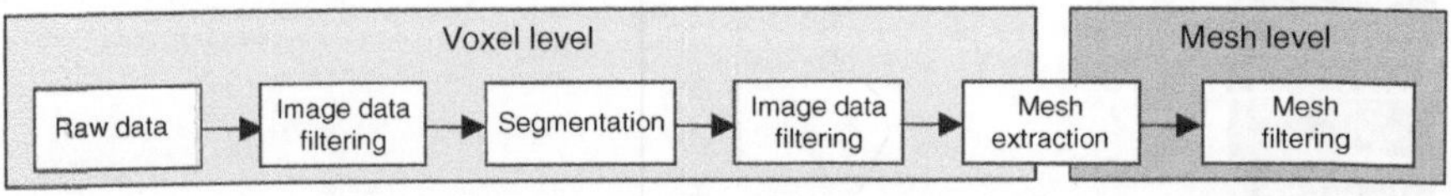

Figure 9.4 Surface extraction pipeline, cuts roughly into a voxel level and a mesh level [13].

(Figure 9.5). To develop the interior structure of the tissue form, many techniques are utilized, including freeform systems, space-filling curves, and implicit surfaces. To create path plans in parametric or Cartesian form as the model is finished, toolpath algorithms are used.

The inside design of the artificial tissue constructs mimics both the internal and outward architecture of the tissue being replaced. For cells to survive, the internal architecture of the tissue needs to have critical elements such as proteins, nutrition, waste elimination, growth factors, mass, and fluid movement. The nutrient flow is influenced by the channel and pore surface chemistry, shape, orientation, size, interconnectedness, and other factors. Additionally, pore connectivity significantly influences cell migration and tissue ingrown. Thus, the survival and growth

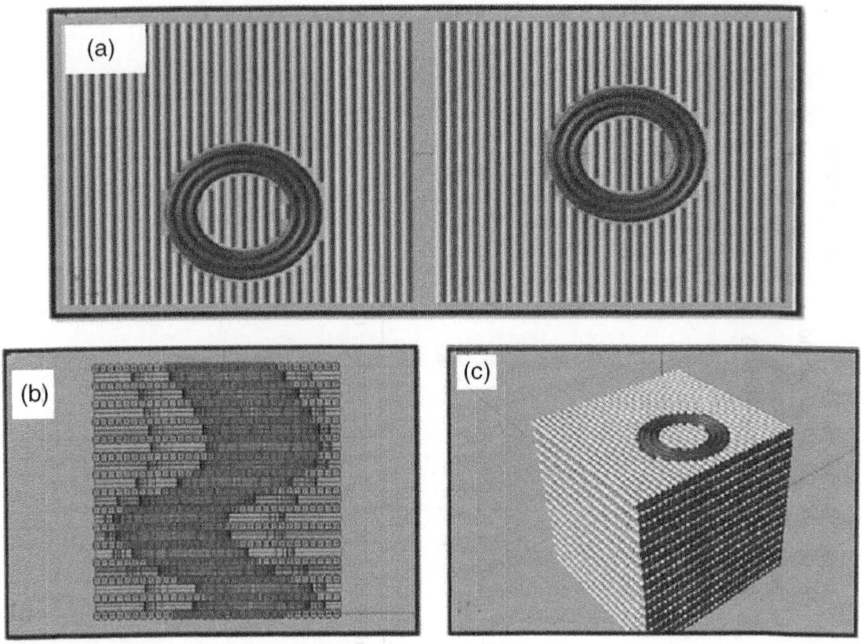

Figure 9.5 (a) Top view of cell aggregates (black) and their support material. (b) Cross-sectional right view of the whole cell and support structure, (c) perspective view of the whole cell and support structure enclosing the aorta model. *Source:* [9] / with permission of Elsevier.

of replacement tissue depend significantly on an internally connected network of channels. This enables the liberation of biomaterials and biomolecules, as well as substantial fluid and nutrition transfer. These characteristics in designed tissue will promote cell proliferation and increase cell growth. The ideal porosity must be established in order for cells to connect and develop. High porosity affects cell development and proliferation, but it also undermines the togetherness and load-carrying capacity of the designed tissue construct. Therefore, a crucial range exists for optimal porosity. The distribution of porosity must concurrently suit the biological and mechanical requirements. Due to manufacturing and design restrictions, the majority of artificial tissues are manufactured with a constant porosity level; nevertheless, porosity may vary geometrically depending on biological, mechanical, and functional necessities. Cell development increases throughout the bone's core and begins on the outside side of the bone. Therefore, there should be more porosity on the outer and less on the interior. Based on each region's functional requirements, the porosity density is chosen. To develop the material and microstructural properties of the artificial tissue, replicating the porosity and elastic properties of bone tissue, an image mask-based homogenization method is used for integration with topology optimization approaches. Porosity with stiffness characteristics was imitated in natural tissue. Porosity is necessary to allow optimal permeability for the transport of genes or cells with the appropriate stiffness property of human bone tissue. It can be accomplished by a selection of appropriate biomaterial with unit cell architecture. By considering unit cells amidst a library, as the component and mapping using CT scan pictures and desired porosity, each layer's porosity may be determined. Another attempt to add pores to existing channels is biomimicry of the microvasculature. The patterning channels of the microvasculature network can be created using one of two methods. The first method involves directly bioprinting the circulatory system with a coaxial nozzle network. Another method is indirect bioprinting. It involves the printing of a vascular network on a sacrificial polymer that may be eliminated by thermally induced de-crosslinking.

The conduct of tissue build-up after planting is significantly impacted by design parameters like mechanical, biological, geometric, transport, and fabrication. Scaffold design and manufacturing should take into account the choice of biomaterial, internal structure, porosity level with connection, external geometry, permeability, and a procedure that is compatible with bioprinting. In addition to the fact that tissue fabrication is a crucial phase in the tissue engineering process, the environment, process control, resolution, and bioink choice all have a big impact on this operation. Bioinks selection is crucial for scaffolds designed specifically for tissues. These factors affects the design and construction of the scaffolds. Bioactive ceramics like hydroxyapatite (HA) along with binders like poly(vinyl alcohol) (PLA) should be present in hard bone tissue bioink. Similar soft tissue, like the skin, calls for a hydrogel-based bioink (collagen and hyaluronic acid). Different bioink materials might need various fabrication techniques. The mechanical strength is influenced by the fabrication technique based on rheological properties and crosslinking mechanisms. Failure of the scaffolding may occur if the bioink droplets are not deposited in the intended locations. Based on the rheological characteristics of the bioink, the software program should strategically plan the toolpath and bioprinting process parameters.

Geometric features of the construct should produce a realistic and anatomically correct approximation of biological tissue and organs when designing tissue constructions, such as tissue scaffolds, tissue, and organ substitutes. For imaging of live tissue and organs, non-invasive procedures like CT and MRI are frequently employed. Ultrasound, nuclear medicine, micro-CT, micro-MRI, and other imaging techniques are also employed to capture images of the regenerating tissue and organs. Based on CT and MRI data, the National Institutes of Health (NIH) developed an online archive of selected sample medical images. Additionally, the 3D Print Exchange of NIH is a platform that gives scientists access to higher quality, scientifically accurate 3D models available in a format suited for printing (Table 9.2).

A sensitive technique for tracking structures with functional alterations within tissue, e.g., tissue regeneration and death, is MRI. These changes in images are reflected by local differences in nuclear magnetic resonance timings and tissue hydration, including its physical condition (e.g., free diffusion and protein bound). MRI uses pulsed radio frequency electromagnetic pulses for the excitation of hydrogen atoms, which are abundant in water and fat present in the human body, and produce detectable radio frequency signals. Magnetic field gradients are produced when the pulsed radio wave parameters are altered, which also affects the nuclear spin energy transition of protons. It therefore causes contrast between the resulting images. A magnetic resonance coil picks up the generated radio frequency signal and sends it to image-processing software. MRI is frequently used to scan soft tissue parts of the human body. MRI is a preferred imaging technique because there is no use of ionizing radiation. The highest resolution in 3 T is 250 m × 250 m × 0.5 mm, and the scan time ranges from 5 to 40 minutes. With a magnetism of 7–9 T, an MRI can reach a higher spatial resolution (5–200 m). In this case, the contrast between soft tissue can be enhanced by the addition of magnetic nanoparticles, a contrasting agent. MRI images can be used to obtain information on soft tissue's anatomical, functional, and cellular structures. The average magnetic field that the human body is exposed to is no greater than 3 T, and longer scan times enhance image resolution.

To view the hard tissue parts of the human body, a CT scan is a medically noninvasive scanning procedure. An X-ray tube emits a conic beam of electromagnetic radiation that selectively penetrates the body during a CT scan. Hard tissue, such as bone, absorbs X-rays significantly higher than soft tissues. The tissue is first scanned cross-sectionally using a detector on the scanner's opposite side, and the obtained results are then further processed to create the desired area of the part of

Table 9.2 Program-specific user intervention level at different workflow steps. For the most part of the segmentation process, "automatic" denotes the usage of menu option otherwise built-in "turnkey" capability. Segmentation phases classified as semi-automatic required user interference but made apply of built-in operating control where user parameter had to be there for established. Steps to facilitate all or nearly all manual user interference are labelled as manual. Following export, final models were utilised for further computations [8].

			Intervention Level				
Program	**Segmentation**	**Parts**	**Thresholding**	**Mesh generation**	**Mesh Refinement**	**Export to STL**	**Final Model**
Program 1	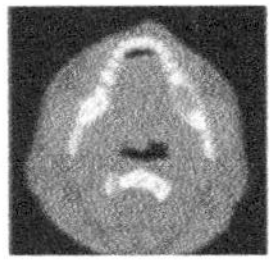	Mandible	Automatic	Automatic	Semi-Automatic	Automatic	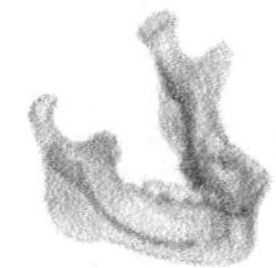
		Tumor	Automatic	Automatic	Manual	Automatic	
		Tumor Nerve	Semi-Automatic	Automatic	Manual	Automatic	
		Healthy Nerve	Semi-Automatic	Automatic	Manual	Automatic	
Program 2	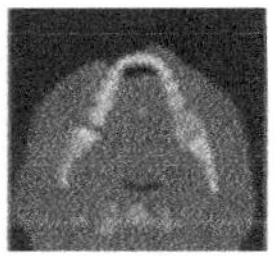	Mandible	Automatic	Automatic	Semi-Automatic	Semi-Automatic	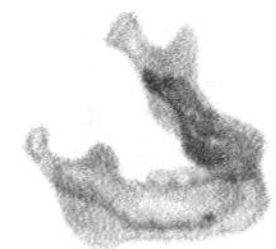
		Tumor	Semi-Automatic	Automatic	Semi-Automatic	Semi-Automatic	
		Tumor Nerve	Semi-Automatic	Automatic	Manual	Semi-Automatic	
		Healthy Nerve	Semi-Automatic	Automatic	Manual	Semi-Automatic	
Program 3	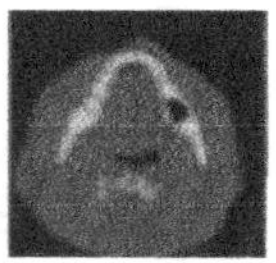	Mandible	Automatic	Automatic	Semi-Automatic	Automatic	
		Tumor	Automatic	Automatic	Manual	Automatic	
		Tumor Nerve	Semi-Automatic	Automatic	Manual	Automatic	
		Healthy Nerve	Semi-Automatic	Automatic	Manual	Automatic	
Program 4	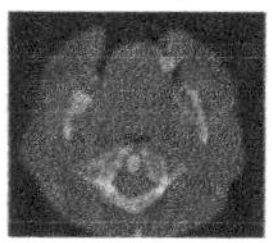	Mandible	Automatic	Automatic	Automatic	Automatic	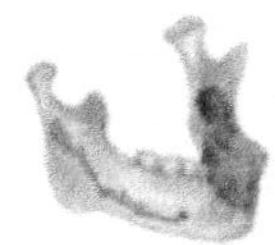
		Tumor	Automatic	Automatic	Automatic	Automatic	
		Tumor Nerve	Automatic	Automatic	Semi-Automatic	Automatic	
		Healthy Nerve	Automatic	Automatic	Semi-Automatic	Automatic	

the human body. A revolving device rotates the scanned body at an angle ranging from 0.25° to 1° and covers a scanned area of 360°, providing 360–1440 images. Using a tomographic reconstruction method, these images have been overlaid. In a short amount of time, a CT scan can produce images with a greater resolution of 0.24–0.3 mm, but this comes at a danger to the patients because it uses ionizing radiation. Only a small dose of individuals were subjected to CT scans at a specified frequency. Medical devices with batteries can be utilized securely during a CT scan because the images produced by the procedure are not sensitive to metal implants. Bone and tumor tissue have both been visualized using CT scans, which can also differentiate between soft tissue and the borders of bone tissue. CT has also been utilized in tissue engineering, notably to track the regeneration of tissue in tiny animals like rats. High resolution in tiny volume samples up to 1–200 m is possible with micro-CT. Iodine and nanoparticles are contrast agents that can enhance picture quality and identify the soft tissue component.

9.2.1 Other Imaging Techniques

Images from MRI and CT scans are used to build scaffolds and evaluate how well they integrate after being implanted. Using ultrasound imaging is another effective method for this. Real-time images of an organ in motion and the flow of blood via a conduit can be recorded during an ultrasound scan. Additionally, local fluid flow is visualized by Doppler ultrasound imaging. Microbubbles are utilized to increase the sensitivity of the microvasculature blood flow visualization system. With the help of ultrasound elastography, it is possible to take precise measurements of the mechanical characteristics of regenerating tissue in images. Additionally, nuclear medicine has been employed as a potent tool for studying the functional and molecular information of target tissue or organs, like using positron emission tomography (PET) and single-photon emission computed tomography (SPECT). In a rabbit critical size deficiency model, the progenitor and the stem cells inside the artificial tissue have been examined for their metabolism and functional activity toward repairing cranial tissue. The type of tissue to be manufactured, as well as the image's resolution and quality, influence the choice of the best imaging modality for bioprinting design purposes. Researchers have recently experimented with multimodal imaging techniques such as MRI with CT, CT with PET, and CT with SPECT (Table 9.3).

9.2.2 Digital Process for STL Generation

In all programs, the method for generating an STL file from DICOM scans may be summarised into five key steps: Importing DICOM, segmentation, mesh/model creation, smoothing and mesh refinement, also exporting to STL file (or another

Table 9.3 Resemblance between medical imaging techniques for regenerative medicine and tissue engineering.

Modality	Resolution	Scan time	Maximum volume	Preferred tissue types	Advantages	Disadvantages
MRI	250 μm × 250 μm × 500 μm	5–40 min	Human body	Soft and hard tissue (bone)	• Radiation exposure risk nil • Images of moving components like heart contractions in 4D can be captured	• Intolerant to body movement during the scanning process and in vivo medical metallic implants • Extended scanning durations for improved photographs • Patients may undergo claustrophobia and anxiety
μMRI	5–200 μm	6–24 h	Full body	Soft and hard tissue (bone)	• Radiation exposure risk nil • Incorporation of contrasting agents helps in molecular imaging	• Higher cost • Lower portability • Long scan times
CT	0.24–0.33 mm	5 min (8–40 s of actual scan time)	Human body	Hard tissue (bone)	• More tolerant of patient body movement during scans • Higher-resolution images of different types of tissue • Tolerable of the in vivo clinical implants (i.e., metallic otherwise battery operated)	• Radiation exposure restricts the frequency and intensity of exposure; breastfeeding mothers and expectant mothers should avoid it • May negatively impact the child's immune system • Allergic reaction may be created by iodine contrast material
μCT	1–200 μm	2–4 h	Whole rat	Hard tissue (bone)	Higher resolution	Exposure to radiation restricts the amount and frequency of exposure
Ultrasound	1 mm × 1.5 mm × 0.2 mm	10–15 min	Neonatal, blood vessels	All tissue types	• Images of the moving bodily parts in real time • Radiation-free operation and ease of use	• Image quality is poor as compared to other techniques • All abnormalities cannot be detected

Source: Refs. [14–16].

file type well-matched with 3D printers) are the steps in the process. Original DICOM imports preserved the inherent picture volume features (i.e., the source data were unchanged). A single user (MF) completed all five steps of the digital workflow used for all conditions in this study, eliminating every inter-user variability. This was done with the intention of isolating and measuring the effect of various algorithms used throughout the software workflows. The phrases automatic, semi-automatic, and manual are used to broadly classify interventions based on how much user involvement is necessary to complete that phase of the process.

9.2.3 Blueprint Modeling

By segmenting an image, a designed 3D model is produced and delivered in stereolithography (STL) file format. The other formats include 3D graphics (X3D) and the virtual reality modeling language (VRML). A popular file format for interchange in finite element analysis (FEA), 3D printing, and scanning is STL. Upon successful 3D modeling in CAD software, a useful tool FEA is used to assess model's fluid flow and mechanical properties. Before bioprinting, a solidly modeled tissue construct's properties can be promptly determined using the potent virtual simulation tool known as FEA [17]. The relevant software system (bioprinter specific), which assists each bioprinting procedure in different process programs and toolpath processes, enables users to adhere to pre-guided steps for successful bioprinting. FDM-based 3D printers operate in a manner similar to extrusion-based printers. Other bioprinting techniques, such as droplet-based or laser-based bioprinting, use various bioprinting methods and call for design software that is specific to the bioprinter. The software package for piezo-inkjet bioprinters, such as the Jetlab of Microfab Inc., enables users to program a droplet pattern [18]. Microsoft PowerPoint slides marked by white and black patterns indicating transparent and opaque parts are used in stereolithography along with the manufacturer's software and a straightforward algorithm for defining masks. Transparent sections allow transmission of ultraviolet (UV) light across the mask and start photopolymerization.

9.2.4 CAD-Based Systems Characteristics

In order to design tissue for bioprinting, CAD software is used. For the construction of tissue constructs, modeling approaches including spatial occupancy enumeration (SOE), boundary representation (B-Rep), and constructive solid geometry (CSG) are applied. Using a combination of building components, such as cubic unit elements to depict large-scale solid things, SOE represents closed objects. Boundary elements like edges, vertices, and faces are crucial to B-Rep.

In order to create various models, CSG uses the solid primitive and Boolean operations. Each of the aforementioned methods has its own drawbacks. For instance, SOE and CSG depend on the union of several solid primitives. This requires a lot of data storage and expensive computation. Boundary elements used by B-Rep enclose the entire object. Complex geometry is more conveniently modeled by SOE and CSG than by the B-Rep approach. Typically, commercially accessible CAD programs like CATIA, SOLIDWORKS, and PTC Creo are used in conjunction with all of the aforementioned methods to create tissue constructs [11]. Pore designing elements from the design library are utilized to produce a porous geometry, which is then removed from the external surface geometry. Similar to this, it is possible to create an internal porous architecture with fixed pore sizes, levels, and geometry [19]. A 3D porous lattice structure is produced by subtracting the negative geometry of the structures that create pores from the surface model (which is constructed in 3D as a periodic element). The designing algorithms took into consideration a number of pores-making models, including the octahedron, tetrahedron, triangular, and square prism by means of adaptable geometry as geometric dimensions. The boolean operation is performed to add primitives for building the pores or structure and the boolean function called from the designed library. Further integrating into bigger scale tissue structures like the femur and spine were these unit cells. To increase the heterogeneity in the intended models, several unit cells can be joined in composite structures. Although CAD software packages are quite capable of creating tissue scaffolds, they perform poorly and inefficiently when trying to simulate porous architecture and have little control over the non-Euclidian solid and biomechanical properties that are generally attributed to tissue constructs [20].

9.2.5 Image-Based Systems

CAD modeling systems are a time- and money-intensive tool for creating and simulating tissue constructs. Another strategy is image-based design. In the image based design approach the defected region in the CT scan and MRI images were identified and filled by stack of 3D images [9]. Here, a pile of two 3D images prepared from binary unit cells was used to highlight and fill the faulty regions in imaging like MRI and CT. Assigned with a random number generator-based method, tissue constructions with regular porous designs and irregular porosity schemes might be created. In this, the flaw generates several voxels at random [21]. This method was first used in the construction of mandibular bone, orbital floor segments, and sample tissue using a selective laser sintering method. Instead of producing a CAD model quickly, different manipulations in 3D design architectures employing the aforementioned image-based approach produce defective models [22]. By assigning various design architectures to specific regions of the defected sites, it also gives the user more ability to vary the porosity network.

9.2.6 Freeform Systems

Both computer-aided methods and image-based systems can build uniformly porous structures. A tissue model with a flexible architecture can be made using a freeform-based methodology. Due to the freeform-based system's ability to be divided into many regions, the population of materials and the architecture in different regions may vary. As an illustration, consider a wound device surface that has been modeled using 2D wound photos. Through the use of Image J software's intensity function, this surface is further transformed into a 3D representation. With respect to the center point of the top surface of the wound, this generated surface is further approximated by non-uniform rational B-Spline (NURBS) surfaces next to lofting. Lofting, also known as skinning, is a method of creating surfaces using a number of curves referred to as generators. This procedure creates a number of zones, each of which is filled with rasters of various materials. Researchers have used different biomaterials and quantities of sodium alginate to 3D bioprint several different kinds of wound devices. It aids in synchronizing the transportation of biomolecules for healing a wound. When designing 3D models of the vertebra, femur bone, and aorta for porous media architecture, the freeform system is also used. Tissue can be created and printed with structural control over the geometric architecture along with material compositions using the freeform design approach [23]. Also comprehensive control over the release mechanism of biomaterials and other biologically agile elements, along with optimization in the build direction and total fabrication time are currently considered for freeform objects.

9.2.7 Designs Using Implicit Surfaces

The inability of the CAD and image-based systems to quickly depict a very complicated structure is one of their primary flaws. The degree of primitive complexity sets a limit on geometric complexity [24]. Therefore, implicit functions are utilized to demonstrate periodic minimum surfaces for tissue scaffolds with complex shapes, which can precisely control the distribution of porosity in space inside an arbitrary-shaped structure. The benefits of minimum surfaces are their architecture's lightweight and strong structural strength. It has been demonstrated that triply periodic minimum surfaces (TPMSs) exist [25]. The space is divided into two interconnected labyrinthine domains by smooth infinite surfaces called TPMSs, which are periodical in all three different latticework directions. For tissue biofabrication, the Diamond and Gyroid designs by Schwartz and Schoen were 3D printed using TPMSs [26]. In order to examine mechanical feedback responses in deformations due to stretching and bending for bone tissue engineering, scaffolds with heterogeneously porous designs were exhibited.

9.2.8 Space-Filling Curves

Solid primitives are used as building blocks in image-based systems or CAD models to develop bigger size tissue construct models. In the majority of 3D printing methods, this is difficult to support. Extrusion-based bioprinting (EBB) is one of the most used methods for 3D bioprinting tissue. It depends on the layer deposition using a 0/90° lay-down raster orientation. Other design approaches in this pattern, like the freeform method approach, are difficult to implement at EBB. The interior architecture of the tissue construct is therefore designed using an alternate method that uses toolpath controlled by space-filling curves. Hilbert and Sierpinsky curves, as well as material deposition patterns with raster angle patterns of 0°/45°/90°/135°, have been employed in creating models by several space-filling curves. The filling curve's design architecture was modified to produce periodic sub-patterns that made it possible to build each layer using EBB in order to provide adequate support for structures with structural integration, also the direction of space-filling curves for the subsequent layers was altered according to raster orientation considered. For instance, Hilbert curves were further studied for the purposes of discriminating the whole region by unit cells and allocating various Hilbert scales to every sector. It is possible to implement a continuous toolpath plan by combining the Hilbert curves of various regions. As a result, the purpose of filling curves offer a platform that is harmonious with EBB as well as its associated methods (such as mechanical, pneumatic, and solenoid micro extrusion). When branched structures were bioprinted using EBB, a similar method pertaining to the Lindenmayer method was used to imitate various vascular networks (Table 9.4).

9.2.9 Planning of Toolpath for Bioprinting

Currently, bioprinting of organ and tissue structures uses two different types of toolpath planning strategies. The purpose of bioprinting heavy tissue constructs is to fill simple, elementary geometry, like cubic tissue constructs. Therefore, highly specialized tool path planning is not necessary, and the geometry can be simply filled using typical toolpath plans without much need for accuracy. Examples of this include bioprinting non-porous bulky structures for pharmaceutics and regenerative medicine, namely tissue constructs [32] and tissue replicas [33, 34].

Two distinct toolpath planning strategies, comprising bioprinting raster orientation and Cartesian and parametric configuration, have been put forth in the literature. The authors here omit toolpath planning in LBB and DBB modalities because their adoption in bioprinting is significantly lower as compared to EBB and because bioprinters related to these processes use manufacturer-directed toolpath planning procedures (in Cartesian configuration).

Table 9.4 Comparison of blueprint modeling methods worked in tissue construct bioprinting.

Blueprint modeling techniques	Advantages	Disadvantages	Compatible bioprinting modalities
CAD-based systems	Faithfully depict the simple forms; simple to model.	Computationally costly; processing is sluggish; complication is constrained to unit cell resolution.	LBB and DBB procedures that rely on cell transfer
Image-based systems	Utilizes clinical imagery; does away with 3D segmentation; makes anatomical architecture easier.	Not compatible with LBB and DBB; low accuracy.	LBB processes are based on photopolymerization
Freeform system	Facilitate heterogeneous design environments; computationally efficient; supports various geometric complexities.	Incompatible with most CAD programs on the market.	DBB, LBB, and EBB
Implicit surfaces	Produces structurally sound design models, supports a variety of geometric complexity, is computationally efficient, and promotes a heterogeneous design environment.	Not working with most CAD programs on the market.	LBB
Space-filling curves	Supports the majority of 3D bioprinters with FDM-based 3D printers; it is computationally competent and enables heterogeneous design environments.	Does not assist models with complex geometry.	EBB

Source: Refs. [27–31].

9.2.10 Cartesian Form Toolpath Planning

Due to its ease of use in producing raster layers stacked up in 3D, Cartesian (x, y, z) toolpath planning has been extensively employed in most of the bioprinting techniques, including EBB, DBB, and LBB. Researchers have been drawn to toolpath planning of EBB since it is the most widely used bioprinting technique. In EBB, deposition is rotated to produce a 0/90° laydown pattern at each layer. Even though various angles (such as 0/45° and 0/135°) can be utilized, 0/90° provides superior physical properties and structural strength as compared to other angles, which is why it is frequently employed in 3D printing process based on FDM and also bioprinting. For instance, Figure 9.6a, b shows a toolpath intended for constant downfall, low fluid clogging, but zigzag pattern has an issue with excessive bioink accumulation at the site of directional shift. Deposition stops when severe curves approach, the direction changes, and feed rate increases in the new command. Although continual deposition is visible, the homogeneity of lay-down filament thickness has significantly deteriorated. It can be improved by choosing a high feed rate as you approach sharp curves, but it cannot entirely be avoided. Tangential continuity (C1) across toolpath is a logical solution to such a situation [35].

Cartesian toolpath planning may also present challenges if functionally grouped tissue constructs are intended. For instance, a Cartesian form toolpath design was given to create multi-functional injury repair appliances (refer to Figure 9.6c, d). 3D blending method used to produce functionally graded wound devices having various biomaterial as well as biomolecular concentrations stacked in various locations. This created toolpath program in the Cartesian form, however, is simply an approximation method of genuine incline because the blending direction and toolpath deposition direction are independent together. The toolpath in Figure 9.6e adheres to Cartesian coordinates rather than blending direction in between the two features. Because the toolpath is unable to follow the geometry, a smaller size raster with closer spacing with neighboring rasters is required to improve accuracy. Motion or jump in the absence of deposition is also important because of how the toolpath is created, which is an independent feature.

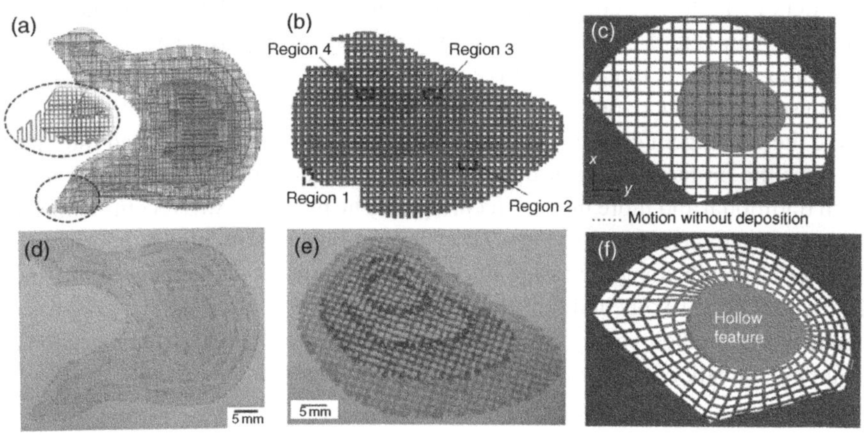

Figure 9.6 Toolpath preparation at Cartesian coordinates: (a) toolpath outline for an uninterrupted path with varying porosities, (b) bilayer bioprinted vertebrae design, (c) toolpath plan for functionally graded wound devices and variable bioink arrangement, (d) toolpath plan outline for functionally graded tissue construct compared with (e) Cartesian Coordinates and (f) parantric modeling. *Source:* Ibrahim et al. [7] / with permission from Elsevier.

Certain hollow attributes, such as blood vessel lumen, are introduced to a tissue design, and toolpath planning in Cartesian coordinates becomes more problematic. Nearly every toolpath planning method described in the articles recommends a close deposition model in Cartesian coordinates while hollow structures are applied to the object design [10]. Joana et al. [36] give a disclosure to this problem, although their offered solution was really just a better technique to occupy the available space. In that investigation, the gap was closed using an adaptive road width toolpath generation technique. However, the deposition began and stopped far too frequently due to the introduction of numerous hollow shapes. Once more, the deposition process ignored the geometry of the hollow feature and used Cartesian coordinates. As a result, toolpath-compatible parametric modeling of this geometric area may be the answer to this problem. The toolpath plan adheres to the bioink's directional inclination in Figure 9.6f, which provides an example.

9.2.11 Parametric Form in Toolpath Planning

To solve challenges with toolpath design in Cartesian coordinates, researchers have been interested in toolpath plan in the parametric (u,v) coordinates. Ozbolat et al. [37–39] recommended working with parametric regulating lines for toolpath planning in complicated geometric areas, which determines composition and porosity of material. First, a CAD file made from a medical image was split up into layers (Fig 9.7 a,e), and each layer's curves were matched with NURBS splines (Fig 9.7 b,f). After the splines were assigned with required parameters and sampled in points, a parametric ruling lines produced between the NURBS splines. Further, two tiers of toolpath generating algorithms have been deployed to produce the toolpath plan (Fig 9.7 c,d). The first layer consists of a zigzag pattern based on governing lines was fitted with curves at their intersections. Arc fitting technique then used to eliminate two consecutively steep turns within zigzag pattern to preserve the C1 tangent continuity maintaining a invariant deposition feed rate. This resolves the problems discussed previously with hollow object prototyping using both traditional and more modern techniques. The next layer was recommended to have a spiral toolpath design (Fig 9.7 d), which enhanced structural support for the zigzag pattern all along the ruling lines and was nearly at right angles to the lines in the junction. Two benefits of the spiral toolpath over the classic zigzag-based space-filling method are (1) continuous deposition over hollow structures and (2) seamless transition across material composition properties.

9.2.12 Bioprinting Methods

9.2.12.1 Extrusion Bioprinting

The most promising method is extrusion bioprinting (Figure 9.8), which enables the production of organized structures in sizes that are clinically relevant in a reasonable amount of time. This method creates a constant bioink filament trace that emerges

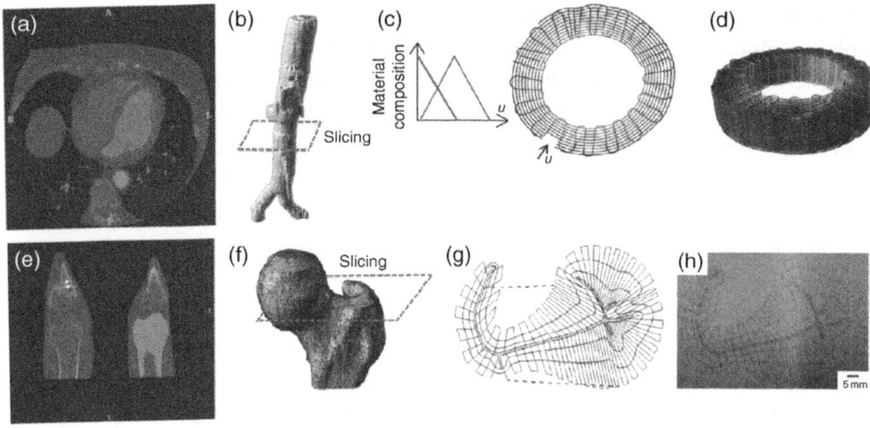

Figure 9.7 Designing a toolpath in parametric coordinates. (a,b) A toolpath plan was created with controlled material composition along the parametric distance *u* using the medical image processing described in (c,d) to create a CAD model of the aorta. *Source:* Ozbolat et al. [40], with permission from American Society of Mechanical Engineers. A femur model was created using medical image processing (e), and it was then cut into slices for toolpath planning with controlled porosity (f), which increased the parametric distance *u* (g), a sample doublelayer structure was bioprinted using sodium alginate hydrogel (h). *Source:* Ozbolat et al. [40], with permission of Elsevier.

Figure 9.8 A graphical presentation of extrusion bioprinting of stem cells. A cell laden bioink is forced from a syringe container and deposited as a stretched trace on the collector plate. *Source:* [1] / MDPI / Licensed Under CC BY 4.0.

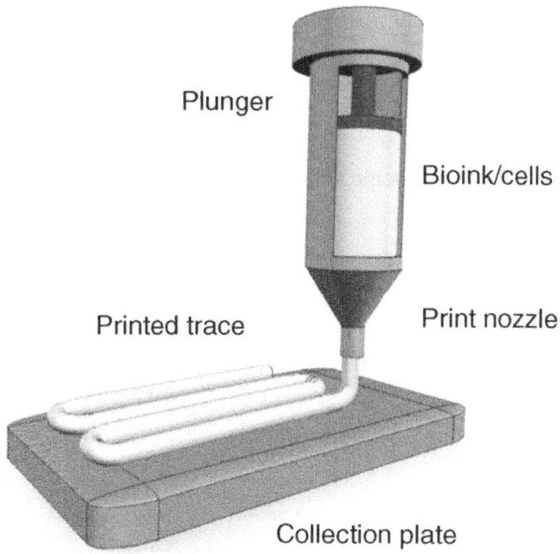

out of a syringe, resembling toothpaste coming out of a squeezed tube. Using a mechanical screw plunger or air (pneumatic) back pressure, hydrogel bioink is expelled from the syringe. For the deposition of hydrogel in 3D patterns, a syringe is coupled to a print arm that moves in the *z*–*y* direction through a collector that moves in the *x*-axis (this direction planning may differ between the bioprinter models) [12, 41].

For extrusion bioprinting, a variety of biocompatible hydrogel polymers can be used as bioinks since they can supply a significant amount of cell concentration. For printing more viscous bioinks, back pressure device works better; however, when trying to stop bioink flow from the nozzle, a delay in control has been observed [42]. The mechanical screw plunger improves patterning by giving the bioink flow more control. Mechanical screw extrusion, on the other hand, might result in significant pressure drops along the nozzle for cell printing. The deformation and apoptosis of the cells enclosed in bioink are linked to the pressure drop [43].

9.2.12.2 Inkjet Printing

The first method, inkjet-type bioprinting, was created by modifying industrial inkjet printers [44], and as a result, it shares many similarities with the method used for standard inkjet printing of papers. There are two types of inkjet printing: drop-on-demand (DOD) and continuous inkjet (CIJ). CIJ deposits individual ink droplets on carefully synchronized points, printing up in the modular droplet units, while the former creates an uninterrupted trail [45]. Only the DOD method is thought to be appropriate for bioprinting as the CIJ approach necessitates conductive ink. The careful control of a thermal or else piezoelectric actuator produces well-placed droplets from a print head that are formed from a cartridge that has been filled with cells in the hydrogel bioink (Figure 9.9). Although the actuator's element temperature can approach 200 °C, the thermal system is more frequently used for cell

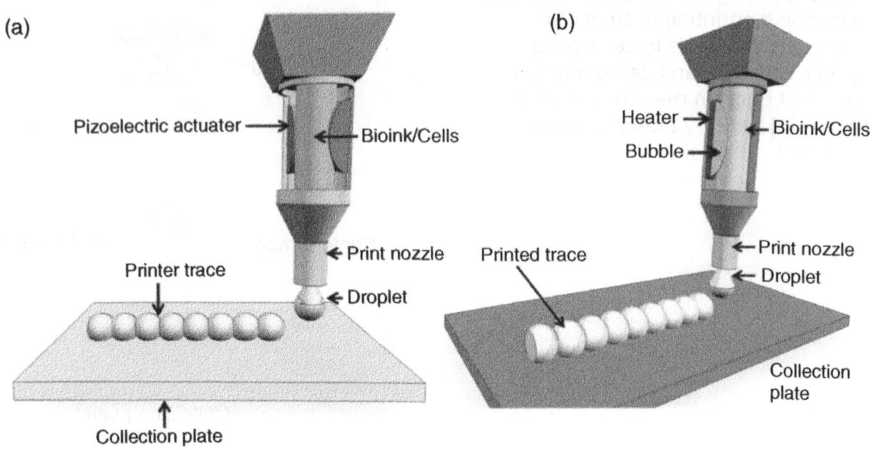

Figure 9.9 A graphical presentation of inkjet bioprinting of stem cell. Droplets of a stem cell carrying bioink are formed in the reservoir by the action of a piezoelectric (a) or thermal actuator (b) to be deposited on a collection plate. *Source:* [3] / MDPI / Licensed Under CC BY 4.0.

printing since the cells only undergo small, transient temperature fluctuations. Currently, inkjet print heads frequently use micro-electro-mechanical systems (MEMS), which have trouble printing materials with high viscosities or dense cell populations.

9.2.12.3 Laser-Assisted Printing

This approach of laser-assisted printing was developed from the laser direct writing with laser-induced transport technologies. The method works on focused laser stimulation of the upper surface of the energy-absorbing sacrificial layer. Cells containing biopolymer or bioink have a coating on the lower surface of the energy-absorbing layer. Sacrificial material vaporizes by laser stimulation, producing a pressure bubble that drives a bioink droplet onto the collecting/receiving substrate with high-precision placement (Figure 9.10). A dispensing nozzle is not necessary for laser-aided bioprinting, in contrast to inkjet printing. This eliminates nozzle clogging [46, 47] and lowers the shear stress that cells are subjected to during deposition [12]. Because it can print higher cellular densities with strong post-deposition feasibility and can use viscous bioinks, this technique is thought to have good cell printing qualities. Additionally, the resolution might drop to 10 m [48]. The method is very expensive and complex [1, 12, 49] (Table 9.5).

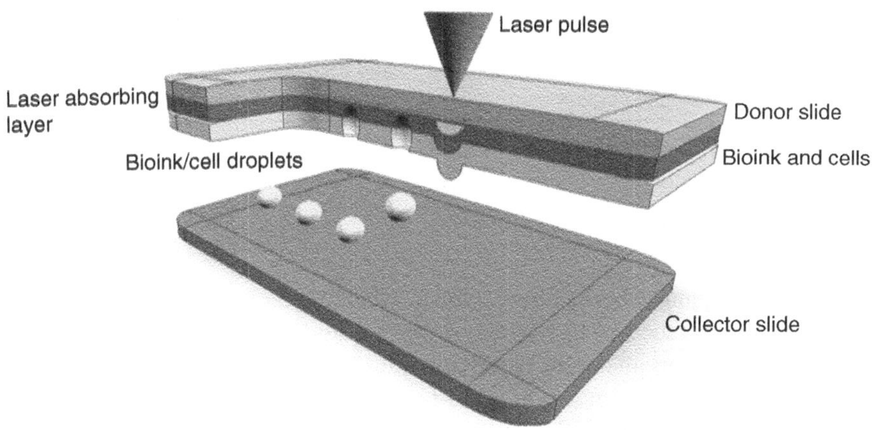

Figure 9.10 A representation of laser-assisted bioprinting of cells. The laser absorbing layer is stimulated by the laser pulse to generate a pressure bubble, thus propelling the cells onto the collector slide. *Source:* [3] / MDPI / Licensed Under CC BY 4.0.

Table 9.5 A comparison of bioprinting methods. The table is a composite of bioprinting methodology comparison tables prepared from reference.

	Bioprinting methodology		
Property	**Extrusion**	**Laser assisted**	**Inkjet**
Costs	Moderate	High	Low
Material viscosity	High	Medium-high	Medium
Resolution	Medium (100 μm–1 mm)	High (≥10 μm)	Medium to high (10 μm–1 mm)
Print speed	Continuous/slow	200–1600 mm/s	1–10 k droplets/s
Fabrication time	Short	Long	Medium
Cell viability	Medium-high, 40–95%	High, 95%	High, 85%
Cell density	High	Medium	Low
3D build-up capability	Good	Medium	Low
Benefits of stem cell bioprinting	Prints ECM-like hydrogels and cells at physiologically relevant density.	Low shear stress levels on cells during delivery, which may affect phenotype.	Nozzle free and exerts low shear stress on cells during deposition. High resolution.
Disadvantages of stem cell bioprinting	Prone to higher shear stress that can affect viability and phenotype.	Struggles with viscous polymers and high cell densities.	Can cause mechanical deformation of the cells.

Source: [1] / MDPI / Licensed Under CC BY 4.0.

9.3 Conclusion

Bio-printing is going to be the future trend to develop and repair of bio-organs. This chapter concludes the step by step description of; study of the organ, model development of the organ, materials requirement for the printing of the organ, slicing and editing of the solid model to reach towards real, physical organ and finally printing of the finished organ. Particularly the emphasis has been given on processing of medical images, segmentation and smoothing techniques, creation of solid model and slicing methods suitable for bioprinting. Further brief report some commonly available bioprinters, their applications and limitations are mentioned.

References

1 Irvine, S.A. and Venkatraman, S.S. (2016). Review: bioprinting and differentiation of stem cells. *Molecules* 21: 1188. https://doi.org/10.3390/molecules21091188.

2 Murphy, S.V. and Atala, A. (2014). 3D bioprinting of tissues and organs. *Nat. Biotechnol.* 32: 773–785. https://doi.org/10.1038/nbt.2958.

3 Roche, C.D., Lin, H., Huang, Y. et al. (2023). 3D bioprinted alginate-gelatin hydrogel patches containing cardiac spheroids recover heart function in a mouse model of myocardial infarction. *Bioprinting* 30: e00263. https://doi.org/10.1016/j.bprint.2023.e00263.

4 Arslan-Yildiz, A., Assal, R.E., Chen, P. et al. (2016). Towards artificial tissue models: past, present, and future of 3D bioprinting. *Biofabrication* 8: 014103. https://doi.org/10.1088/1758-5090/8/1/014103.

5 Heid, S. and Boccaccini, A.R. (2020). Advancingbioinks for 3D bioprinting using reactive fillers: a review. *Acta Biomater.* 113: 1–22:https://doi.org/10.1016/j.actbio.2020.06.040.

6 Zhang, Z., Chenming, W., Dai, C. et al. (2022). A multi-axis robot-based bioprinting system supporting natural cell function preservation and cardiac tissue fabrication. *Bioact. Mater.* 18 (2022): 138–150. https://doi.org/10.1016/j.bioactmat.2022.02.009.

7 Ozbolat, I. and Gudapati, H. (2016). A review on design for bioprinting. *Bioprinting* 3: 1–14. https://doi.org/10.1016/j.bprint.2016.11.001.

8 Fogarasi, M., Coburn, J.C., and Ripley, B. (2022). Algorithms used in medical image segmentation for 3D printing and how to understand and quantify their performance. *3D Print. Med.* 8: 18. https://doi.org/10.1186/s41205-022-00145-9.

9 Kucukgul, C., Ozler, B., Karakas, H.E., and Gozuacik, D. (2013). 3D hybrid bioprinting of macrovascular structures. *Procedia Eng.* 59: 183–192. https://doi.org/10.1016/j.proeng.2013.05.109.

10 Liu, J., Zhou, Z., Zhang, M. et al. (2022). Simple and robust 3D bioprinting of full-thickness human skin tissue. *Bioengineered* 13 (4): 10090–10100. https://doi.org/10.1080/21655979.2022.2063651.

11 Wang, X. (2019). Review: advanced polymers for three-dimensional (3D) organ bioprinting. *Micromachines* 2019 10: 814. https://doi.org/10.3390/mi10120814.

12 Agarwal, T., Chiesa, I., Presutti, D. et al. (2021). Recent advances in bioprinting technologies for engineering different cartilage-based tissues. *Mater. Sci. Eng. C* 123: 112005:https://doi.org/10.1016/j.msec.2021.112005.

13 Bade, R., Haase, J., and Preim, B. (2006). Comparison of fundamental mesh smoothing. *Sim. Vis.* 6: 289–304. http://dblp.uni-trier.de/db/conf/simvis/simvis2006.html#BadeHP06.

14 Nam, S.Y., Ricles, L.M., Suggs, L.J., and Emelianov, S.Y. (2014). Imaging strategies for tissue engineering applications. *Tissue Eng. Part B Rev.* 21 (1): 88–102. https://doi.org/10.1089/ten.teb.2014.0180.

15 Ballyns, J.J. and Bonassar, L.J. (2009). Image-guided tissue engineering. *J. Cell. Mol. Med.* 13 (8a): 1428–1436. https://doi.org/10.1111/j.1582-4934.2009.00836.x.

16 Janmey, P.A., Winer, J.P., and Weisel, J.W. (2009). Fibrin gels and their clinical and bioengineering applications. *J. R. Soc. Interface* 6: 1–10. https://doi.org/10.1098/rsif.2008.0327.

17 Bishop, E.S., Mostafa, S., Pakvasa, M. et al. (2017). 3-D bioprinting technologies in tissue engineering and regenerative medicine: current and future trends. *Genes Dis.* 4: 185–195:https://doi.org/10.1016/j.gendis.2017.10.002.

18 Rimann, M., Bono, E., Annaheim, H. et al. (2016). Standardized 3D bioprinting of soft tissue models with human primary cells. *J. Lab. Autom.* 21 (4): 496–509. https://doi.org/10.1177/2211068214567146.

19 ALZAID, S., Hammad, N.S., Albalawi, H.I. et al. (2022). Advanced software development of 2D and 3D model visualization for TwinPrint, a dual-arm 3D bioprinting system for multi-material printing. *Mater. Sci. Add. Manuf.* 1 (3): https://doi.org/10.18063/msam.v1i3.19.

20 Ruiz-Cantu, L., Gleadall, A., CallumFaris, J.S. et al. (2020). Multi-material 3D bioprinting of porous constructs for cartilage regeneration. *Mater. Sci. Eng. C* 109: 110578. https://doi.org/10.1016/j.msec.2019.110578.

21 de Toledo Bagatin, J., Camarena, D.E., Osaki, L.H. et al. (2023). Bioprinted and manual human epidermis production: a compared performance for skin irritation tests. *Bioprinting* 29: e00251. https://doi.org/10.1016/j.bprint.2022.e00251.

22 Fang, Y., YuzhiGuo, T.L., RunzeXu, S.M. et al. (2022). Advances in 3D bioprinting. *Chin. J. Mech. Eng.: Addit. Manuf. Front.* 1: 100011. https://doi.org/10.1016/j.cjmeam.2022.100011.

23 O'Connell, C.D., Konate, S., Onofrillo, C. et al. (2020). Free-form co-axial bioprinting of a gelatin methacryloyl bio-ink by direct in situ photo-crosslinking during extrusion. *Bioprinting* 19: e00087. https://doi.org/10.1016/j.bprint.2020.e00087.

24 Robu, A., Robu, N., and Neagu, A. (2018). New software tools for hydrogel-based bioprinting. *SACI 2018, IEEE 12th International Symposium on Applied Computational Intelligence and Informatics* (17–19 May) Timisoara, Romania. https://doi.org/10.1109/SACI.2018.8440971.

25 ÖzgenÖztürk-Öncel, M., Leal-Martínez, B.H., Monteiro, R.F. et al. (2023). A dive into the bath: embedded 3D bioprinting of freeform in vitro models. *Biomater. Sci.* 2023 (11): 5462–5473. https://doi.org/10.1039/D3BM00626C.

26 Admane, P., Ghaghada, K.B., Devkota, L. et al. (2023). Multimodal imaging compatible micro-physiological system. *Bioprinting* 29: e00249. https://doi.org/10.1016/j.bprint.2022.e00249.

27 Wettergreen, M.A., Bucklen, B.S., Starly, B. et al. (2005). Creation of a unit block library of architectures for use in assembled scaffold engineering. *CAD Comput. Aided Des.* 37: 1141–1149. https://doi.org/10.1016/j.cad.2005.02.005.

28 Hollister, S.J., Maddox, R.D., and Taboas, J.M. (2002). Optimal design and fabrication of scaffolds to mimic tissue properties and satisfy biological constraints. *Biomaterials* 23: 4095–4103. https://doi.org/10.1016/s0142-9612(02)00148-5.

29 Ozbolat, I.T. and Koc, B. (2011). Multi-function based modeling of 3D heterogeneous wound scaffolds for improved wound healing. *Comput. Aided Des. Appl.* 8: 43–57. https://doi.org/10.1080/16864360.2010.10738810.

30 Kapfer, S.C., Hyde, S.T., Mecke, K. et al. (2011). Minimal surface scaffold designs for tissue engineering. *Biomaterials* 32: 6875–6882. https://doi.org/10.1016/j.biomaterials.2011.06.012.

31 Starly, B. and Sun, W. (2007). Internal scaffold architecture designs using lindenmayer systems. *Comput. Aided Des. Appl.* 4: 395–403. https://doi.org/10.1080/16864360.2007.10738559.

32 Wang, Y., Yuan, X., Yao, B. et al. (2022). Tailoring bioinks of extrusion-based bioprinting for cutaneous wound healing. *Bioact. Mater.* 17: 178–194. https://doi.org/10.1016/j.bioactmat.2022.01.024.

33 Jessop, Z.M., Al-Sabah, A., Gardiner, M.D. et al. (2017). 3D bioprinting for reconstructive surgery: principles, applications and challenges. *J. Plast. Reconstr. Aesthet. Surg.* 70 (9): 1155–1170. https://doi.org/10.1016/j.bjps.2017.06.001.

34 Xia, Z., Jin, S., and Ye, K. (2018). Tissue and organ 3D bioprinting. *SLAS Technol.* 23 (4): 301–314:https://doi.org/10.1177/24726303187605.

35 Marro, A., Bandukwala, T., and Mak, W. (2016). 3D printing and medical imaging: a review of the methods and applications. *Curr. Probl. Diagn. Radiol.* 45 (1): 2–9. https://doi.org/10.1067/j.cpradiol.2015.07.009.

36 Moura, J.A., Meldrum, K., Doak, S.H., and Clift, M.J.D. (2023). Alternative lung cell model systems for toxicology testing strategies: current knowledge and future outlook. *Semin. Cell Dev. Biol.* 147: 70–82. https://doi.org/10.1016/j.semcdb.2022.12.006.

37 Khoda, A.K.M., Ozbolat, I.T., and Koc, B. (2013). Designing heterogeneous porous tissue scaffolds for additive manufacturing processes. *CAD Comput. Aided Des.* 45: 1507–1523. https://doi.org/10.1016/j.cad.2013.07.003.

38 Derakhshanfar, S., Mbeleck, R., Xu, K. et al. (2018). 3D bioprinting for biomedical devices and tissue engineering: a review of recent trends and advances. *Bioact. Mater.* 3: 144–156. https://doi.org/10.1016/j.bioactmat.2017.11.008.

39 Zhang, B., Gao, L., Ma, L. et al. (2019). 3D bioprinting – review: 3D bioprinting: a novel avenue for manufacturing tissues and organs. *Engineering* 5: 777–794. https://doi.org/10.1016/j.eng.2019.03.009.

40 Ozbolat, I.T. and Khoda, A. (2014). Design of a new parametric path plan for additive manufacturing of hollow porous structures with functionally graded materials. *J. Comput. Inf. Sci. Eng.* 14: 41005. https://doi.org/10.1115/1.4028418.

41 Lee, J., Kim, K.E., Bang, S. et al. (2017). A desktop multi-material 3D bio-printing system with open-source hardware and software. *Int. J. Precis. Eng. Manuf.* 18 (4): 605–612. https://doi.org/10.1007/s12541-017-0072-x.

42 Daly, A.C., Critchley, S.E., Rencsok, E.M., and Kelly, D.J. (2016). A comparison of different bioinks for 3D bioprinting of fibrocartilage and hyaline cartilage. *Biofabrication* 8: 045002. https://doi.org/10.1088/1758-5090/8/4/045002.

43 Pottathara, Y.B. and Kokol, V. (2023). Effect of nozzle diameter and cross-linking on the micro-structure, compressive and biodegradation properties of 3D printed gelatin/collagen/ hydroxyapatite hydrogel, *Bioprinting*. 31: e00266. https://doi.org/10.1016/j.bprint.2023.e00266.

44 Pakhomova, C., Popov, D., Maltsev, E. et al. (2020). Software for bioprinting. *Int. J. Bioprinting* 6 (3): 279. https://doi.org/10.18063/ijb.v6i3.279.

45 Ashammakhi, N., Ahadian, S., Xu, C. et al. (2019). Bioinks and bioprinting technologies to make heterogeneous and biomimetic tissue constructs. *Mater. Today* 1: 100008:https://doi.org/10.1016/j.mtbio.2019.100008.

46 Mao, H., Yang, L., Zhu, H. et al. (2020). Recent advances and challenges in materials for 3D bioprinting. *Prog. Nat. Sci.: Mater. Int.* 30: 618–634. https://doi.org/10.1016/j.pnsc.2020.09.015.

47 Xiang, Y., Zhong, Z., Yao, E.J. et al. (2023). 3D bioprinting of gene delivery scaffolds with controlled release. *Bioprinting* 31: e00270. https://doi.org/10.1016/j.bprint.2023.e00270.

48 Jin, Z., Li, Y., Kang, Y. et al. (2021). 3D printing of physical organ models: recent developments and challenges. *Adv. Sci.* 2021 8: 2101394:https://doi.org/10.1002/advs.202101394.

49 Semba, J.A., Mieloch, A.A., and Rybka, J.D. (2020). Introduction to the state-of-the-art 3D bioprinting methods, design, and applications in orthopedics. *Bioprinting* 18: e00070. https://doi.org/10.1016/j.bprint.2019.e00070.

10

Computational Engineering for 3D Bioprinting: Models, Methods, and Emerging Technologies

Vidyapati Kumar[1], Ankita Mistri[2], Varnit Jain[3], and Manojit Ghosh[3]

[1] *Department of Mechanical Engineering, Indian Institute of Technology, Kharagpur, West Bengal, India*
[2] *Department of Mechanical Engineering, Indian Institute of Technology, Dhanbad, Jharkhand, India*
[3] *Department of Metallurgy and Materials Engineering, Indian Institute of Engineering Science and Technology, Shibpur, West Bengal, India*

10.1 Introduction

Amid the Fourth Industrial Revolution, where constant innovation and customization are imperative for staying competitive, the exploration of unconventional manufacturing processes has become indispensable. Among these processes, 3D bioprinting has emerged as a pivotal player, offering remarkable potential in the manufacturing landscape. Also known as additive manufacturing (AM), rapid prototyping, or free-form fabrication, 3D bioprinting revolutionizes the production by seamlessly layering materials, eliminating traditional machining techniques. With its ability to create intricate and sophisticated three-dimensional geometries using metal powders, 3D bioprinting has become a rapidly growing industry. Recognized as a critical manufacturing priority and a cornerstone of Industry 4.0, it holds the power to reshape the global manufacturing sector. Notably, the demand for 3D bioprinting skyrocketed following the onset of the COVID-19 pandemic, as numerous industries sought innovative solutions amid economic challenges. This groundbreaking technology is already spearheading a design and industrial revolution across diverse sectors, including aviation, power, automobile, health care, tooling, and consumer products. To comprehend the manufacturing process of printing tissues using the 3D bioprinting technique, let us focus on Figure 10.1.

3D Bioprinting from Lab to Industry, First Edition. Edited by Prosenjit Saha, Sabu Thomas, Jinku Kim, and Manojit Ghosh.

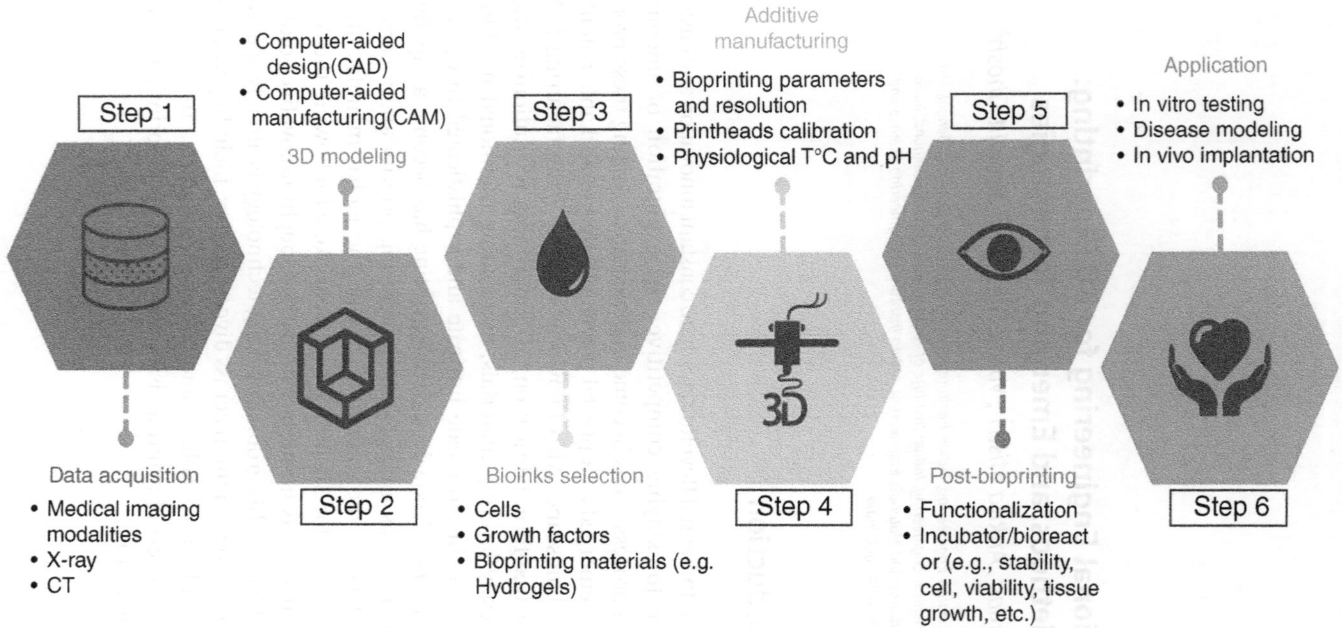

Figure 10.1 Illustration of the sequential workflow in 3D bioprinting for biomedical applications.

It begins with the vital phase of data collection, where information about the specific organ or body part to be printed is gathered through medical imaging methods, like X-ray and computed tomography (CT) scans. Once data are obtained, a three-dimensional model of the desired component takes shape through computer-assisted design (CAD) and computer-aided manufacturing (CAM) processes. Careful selection of bioinks, followed by calibration of bioprinting settings and resolutions, sets the stage for the actual 3D bioprinting process. Subsequently, the bioprinted material undergoes rigorous testing in incubators or bioreactors to confirm stability, cell viability, tissue development, and other crucial factors. Armed with these thorough evaluations, the printed component becomes ready for testing, disease modeling, and even in implantation within the human body, thereby saving countless lives within health-care systems. Considering these perspectives, it becomes essential to recognize the pivotal role of 3D bioprinting in health care and biological sciences. Traditionally, preclinical research in health care heavily relied on animal-based models, particularly rodents [1]. From a regulatory standpoint, conducting animal model experiments has been deemed critical and mandatory to bridge the gap between preclinical and clinical trials and bring forth the latest therapeutic breakthroughs. However, recent scrutiny from humanitarian and scientific considerations has shed light on a pressing issue: Approximately, 80% of promising treatments fail during clinical trials despite demonstrating effectiveness and safety in preclinical investigations [2]. Several underlying factors contribute to this disheartening reality, including inadequate characterization of relevant animal models [3, 4], suboptimal experimental quality in vivo studies, and significant interspecies differences in anatomy, pathophysiology, and immunology compared with humans. Various 3D-printing technologies have emerged to engineer three-dimensional tissue and organ structures, as depicted in Figure 10.2. Bioprinting techniques can be broadly classified into three main categories: extrusion-based, droplet-based, and laser-assisted bioprinting [5]. Each technique has its own capabilities and limitations that determine its suitability for specific applications, as reviewed in detail by Huh et al. [5]. The success of these techniques hinges on the careful selection and functionality of biomaterials.

Moreover, many diseases that manifest in humans are not naturally occurring in animals, necessitating the induction of artificial disease states to simulate human conditions. Even in cases where animals develop similar diseases, the underlying pathophysiological mechanisms may remain largely unknown. For instance, a mouse-based model of SARS corona virus infection has exhibited severe encephalitis, a condition absent in human pathogenesis. This lack of concordance and reproducibility in preclinical animal models, as depicted in Figure 10.2 [6], has cast doubt on their translational utility. Figure 10.3 also

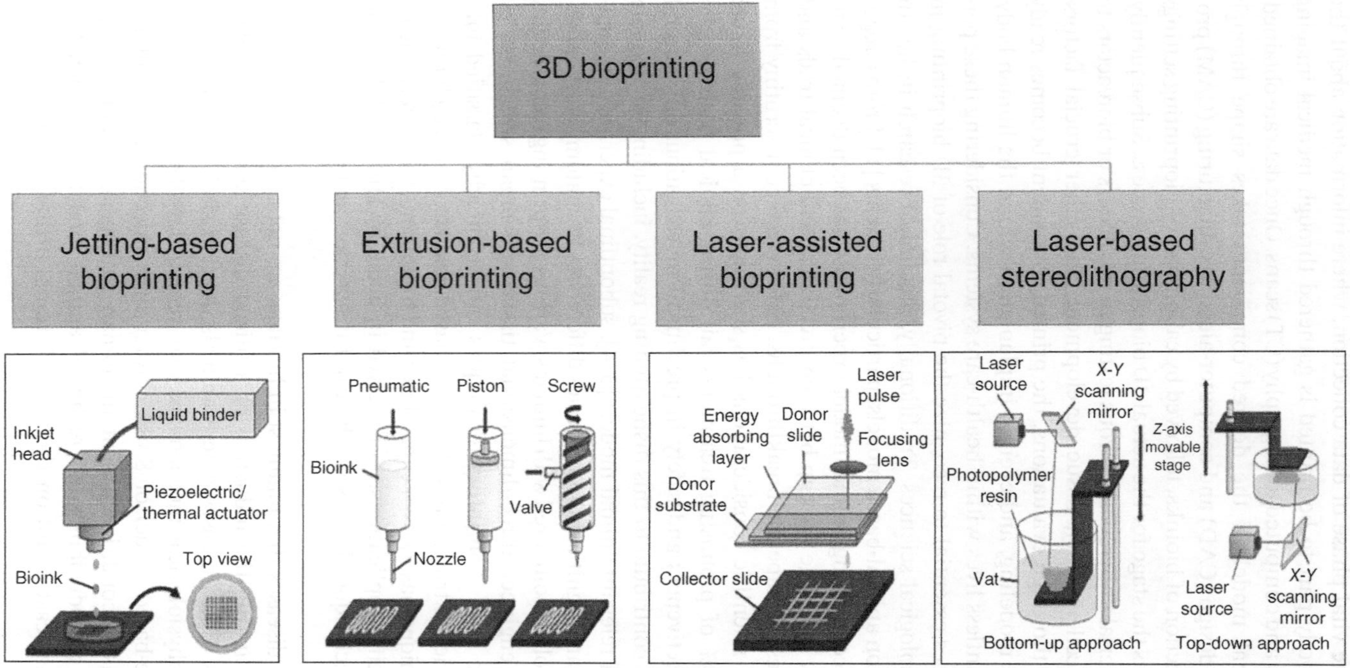

Figure 10.2 Types of 3D bioprinting. *Source:* Huh et al. [5] / Elsevier.

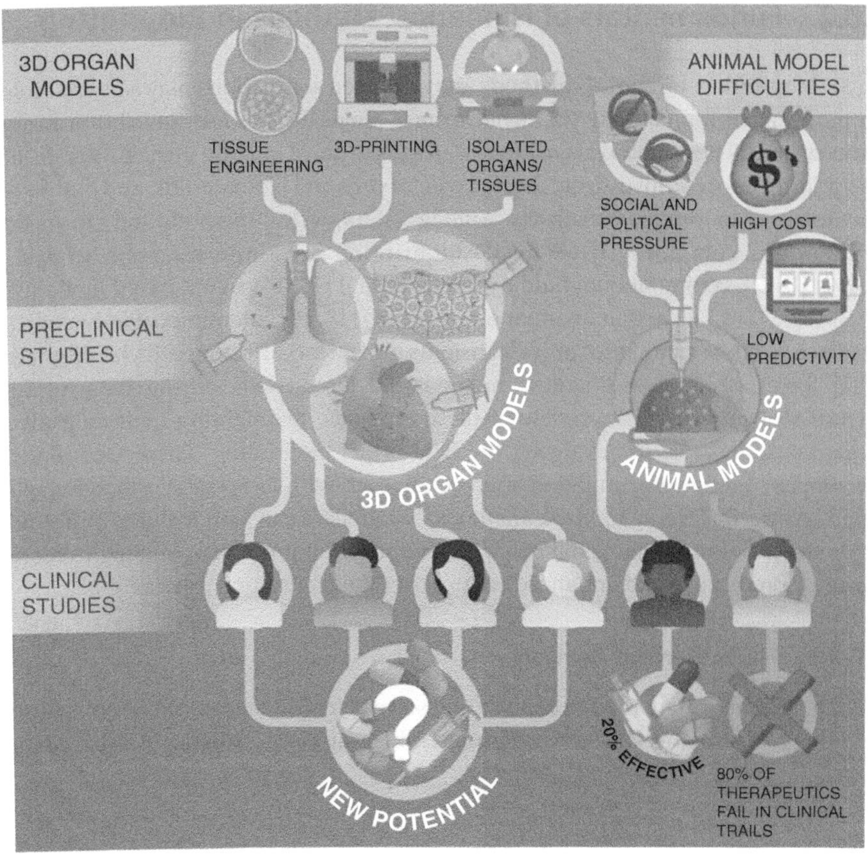

Figure 10.3 The potential of 3D bioprinting technologies for producing 3D organ models. *Source:* Weinhart et al. [6] / Elsevier.

highlights the exorbitant costs of drug development and the relatively high failure rates observed during the clinical phase, raising concerns about the current applicability of existing research methodologies. To address these limitations, the emergence of 3D bioprinting technologies has assumed a critical role. This revolutionary technology offers an invaluable tool for fundamental and preclinical research by enabling the creation of three-dimensional organs or human-based models. 3D bioprinting has garnered significant attention due to its potential to produce human-based models that can serve as alternatives to animal models for scientific purposes, offering new avenues for exploration and breakthroughs in biomedical research.

10.2 Fundamentals of Numerical Methods in Bioprinting

In 3D bioprinting, numerical methods and advanced simulations play a crucial role in modeling and predicting the behavior of biological materials and the printing process itself. These methods enable researchers to delve into the intricate details of the printing process, investigate the behavior of cells and biomaterials, and optimize the design of bioprinted constructs. Among the numerical methods and advanced simulations commonly used in 3D bioprinting are finite element analysis (FEA) [7–12], computational fluid dynamics (CFD) [13, 14], agent-based modeling (ABM) [15–17], lattice Boltzmann method (LBM) [18, 19], and molecular dynamics (MD) [20]. These techniques provide insights into factors affecting cell viability, like fluid forces, scaffold mechanics, and material interactions during bioprinting. Figure 10.4 provides an overview of these methods, highlighting their collective importance and role in advancing the field of 3D bioprinting. These techniques encompass a range of simulation approaches, each with its specific principles and applications. By leveraging these methods, researchers can gain insights into fluid flow, structural mechanics, cell behavior, and material properties, leading to more accurate predictions and enhanced control over the bioprinting process.

10.2.1 Finite Element Analysis

FEA is a numerical method to analyze and solve complex engineering problems by dividing a continuous structure into more minor, discrete elements. It is based on

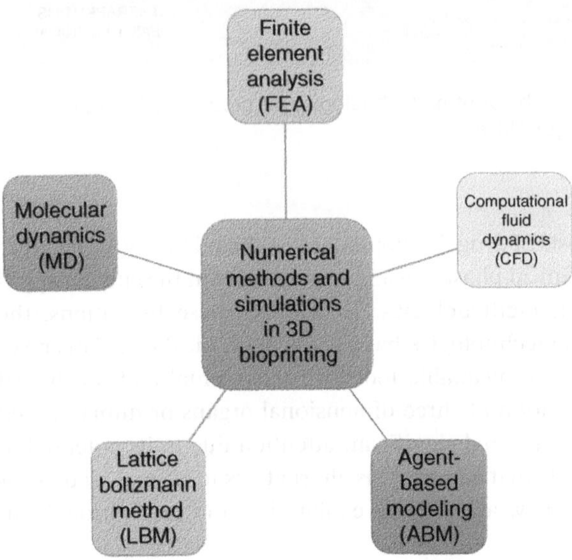

Figure 10.4 Numerical methods and simulations in 3D bioprinting.

the principle of discretization, where the structure is represented as a mesh of interconnected nodes and elements [7]. FEA involves solving the equations of motion and equilibrium for each element to determine the system's overall behavior.

In 3D bioprinting, FEA simulates and analyzes the mechanical properties and behavior of bioprinted structures. It helps optimizing the printed constructs' design and structural integrity by predicting factors, such as deformation, stress distribution, and mechanical stability. FEA can also assist in identifying potential failure points, optimizing material properties, and improving the overall performance of bioprinted tissues and organs [8]. FEA can simulate the shear stresses experienced by cells during extrusion through the print nozzle to predict potential cell damage.

Several software packages are commonly used for FEA in 3D bioprinting research. Some popular software options include the following:

1) *Abaqus*: Developed by Dassault Systèmes, Abaqus offers advanced FEA capabilities and widely used in various engineering fields, including 3D bioprinting.
2) *ANSYS*: ANSYS is a comprehensive simulation software suite that provides FEA functionalities and has modules specifically designed to analyze structures' mechanical behavior [21].
3) *COMSOL Multiphysics*: COMSOL Multiphysics is a versatile simulation platform allowing multiphysics simulations, including FEA, fluid dynamics, heat transfer, and more. It is often used for complex simulations in 3D bioprinting [21].

Examples of FEA applications in 3D bioprinting include the following:

1) *Structural analysis of bioprinted scaffolds*: FEA can be used to evaluate the mechanical stability and deformation behavior of bioprinted scaffolds, ensuring they can support cell growth and tissue formation [7, 8].
2) *Stress analysis in tissue constructs*: FEA helps to predict the stress distribution within bioprinted tissue constructs, enabling the identification of regions that may be subjected to excessive mechanical stress, which could affect cell viability and tissue functionality [9].
3) *Optimization of printing parameters*: FEA simulations can aid in optimizing printing parameters, such as nozzle size, printing speed, and material properties, to achieve desired structural integrity and minimize defects in the printed constructs [11].
4) *Simulation of fluid flow*: FEA combined with CFD can simulate the behavior of bioinks during extrusion [10].

10.2.2 Computational Fluid Dynamics

CFD is a simulation technique used to analyze fluid flow patterns, heat transfer, and other related phenomena. It employs numerical methods and algorithms to solve the governing equations of fluid motion, such as the Navier–Stokes equations, and to predict and visualize fluid behavior within a defined geometry [13].

In 3D bioprinting, CFD plays a vital role in understanding and optimizing fluid dynamics during printing.

The principle of CFD is based on discretizing the domain into smaller computational cells, typically a grid or mesh, and solving the governing equations for fluid flow and heat transfer within each cell. These equations, which describe the conservation of mass, momentum, and energy, are solved iteratively to obtain numerical solutions that represent the flow behavior and distribution of variables (e.g., velocity, pressure, and temperature) throughout the domain [14]. Several software tools are commonly used for CFD simulations in 3D bioprinting, such as ANSYS Fluent and COMSOL Multiphysics [21]. In bioprinting, CFD helps in minimizing the cell damage during the printing process by enabling the optimization of nozzle geometry, applied pressures, and printing speed to reduce shear stresses on suspended cells.

In 3D bioprinting, CFD is used to study and optimize various aspects of the bioprinting process, which are as follows:

1) *Bioink flow and extrusion*: CFD simulations are employed to analyze the behavior of bioinks as they are extruded through the printing nozzle. By studying nozzle geometry, bioink rheology, and extrusion parameters, CFD can predict and optimize the printed construct's flow characteristics, velocity profiles, and pressure distribution. This helps in ensuring uniform deposition, prevent clogging, and improve print accuracy.
2) *Bioreactor design and optimization*: CFD models fluid dynamics and mass transports within bioreactors used for tissue culture and cultivation. By simulating fluid flow patterns, nutrient distribution, and oxygen transport, CFD helps in optimizing bioreactor design and operational parameters. It aids in achieving efficient nutrient supply to cells, maintaining optimal shear stress levels, and promoting cell growth and tissue development.
3) *Microfluidic systems*: CFD plays a significant role in designing and analyzing microfluidic devices used in 3D bioprinting. It enables the study of fluid flow behavior, mixing, and cell behavior within microchannels, facilitating the precise control of material deposition and the creation of complex structures. CFD simulations aid in optimizing channel geometries, flow rates, and cell-loading strategies to achieve desired printing outcomes.

10.2.3 Agent-Based Modeling

ABM is a simulation technique used to study complex systems consisting of individual agents interacting with each other and their environment. In ABM, each agent represents an autonomous entity with its own rules and behaviors, allowing for the simulation of emergent behaviors and patterns at a collective level [15]. In the field of 3D bioprinting, ABM is employed to model and simulate the behavior of individual cells within bioprinted constructs.

The principle of ABM is based on defining agents as autonomous entities with specific characteristics, behaviors, and interaction rules. Agents can represent individual cells, biomolecules, or other components involved in the bioprinting process. ABM simulations proceed by iteratively updating the states and behaviors of respective agents based on their local interactions and the rules governing their behavior [16]. Through the collective behavior of individual agents, ABM can model and predict complex phenomena in 3D bioprinting.

In 3D bioprinting, ABM is used to study and understand various aspects of the bioprinting process, which are as follows [22]:

1) *Cell behavior and migration*: ABM allows for the modeling of individual cell behavior, including migration, proliferation, differentiation, and interactions with other cells and the surrounding environment. By incorporating biological rules and properties into the simulation, ABM can predict the spatiotemporal distribution of cells within bioprinted constructs.
2) *Tissue development and organization*: ABM simulates the self-organization and tissue development processes within bioprinted constructs. It can capture the emergence of complex tissue structures, such as blood vessel networks or cell aggregates, by modeling the interactions between cells, extracellular matrix, and biochemical signaling [23].
3) *Optimization of printing strategies*: ABM simulations aid in optimizing bioprinting strategies by exploring different printing parameters, such as cell density, scaffold geometry, and bioink properties. The simulations can predict the effects of these parameters on cell viability, cell distribution, and tissue functionality, facilitating the design of more effective printing strategies.

Several software tools are commonly used for ABM simulations in 3D bioprinting, such as the following:

1) *NetLogo*: An open-source software platform designed explicitly for ABM and simulation. It provides a user-friendly environment for developing and implementing ABM simulations, making it widely used in various research fields [24].
2) *Repast*: Another popular open-source platform for ABM. It offers a range of tools and libraries for developing and analyzing ABM simulations, allowing for flexibility and customization [25].

10.2.4 Lattice Boltzmann Method

The LBM is a numerical technique used to simulate fluid flow at the mesoscopic level. It is well-suited for simulating complex fluid behaviors, including multiphase flows, fluid–structure interactions, and non-Newtonian fluid flow.

In 3D bioprinting, LBM is employed to model and understand fluid dynamics during the bioprinting process [18].

The principle of LBM is based on representing the fluid flow using a set of fictitious particles, known as "lattice Boltzmann particles," which move on a discrete lattice structure. Instead of directly solving the Navier–Stokes equations, LBM simulates these particles' streaming and collision processes to compute the macroscopic fluid properties [19]. The particles' distribution functions are updated through successive iterations, resulting in a numerical solution that describes the fluid flow and related phenomena.

In 3D bioprinting, LBM is used to study and optimize various aspects of the bioprinting process, which are as follows:

1) *Bioink rheology and extrusion*: LBM simulations allow for the investigation of bioink rheology and flow behavior during the extrusion process. By incorporating relevant fluid properties, such as viscosity, shear-thinning behavior, and yield stress, LBM can predict flow patterns, pressure distributions, and shear forces experienced by the bioink, as it passes through the printing nozzle.
2) *Fluid flow in microchannels and capillaries*: LBM models fluid flow in microchannels, capillaries, or small-scale vessels within bioprinted constructs. By simulating the fluid flow behavior at this microscale, LBM can help understand the transport of nutrients, oxygen, and waste products within the printed tissues, aiding in the design of vascularized constructs.
3) *Aerosolization and droplet formation*: LBM simulations can be used to study the aerosolization process, particularly in cell encapsulation or bioprinting with high cell density. LBM helps analyze the dynamics of droplet formation, droplet size distribution, and jet breakup, enabling optimization of printing parameters and improving cell viability and distribution.

Several software tools that are commonly used for LBM simulations in 3D bioprinting are as follows:

1) *Palabos*: It is an open-source software library designed explicitly for LBM simulations. Palabos provides a range of modules and functionalities for various fluid flow problems, including complex geometries and multiphase flows [26].
2) *OpenLB*: It is another popular open-source LBM software package offering a versatile framework for simulating fluid flow and related phenomena. OpenLB supports various boundary conditions and handles both 2D and 3D simulations [27].

10.2.5 Molecular Dynamics

MD is a computational simulation technique used to study the behavior and interactions of atoms and molecules over time. MD simulations simulate the motion of individual atoms or molecules based on their interactions, such as interatomic forces, to understand biomaterials' dynamics, structure, and properties [20]. In the field of 3D bioprinting, MD is employed to study the behavior of biomaterials at the molecular level and optimize bioink formulations.

The principle of MD is based on Newton's equations of motion. The simulations numerically solve these equations to calculate the positions and velocities of atoms or molecules over time. MD simulations use force fields describing the potential energy and forces between atoms or molecules to model their interactions accurately. MD provides insights into molecular behavior, structure, and properties by integrating the equations of motion and iteratively updating atomic positions. In 3D bioprinting, MD simulations of bioink components help understand their interactions with cells at the molecular level, providing design guidelines for bioinks with improved cytocompatibility.

In 3D bioprinting, MD is used to study and optimize various aspects of biomaterials, which are as follows:

1) *Material properties*: MD simulations aid in understanding the mechanical, thermal, and rheological properties of biomaterials used in 3D bioprinting [28]. By simulating the behavior of molecules, MD provides insights into the material's response to external stimuli, such as stress, temperature, or deformation, helping optimize bioink formulations for desired mechanical and rheological properties.
2) *Interactions with cells and biomolecules*: MD simulations enable studying biomaterial interactions with cells, proteins, and other biomolecules [29]. These simulations help elucidate the binding mechanisms, molecular recognition, and stability of biomaterial–cell or biomaterial–protein complexes. Understanding these interactions aids in designing bioinks with enhanced biocompatibility and biofunctionality.
3) *Optimization of bioink formulations*: MD simulations aid in the optimization of bioink formulations by exploring different molecular compositions, concentrations, and interactions [30]. These simulations provide insights into the self-assembly, aggregation, and gelation behavior of bioinks, facilitating the design of bioinks with suitable rheological properties, printability, and cell encapsulation capabilities.

Several software tools are commonly used for MD simulations in 3D bioprinting, which are as follows:

1) *GROMACS*: A widely used and highly optimized software package for MD simulations, GROMACS offers a range of force fields, simulation protocols, and analysis tools specific to biomolecular simulations. It is known for its efficiency and scalability on various computing architectures [31].
2) *AMBER*: Another popular software suite for MD simulations, which provides extensive features for studying biomolecular systems. AMBER offers diverse

force fields, advanced sampling techniques, and analysis capabilities, allowing for detailed investigations of biomaterial behavior [32].

In recent years, computational approaches have been developed to simulate and provide virtual insights into bioprinting [33, 34]. However, modeling the entire bioprinting process requires the integration of multiple modeling fields and computational technologies. The extrusion of bioink involves fluid dynamics, where the non-Newtonian behavior of the bioink and its nonlinear mechanical response need to be considered. Bioink itself is a heterogeneous material, consisting of hydrogel and solid cellular components, which interact during the extrusion process. In addition, the interaction between the bioink and air at the interface plays a role in droplet or fiber formation during bioprinting [34, 35]. After deposition, cross-linking stabilizes the bioink and significantly alters its mechanical properties [36]. At this stage, describing the bioink based on solid mechanics rather than fluid dynamics becomes more appropriate. Subsequently, numerical simulations should incorporate cell activity and tissue generation to understand the process comprehensively. Table 10.1 provides an overview of different references and their corresponding objects of study, aims of the study, rheological models used, governing equations, and numerical techniques applied for fluid mechanics modeling in bioink extrusion.

After printing, bioprinted constructs must solidify and integrate to form a cohesive structure. This depends on balancing the forces between the bioink components. Furthermore, tissue development and remodeling from embedded cells also change the construct's composition over time. Predicting these spatiotemporal changes requires solving solid mechanics problems that satisfy fundamental physical laws—conservation of mass, momentum, and energy. Although limited literature exists for modeling bioprinted constructs specifically, noteworthy examples for tissue engineering scaffolds broadly include the studies listed in Table 10.2. This table summarizes various computational solid mechanics models applied to understand bioprinted constructs, including the analysis, research object, aims, constitutive model, and numerical technique. Overall, computational modeling can provide insight into the evolving mechanical behavior of bioprinted tissues based on their changing material composition and microstructure.

10.3 Application of Machine Learning for 3D Bioprinting

ML is a cutting-edge field of artificial intelligence that empowers computer systems to learn from data and improve their performance without explicit programming. ML techniques have gained widespread recognition and application in diverse domains, including AM. In the context of AM, ML offers valuable tools to optimize processes, predict material properties, enhance design capabilities, and accelerate product development.

Table 10.1 Models for bioink extrusion in fluid mechanics.

Study objective	Computational method	Material investigated	Rheological model	Governing equations	Research reference
Evaluation of printing parameters and cross-linking effects	Finite element method	GelMA	Power law	Navier–stokes (NS)	Billiet et al. [7]
Prediction of biomaterial extrudability	Finite element method	Self-assembling peptide hydrogel (SAPH)	Newtonian and power law	Navier–stokes (NS)	Chiesa et al. [8]
Analysis of pressure distribution in the nozzle	Finite element method	Viscosity variations	Newtonian	Navier–stokes (NS)	Gómez Blanco et al. [9]
Study of extrusion and fluid behavior	Finite volume method	Viscoelastic fluid extrusion	Carreau-viscoelastic	Navier–stokes (NS)	Ingelsten et al. [10]
Prediction of printing speed line–width relationship	Volume of fluid (VOF)	Cellulose hydrogels and alginate	Power law	Navier–stokes (NS)	Leppiniemi et al. [11]
Examination of printing parameters on cell viability	Finite element method	Alginate hydrogel	Power law	Navier–stokes (NS)	Ning et al. [12]
Analysis of printability and cell viability	Not specified	Alginate hydrogel	Power law	Navier–stokes (NS)	Paxton et al. [35]
Rheological behavior analysis	Not specified	Alginate hydrogel	Power law	Hagen–poiseuille	Rezende et al. [37]
Investigation of printing performance and shape fidelity	Computational fluid dynamics	Jammed gelatin microgel composite ink	Bingham and Herschel–Bulkley	Navier–stokes (NS)	Song et al. [13]
Comparison of miscible and immiscible models	Computational fluid dynamics	Simultaneous flow of alginate within $CaCl_2$ solutions along the same axis	Carreau–Yasuda and Carreau	Navier–stokes (NS)	Williams et al. [14]

Table 10.2 Solid mechanics models for bioprinted constructs.

Bioprinted construct	Constitutive model	Numerical technique	Study objective	Research reference
PCL scaffolds obtained via extrusion	Volumetrically hardened collapsible foam	FEM	Exploration of the influence of mechanics on scaffold pore dimensions	Almeida and Bartolo [38]
Bulk hydrogels (nonprinting technology)	Poromechanics with significant deformation	Time integration using an explicit method combined with a finite volume scheme in spatial discretization	Dynamics of swelling and drying	Bertrand et al. [39]
Self-assembling peptide hydrogels (SAPH)	Linear viscoelastic	FEM	Evaluation of mechanical stability over time	Chiesa et al. [8]
Alginate hydrogels	Large deformation chemo-poromechanics	FEM	Modeling reactive–diffusive mechanisms during cross-linking	Hajikhani et al. [40]
Polyelectrolyte hydrogels	Multiphasic or transport poroelastic model	FEM	Analysis of swelling and deformation behaviors	Feng et al. [41]
Polylactic acid (PLA) scaffolds obtained via FDM	Linear elastic	FEM	Investigation of layer penetration and postheating effects	Naghieh et al. [42]
Alginate scaffolds obtained via FDM	Linear elastic	FEM	Assessment of interlocked strands penetration effects	Naghieh, Sarker, et al. [43]
PCL scaffolds obtained via FDM	Plasticity with von Mises yield criterion and isotropic reinforcement	FEM	Study of scaffold geometrical features on mechanical effects	Schipani et al. [22]

Table 10.2 (Continued)

Bioprinted construct	Constitutive model	Numerical technique	Study objective	Research reference
PCL scaffolds obtained via extrusion	Finite-strain hyperelastic framework	FEM	Analysis of scaffold microstructure and its mechanical effects	Soufivand et al. [44]
Phase transition of physical hydrogels	Thermo-chemical–mechanical model (polymer network model, chemical blending, ionic bonding)	Time integration using the fourth-order Runge–Kutta method combined with FEM in space	Investigation of phase transition effects in physical hydrogels	Wu and Li [45]

Table 10.3 [46–57] showcases various ML techniques applied in the realm of AM, highlighting their specific features and contributions. The featured ML methods include linear regression, convolutional neural networks (CNN), genetic algorithm, support vector machines (SVM), self-organizing maps, long short-term memory (LSTM) networks, classification trees (C4.5), and more. Each entry in the table corresponds to a distinct application area within bioprinting, such as composite material design, process optimization, stress prediction, geometric compensation, and efficient numerical modeling.

10.4 Summary

This chapter provides a comprehensive overview of the computational engineering approaches and technologies being applied to advance bioprinting. It introduces bioprinting and its ability to fabricate biological structures layer-by-layer using cells, biomaterials, and biomolecules. The main bioprinting techniques, such as extrusion, jetting, laser-assisted, and stereolithography, are described along with their respective strengths and limitations. This chapter then delves into the computational modeling methods utilized to simulate and optimize different aspects of the bioprinting process. These include FEA for evaluating scaffold mechanics, CFD for modeling bioink extrusion dynamics, ABM for cell behavior, lattice Boltzmann techniques for fluid flows, and MD for biomaterial interactions.

Table 10.3 ML advancements in bioprinting: techniques and key findings.

Feature	ML technique	Key findings/advantages	Reference
Composite material design	Linear regression and CNN	Accurate prediction of mechanical properties	Gu et al. [46]
Process optimization	Genetic algorithm and gradient-based schemes	Addressing nondifferentiable objective functions	Chaparro et al. [47]
Design recommendations	Hierarchical clustering and SVM	Assisting novice designers in exploring design possibilities	Yao et al. [48]
Microstructure and microhardness tuning	Self-organizing maps	Precise tuning using physics-based models and data mining	Gan et al. [49]
Build orientation optimization	CNN and linear regression	Improved precision in estimating factors related to build orientation	Williams et al. [50]
Flatness perception analysis	Classification tree (C4.5)	Identifying perceptual differences in flatness quality	Petrov et al. [51]
Geometric compensation	Feed-forward ANN	Simulating and compensating for deformations in AM parts	Chowdhury and Anand [52]
Manufacturability assessment	Heat kernel signature (HKS)	Accelerating product development and reducing human errors	Shi et al. [53]
Part estimation	Knowledge-based ANN	Incorporating topological zones and modeling missing knowledge	Nagarajan et al. [54]
Efficient numerical modeling	ANN with fully connected layers and LSTM	Drastically reducing computational time with accurate results	Koeppe et al. [55]
Stress prediction	Two-stream CNN	Outperforming simple neural networks in stress prediction	Khadilkar et al. [56]
Composite material design	CNN	Coarse-graining materials for analysis and design	Gu et al. [46]
Designing surrogate systems	ANN	Replicating dynamic characteristics of targets	Sarlo and Tarazaga [57]

Figure 10.5 presents a concise flowchart depicting the procedural steps of deploying computational engineering in bioprinting. It begins with gathering geometrical and material data, selecting a numerical method, and creating a mesh or grid. Boundary conditions, material models, and governing equations are defined. The numerical solver performs simulations, and results are analyzed, optimized, and validated against experimental data. The flowchart concludes with successful bioprinting deployment using computational engineering.

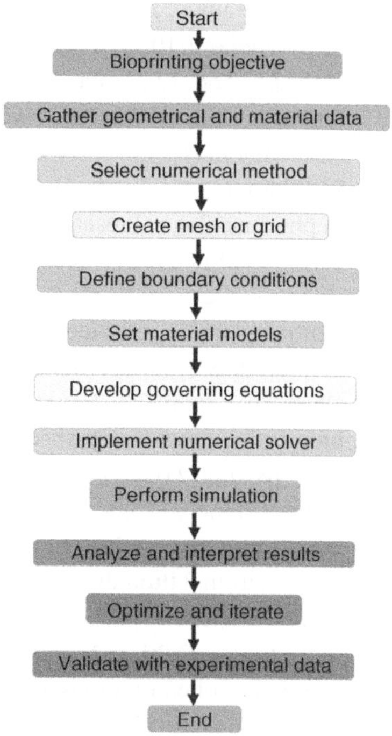

Figure 10.5 Sequential stages involved in implementing computational engineering for bioprinting deployment.

Such simulations provide critical insights into the physics, material properties, and dynamics involved in bioprinting. The emerging role of machine learning in bioprinting is also highlighted, with examples, like composite design, process optimization, geometry compensation, and efficient numerical modeling. Overall, integrating advanced computational engineering technologies and data-driven techniques enables greater understanding and control over the bioprinting process, helping drive progress in areas, like regenerative medicine and tissue engineering.

References

1 Osuchowski, M.F., Remick, D.G., Lederer, J.A. et al. (2014). Abandon the mouse research ship? Not just yet! *Shock* 41 (6): 463–475. https://doi.org/10.1097/SHK.0000000000000153.

2 Pound, P. and Bracken, M.B. (2014). Is animal research sufficiently evidence based to be a cornerstone of biomedical research? *BMJ* 348: https://doi.org/10.1136/bmj.g3387.

3 Perrin, S. (2014). Preclinical research: make mouse studies work. *Nature* 507 (7493): 423–425: https://doi.org/10.1038/507423a.

4 Couzin-Frankel, J. (2013). When mice mislead. *Science* 342 (6161): 925–923: https://doi.org/10.1126/science.342.6161.922.

5 Huh, J.T., Yoo, J.J., Atala, A., and Lee, S.J. (2020). Three-dimensional bioprinting for tissue engineering. In: ***Principles of Tissue Engineering***, 1391–1415. Academic Press.

6 Weinhart, M., Hocke, A., Hippenstiel, S. et al. (2019). 3D organ models – revolution in pharmacological research? *Pharmacol. Res.* 139: 446–451: https://doi.org/10.1016/j.phrs.2018.11.002.

7 Billiet, T., Gevaert, E., De Schryver, T. et al. (2014). The 3D printing of gelatin methacrylamide cell-laden tissue-engineered constructs with high cell viability. *Biomaterials* 35 (1): 49–62: https://doi.org/10.1016/j.biomaterials.2013.09.078.

8 Chiesa, I., Ligorio, C., Bonatti, A.F. et al. (2020). Modeling the three-dimensional bioprinting process of β-sheet self-assembling peptide hydrogel scaffolds. *Front. Med. Technol.* 2 (October): 1–16: https://doi.org/10.3389/fmedt.2020.571626.

9 Gómez Blanco, J.C., Mancha Sánchez, E., Torrejón Martín, D. et al. (2020). Nozzle pressure analysis of several hydrogel on extrusion-based bioprinting using computational fluid dynamics. *Jornadas de Automática* 737–741: https://doi.org/10.17979/spudc.9788497497565.0737.

10 Ingelsten, S., Göhl, J., Mark, A., and Edelvik, F. (2018). A virtual framework for simulation of complex viscoelastic flows. *Procedia CIRP* 72: 392–397: https://doi.org/10.1016/j.procir.2018.03.226.

11 Leppiniemi, J., Lahtinen, P., Paajanen, A. et al. (2017). 3D-printable bioactivated nanocellulose-alginate hydrogels. *ACS Appl. Mater. Interfaces* 9 (26): 21959–21970: https://doi.org/10.1021/acsami.7b02756.

12 Ning, L., Betancourt, N., Schreyer, D.J., and Chen, X. (2018). Characterization of cell damage and proliferative ability during and after bioprinting. *ACS Biomater. Sci. Eng.* 4 (11): 3906–3918: https://doi.org/10.1021/acsbiomaterials.8b00714.

13 Song, K., Zhang, D., Yin, J., and Huang, Y. (2021). Computational study of extrusion bioprinting with jammed gelatin microgel-based composite ink. *Addit. Manuf.* 41: https://doi.org/10.1016/j.addma.2021.101963.

14 Williams, H., McPhail, M., Mondal, S., and Münch, A. (2018). Modeling gel fiber formation in an emerging coaxial flow from a nozzle. *J. Fluids Eng. Trans. ASME* 141 (1): https://doi.org/10.1115/1.4040833.

15 Ramírez López, D.V., Peña-Reyes, C., and Rojas, Á.J. (2019). Agent-based modeling of mesenchymal stem cells on a 3D-printed bio-device for the regenerative treatment of the infarcted myocardium. *Proceedings – 2018 IEEE International Conference on Bioinformatics and Biomedicine, BIBM 2018*, pp. 2033–2040 (3–6 December 2018), Madrid. https://doi.org/10.1109/BIBM.2018.8621314.

16 Keshavarzian, M., Meyer, C.A., and Hayenga, H.N. (2019). In silico tissue engineering: a coupled agent-based finite element approach. *Tissue Eng. – Part C Methods* 25 (11): 641–654: https://doi.org/10.1089/ten.tec.2019.0103.

17 Nolan, D. and Lally, C. (2018). Coupled finite element-agent-based models for the simulation of vascular growth and remodeling. In: *Numerical Methods and Advanced Simulation in Biomechanics and Biological Processes* (ed. M. Cerrolaza, S.J. Shefelbine, and D. Garzón-Alvarado), 283–300. Academic Press.

18 Cristea, A. and Neagu, A. (2016). Shape changes of bioprinted tissue constructs simulated by the Lattice Boltzmann method. *Comput. Biol. Med.* 70: 80–87: https://doi.org/10.1016/j.compbiomed.2015.12.020.

19 Cristea, A., Horhat, R., and Neagu, A. (2021). Lattice Boltzmann simulations of morphogenesis in artificial tissues. *Rom. J. Physiol.* 66 (3–4): 185.

20 Wei, Q., Yang, R., Sun, D. et al. (2022). Design and evaluation of sodium alginate/polyvinyl alcohol blend hydrogel for 3D bioprinting cartilage scaffold: molecular dynamics simulation and experimental method. *J. Mater. Res. Technol.* 17: 66–78: https://doi.org/10.1016/j.jmrt.2021.12.130.

21 Brunner, D., Khawaja, H., Moatamedi, M., and Boiger, G. (2018). CFD modelling of pressure and shear rate in torsionally vibrating structures using ANSYS CFX and COMSOL multiphysics. *Int. J. Multiphys.* 12 (4): 349–358: https://doi.org/10.21152/1750-9548.12.4.349.

22 Schipani, R., Nolan, D.R., Lally, C., and Kelly, D.J. (2020). Integrating finite element modelling and 3D printing to engineer biomimetic polymeric scaffolds for tissue engineering. *Connect. Tissue Res.* 61 (2): 174–189: https://doi.org/10.1080/03008207.2019.1656720.

23 Babbar, A., Jain, V., Gupta, D. et al. (2022). Additive manufacturing for the development of biological implants, scaffolds, and prosthetics. *Addit. Manuf. Process. Biomed. Eng.* 27–46: https://doi.org/10.1201/9781003217961-2.

24 Hjorth, A., Head, B., Head, B., and Wilensky, U. (2020). Levelspace: a netlogo extension for multi-level agent-based modeling. *JASSS* 23 (1): https://doi.org/10.18564/jasss.4130.

25 Größler, A., Stotz, M., and Schieritz, N. (2003). A software interface between system dynamics and agent-based simulations – linking vensim and repast. *Proc. 21st Int. Conf. Syst. Dyn. Soc.* 8: 20–24.

26 Latt, J., Malaspinas, O., Kontaxakis, D. et al. (2021). Palabos: parallel lattice Boltzmann solver. *Comput. Math. Appl.* 81: 334–350: https://doi.org/10.1016/j.camwa.2020.03.022.

27 Krause, M.J., Kummerländer, A., Avis, S.J. et al. (2021). OpenLB – open source lattice Boltzmann code. *Comput. Math. Appl.* 81: 258–288: https://doi.org/10.1016/j.camwa.2020.04.033.

28 Kumar, V., Prakash, C., Babbar, A. et al. (2022). Additive manufacturing in biomedical engineering. *Addit. Manuf. Process. Biomed. Eng.* 143–164: https://doi.org/10.1201/9781003217961-8.

29 Kumar, V., Babbar, A., Sharma, A. et al. (2022). Polymer 3D bioprinting for bionics and tissue engineering applications. In: *Additive Manufacturing of*

Polymers for Tissue Engineering (ed. A. Babbar, R. Kumar, V. Dhawan, et al.), 17–39. Boca Raton: CRC Press.

30 Babbar, A., Tian, Y., Kumar, V., and Sharma, A. (2022). 3D bioprinting in biomedical applications. In: *Additive Manufacturing of Polymers for Tissue Engineering* (ed. A. Babbar, R. Kumar, V. Dhawan, et al.), 1–16. Boca Raton: CRC Press.

31 Páll, S., Zhmurov, A., Bauer, P. et al. (2020). Heterogeneous parallelization and acceleration of molecular dynamics simulations in GROMACS. *J. Chem. Phys.* 153 (13): https://doi.org/10.1063/5.0018516.

32 Mermelstein, D.J., Lin, C., Nelson, G. et al. (2018). Fast and flexible gpu accelerated binding free energy calculations within the amber molecular dynamics package. *J. Comput. Chem.* 39 (19): 1354–1358: https://doi.org/10.1002/jcc.25187.

33 Emmermacher, J., Spura, D., Cziommer, J. et al. (2020). Engineering considerations on extrusion-based bioprinting: interactions of material behavior, mechanical forces and cells in the printing needle. *Biofabrication* 12 (2): https://doi.org/10.1088/1758-5090/ab7553.

34 Göhl, J., Markstedt, K., Mark, A. et al. (2018). Simulations of 3D bioprinting: predicting bioprintability of nanofibrillar inks. *Biofabrication* 10 (3): https://doi.org/10.1088/1758-5090/aac872.

35 Paxton, N., Smolan, W., Böck, T. et al. (2017). Proposal to assess printability of bioinks for extrusion-based bioprinting and evaluation of rheological properties governing bioprintability. *Biofabrication* 9 (4): https://doi.org/10.1088/1758-5090/aa8dd8.

36 Ozbolat, I.T. and Hospodiuk, M. (2016). Current advances and future perspectives in extrusion-based bioprinting. *Biomaterials* 76: 321–343: https://doi.org/10.1016/j.biomaterials.2015.10.076.

37 Rezende, R.A., Bartolo, P.J., Mendes, A., and Filho, R.M. (2009). Rheological behavior of alginate solutions for biomanufacturing. *J. Appl. Polym. Sci.* 113 (6): 3866–3871: https://doi.org/10.1002/app.30170.

38 Almeida, H.A. and Bártolo, P.J. (2013). Numerical simulations of bioextruded polymer scaffolds for tissue engineering applications. *Polym. Int.* 62 (11): 1544–1552: https://doi.org/10.1002/pi.4585.

39 Bertrand, T., Peixinho, J., Mukhopadhyay, S., and MacMinn, C.W. (2016). Dynamics of swelling and drying in a spherical gel. *Phys. Rev. Appl.* 6 (6): https://doi.org/10.1103/PhysRevApplied.6.064010.

40 Hajikhani, A., Wriggers, P., and Marino, M. (2021). Chemo-mechanical modelling of swelling and crosslinking reaction kinetics in alginate hydrogels: a novel theory and its numerical implementation. *J. Mech. Phys. Solids* 153: https://doi.org/10.1016/j.jmps.2021.104476.

41 Feng, L., Jia, Y., Li, X., and An, L. (2011). Comparison of the multiphasic model and the transport model for the swelling and deformation of polyelectrolyte hydrogels. *J. Mech. Behav. Biomed. Mater.* 4 (7): 1328–1335: https://doi.org/10.1016/j.jmbbm.2011.05.001.

42 Naghieh, S., Karamooz Ravari, M.R., Badrossamay, M. et al. (2016). Numerical investigation of the mechanical properties of the additive manufactured bone scaffolds fabricated by FDM: the effect of layer penetration and post-heating. *J. Mech. Behav. Biomed. Mater.* 59: 241–250: https://doi.org/10.1016/j.jmbbm.2016.01.031.

43 Naghieh, S., Sarker, M.D., Karamooz-Ravari, M.R. et al. (2018). Modeling of the mechanical behavior of 3D bioplotted scaffolds considering the penetration in interlocked strands. *Appl. Sci.* 8 (9): https://doi.org/10.3390/app8091422.

44 Soufivand, A.A., Abolfathi, N., Hashemi, S.A., and Lee, S.J. (2019). Prediction of mechanical behavior of 3D bioprinted tissue-engineered scaffolds using finite element method (FEM) analysis. *SSRN Electron. J.*: https://doi.org/10.2139/ssrn.3431851.

45 Wu, T. and Li, H. (2018). Phase-field model for liquid–solid phase transition of physical hydrogel in an ionized environment subject to electro–chemo–thermo–mechanical coupled field. *Int. J. Solids Struct.* 138: 134–143: https://doi.org/10.1016/j.ijsolstr.2018.01.005.

46 Gu, G.X., Chen, C.T., and Buehler, M.J. (2018). De novo composite design based on machine learning algorithm. *Extrem. Mech. Lett.* 18: 19–28: https://doi.org/10.1016/j.eml.2017.10.001.

47 Chaparro, B.M., Thuillier, S., Menezes, L.F. et al. (2008). Material parameters identification: gradient-based, genetic and hybrid optimization algorithms. *Comput. Mater. Sci.* 44 (2): 339–346: https://doi.org/10.1016/j.commatsci.2008.03.028.

48 Yao, X., Moon, S.K., and Bi, G. (2017). A hybrid machine learning approach for additive manufacturing design feature recommendation. *Rapid Prototyp. J.* 23 (6): 983–997: https://doi.org/10.1108/RPJ-03-2016-0041.

49 Gan, Z., Li, H., Wolff, S.J. et al. (2019). Data-driven microstructure and microhardness design in additive manufacturing using a self-organizing map. *Engineering* 5 (4): 730–735: https://doi.org/10.1016/j.eng.2019.03.014.

50 Williams, G., Meisel, N.A., Simpson, T.W., and McComb, C. (2019). Design repository effectiveness for 3D convolutional neural networks: application to additive manufacturing. *J. Mech. Des. Trans. ASME* 141 (11): https://doi.org/10.1115/1.4044199.

51 Petrov, A., Pernot, J.P., Giannini, F., and Falcidieno, B. (2016). Mapping aesthetic properties to 3D free form shapes through the use of a machine learning based framework. *IMATI Report*.

52 Chowdhury, S. and Anand, S. (2016). Artificial neural network based geometric compensation for thermal deformation in additive manufacturing processes. *ASME 2016 11th International Manufacturing Science and Engineering*

Conference, MSEC 2016, vol. 3, 27 June–1 July 2016. Blacksburg, Virginia, USA. https://doi.org/10.1115/MSEC20168784.

53 Shi, Y., Zhang, Y., Baek, S. et al. (2018). Manufacturability analysis for additive manufacturing using a novel feature recognition technique. *Comput. Aided. Des. Appl.* 15 (6): 941–952: https://doi.org/10.1080/16864360.2018.1462574.

54 H.P. Nagarajan, H Jafarian, A Hamedi *et al.* (2018). Knowledge-based optimization of artificial neural network topology for process modeling of fused deposition modeling. *Proceedings of the ASME Design Engineering Technical Conference*, vol. 4, https://doi.org/10.1115/DETC2018-85187.

55 Koeppe, A., Hernandez Padilla, C.A., Voshage, M. et al. (2018). Efficient numerical modeling of 3D-printed lattice-cell structures using neural networks. *Manuf. Lett.* 15: 147–150: https://doi.org/10.1016/j.mfglet.2018.01.002.

56 Khadilkar, A., Wang, J., and Rai, R. (2019). Deep learning–based stress prediction for bottom-up SLA 3D printing process. *Int. J. Adv. Manuf. Technol.* 102 (5–8): 2555–2569: https://doi.org/10.1007/s00170-019-03363-4.

57 Sarlo, R. and Tarazaga, P.A. (2016). A neural network approach to 3D printed surrogate systems. *Conf. Proc. Soc. Exp. Mech. Ser.* 10: 215–222: https://doi.org/10.1007/978-3-319-30249-2_18.

11

Controlling Factors of Bioprinting

Mridula Sreedharan[1], *D.A. Gouripriya*[2], *Ankita Deb*[2], *Yves Grohens*[3], *Nandakumar Kalarikkal*[1,4], *Prosenjit Saha*[2], *and Sabu Thomas*[5,6,7]

[1] *International and Inter University Centre for Nanoscience and Nanotechnology, Mahatma Gandhi University, Kottayam, Kerala, India*
[2] *Centre for Interdisciplinary Sciences, JIS Institute of Advance Studies and Research, Kolkata, West Bengal, India*
[3] *Laboratoire d'Ingénierie des Matériaux de Bretagne, Université Bretagne Sud, Lorient, France*
[4] *School of Pure and Applied Physics, Mahatma Gandhi University, Kottayam, Kerala, India*
[5] *School of Energy Materials, School of Nanoscience and Nanotechnology, School of Polymer Science and Technology, School of Chemical Science and International and Inter University Centre for Nanoscience and Technology (IIUCNN), Mahatma Gandhi University, Kottayam, Kerala, India*
[6] *Department of Chemical Sciences, University of Johannesburg, Doornfontein, Johannesburg, South Africa*
[7] *TrEST Research Park, Sreekariyam, Trivandrum, Kerala, India*

11.1 Introduction

Three-dimensional (3D) bioprinting is an emerging technology used mainly for tissue engineering and regenerative applications. The material with which 3D constructs are made is called inks for bioprinting. Bioprinting can be done with or without seeding cells and it is called bioink and biomaterial ink, respectively. A wider range of biomaterials show desirable properties to be used for bioprinting. For each material, the influence of the printing process and parameters will be different. So careful selection and optimization of parameters should be done to develop scaffolds with respect to their application. Mechanical stability and cytocompatibility are necessary for the materials used for printing. Biological cells are very sensitive to their growth medium and environment. Therefore, while printing, the effects of different factors on the viability of cells in the scaffolds should be taken care as the major application area is tissue regeneration. Each printing techniques work upon different factors and methods of ink deposition, therefore, the behavior of cell growth in each environment has to be well

3D Bioprinting from Lab to Industry, First Edition. Edited by Prosenjit Saha, Sabu Thomas, Jinku Kim, and Manojit Ghosh.

studied. Bioprinting methods are mainly classified into droplet-based, extrusion-based, and light-based bioprinting. While in droplet-based bioprinting droplet size, velocity, and heat determine the plight of cells, in extrusion-based bioprinting, the pressure and nozzle parameters influence more [1]. In light-based bioprinting, the crosslinking efficiency and properties of light affect the resulting structures. External factors such as topography and patterning of the printed surfaces can control and promote the directional growth of cells in the printed constructs. Bioprinting has given the freedom to control the spatial resolution and morphology of structures and there are so many choices to modulate the mechanical stability of scaffolds by different crosslinking techniques. Still, each technique lacks the efficiency to provide both mechanical robustness and a high viability percentage in one go. So, studies are progressing to achieve maximum desirable properties by careful selection of materials and methods and printing parameters [2].

11.2 Factors Influencing the Printability of Hydrogel Bioink

Bioprinting for fabrication of various scaffolds for tissue engineering depends on numerous factors that affect printability and cell response. Extrusion-based bioprinting mainly depends on the bioink properties that make the scaffold perfect for the cells to grow and form desired tissue to mimic the damaged or replaced organs. The hydrogel precursor or the bioink needs to be designed so that the printability aspects are matched to fit the demand. Even though there are many printing techniques, extrusion-based printing is the most widely used method in printing cells along with biomaterials [3]. The bioinks should circumvent various criteria before they can be used to encapsulate and support cells to form tissues. The chemical building of the bioink, rheological properties, and printing parameters are the main factors to be considered while designing a bioink for the 3D printing of tissues [4, 5].

In order for a bioink to be successfully printed, we need to consider a lot of designing functions, some of which are listed below:

1) Cyto non-toxicity, hydrogel nature, and non-triggering of the host immune response.
2) The material of the bioink should have shear-thinning properties such that it should be able to homogenously mix the cells and prevent them from settling. For this, the viscosity of the bioink should be high, and a strong sheer thinning behavior will reduce the stress on the cells.
3) The material should have cell attachment sites for the cells to proliferate.
4) It should have mechanical properties that match the organ to be implanted onto and also withstand stress during handling.
5) The swelling and degradation of the bioink should match the application.
6) The material should be permeable for the nutrients and oxygen to reach all sites inside and outside the 3D-printed constructs.

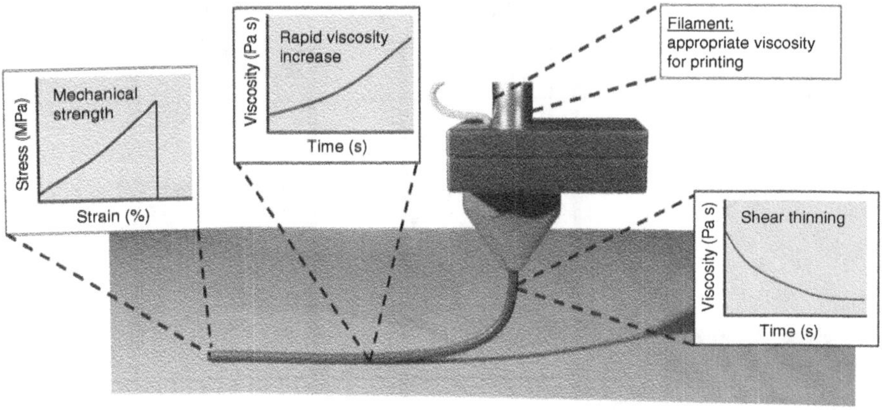

Figure 11.1 Key properties required for 3D extrusion printing of polymers. *Source:* Jiang et al. [3] / John Wiley & Sons/ CC by 4.0.

7) Most importantly, the material of the bioink should be able to mimic the extracellular matrix (ECM) for the cells to behave naturally [3, 6].

The set of printing conditions, which we call printability, depends on many physical and chemical properties of the bioink (Figure 11.1) and also the conditions needed during and after printing.

11.2.1 Extrudability

The material or the bioink is forced through a very small or micrometer nozzle and is pushed out to form the desired shape. This is known as extrudability. The amount of force used to extrude the bioink should not be too high such that the cells get damaged permanently. The processing time and cell viability are influenced by extrudability. The cell viability during extrusion depends on the time cells are present in the syringe, shear stresses on the cells in the nozzle, and the pressure they experience in the syringe. The extrudability can be measured by

1) live/dead assay immediately after printing,
2) measuring extrusion pressure needed to achieve the desired flow rate or the least pressure needed for consistent flow,
3) measuring shear rate–viscosity association, and
4) power law constants.

To achieve a decent print duration and cell viability, bioinks must be extruded above a minimum flow rate and under maximum pressure/shear stress. The pressure required to extrude a specific amount of bioink per construct can be measured and can be related to the storage and loss moduli of that bioink [7, 8].

For easy extrusion of bioinks, the diameter of the nozzle can be increased, the length of the nozzle can be decreased, and a conical-shaped nozzle can be used.

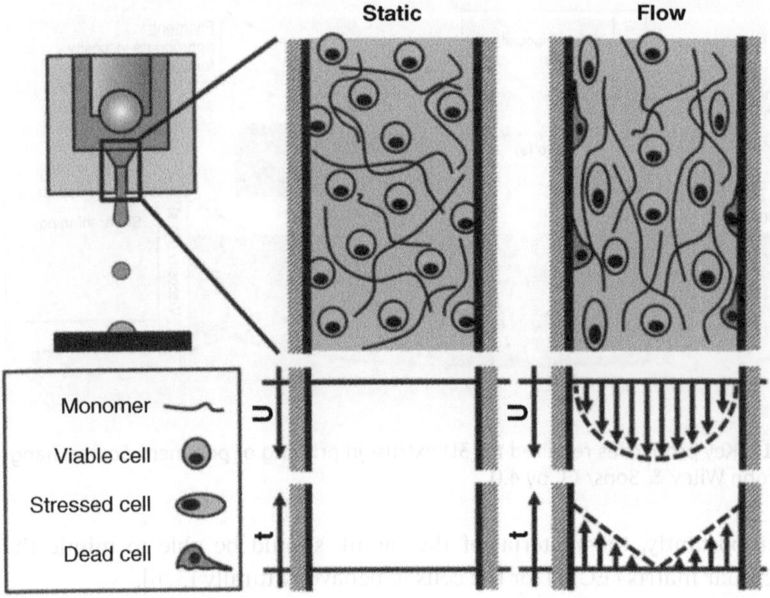

Figure 11.2 Schematic illustration of the velocity and shear stress distribution as well as the stress impinged on cells inside a bioprinter nozzle. *Source:* Blaeser et al. [16] / Reproduced from John Wiley & Sons.

Extrusion forces and how bioink resists them can be studied using the viscosity measurements (Figure 11.2). Shear thinning, which occurs when the viscosity of the bioink reduces with an increase in shear strain, is the most important capacity that the printing material should have. However, certain materials show thixotropic behavior where the viscosity decreases under shear [7].

11.2.2 Filament Type

After extrusion, the bioink forms into different types of filaments. The filaments are classified as (i) continuous filaments that are smooth (uniform) or curvy (nonuniform) and (ii) droplets. Filament type can be quantitatively measured using uniformity ratio, which refers to the perimeter of the filament normalized by length or pore perimeter normalized by pore area. The filament type can also be tested using the filament drop test to understand whether it is forming droplets, filament, or over-gelled filament. Bioprinters can easily control smooth filament deposition, but nonuniform filaments stay away from the path of printing and will have different cross-sections along their length. Alginate and Pluronic can form smooth filaments. The concentrations and level of crosslinking determine the type of filaments, whereas increasing the concentration and degree of alginate crosslinking can determine the filament type. Addition of alginate to the bioink can make the transition from droplet formation to smooth filament formation. Thus, the rheology of the material used influences the filament formed [7].

11.2.3 Shape Fidelity

3D printing makes multiple layers on top of each other, and thus maintaining the shape of the scaffold is a very important factor in printing. This precision in shape or shape fidelity is to be maintained, which determines the final size and shape of the printed construct. It is the capability of a bioink to preserve its structure upon deposition, which can be measured by the width or height of a single filament. The amount of the substance deposited and its interaction with the substrate controls the shape of the printed construct. Shape robustness can be measured using various parameters like spreading ratio, aspect ratio, pore perimeter, height of the structures (cylinder or tubular), pore area, and angle of deflection of unsupported filament [8].

11.2.4 Optimization of Printing Parameters

In order to achieve the desired resolution and repeatability, various printing parameters like printing speed, extrusion pressure, nozzle temperature and diameter, printing time, dispensing pressure, and z-stepping height between layers should be optimized [9].

11.3 Bioink Formulation

Currently available bioinks have limitations that indicate the need for a novel combination of materials to be used as bioink. The bioink palette should be expanded to achieve desired functions. Certain hydrogel like Pluronic F127 shows good printability and shape but have poor compatibility with cells. Decellularized membranes are good cytocompatible material but do not form desired shape as they develop into a weak gel. Bioinks that hold good shape need to be extruded under high pressure, which can affect the cells. Thus, formulation of bioink is still under research to achieve desired functions. Factors that influence printability have direct relation with bioink formulation. Numerous parameters influence printability that are interrelated and complex. Many relationships still remain unstudied, and thus the researchers have to explore the relationship between the various parameters and eventually contribute to formulation of bioinks that can achieve the purpose [7, 10, 11].

11.4 Influence of Printing Process on Cell Behavior

During and after bioprinting, the cells will experience a transition from their reservoir position to arranged position that is affected by various printing parameters. The interplay between cells in bioink and printing process has to be assessed to understand the success of bioprinting. The generation of viable tissue constructs in 3D bioprinting is achieved based on a plethora of factors prior to, during, and after printing. 3D bioprinting stands out as the most in-demand technique in tissue engineering because it aids in fabricating multicellular tissue structures that are complex but the drawback lies in the negative effects of various stresses on cell behavior.

In the extrusion-based 3D bioprinting process, the bioink will be in an initial resting state. Then this bioink is subjected to a shear stress, and when it passes out through the nozzle for printing, it either regains its original state or attains a new state. The important features that should be considered to get an idea about the transition are the viscosity, shear moduli, shear stress, elastic recovery, etc. There should be a perfect balance of viscosity and elasticity of the bioink to print with high fidelity. Generally, materials having high viscosity give good printability, but they can also cause the opposite result beyond a particular limit. High viscosity is mainly concerning as it can produce high shear stress while extruding through a nozzle, which can affect the cell viability of the bioink. One of the desirable properties of the bioink is the shear-thinning property, which is a characteristic feature of non-Newtonian fluids where viscosity is decreased on increasing shear rates. It makes the process of printing easier as the decrease in viscosity enables the smooth flow of ink through the nozzle and it can regain its original state after printing, which maintains the shape fidelity of the constructs. This property is shown by certain polymer blends, polymer solutions, partially crosslinked hydrogels, etc., due to the disentanglement of long polymer chains upon application of shear stress [12]. The printed structures retain their mechanical stability by regaining the interactions upon release of shear-induced stress. In some cases, the shape fidelity is maintained for a long term by post-crosslinking of the constructs [13].

Another desirable trait of bioink is viscoelasticity where it shows both viscous and elastic nature. The viscoelasticity is mainly determined by the storage or elastic modulus G' and loss or viscous modulus G'' and the yield point. Yield point is the stress above which the deformation takes place. Generally, high storage modulus and yield point can cause better filament formation. But in the case if cells are encapsulated in the printing material, high shear will result in reduced cell viability. Biomaterials such as gellan gum, hyaluronan, or carrageenan are used to increase the yield stress of a given ink [14] Vivian H M Mouser et al. studied the effect of additives in the printability of hydrogels by adding gellan gum to GelMA. It was found that the addition of gellan gum improved deposition of filament by inducing yielding behavior, increased stiffness, and supported chondrogenesis. Yield stress determines bioprintability of hydrogels based on gelatin methacryloyl and gellan gum for cartilage bioprinting [14].

The printability determining factors vary between each printing technique. In extrusion-based bioprinting, if shear-thinning behavior and high modulus are considered desirable, in lithography-based bioprinting much less viscous solutions are preferred as they can flow in the platform even in the absence of mixing mechanisms [15].

The printing parameters like nozzle diameter, pressure, and viscosity affect the shear stress level, which is one of the main factors controlling cell behavior in the printed construct. Shear stress on moderate levels aids in cell differentiation and maturation, but if it is high, cell disruption can occur. To maintain cell integrity, the shear stress-affecting factors should be controlled. Viscosity of the bioink and the nozzle structure determines the level of shear stress and thus cell behavior in a hydrogel that is 3D printed [16].

Low rates of shear stress (<5 kPa) do not have any effect on cell viability and membrane integrity as more than 95% of cells remain viable. For nozzle shear >10 kPa, cell viability will decrease to less than 75% showing that the cell integrity gets affected by shear rates >10 kPa. Moderate shear stress over a long period of up to seven days post-printing can uplift the intracellular calcium concentration thereby serving in stem cell proliferation. The shear stress can also affect the phenotypic alterations in mesenchymal cells. Moderate shear stress does not alter the phenotypic characteristics of the cells even for long period of time. Even when exposed to high shear stress of 10 kPa, for short time period, the cell phenotypic expression does not change and has negligible effect on cell proliferation. Immunocytochemical techniques like staining for markers like CD 34 and vimentin can confirm the effect of phenotype of stem cells and can find out the effect of shear stress on cell phenotypic markers. However, the results show negligible or no effect on change in phenotypic characteristics due to shear stress. A deep understanding about shear stress and fluid dynamics is necessary to augment cell response in 3D printing. Major factors that influence drop formation in printing can have profound effects on cell behavior and should be optimized. The effect of shear stress on cells immediately and in the long term should be understood for effective 3D bioprinted scaffolds. Shear stress should be studied in relation with nozzle size, hydrogel viscosity, and printing pressure [16, 17].

The articular chondrocytes can change their morphology and metabolic activity at around 1 Pa of shear stress while mesenchymal stem cells can withstand approximately 10^{-4} Pa. At this shear stress, the upregulation of messenger RNA expressions will also be triggered. In order to modify the rheology of the bioink, nanoparticles will be added, which can impart toxicity to cells. Interaction of cells and polymers has already been well studied, but the effect of nanoparticles in bioink on the cells has to be analyzed. The nanoparticles intermingle with cytosolic proteins and mitochondria and trigger the generation of reactive oxygen species. The concentration of nanoparticles should be optimized as it can influence the cell behavior in the printed hydrogel structures [16, 17].

Apart from the flow shear stress, internal shear stress also affects cell behavior. Most commonly and widely used bioink polymers like alginate, GelMA, and PEG composites with peptides and other polymers have been utilized to understand the interaction between cell and printing parameters. Pre-printing parameters like bioink compositions and their concentrations also affect the cell behavior greatly. The mass flow rate has a linear relationship with cell viability. An increased hydrogel robustness can damage the cells encapsulated in it. Eventually, bioink shear rate distribution can affect the cell viability, and it is vital to determine the shear rate distribution before printing [16].

Another factor that can make an impact while printing is the temperature. The influence of temperature is mainly significant in printing thermoresponsive polymer biomaterial. Thermoresponsive polymers exhibit sol–gel phase transition upon change in temperature. So it can be used for extrusion bioprinting by choosing polymers having transition temperature that comes in the range of printing temperature. Thermoresponsive polymers can exhibit upper or lower critical solution temperature; therefore, they can convert from liquid state to stable gels when the temperature is raised above or below their critical point, respectively. They can be beneficial for printing purposes, and the printed construct can be made stable by heat treatment [18]. Anna Abbadessa et al. developed hydrogels based on triblock copolymers of polyethylene glycol and partially methacrylated poly[*N*-(2-hydroxypropyl) methacrylamide mono/dilactate] with tunable thermoresponsive and mechanical properties for cartilage repair applications [19]. Fu-Yu Hsieh et al. in their study synthesized two thermoresponsive biodegradable polyurethane dispersions that form gel near 37 °C without any crosslinker. Neural stem cells were incorporated in the dispersions, and the gels were printed at 37 °C [20]. Jongbeom Ku et al. worked on thermoresponsive PEG-based hydrogels for 4D bioprinting applications using temperature as one of the variables. Use of thermoresponsive polymers for producing mechanically stable constructs upon heat treatment can avoid the use of crosslinkers that cause damage to cells [21]. Miranda Torre et al. analyzed the printability and shape retention ability of thermoresponsive Pluronic combined with photocrosslinkable GelMA hydrogel systems that turn out to show very good results [22].

11.5 Importance of Patterning and Surface Topography

11.5.1 Importance of Patterning with 3D Bioprinting

Patterning with 3D bioprinting affects the functions of cells. Patterning plays a crucial and irreplaceable role in 3D bioprinting, which is important for creating detailed and functional tissue structures. By carefully arranging cells, biomaterials, and bioinks, patterning allows us to make tissue structures that closely resemble natural tissues. This is very significant for various areas like tissue engineering, regenerative medicine, and drug discovery. Bioprinting provides flexible positioning to the cells with desired position, which comes with patterning. With the help of bioprinting, positioning of cell spheroids can be done, which offers closeness of spheroids cells with one and all. Construction of complex geometries interrelated networks porous structures, improvement of pore arrangements, and variation of pore density to construct gradient-based structures are enabled by 3D bioprinting. The scaffold geometries with pores are determined during the alignment of filaments or struts [23].

One major benefit of using patterning in 3D bioprinting is the ability to make organized tissue structures with clear spatial features. For example, creating blood vessel networks in bioprinted tissues is important for supplying nutrients, removing waste, and helping the tissues fit together well. Patterning techniques let researchers place cells and biomaterials strategically to build complex blood vessel structures, which improves the overall function and health of the printed tissues [24]. Patterning can also help determine how different cell types are distributed in printed structures. The way cells are arranged spatially is crucial for making tissues that mimic the complex cellular environments found in natural tissues. Patterning can create separate sections within the printed structures, making it easier to grow multiple types of cells together or include specific cell environments. This helps researchers make tissue models that more accurately represent how cells are organized and interact in real tissues, which is important for research and drug testing [23, 24]. Additionally, patterning can be used to include gradual changes in the concentration of certain molecules or growth factors in 3D bioprinted tissues. By carefully controlling how specific molecules are placed in the bioinks used for printing, researchers can create gradual changes in concentration, similar to what happens in living tissues. These changes can strongly influence how cells behave, guiding their movement, growth, and specialization. Being able to create such changes through patterning opens up new opportunities for studying tissue development, cell communication, and tissue regeneration in a controlled and realistic way [24].

11.5.2 Effects of Patterning

Patterning with bioprinting leads to interconnectivity of the pores, which ensures the transfer of nutrient and oxygen diffusion. Intended porosity helps to support vascularization for thick tissues. Mechanical properties like elastic modulus of scaffold can be adjusted with porosity, like the elastic modulus is increased with the reduction of porosity.

Different types of designs with different patterns give various kinds of properties to the scaffold. According to finite element analysis (FEA), designing the scaffold with oriented filaments gives a solidified polymer design because of the similar point of intersection and brings great resistance to compression. On the other hand, the zig-zag configuration of filaments exerts failure of the structures in a concerted way, which generates high stress at the hinges. The scaffold with high elastic modulus can be tailored with filament orientation by patterning with bioprinting in approximately 0/90 or 0/60/120-degree rotations, which lead the filament to traverse at an identical position and create a strong polymer column. Besides that, traverse takes place at different positions with approximately 0/45/90/135-degree orientation and gives not so much elastic modulus than the previous orientation. Hexagonal patterns, among curved, rectangular, and zigzag patterns, offer a very large contact area with the adjoin layers and show superior mechanical properties than the other designs [25].

Cell–cell interactions, cell morphology, migration, and proliferation rate are influenced by the geometries of the substrate that comes from patterning by 3D bioprinting. When the length of the gap is 15 μm or broad, cell bodies become confined inside the grooves, whereas for gap sizes of 5 μm and smaller, cells stay on top of the groove. On fine gaps with square posts, cells have a tendency to elongate in the path of the gaps. Comparatively, square patterns result in higher aspect ratios of cell extension when compared to the geometry of hexagonal patterns. Square patterns have an advantage as they feature continuous grooves, while hexagonal patterns lead to periodic changes in the cell elongation pathway [26].

Electrohydrodynamic (EHD) printing is used to create scaffolds with different fiber sizes ranging from 3 to 22 μm. By arranging thick and thin fibers in a certain way, it can be controlled how cells grow in a grid pattern. When the pores are small, cells fill up the spaces, and in broad pores, cells stick to the fibers (Figure 11.3a). In one part of the scaffold where both thick and thin fibers are present, cells propagate along the thick fibers. In another part with only thick fibers, cells grew in a circular manner along the pores. In the area with thin fibers, cells grew randomly. Cells that were similar in size to the thick fibers couldn't wrap around them but stuck to their surface. However, with thin fibers, cells wrapped around them and formed bridges. When the thick fibers were aligned in the x-direction and thin fibers were aligned in the y-direction, cells grew in a specific direction, creating a structured pattern due to the combination of these two properties (Figure 11.3b).

In summary, the importance of patterning in 3D bioprinting cannot be overstated. It allows us to make intricate and complex tissue structures that faithfully reproduce the natural organization of cells and tissues. By incorporating blood vessel networks, creating separate sections for different cell types, and generating molecular gradients, patterning techniques improve the function and relevance of bioprinted tissues. The continuous advancements in patterning methods drive the field of 3D bioprinting forward, getting us closer to personalized medicine and the evolution of fully operative tissue replacements [25, 26].

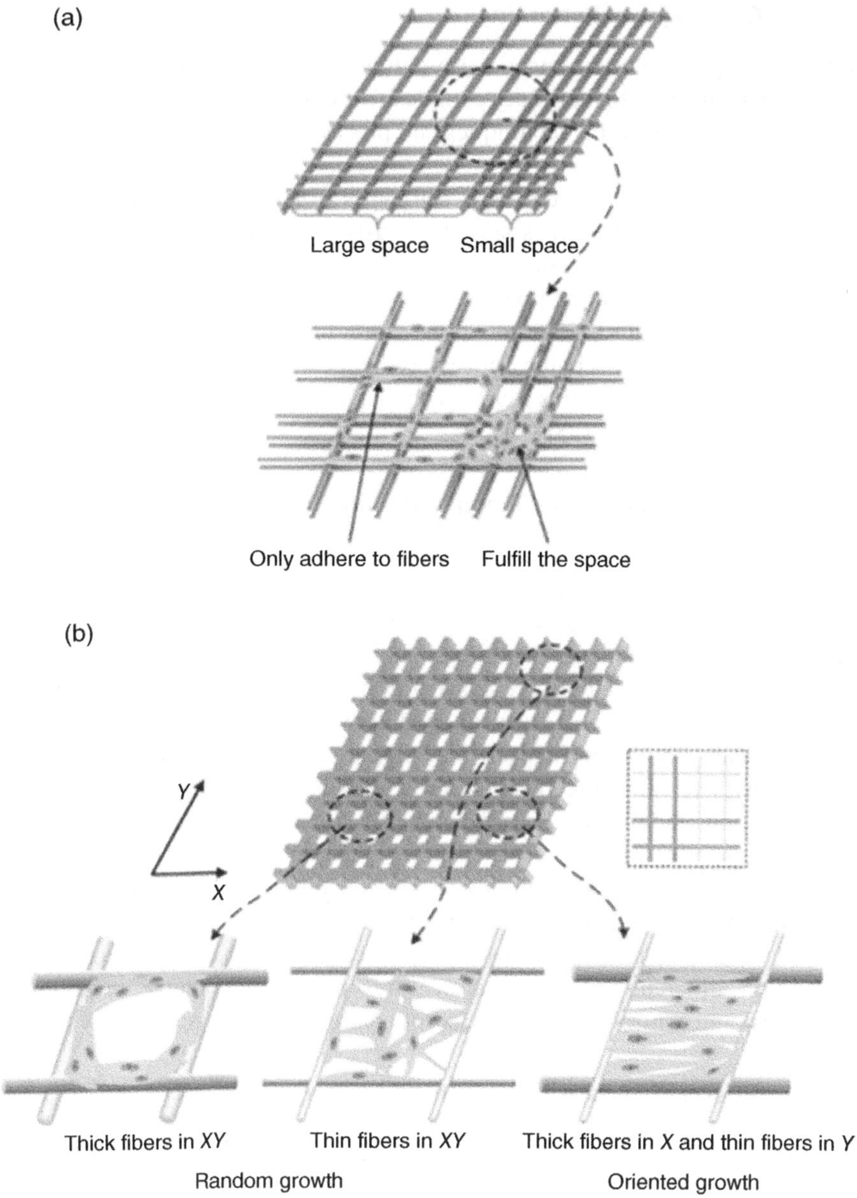

Figure 11.3 (a) Consequences of the size of the pore by patterning on cell growth, (b) consequences of the change in the morphology by different fiber diameters by precise patterning and its effect on growth of the cell. *Source:* [25] / Royal Society of Chemistry / Licensed Under CC BY 4.0.

11.5.3 Challenges of Patterning

Achieving high-resolution and precise patterning is a notable challenge in 3D bioprinting. The need to reproduce complex tissue structures with intricate details requires advanced printing techniques and precise control over the deposition of bioinks.

Many tissues consist of different cell types organized in specific patterns. Bioprinting multiple cell types simultaneously while maintaining their spatial arrangement is a complex task, as it involves ensuring cell viability and functionality throughout the printing process.

Creating functional vascular networks within bioprinted tissues is essential to support cell survival and enable nutrient exchange. Patterning blood vessels along with tissue structures remains a critical challenge to replicate the intricate nature of the native vasculature [25].

11.5.4 Influence of Surface Topography with 3D Bioprinting

Surface topography plays a crucial role in 3D bioprinting, an advanced technology that has the potential to revolutionize the field of regenerative medicine. In recent years, 3D bioprinting has garnered significant attention due to its ability to fabricate intricate and functional 3D tissues and organs using living cells as building blocks. The success of bioprinted constructs heavily relies on the careful consideration and optimization of surface topography, which refers to the microscale and nanoscale features present on the surface of the printed structures. These surface characteristics directly influence cell behavior, adhesion, proliferation, differentiation, and overall tissue functionality, making them a key determinant of the success of bioprinted constructs in vivo (Figure 11.4).

The native ECM portrays nanometer range of different characteristics. Hence the functionality of native ECM is mimicked by the nanoscale topography of the scaffold and stimulates responses of the stem cells. Different cell types react differently with different topographical features at the nanometer scale range. Wettability and controls of protein adsorption are dictated by topography [25].

11.5.5 Effects of Surface Topographical Features

Surface roughness, which refers to the texture on a surface, affects the structure inside a cell known as the cytoskeleton. The way the cytoskeleton is positioned can be directly influenced by the surface texture in a way that requires less energy for the cell. Cells prefer to align themselves in a way that minimizes the deformation of their stiff cytoskeleton, see figure 11.5.

The surface texture also affects how cells function by causing receptors to gather in specific locations, making the cells polarized. Polarized cells behave differently

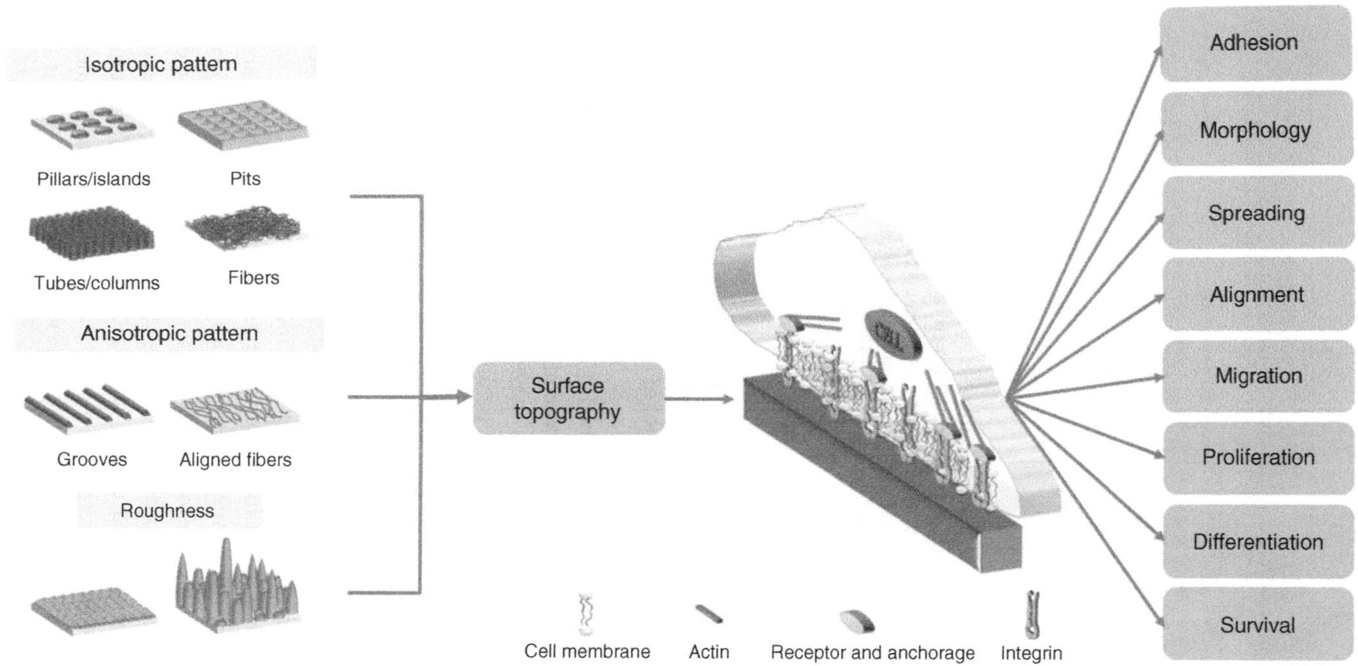

Figure 11.4 Schematic diagram of the effect of different 2D surface features on cell processes. *Source:* [29]. American Chemical Society / Licensed Under CC BY 4.0.

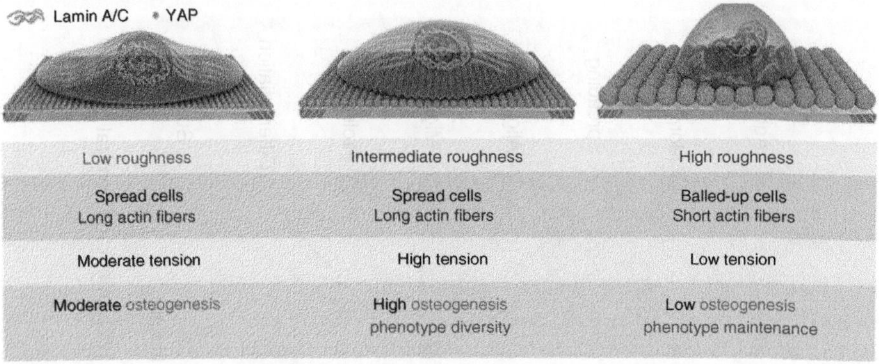

Figure 11.5 Link between surface roughness and cell behavior. Increased cell tension helps with bone formation. The protein YAP, which regulates gene activity, moves into the cell's nucleus more as the surface becomes rougher, up to a certain point, but beyond that point, it decreases. This gene activity directly affects how the cell behaves and looks. *Source:* [25] / Royal Society Of Chemistry/ Licensed Under CC BY 4.0.

compared to cells on a smooth surface with no texture. This change in the shape of cell could also be the reason why the cell nucleus becomes distorted. When the cell nucleus is distorted, it can lead to changes in the way genes are expressed in the cell. This means that the surface texture can impact how genes are activated or deactivated, influencing cell behavior at the genetic level.

Topographical features are preferred over biochemical cues because they are more consistent and stable in the body's environment. In studies with fibroblast cells, it has been found that the speed at which cells move is also influenced by these topographical features. Cells move the slowest on flat surfaces. However, they travel faster when they move parallel to the guiding structures and slower when they move perpendicular to them. With osteoblast cells, the presence of irregular features on the surface enhances the cells' directionality and migration speed. Sometimes, cells even change their direction on bends, with a more noticeable effect when the bend is at 45° compared to 135°. The structures called filopodia and lamellipodia, which are part of the cell's actin filaments, help the cell sense its environment and control its movement.

Surface topography can influence how monocytes/macrophages are activated and how cytokines (inflammatory molecules) are released during inflammation. Scaffolds made with chitosan, having enormous pores and broad angles, encourage higher secretion of pro-inflammatory cytokines, leading to rapid healing. Implants that are inserted during healing can disrupt the formation of macrophages. Cells tend to stretch and elongate along straight and parallel channels on the surface. This effect is more noticeable with narrower ridges that are smaller than 10 μm [25, 27].

There are three commonly used microscale topographies – grooves, pillars, and pits – that affect how cells stick to the surface, grow, and transform into different cell types. These topographies play a significant role in adhesion, proliferation, and differentiation of cells. When pillars are closely packed together, they encourage the formation of strong connections between cells and their surroundings, especially thick actin stress fibers. When the walls of the pores have rough textures due to the surface's features, it becomes easier for cells to stick to them. By adjusting the tiny structures on the surface of the material, we can change the surface's chemical properties and energy, which, in turn, affect how cells attach and stick to it [25, 28].

In conclusion, surface topography plays a pivotal role in 3D bioprinting and significantly impacts the success and functionality of bioprinted tissues and organs. Careful design and optimization of surface characteristics can enhance cell adhesion, proliferation, and differentiation, thus promoting tissue integration and functionality. Additionally, surface topography influences vascularization, which is crucial for the survival of larger constructs, and can also affect the mechanical properties of the printed structures. Understanding the importance of surface topography in 3D bioprinting will undoubtedly lead to advancements in regenerative medicine, enabling the progress of more complex and functional bioprinted tissues and organs [25].

11.6 Contact Guidance and Directional Growth of Cells

Topographic cues in tissue-engineered scaffolds are very crucial for cellular adhesion, migration, and proliferation. The conventional techniques of scaffold fabrication give very less topographic cues, whereas 3D-printed hydrogels and constructs provide the topographic cues needed for cell growth. The ECM inside the body contains several topographic cues needed for the cells to attach and proliferate. Similar conditions are provided by 3D-printed scaffolds so as to mimic the ECM. This mechanism is known as contact guidance. Predesigned topographic cues on 3D-printed scaffolds can provide contact guidance and thus guide in the cell growth facilitating fast tissue regeneration. In order to understand cellular responses and the mechanism of contact guidance, 3D-printed topographic cues are designed. The phenomenon of contact guidance was given importance only after tissue engineering researchers identified its importance as it influences the morphology and organization of cells. The presence of nano- or micro-grooves in scaffolds or the presence of collagen fibers represent topographic cues needed for contact guidance. Focal adhesion, stress fibers, and cellular contraction are the main mechanisms of contact guidance. The signals between extracellular matrix and cells are transmitted by large macromolecular assemblies like focal adhesion. When cells adhere to ECM, certain signaling events, mechanical forces, or regulatory effects are developed and transmitted through focal adhesions. This shows the need for incorporating topographic cues in tissue-engineered scaffolds [29].

11.6.1 2D Topographic Cues

Figure 11.6 shows two-dimensional topographic cues where only a single layer of cells can grow. Straight lines, curves, and lattice points are topographic cues where cells grow differently in each cue. Straight lines can be uniform or multiscale. The alignment and diameter of printed fibers influence cell growth and migration.

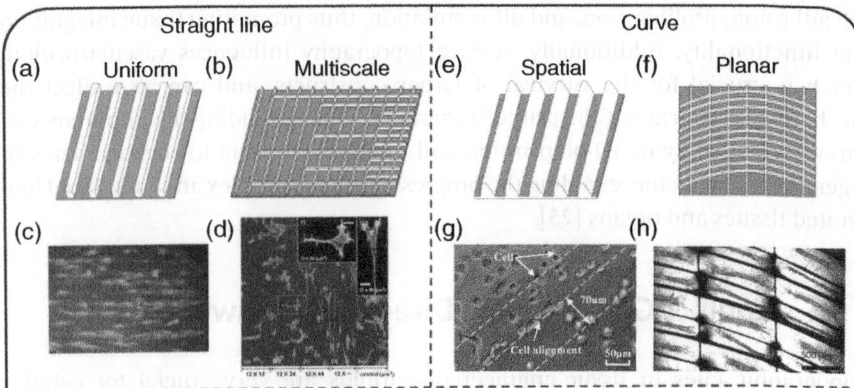

Figure 11.6 Different 2D topographic cues. (a) Uniform straight line; (b) multiscale straight line; (c) fluorescence microscope image of hMSCs aligned on straight grooves printed by DIW; (d) fluorescence microscope image of NIH3T3 cells aligned on multiscale straight line; (e) spatial curve; (f) planar curve; (g) microscope image of L929 cells aligned on spatial curve printed by vat polymerization-based printing, dashed line: trough; (h) Fluorescence image of C2C12 oriented on planar curve printed by electro-printing; (i) Lattice pattern; (j) Microscope image of preosteoblast cells grew on convex lattice pattern printed by vat polymerization-based printing. *Source:* Lai et al. [30] / Reproduced with permission from Elsevier.

Cells migrate differently in aligned or random fibers where cell viability is more on random fibers and the most active cells were found in small diameter fibers but the most aligned cells were found on medium diameter fibers. 3D-printed spatial curves provide good platform for cells to migrate and differentiate. Cells show gathering at minimum curvature, while differentiation can be seen in saddle-like curves. Different shaped curves like arc, spiderweb, and planar influence cell alignment, cell migration, and differentiation. Convex or concave lattice patterns form shapes like cylinders and ellipses. Inkjet printing method can create high-precision lattice patterns. Human mesenchymal stem cells show higher metabolic activity in scaffolds with cylindrical pillar arrays with 100–300 μm spacing [29, 30].

11.6.2 3D Topographic Cues

Multilayer cell growth can occur in 3D scaffolds like shape-based and typical scaffolds. Typical 3D-printed scaffolds will have multilayer parallel fibers that have curves or grooves on the surface. Immune cells will be guided to these surfaces due to these cues that aid in cell maturation, proliferation, and differentiation. 3D printing using fused deposition modeling can create nanocomposite scaffolds that aid in seeding efficiency, proliferation, and differentiation of stem cells. Bone tissue formation was achieved faster on wavy scaffolds than on orthogonal straight-line scaffolds. Figure 11.7 shows 3D topographic cues containing scaffolds.

Shape-based scaffolds are classified into the following as shown in Figure 11.8:

a) Plane-based scaffolds.
b) Cylinder-based scaffolds.
c) Cube-based scaffolds.
d) Plane-based scaffolds are created from planes in order to contact guidance cells.

When the pattern width is smaller than the focal adhesions, constrained focal adhesion alignment plays a critical role in contact guiding. It suggests that cell regulation and differentiation are feasible by the presentation of designer topographies, which might be used to further control tissue creation on regeneration efforts. However, ideas for extrapolating these topographical cues to 3D-printed systems are still in their infancy [30].

11.7 Cell Viability and Mitigation Process

When you say that a 3D bioprinted construct is biologically similar to native tissues or organs, a very high cell viability should be maintained by retaining cellular activities like proliferation and differentiation. The major factors that affect cell viability in various kinds of 3D printing are outlined in Table 11.1.

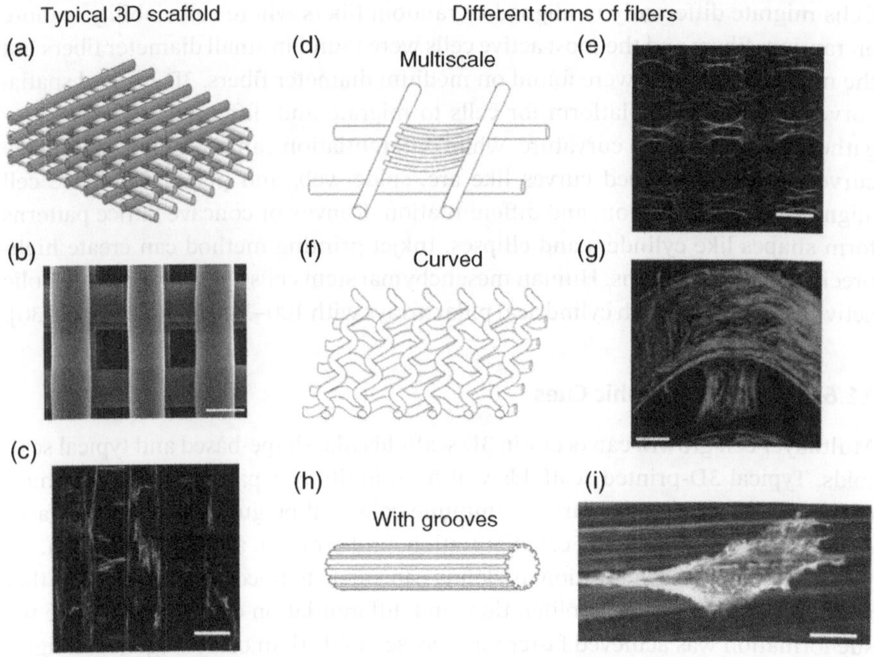

Figure 11.7 Schematic diagrams of typical 3D scaffolds for cell culture. (a) Schematic diagram of a typical 3D scaffold; (b) microscope image of PCL scaffold printed by FDM (scale bar = 300 μm); (c) fluorescence microscope of actin staining of aligned human adipose-derived stem cells; (d) schematic diagram of a scaffold with multiscale fibers; (e) fluorescence microscope of human adipose-derived stem cells aligned on multiscale fibers (scale bar = 50 μm); (f) schematic diagram of a scaffold with curved fibers; (g) multiphoton confocal images of hMSCs cultured on wavy scaffolds printed by FDM. Cells were stained for F-actin and nuclei (scale bars = 200 μm); (h) schematic diagram of scaffold consisted of fibers with grooves; (i) fluorescence microscope image of fibroblast elongated on grooved fiber printed by DIW (scale bar = 10 μm). *Source:* Lai et al. [30] / Reproduced with permission from Elsevier.

11.7.1 Inkjet-Based Bioprinting

Shear and thermal stress are the main factors that affect cell viability in inkjet printing. In order to avoid clogging in nozzle and to reduce shear stress to protect cells from injury, bioinks are created with low viscosity (ranges from 3 to 30 mPa s) and concentration. Thus the cell viability comes to above 85% [31, 32].

11.7.2 Extrusion-Based Bioprinting

This is similar to inkjet printing but with larger nozzle size and thus the viscosity can range from 30 to 6×10^8 mPa s. Due to high viscosity, the shear stress on the cells is more and thus the cell viability is less (40–80%). The only way to achieve cell viability is to optimize the printing parameters and operating conditions.

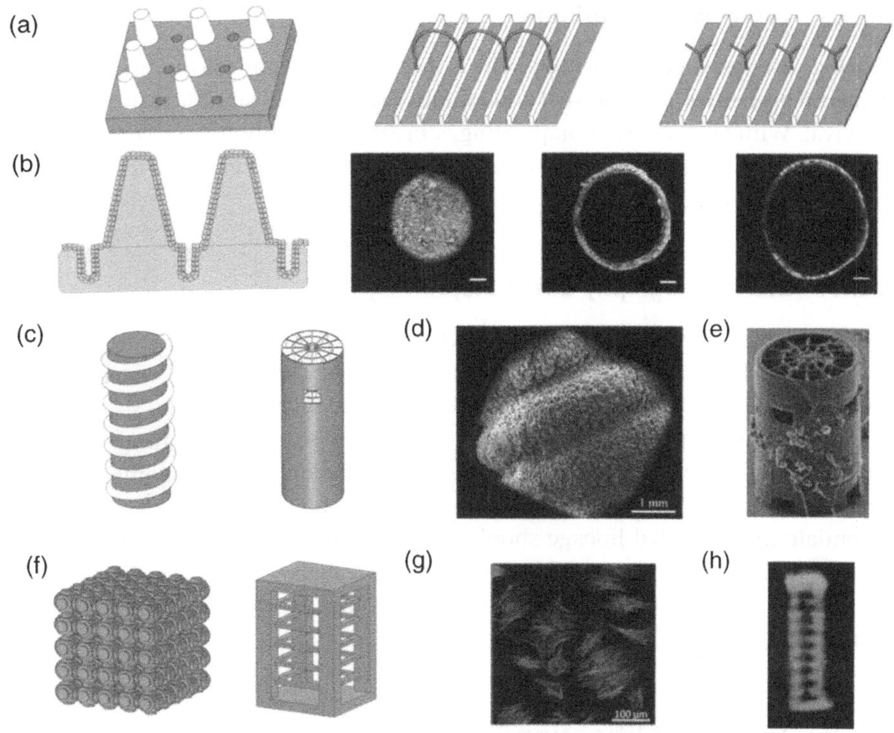

Figure 11.8 Schematic diagrams of representative 3D scaffolds for cell culture along with Fluorescent and SEM images. (a) Plane-based scaffolds; (b) schematic and fluorescent images (top, middle, and base of the villi) of Caco-2 cells seeded on intestinal scaffolds formed by DLP; (c) cylinder-based scaffolds; (d) SEM image of osteoblasts on cylinder-based scaffold printed by SLS; (e) microscope image of neuronal cells grew on a cylinder-based scaffold; (f) cube-based scaffolds; (g) fluorescent images of rBMSCs cultured on cube-based scaffold printed by DLP. (h) Image of a printed cube-based scaffold. *Source:* Lai et al. [30] / Reproduced with permission from Elsevier.

Table 11.1 Major stress factors affecting cell viability in different 3D printing techniques.

Inkjet-based printing	Extrusion-based printing	Laser-assisted printing	Stereolithography-based printing
Shear stress	Shear stress	Shear stress	Thermal stress
Thermal stress	Thermal stress	Material property	Radiative stress
	Radiative stress	Thermal stress	
		Radiative stress	

11.7.3 Laser-Based Bioprinting

During laser-assisted bioprinting, the shear stress during jetting and landing, material properties, and thermal and radiative stressors all significantly affect cell survival. With laser-assisted bioprinting, a broad range of bioink viscosities up to 300 mPa s can be used while preserving more than 90% cell viability thanks to its nozzle-free mechanism.

11.7.4 Stereolithography-Based Bioprinting

Due to the high printing resolution and cell viability of more than 90%, this method is in demand for creating tissue-engineered constructs. The cell viability depends on the thermal and radiative stress that can be controlled more easily than shear stress.

Maintaining the cellular activity is the next crucial step in bioprinting. The proliferation and differentiation capacity of the cells should be maintained during and after printing. When stem cells are used, the potency of the stem cells to differentiate into expected lineage should be maintained. The related gene expression should also be maintained post-printing. The cells that survive the printing process are expected to adhere, migrate, and proliferate in the scaffold materials that mimic the ECM [31–33].

11.8 Possible Mitigation Techniques

In this section we are briefing the various effects on cells during the printing process and how these can be circumvented. In extrusion-based bioprinting, cells get clogged at the print head and this can be overcome by optimizing the cell concentration with 10^7 cells ml^{-1}. The mechanical impact can be overcome by using cell-adhesive coating on the substrate. Shear stress due to low nozzle diameter can be mitigated by optimizing the diameter without affecting the print resolution. In thermal inkjet, the heat and shear stress can be reduced by lowering the time of printing and utilizing less energy. In microvalve-assisted bioprinting, a sudden release of cells can occur, which can be eliminated by calculating the shear stress [1].

11.9 Conclusion

3D bioprinting is a field that is still under research to construct scaffolds and implants for tissue regeneration. While considering cell-based 3D printing, the factors affecting cell growth and proliferation should be given prior importance. The bioink formulation is the key step in 3D bioprinting where the fate of the cell is determined by the proper designing of bioink and the process conditions during printing. The bioink design should focus on minimizing the stress or pressure applied to cells. The rheological property characterization is very important when considering a successful 3D-printed construct. In order to achieve high cell viability, the factors affecting printability should be given focus. In order to have the largest influence in the field of regenerative engineering, next-generation bioinks must simultaneously exhibit good printability, high cell viability, and a wide range of material qualities.

References

1 Adhikari, J., Roy, A., Das, A. et al. (2021). Effects of processing parameters of 3D bioprinting on the cellular activity of bioinks. *Macromol. Biosci.* 21 (1): 2000179.

2 Williams, D., Thayer, P., Martinez, H. et al. (2018). A perspective on the physical, mechanical and biological specifications of bioinks and the development of functional tissues in 3D bioprinting. *Bioprinting* 1 (9): 19–36.

3 Jiang, Z., Diggle, B., Tan, M.L. et al. (2020). Extrusion 3D printing of polymeric materials with advanced properties. *Adv. Sci.* 7 (17): 2001379.

4 Shah, J., Snider, B., Clarke, T. et al. (2019). Large-scale 3D printers for additive manufacturing: design considerations and challenges. *Int. J. Adv. Manuf. Technol.* 104 (9–12): 3679–3693.

5 Nadgorny, M. and Ameli, A. (2018). Functional polymers and nanocomposites for 3D printing of smart structures and devices. *ACS Appl. Mater. Interfaces* 10 (21): 17489–17507.

6 Duty, C., Ajinjeru, C., Kishore, V. et al. (2018). What makes a material printable? A viscoelastic model for extrusion-based 3D printing of polymers. *J. Manuf. Process.* 1 (35): 526–537.

7 Gillispie, G., Prim, P., Copus, J. et al. (2020). Assessment methodologies for extrusion-based bioink printability. *Biofabrication* 12 (2): 022003.

8 Lee, S.C., Gillispie, G., Prim, P., and Lee, S.J. (2020). Physical and chemical factors influencing the printability of hydrogel-based extrusion bioinks. *Chem. Rev.* 120 (19): 10834–10886.

9 Ramesh, S., Harrysson, O.L., Rao, P.K. et al. (2021). Extrusion bioprinting: recent progress, challenges, and future opportunities. *Bioprinting* 1 (21): e00116.

10 Theus, A.S., Ning, L., Hwang, B. et al. (2020). Bioprintability: physiomechanical and biological requirements of materials for 3d bioprinting processes. *Polymers* 12 (10): 2262.

11 Groll, J., Burdick, J.A., Cho, D.W. et al. (2018). A definition of bioinks and their distinction from biomaterial inks. *Biofabrication* 11 (1): 013001.

12 Mezger, T.G. (2011). *The Rheology Handbook*, 3e. Vincentz Network.

13 Ouyang, L., Highley, C.B., Rodell, C.B. et al. (2016). 3D printing of shear-thinning hyaluronic acid hydrogels with secondary cross-linking. *ACS Biomater. Sci. Eng.* 2: 1743–1751.

14 Mouser, V.H., Melchels, F.P., Visser, J. et al. (2016). Yield stress determines bioprintability of hydrogels based on gelatin-methacryloyl and gellan gum for cartilage bioprinting. *Biofabrication* 8: 035003.

15 Schwab, A., Levato, R., D'Este, M. et al. (2020). Printability and shape fidelity of bioinks in 3D bioprinting. *Chem. Rev.* 120 (19): 11028–11055.

16 Blaeser, A., Duarte Campos, D.F., Puster, U. et al. (2016). Controlling shear stress in 3D bioprinting is a key factor to balance printing resolution and stem cell integrity. *Adv. Healthcare Mater.* 5 (3): 326–333.

17 Lepowsky, E., Muradoglu, M., and Tasoglu, S. (2018). Towards preserving post-printing cell viability and improving the resolution: past, present, and future of 3D bioprinting theory. *Bioprinting* 1 (11): e00034.

18 Santoni, S., Sponchioni, M., Gugliandolo, S.G. et al. (2022). Preliminary tests on PEG-based thermoresponsive polymers for the production of 3D bioprinted constructs. *Procedia CIRP* 1 (110): 348–353.

19 Abbadessa, A., Mouser, V.H., Blokzijl, M.M. et al. (2016). A synthetic thermosensitive hydrogel for cartilage bioprinting and its biofunctionalization with polysaccharides. *Biomacromolecules* 17 (6): 2137–2147.

20 Hsieh, F.Y., Lin, H.H., and Hsu, S.H. (2015). 3D bioprinting of neural stem cell-laden thermoresponsive biodegradable polyurethane hydrogel and potential in central nervous system repair. *Biomaterials* 1 (71): 48–57.

21 Ku, J., Seonwoo, H., Park, S. et al. (2020). Cell-laden thermosensitive chitosan hydrogel bioinks for 3D bioprinting applications. *Appl. Sci.* 10 (7): 2455.

22 Torre, M., Giannitelli, S.M., Mauri, E. et al. (2023). Printability assessment workflow of a thermosensitive photocurable biomaterial ink for microextrusion bioprinting. *Bioprinting* 1 (30): e00262.

23 Hinton, T.J., Lee, A., and Feinberg, A.W. (2017). 3D bioprinting from the micrometer to millimeter length scales: size does matter. *Curr. Opin. Biomed. Eng.* 1 (1): 31–37.

24 Dey, M. and Ozbolat, I.T. (2020). 3D bioprinting of cells, tissues and organs. *Sci. Rep.* 10 (1): 14023.

25 Adhikari, J., Roy, A., Chanda, A. et al. (2023). Effects of surface patterning and topography on cellular functions of tissue engineered scaffolds with special reference to 3D bioprinting. *Biomater. Sci.* 11 (4): 1236–1269.

26 Sari, D.P., Bang, S., Nguyen, L. et al. (2016). Micro/nano surface topography and 3D bioprinting of biomaterials in tissue engineering. *J. Nanosci. Nanotechnol.* 16 (9): 8909–8922.

27 Fernandes, C., Suares, D., Dhawan, V., and Prabhu, P. (2018). Nanostructured polymer scaffolds for tissue engineering technology. In: *Fundamentals of Nanoparticles* (ed. A. Barhoum and A.S.H. Makhlouf), 451–483. Elsevier.

28 Magin, C.M., Alge, D.L., and Anseth, K.S. (2016). Bio-inspired 3D microenvironments: a new dimension in tissue engineering. *Biomed. Mater.* 11 (2): 022001.

29 Iturriaga, L., Van Gordon, K.D., Larrañaga-Jaurrieta, G., and Camarero-Espinosa, S. (2021). Strategies to introduce topographical and structural cues in 3D-printed scaffolds and implications in tissue regeneration. *Adv. NanoBiomed Res.* 1 (12): 2100068.

30 Lai, H., Gong, B., Yin, J., and Qian, J. (2022). 3D printing topographic cues for cell contact guidance: a review. *Mater. Des.* 218: 110663.

31 Xu, H.Q., Liu, J.C., Zhang, Z.Y., and Xu, C.X. (2022). A review on cell damage, viability, and functionality during 3D bioprinting. *Mil. Med. Res.* 9 (1): 70.

32 Panda, S., Hajra, S., Mistewicz, K. et al. (2022). A focused review on three-dimensional bioprinting technology for artificial organ fabrication. *Biomater. Sci.* 10 (18): 5054–5080.

33 Bhattacharyya, A., Janarthanan, G., and Noh, I. (2021). Nano-biomaterials for designing functional bioinks towards complex tissue and organ regeneration in 3D bioprinting. *Addit. Manuf.* 1 (37): 101639.

12

In Situ Bioprinting

Mina Mina[1], Kevin Y. Wu[2], Ananda Kalevar[2], and Simon D. Tran[3]

[1] *Department of Mechanical and Manufacturing Engineering, University of Calgary, Calgary, Alberta, Canada*
[2] *Department of Surgery, Division of Ophthalmology, University of Sherbrooke, Sherbrooke, Quebec, Canada*
[3] *Faculty of Dental Medicine and Oral Health Sciences, McGill University, Montreal, Quebec, Canada*

12.1 Introduction

In the realm of cutting-edge biomedical research, the concept of in situ bioprinting often evokes a sense of science fiction, seemingly distant from our current realities. However, its extraordinary potential and transformative capabilities have captured the attention of researchers worldwide. In situ bioprinting is a technology that permits the direct deposition of biomaterials onto a target site, whether external, like the skin or internal, through minimally invasive surgery (MIS). This technology minimizes the risk of tissue contamination during implantation and frees medical personnel, facilitating high-throughput repair directly onto patients' bodies.

Despite its promising potential, in situ bioprinting does not come without challenges. One significant hurdle in the transition from lab to industry is the difficulty in controlling functional vascularization development, which is essential for engineered tissue thicker than the diffusion limit of oxygen. Additionally, the high costs of commercially available in situ bioprinters, the absence of crosslinking guidance databases, and the overall deficiency in tissue-based three-dimensional (3D) models present issues that require resolution. Furthermore, the precise control over the deposition of different cell types in a complex pattern poses a substantial challenge, particularly for complex organ formation.

In this chapter, the authors aim to provide a comprehensive exploration of in situ bioprinting, a transformative technology poised to become a crucial element

3D Bioprinting from Lab to Industry, First Edition. Edited by Prosenjit Saha, Sabu Thomas, Jinku Kim, and Manojit Ghosh.

in the future of healthcare. The chapter will start by defining in situ bioprinting and discussing its advantages, guiding readers through the potential applications of these printers. It will then explore the most current bioprinting technologies, including hardware, bioink, and biomaterials, that have been adapted for in situ use. Subsequently, the chapter will highlight in situ approaches for tissue regeneration, supported by the latest studies, and conclude by addressing the current challenges and future directions of this technology.

12.2 Advantages of In Situ Bioprinting

12.2.1 Precision in Cellular Deposition

The conventional approach for tissue engineering faces challenges in achieving precise cell deposition, particularly in the context of complex organ formation. For instance, a kidney's single glomerulus, which is crucial for its filtering function, consists of three interdependent cell types: endothelial cells, podocytes, and mesangial cells [1, 2]. Similarly, the heart incorporates specialized myocardial cells, forming a complex conduction system essential for proper organ function [3]. Replicating such intricate organs requires technological advancements that enable precise control over the deposition of different cell types in a complex pattern. As will be explored further in this chapter, in situ bioprinting uses multiple platforms to, automatically or manually, precisely place cells layer by layer and reconstruct a complex biological tissue along a predefined path with high resolution. Newer bioprinting technologies offering even higher resolutions, such as multiphoton lithography (<1 μm) and electrowriting (5 μm), are opening doors for the fabrication of even more complex structures in the future [4].

12.2.2 Enhanced Integration with Host Tissue

One of tissue engineering's great challenges in the transition from lab to industry remains the inability to control functional vascularization development [5]. An early functional vasculature is a prerequisite for engineered tissue thicker than the diffusion limit of oxygen (around 100–200 μm), which explains why hollow and thin constructs are those that are most successfully engineered currently [6]. Without it, more complex constructs cannot integrate with host tissue and eventually undergo necrosis [7]. Attempts to tackle this problem using conventional TE methods involve induction of cultured endothelial cells to grow onto a synthetic scaffolding platform creating a prevascular network that will subsequently be anastomosed with the host's vascular network once implanted [8–12]. Unfortunately, it usually takes weeks for such a network to develop into an implant of several millimeters in size, and the anastomosis process poses a challenge of its own [5].

With the advent of 3D bioprinting, seeding endothelial cells directly unto scaffolds made up of various biological and synthetic materials has gained much interest, though the issue of transplanting the final construct remains [13–16]. In fact, the smallest capillaries in human tissues reach about 5 μm [8]. However, extrusion-based printing method produces filaments typically of the 100 μm scale, with successes down to 20–30 μm using pH-controlled collagen gelation, making the printing of capillaries a challenge [17]. While current 3D bioprinting has seen the emergence of new technologies offering better resolution, these high-resolution techniques are also self-limited by their slow printing speed resulting from low throughput, which potentially compromises tissue viability if the speed of printing cannot be completed in a timely manner [18–21].

In situ bioprinting provides an interesting alternative to these problems. By either incorporating angiogenic factors into the bioink or directly constructing scaffolds for microvascular growth, it allows host vascularization to be directly stimulated to produce capillaries around the tissue construct without the need for transplantation [22]. Successes in inducing host vasculature have been obtained in osteogenic tissues wherein high vascular densities and kinetics were obtained by implanting microsurgically constructed arteriovenous loops within a bone matrix in rats [23]. As will be further explored in this chapter, combined osteogenesis and angiogenesis have achieved success in providing in situ bone defect repair using 3D printing technology [24]. Therefore, it is not far-fetched to imagine this technique being utilized in vivo to generate tissues with vascularization that develops concomitantly.

12.2.3 High-Fidelity Replication of the Extracellular Environment

In situ bioprinting enables precise replication of the extracellular environment of native tissues. In the native organism, cellular interactions in the form of cytokines and chemokines have been shown to regulate cell growth, replication, apoptosis, and differentiation in neighboring tissues and even faraway organs [25]. A notable example is that of stromal–epithelial interactions wherein paracrine signaling between epithelial and mesenchymal tissues regulate each other's function and development [26].

Conventional 3D bioprinting has attempted to emulate the epithelial–extracellular matrix (ECM) interaction ex vivo by integrating epithelial cells – such as keratinocytes – in the printing of human skin with fibroblast loaded inside hydrogels composed of a variety of biomaterials intended to mimic the functional aspect of native ECM such as collagen, fibrin, alginate, and hyaluronic acid (HA) [27,

28]. Compared to synthetic polymers, these natural biopolymers offer enhanced cell response and better bioactivity [29]. Despite these attempts, cellular interactions inside living organisms are dynamic and complex. The sheer volume of signaling molecules poses a significant challenge for tissue engineering ex vivo to accurately reproduce the in vivo environment. By directly depositing cells onto their natural milieu, in situ bioprinting offers an attractive solution to allow cell differentiation without additional in vitro manipulation.

12.2.4 Reduced Immune Response

One major ethical issue concerning clinical research on tissue engineering surrounds its cell sources and processing methods [30]. Due to their ability to divide infinitely and differentiate into any cell lineage, stem cells have wide therapeutic purposes. However, the use of embryonic stem cells extracted from blastocysts in living or aborted fetuses has been considered unethical in some countries [31]. Furthermore, long-term studies raise concerns that HLA-mismatches and ABO disparities are associated with higher graft rejection and graft failure rates, reaching as much as 50 times in leukemia patients receiving myeloablative conditioning [32, 33]. Patient-specific stem cells avert the issues surrounding the harvesting of and immune response caused by allogeneic stem cells. However, their processing is challenging and expensive [34]. In situ bioprinting and its application of small-scaled personalized constructs may reduce the cost of using the patient's own cells for therapeutic regeneration [31].

12.2.5 Regulatory Requirements

Legislative requirements for bioprinted tissue engineering products are evolving and vary by region [30]. In Canada, most tissue-based products are classified as biologic drugs, regulated by Health Canada's Biologic and Radiopharmaceutical Drugs Directorate [30]. Sponsors of biologic drugs must submit a Clinical Trial Application (CTA) prior to commencing clinical trials and a New Drug Submission (NDS) for premarket approval. This 300-day process ensures that the product fulfills rigorous safety, effectiveness, and quality criteria [35]. Similarly, in the United States, tissue engineering constructs are regulated as high-risk biologic drugs and must always undergo a 12-month premarket approval review by the Office of Tissues and Advanced Therapies [36].

Conventional tissue engineering approaches require the tissue construct – biomaterials and cells – to be implanted into the host. The storage and handling conditions of these fully functional organs must be submitted to tight regulatory requirements. Without a transplantation requirement, in situ bioprinting may face lower barriers to clinical translation than conventional tissue engineering methods.

12.2.6 Potential for Personalized Medicine

Personalized medicine – or precision medicine – uses a person's own biological information, genes, proteomics, and metabolomics, to target diagnosis and treatment [37]. Along with traditional 3D bioprinting, in situ 3D bioprinting is a promising technology with versatile clinical applicability that will allow the accurate reproduction of individual organ defects. Using computer-aided design, defects are modeled and constructs can be printed in real time without the delay needed by traditional 3D bioprinting constructs [38]. As will be explored in later sections, in situ bioprinting has achieved success in regenerating tissues such as skin, bones, and cartilage [39–43].

The potential of in situ bioprinting for topical or localized drug delivery is readily apparent, thanks to its high-resolution printing system and precise deposition capability at micron-level resolutions [44]. Currently, 3D printing has been applied in the pharmaceutical field to build oral drug delivery systems with enhanced absorption, transdermal patches securing sustained delivery, wound dressings, and various types of drug-encapsulated microneedle delivery systems [45–53]. While other drug delivery systems are gradually entering the realm of in situ printing, the printing of wound dressings has garnered remarkable attention, leading to significant advancements [54]. These noteworthy progressions will be further investigated in subsequent sections of this chapter. With the ongoing advancements, the realization of personalized medicine is gradually approaching.

Furthermore, in situ bioprinting eliminates the risks of tissue contamination during implantation and liberates medical personnel, allowing high-throughput repair directly unto patients' bodies. In the future, one can envision a continuum of robotic in situ bioprinters operating tirelessly in conflict zones or isolated regions, surpassing human limitations, under the careful supervision of a surgeon [55].

12.3 In Situ Bioprinting Technologies

12.3.1 Brief Review of 3D Bioprinting Methods

Several key features distinguish in situ bioprinting from in vitro. These differences – such as its "on the spot" printing functionalities – make in situ bioprinting desirable and enable a larger therapeutic potential. However, in situ 3D bioprinters require mechanical engineering modifications to the existing bioprinting systems to incorporate such differences and reap the advantages. These adaptations allow them to print directly on the targeted site, whether that target is an external surface such as the skin or an internal organ such as the lumen of the colon. Depending on the clinical context, the standard bioprinter is modified to fulfill the needs of the application. Currently, there are three main types of printers in use: automated bioprinters, semiautomated bioprinters, and handheld bioprinters. These printers and their subcategorizations have been summarized in Figure 12.1.

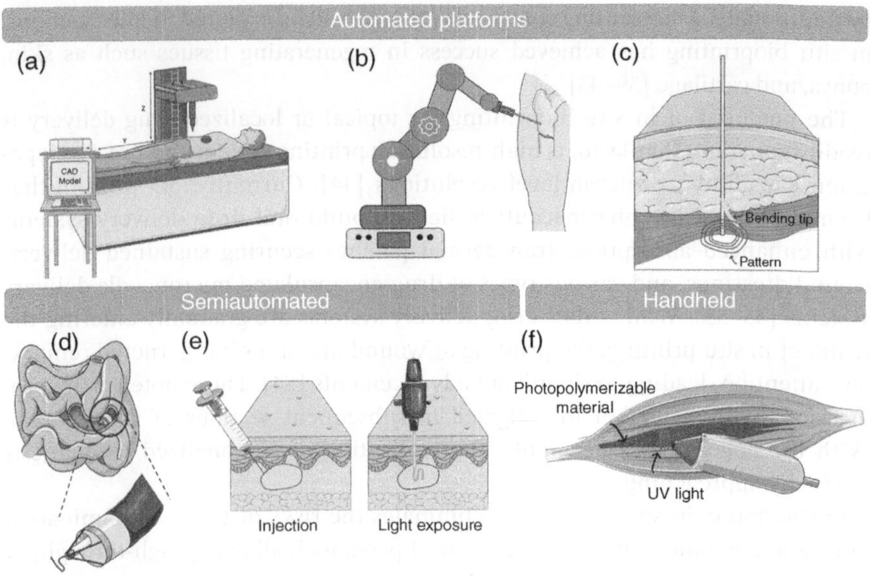

Figure 12.1 In situ bioprinting platforms and innovative methods of additive manufacturing within each category. (a) *Frame-based/bedside-mounted bioprinter:* conventional bioprinter that requires 3D scanning/imaging of the defect, computer system to run the computer-aided design (CAD) file and provide directions to the printer head, and a bioprinting frame, (b) *articulated robotic arm:* bioprinters adapted to perform minimally invasive surgery and address internal defects, through an incision point on the skin, (c) *ferromagnetic soft catheter:* using an external magnetic field, a ferromagnetic soft catheter nozzle is controlled to deposit layers of the bioink at the target site, (d) *endoscopic robotic system:* endoscopic printing heads mounted on existing equipment can enable internal organ bioprinting potentially paving for a future where it is possible to repair gastric or intestinal defects without open surgical intervention, (e) *intravital bioprinting:* in this two-step process, a photopolymerizable bioink is first injected at the defect site; next, the bioink is crosslinked to form the desired matrix by exposure to a particular intensity and wavelength of light, (f) *light-activated handheld bioprinters*: similar to intravital bioprinting, these printers use light to photopolymerize a bioink; however, light-activated printers are designed to be handheld, allowing for a quick defect filling time. *Source:* Parts of the figure were drawn by using pictures from Servier Medical Art. Servier Medical Art by Servier is licensed under a Creative Commons Attribution 3.0 Unported License (https://creativecommons.org/licenses/by/3.0/).

12.3.2 Adaptations for In Situ Applications

Three main types of in situ bioprinters have thus far been designed. These printers vary in their level of automaticity. Fully automated printers are those that, once the design is complete and the printer is prepared with the bioink, autonomously print in additive, successive layers to fabricate the target tissue (Figure 12.1a–c). These printers deposit bioink through a predefined printing path, guided by a computer system. Semiautomated printers are those that require human intervention during the bioprinting process (Figure 12.1d, e). One such system is intravital 3D bioprinting. In it, a qualified healthcare administrator first injects a photosensitive biopolymer, which is then polymerized through a bioprinter that emits light according to the planned path. Finally, handheld bioprinters are those systems that can be brought to the target site, to the patient's bedside. These systems are mobile and rely predominantly on healthcare administrators to deliver the bioink in an appropriate formation (Figure 12.1f). Each of these types of bioprinters is described in their respective sections.

12.3.2.1 Automated In Situ Bioprinters

Automated bioprinters offer many advantages, like higher printing resolution and accuracy. Resolution primarily relies on the printing modality and the selected bioink, whereas the accuracy is dependent on the constraints and calibration of the printer. Due to their automated capability, these bioprinters can permit a larger range of printing modalities as compared to their handheld counterparts. The majority of handheld bioprinters are extrusion-based systems since other printing modalities such as stereolithography (SLA) and laser-assisted bioprinting (LAB) pose technical challenges to the miniaturization needed as part of their handheld design [56]. On the other hand, automated bioprinters, which permit printing with SLA and LAB, can generally facilitate higher printing resolution. Such a finding is unsurprising since printing with laser would generally produce higher resolution than extrusion-based methods. Automated bioprinters also enable high printing accuracy because of their predefined and mechanically well-constrained printing path. Since these printers are controlled via motorized systems, their accuracy is superior to that of handheld bioprinters. This combination between improved resolution and accuracy creates therapeutic potential for in situ tissue engineering and biofabrication that handheld systems would be unable to fulfill. Specifically, automated bioprinters enable printing in "high definition" (HD). HD bioprinting is the ability to produce objects with features that are less than 50 μm in size. This capability will allow delicate tissue repair like that of nephrons, intestinal crypts, or pulmonary alveoli [57]. In addition to the improved resolution and accuracy, automated bioprinters can incorporate closed-loop systems. Since automated bioprinters are connected to a computer, they utilize the processor to receive external information from the printing site and feed it back to the printer. Through such closed-loop control systems, it is possible for automated in situ applications to respond to movements on printing sites adaptively. Printing on the human body, with its dynamic and physiologic actions such as respiratory function or cardiac rhythm, may result in reduced accuracy of bioprinters if real-time tracking and printer compensation are not incorporated [58].

Frame-Based Bioprinting Platforms

Conventional, frame-based bioprinter systems are the most common type of automated bioprinters. They consist of a 3D imaging or scanning modality, a computer control system that directs the bioprinter, and a bioprinting unit that moves on multiple axes. The following is a review of these components:

3D scanning: Frame-based systems, as with the other automated in situ platforms, make use of a 3D scanner or other imaging modality to obtain an impression of the geometry at the defect or target site. Several imaging methods have been used to achieve this function, including magnetic resonance imaging (MRI), computed tomography (CT), and optical 3D scanners [38, 59, 60]. MRI and CT imaging are valuable in that these imaging techniques are already integrated in the healthcare system and many patients may have such imaging done as part of their admission. MRI is a well-established tool in the imaging of soft tissues; in comparison, CT imaging is usually indicated for viewing bone structures. Both these modalities can image internal structures of the body, which can enable bioprinting of internal structures through MIS. However, while advantageous, these systems are expensive and generally require operational expertise. On the other hand, optical 3D scanner systems are cheaper and are generally user-friendly, requiring minimal training for operation. However, optical 3D scanners can only capture the topography of external surfaces that can be subject to visible light. All in all, the selection of a scanning modality should be based on the defect location (external vs internal organs), the tissue properties (soft vs hard tissue), and the clinical context of the patient (i.e., if imaging is already indicated as part of the medical management). After image acquisition of the defect site and its features, the scanned data are translated to standard tessellation language (STL) format. STL files are a 3D construction of the surface geometry, which enables visualization of the model attributes. This STL file is then read by computer-aided design (CAD) software where the 3D shape can be edited as desired to optimize the printing. Finally, once the file is prepared, it is exported as a G-Code in the printer's coordinate system. G-Code files take the printer's axes and create a series of instructions for the printer, including the direction, speed, and path of movement. Through this, the G-Code relays to the printer its planned path, in coordinates, directing it to serially build the desired tissue [61]. Here, the registration of the printing path to the patient reference frame is important, ensuring that the printer is aware of the patient's location, and maps those coordinates within its own reference frame [38]. A transformation matrix is often needed to align both reference frames.

Computer control system: In addition to a 3D scanning modality, frame-based systems make use of computer control systems. Such a control system is important to regulate and adaptively respond to movements of the printing surface during in situ printing. To achieve adaptive control, closed-loop feedback is required. Closed-loop control is possible through continuous, real-time monitoring of spatial coordinates placed on the printing body via a tracking camera/scanner [62]. The recorded information from the monitoring camera is fed into software that can calculate the required compensation to maintain the planned trajectory. Overall, this process enables adaptive 3D bioprinting on moving freeform surfaces [62].

Bioprinting frame: Finally, frame-based systems require the movement of the bioprinting unit on multiple axes. Most of the existing frame-based bioprinters are limited to three degrees of freedom (DoF), allowing them to move in the *X*, *Y*, and *Z* planes. However, with three DoF, the printheads of these bioprinters are typically oriented in a vertical-downward direction confining the printer to a horizontal printing plane. Since surfaces of the human body have complex curvatures, a three-DoF printing system can be limiting to the quality or functionality of the printer (Figure 12.2) [62]. In addition to the complex geometry,

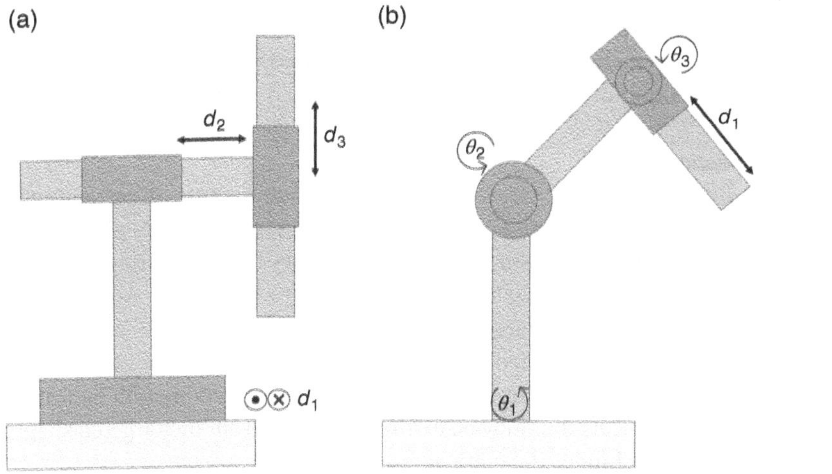

Figure 12.2 Kinematic configurations of different degree of freedom (DoF) systems. (a) *Three-DoF system:* In a three-DoF system, the printheads are typically oriented in a vertical-downward (d_3) direction confining the printer to a horizontal printing plane, which can have limitations when printing in situ. (b) *Four-DoF system:* Higher DoF systems have increased printing freedom, allowing for applications that may have not been possible with the constraints of lower DoF systems. Figure 12.3 demonstrates the benefits of increasing the DoF of bioprinters.

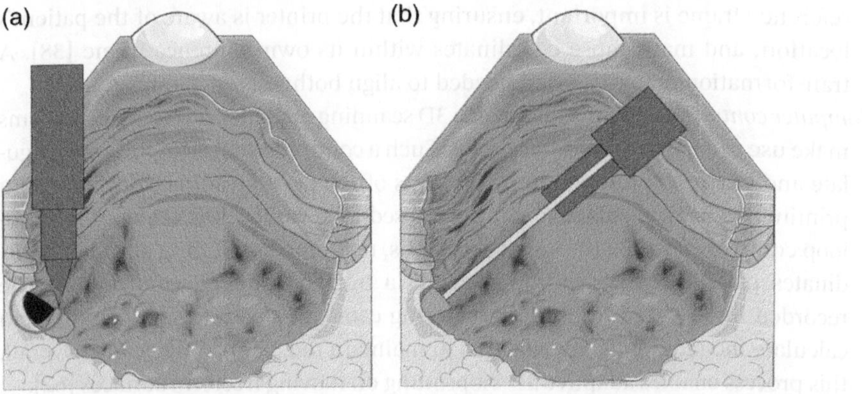

Figure 12.3 (a) Underfilling (underfilled area demonstrated by the encircled marking) that can be seen with three-DoF systems. (b) In comparison, higher DoF articulated robots have superior dexterity, which enables better filling of the defect site. Enhancing the overall geometry of the manufactured scaffold more closely resembles natural tissue and supports the scaffold integration and cellular proliferation. *Source:* Parts of the figure were drawn by using pictures from Servier Medical Art. Servier Medical Art by Servier is licensed under a Creative Commons Attribution 3.0 Unported License (https://creativecommons.org/licenses/by/3.0/).

living tissues – unlike many other 3D printing surfaces – are delicate, have low rigidity, and are susceptible to injury. Consequently, higher DoF systems will allow increased dexterity, which, when amalgamated with meticulous closed-loop control, can be effective at bioprinting of tissues without causing secondary injury to existing defects.

Articulated Robotic Arm

Defect sites due to skin burns, tissue ulceration, or bone fracture all present with unique geometry that extends beyond a single two-dimensional (2D) plane. With such complex curvatures, stacking of planar 2D layers through cartesian three-DoF bioprinters risks underfilling the defect site (Figure 12.3a) [63]. Additionally, with sophisticated profiles and larger angles, a specific phenomenon, namely, the stair-stepping effect, is exacerbated. In the stair-stepping effect, there is distortion and deviation between layers overlying each other. This phenomenon can often result in bioprinted objects that may be distinctly altered from the targeted shape, and thus may possibly hinder the clinical outcomes.

To overcome many of these challenges, articulated robotic arms have been considered for in situ bioprinting (Figure 12.1b). These systems have multiple 360° rotating joints that enable the printer head to be oriented in many different directions for printing complex shapes onto unique curvatures. Because of their multiple axes, these robotic arms are more adaptable with anthropomorphic qualities. Moreover, with the ability to design a variety of kinematic joints, larger DoF systems have been implemented that facilitate applications that even six-axis systems could not perform due to dexterity limitations. One such example is the use of a seven-axis serial robot for the repair of fetal membranes, which, if implemented clinically, can be used to avoid premature rupture of membranes in pregnancies [64]. Similarly, because of their increased flexibility, these systems can perform MIS, which includes inserting long and narrow surgical tools through small incisions in the skin. The reason that MIS requires high dexterity of the robotic arm is because of the remote center of motion (RCM) constraint. As seen in Figure 12.4, the printer end effector must pivot around the incision point and align well with the defect site. It is this flexibility and redundancy, which is intrinsic to articulated robots, that provides them with their advantages.

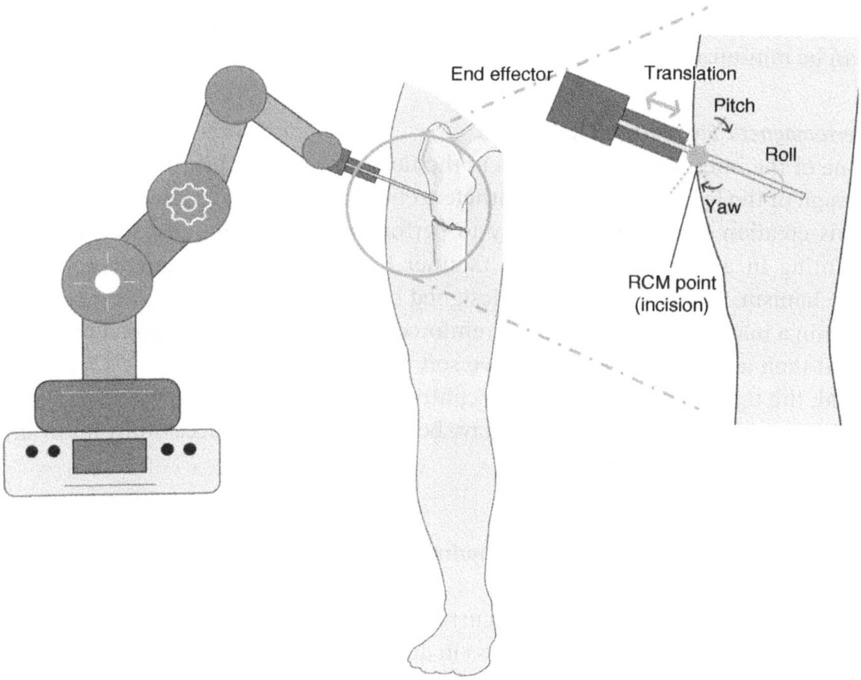

Figure 12.4 *Schematic diagram of a 3D bioprinter end effector, mounted on an articulated robotic arm, as it moves under the remote center of motion (RCM) constraint.* The printer end effector must pivot around the incision point and align well at the target defect site. *Source:* Parts of the figure were drawn by using pictures from Servier Medical Art. Servier Medical Art by Servier is licensed under a Creative Commons Attribution 3.0 Unported License (https://creativecommons.org/licenses/by/3.0/).

However, their flexibility comes at a cost – namely, an increased programming complexity of the driving kinematic models. To be successful, the robotic arm ought to perform dexterous movements, including movements of multiple joints, while avoiding collisions during the operation. Moreover, articulated robots require advanced algorithms for path planning that dynamically manipulate the end effector such that it is always perpendicular to the printing surface. To address safety concerns, both software and hardware advancements have been implemented. Software safety mechanisms include complex failure prevention models that should ensure that every movement of the robot stays clear of the surrounding human tissue, accounting for slight body movements in the process. Such algorithms have been designed and tested including safety mechanisms where the printing pattern is projected onto the printing surface prior to printing [65]. As for hardware, encoders can, and have been introduced on axes of articulated robots that allow for real-time monitoring of the printing head [65]. If the encoder positioning value differs from the expected printing path, the printing code is terminated [65]. Through such a regulatory process, any harm from possible collisions can be minimized.

Ferromagnetic Soft Catheter Robot
One of the interesting innovations in the field of in situ bioprinting has been the design of the ferromagnetic soft catheter robot (FSCR) system (Figure 12.1c) [66]. This creation possesses the ability to perform in situ computer-controlled bioprinting in a minimally invasive manner by employing a magnetic actuation mechanism. The FSCR system is designed by integrating ferromagnetic particles within a matrix composed of fiber-reinforced polymer. A superimposed magnetic field then acts on the magnetoactive soft printer nozzle. Through this magnetic field, the tip is steered as needed to control the printing process. While such systems are still in their infancy, they have been demonstrated successfully for in situ bioprinting of liver in rats [67].

12.3.2.2 Semiautomated In Situ Bioprinters

Endoscopic Robotic System
While most in situ bioprinters are currently designed to repair defects located on outer tissues, there has been progress in designing bioprinters capable of printing intracorporeally. Such systems have been demonstrated in minimally invasive repairs of internal organs including endoscopic intestinal defects [68] and gastric wall repair [69]. Such printers have been predominantly semi-automated. One such system, called the flexible 3D bioprinter (F3DB), has been designed employing a master–slave model, wherein the user stationed at the master console controls a slave manipulator (Figure 12.1d) [68]. This manipulator comprises a pliable

robotic arm and a 3D printing apparatus. Once the 3D printing head is directed by the master to its desired destination, an automatic control algorithm, rooted in an inverse kinematic model, is engaged to initiate movement in three directions for both the soft robotic arm and the printing nozzle. This orchestrated motion enables the deposition of multilayered biomaterials onto the surfaces of internal organs or tissues across multiple locations. Employing predefined trajectories, the 3D printing head then autonomously executes the program. Once a printing task at one site is complete, the soft robotic arm is guided toward other locations, and the printing process resumes. This dynamic workflow enables the system to perform multisite printing, allowing for the creation of complex structures across different areas of interest.

However, while these systems facilitate a wider range of applications, their complex engineering design can be a hindrance in their market adoption. Particularly, translating benchtop in situ bioprinters to an endoscopic form involves designing a micro 3D bioprinting platform that can be incorporated with current medical devices. With such miniaturization, it has been observed that the extrusion pressure applied on the bioink increases dramatically, even upwards by more than a factor of 10 due to area reduction [70]. Such an increase, however, can impact the cell viability of bioinks, which can hinder the printer's application. As such in the design of these systems, specific engineering design considerations need to be regulated, including both kinematics of the device and fluid dynamics. From a kinematics standpoint, endoscopic systems are designed to have both extracorporeal and intracorporeal joints. As with regard to fluid dynamics, several factors are considered. These include the bioink itself, with its rheological properties, as well as the flow behavior induced by the printer tubing. All in all, while this form of bioprinting poses technical challenges, its application would create enormous therapeutic potential. Thus, it is the role of engineers and scientists to collaborate in order to overcome these challenges and make these printers a reality.

Intravital Bioprinting

Intravital bioprinting is the process by which photoactive polymers are injected directly at the desired site, which may be at or below the skin surface. After injection, the polymer is crosslinked using a multiphoton microscope to form the designed 3D object (Figure 12.1e) [71]. One of the advantages of this in situ bioprinting technique is its ability to print through live tissues without inducing damage to them. This is advantageous in that it allows 3D bioprinting without open surgery, thereby reducing the risk of any infection caused at the site of incision if it were to be completed through MIS, or open surgery. Additionally, through this technique, it is possible to preserve the pre-existing cellular structure that should support the integration of the bioprinted 3D object. However, one limitation of intravital bioprinting is that the crosslinking light might not be able to penetrate deeply without causing damage. Thus, intravital bioprinting may be constrained – at least for now – to more surface organs.

12.3.2.3 Handheld In Situ Bioprinters

In comparison to automated printers, handheld printers enable direct writing and crosslinking of biomaterials by surgeons or physicians. These devices are simple – consisting of a handle, a nozzle, an extrusion system, a crosslinking device, and select biomaterials. Their simplicity is central to making these devices affordable, with the total prices being reported as low as $100 [43]. In addition to their price advantage, when compared to automated bioprinters, handheld systems are superior in several facets. First, these systems avoid the complexities inherent to automated methods; they do not require many of the pre-processing steps. For instance, handheld devices do not require 3D scans or imaging of the defect site [72]. More importantly, they do not require the preparation of 3D files and G-Code to enable computer-directed printing, which can both be time-consuming and often demanding from a technical standpoint. Secondly, because of their size and portability, handheld bioprinters are typically easier to sterilize. The convenience they offer is also valuable in increasing the accessibility to treatments that may be offered with the implementation of 3D bioprinting. That is, once these devices are adopted within healthcare systems, handheld bioprinters will be more accessible in remote communities and at point of care. This portability is important in breaking down barriers to access and in expanding the reach of the healthcare system to areas impacted by natural disasters, wars, or otherwise, where there is lack of infrastructure. Thirdly, handheld bioprinters give the user the ability to flexibly deposit biomaterial based on the wound geometry, with real-time tuning by a person having analytical acumen. To reap the many advantages of handheld bioprinters, however, many engineering design constrains are to be considered and implemented. These include: (i) producing an ergonomic design that supports the user in efficiently and safely operating the device; (ii) controlling the bioprinting pressure and flow rate, which is usually programmed, allowing the user to focus on the printing location and overall speed; and (iii) designing the nozzle head to control for bioink shear stress and viscosity [73]. Of the handheld bioprinters, the Biopen is widely known in the field [74]. The Biopen, designed in 2016, is an extrusion-based, pneumatic device that enables the printing of multiple inks in a collinear geometry [55]. The device went through design changes and was demonstrated to be an effective tool for in situ bioprinting with cells and scaffold to regenerate articular cartilage in animal models [75, 76]. Several types of handheld printers have been discussed in the literature. These, however, predominantly fall into two major categories – those that are light-activated and those that are ionically activated.

Light-Activated Bioprinters

To treat volumetric muscle loss (VML) – a complication often resulting from traumatic injury to skeletal muscle (SM) – Russel et al., in 2020, pursued in situ 3D bioprinting when looking for a solution. The use of a handheld bioprinter, in this scenario, was of particular importance since the treatment of VML injuries requires rapid intervention to avoid inflammatory reactions that prevent functional muscle restoration [77]. Thus, the use of handheld bioprinters can be integral in avoiding any time losses associated with defect imaging, as well as 3D design and printing path generation. In their solution, the authors developed a handheld bioprinter that is equipped with ultraviolet (UV) light capable of photocrosslinking hydrogel-based printed scaffolds (Figure 12.1f) [77]. Their solution, after optimization of the nozzle parameters such as diameter and extrusion rate, was capable of depositing myoblasts that were viable and capable of proliferation. The light-activated handheld printer was deemed to be an effective solution that should be implemented in cases of traumatic injuries where response time is essential and dexterity is indispensable.

While Russel et al. in 2020 sustained cell viability after using UV light to crosslink their polymer, it is generally accepted that visible light is more biocompatible, and thus more suitable for bioprinting purposes [78]. Al-Abboodi et al. in 2019 were able to design a handheld bioprinter, using off-the-shelf syringes, to print in situ tissue sealant with a bioink that can be crosslinked with visible light. Such a sealant is used for a variety of reasons including as an acute intervention to stop bleeding [79]. Along with its sealing properties, the bioink can be incorporated with proteins, drugs, or metal nanoparticles to promote cell proliferation, inhibit inflammatory reactions, or prevent infections. Due to its visible light crosslink properties, the matrix photocrosslinks within 30 s and is thus an ideal candidate to be combined with handheld bioprinters for use in acute settings – in surgeries or trauma. Such innovation demonstrates the spectrum of applications that is possible with handheld bioprinters, especially in acute, in situ purposes.

Ionically Activated Bioprinters

Unlike light-activated bioinks, ionically activated bioprinting systems require different design considerations. These systems work by allowing the integration of two different materials, one being a crosslinker material and the other being a bioink. These differences are predominately incorporated through innovative extruder designs. The extruder system ought to combine the crosslinker with the bioink at specified amounts to ensure that the latter is sufficiently cured at the conclusion [80]. Typically, this design includes a microfluidic cartilage that separates the two distinct solutions. At the apex of the bioprinter nozzle, the cartilage channels merge allowing the bioink to contact the crosslinker immediately prior to the deposition surface.

12.4 Bioinks and Biomaterials for In Situ Bioprinting

12.4.1 Overview

A bioink – defined as a bioprintable compound that consists of cells, biocompatible materials, and functional elements – is central to the success of in situ bioprinters. Not only do bioinks determine the initial structural integrity of the printed object, but they are also fundamental for the proliferation of cells at the defect sites in the days and weeks following the structure manufacturing. When designing bioinks for in situ bioprinting, careful consideration should be given to the following design criteria. (i) Since in situ printing takes place at the patient's wound site, bioinks used in this method should have rapid crosslinking to avoid collapse of the structure [81]. The temperature, surrounding blood or bodily fluids, and physiologic movements are all factors that can impact the structural integrity of the printed form prior to its crosslinking. Moreover, unlike in other forms of 3D printing, in situ bioprinting does not always enable the printing of support structures in the initial, base layers to support higher layers in the structure [82]. It is also worthy of note that, in general, the crosslinking of bioinks can be limited for in situ applications because of the surrounding environment. Since the printing platform is constantly changing due to fluids at the defect site, the concentration of the crosslinking reagent, which ought to be strictly controlled, can be susceptible to change. Such a change in the ratio of bioink to crosslinking reagents can influence the final properties of the structure. Thus, specific care should be taken to ensure that the crosslinking method is not impacted in a way that can bear negative consequences to the outcome. Therefore, non-chemical crosslinking methods, such as light crosslinking, are more commonly used for in situ applications [82]. (ii) The bioink must create an environment that facilitates cell growth and induces desired cellular responses. It must then, with certainty, be biocompatible – not provoking cell apoptosis or instigating immunogenic responses in the host. In addition, bioink-generated structures should degrade at a rate similar to tissue maturation. Different bioinks should be used for varying tissue types depending on the native tissue growth rate [62]. (iii) The designed bioink should maintain a delicate balance between having strong and weak mechanical properties. While these bioink structures should exhibit high mechanical strength to be able to match the defect site and uphold cells during repair, they should also exhibit low mechanical modulus to allow cells in the matrix to exert their therapeutic roles [81]. Many materials have been considered for incorporation with in situ bioprinting including gelatin, alginate, chitosan, HA, and collagen [83–89]. Each of these bioinks presents its advantages. For instance, HA has been shown to act as a natural signaling molecule capable of increasing cell proliferation at the defect site [90].

12.4.2 Engineering Consideration

As described in the previous section, bioinks used for in situ bioprinting require specific engineering considerations. Each of these is described in further detail in the following sections.

12.4.2.1 Photocrosslinking

As mentioned, photocrosslinking is the most used method for in situ bioprinting [91]. UV, laser, and visible light are light radiation sources used for photocrosslinking; of these, UV light is the most common. Many parameters are controlled with photocrosslinking, including the intensity and the exposure time, which should be adjusted based on the bioink used and the desired construct stiffness. However, it would be cumbersome to study every bioink, testing the mechanical characteristics of each based on multiple variable crosslinking parameters. Instead, computational modeling offers a time- and cost-effective solution that can be contextualized to various clinical problems. One such computation model has already been developed, which studied the crosslinking kinetics of alginate-based hydrogels [92]. Ultimately, validated models will facilitate optimization and facilitate creation of registries for future research.

12.4.2.2 Biocompatibility and Biomimicry

Beyond crosslinking, in situ bioprinting should ideally produce constructs that can guide native tissues to repair and replace damaged tissue at the defect site [93]. Often, that is made possible through constructing biomimetic structures that mimic the natural microenvironment of the defect sites [57]. For instance, nanostructured scaffolds that have matrix configurations of similar size to native ECM are known to promote migration, integration, cellular attachment, and differentiation of mesenchymal stem cells (MSCs) [94]. One example of such a structure was the in situ bioprinting of nanohydroxyapatite/collagen bioink onto a calvaria defect model in mice [95]. This study incorporated mesenchymal stromal cells, and through it, the authors demonstrated that varying the cellular arrangements of the bioprinted scaffold can influence native bone tissue regeneration.

12.4.2.3 Mechanical Properties

Degradation Properties

The degradation of bioprinted scaffolds is an important engineering design challenge. Ideally, the degradation rate should be long enough to provide structural support and promote cell growth and tissue formation. At the same time, however, the degradation rate should also correspond to the growth of native tissue at the defect site. Striking this balance depends on understanding the scaffold material, its porosity, and the specific environment in which the scaffold is placed.

Scaffolds with higher porosity, for instance, tend to degrade more quickly as they provide more surface area for enzymatic and hydrolytic processes. Cellular activity at the implant site can also contribute to scaffold degradation. When cells are seeded onto the scaffold, they can produce enzymes or secrete acidic byproducts that can accelerate the degradation process. Additionally, cells can physically remodel the scaffold by exerting forces and causing structural changes.

Carefully tuning bioinks to fabricate an ideal scaffold is challenging. Barcelo et al. in 2022 investigated the impact that different ratios of raw and partially oxidized alginate (OA) have on the rate of bioink degradation [96]. Their results showed that the degradation rate is highly dependent on the OA content of the bioink [96]. While such a result is positive since it indicates an ability to fine-tune bioinks, it is also burdensome. Knowing that minor variations directly and easily influence degradation properties of scaffolds in bioinks creates a challenge in regulating the exact amounts of every compound in the bioink.

Rheological Properties

When passing through the diameter of a 3D bioprinter nozzle, the bioink experiences pressure that may impact its properties. As such, it is crucial that a functional bioink is tested under such conditions and demonstrates shear-thinning properties [97]. Shear thinning is characterized by a decrease in the bioink resistance to flow when subjected to shear stress, such as within the nozzle. Another important rheological property of a functional bioink is its ability to maintain its structural integrity by exhibiting a viscoelastic response [98]. Through viscoelastic responses, bioinks can return to their original shape following an application of force, such that one experienced at extrusion. These rheological properties of bioinks – including the shear thinning and viscoelastic responses, correlate to the bioink's integrity, consistency, and extrusion rate. As such, because of the significant impact these rheological properties have on the quality of the bioproduct, it is important to control the composition of the bioink carefully.

12.5 In Situ Approaches for Tissue Regeneration

12.5.1 Cartilage and Bone Regeneration

Osteoarthritis (OA) is a chronic degenerative condition that causes cartilage loss and bone remodeling in joints. Symptoms typically involve joint pain, stiffness and reduced daily mobility, and general functionality. Depending on the nature of the condition, treatment may involve pharmacological intervention to alleviate pain and inflammation or may require more aggressive surgical intervention [99]. Importantly, cartilage is aneural; it lacks a direct nerve supply, and consequently, it lacks sensation of pain and other stimuli. While this quality contributes to its uniquely designed structure and function, the downside is that early physical symptoms are left undetected; with the added challenge of its instability to self-repair, damage to this tissue can quickly compound [100]. While OA is a significant source of disability worldwide, current treatments are lacking in many domains; namely, they are unable to restore normal cartilage in the presence of chondral defects [74].

Early stage cartilage damage is not a significant cause for disability, so non-operative management, like patient education and lifestyle changes, is the first course of action. This can include developing a full-range mobility and strength training regimen, making dietary changes to promote healthy weight loss or prosthetic bracing for external support. The natural progression is pharmacological interventions: oral supplements, like glucosamine, which enable the synthesis of pro-healing proteins like glycosaminoglycan and glycoprotein; steroid injections that can reduce pain and offer an anti-inflammatory benefit; and injectable viscosupplementation, which is a high-molecular-weight HA injection to combat the commonly observed reduction in viscoelasticity of synovial fluid. Orthobiologics are also becoming an emerging treatment modality. Treatments like platelet-rich plasma (PRP), stromal vascular fraction injections (SVF), or even stem cells have shown evidence for countering cartilage erosion and improving function. However, the evidence and reliability for the abovementioned treatments are scattered and inconclusive. Moreover, they can be very expensive, and results are often short-term. In more severe cases, surgical interventions may be recommended: mosaicplasty, microfracture, autologous chondrocyte implantation, and hyaluronan-based scaffolds to name a few. However, like the non-operative approaches, clinical evidence for hyaline cartilage regeneration or long-term symptom reduction is inconclusive [99].

In recent years, regenerative medicine approaches have been explored as possible alternatives to conventional therapies. Some investigations have been conducted involving the implantation of engineered biomaterials to damaged areas to promote tissue regeneration [101–103]. Other groups have explored the potential of 3D bioprinting constructs by layering therapeutic materials for targeted and personalized treatment. However, one challenge with pre-engineering biomaterials in the context of chondral lesions is that the standard protocol for injury repair requires the surgical team to first conduct an initial debridement of the damaged area. This step is to ensure that scarred or fibrotic tissue is removed from the area prior to the intervention. However, the amount of debridement required and the morphology of the final defect cannot be accurately predicted pre-surgery. Consequently, it is incredibly challenging to precisely pre-fabricate constructs. Therefore, a modality that would allow for on-demand autologous deposition of biomaterials would be ideal [74].

Some studies have examined the use of LAB and shown efficacy in murine models. They offer excellent precision and control, high-throughput potential, and reliable reproducibility. One group demonstrated its in situ potential by demonstrating successful printing of mammalian cells to treat bone defects [95]. While robotically controlled apparatuses with this functionality could be installed in operating theaters, they require large, space-consuming equipment that requires frequent sterilization and is separated from the surgical site; consequently, it has limited adaptability when it comes to time-sensitive changes during surgery and distances the surgical team from the patient during the procedure [74].

To address these concerns, O'Connell and team in 2016 developed a handheld biofabrication device prototype called the Biopen. Not only does their prototype offer precise deposition of biomaterials, but it also offers the ability to surgically sculpt tissue ad hoc, improve dexterity for in situ fabrication on difficult-to-map areas, easy sterilizability and portability, and is significantly more cost-effective than other surgical bioprinters due to its simple design [74]. The Biopen was tested on ovine models. The authors collected MSCs from the adipose tissue beneath the patella of a sheep, known as the Hoffa's fat pad. These were then incorporated into an HA-GelMA hydrogel to form a bioscaffold core with an additional outer layer of HA-GelMA coupled with a photoinitiator. When the photoinitiator interacts with UV light, a chemical reaction is triggered, releasing reactive chemical species; subsequently, photo-curing facilitates the formation of chemical bonds and solidifies the gel-like scaffold into a hardened form. Six sheep with bilateral chondral defects at the patellar joint were then randomly placed in four treatment groups: (i) an untreated group, (ii) a microfracture surgery group, (iii) a prefabricated 3D-printed bioscaffold group, and (iv) a Biopen-fabricated bioscaffold group. A standardized medial parapatellar arthrotomy was performed across all groups and the interventions were distributed accordingly [75].

The perioperative and postoperative outcomes were positive: there was an absence of complications, and the Biopen was user-friendly, provided a large range of user control, and was overall flexible to the needs of the surgeon. Moreover, the treatment was well-tolerated by the sheep after their procedure, marked by the successful completion of an eight-week postoperative observation period. Macroscopic assessments revealed limited inflammation despite a preoperative Grade-IV – severe chondral defect – diagnosis across most of the groups. Histologically, the Biopen-treated group demonstrated a greater amount of chondral regeneration, columnar alignment, and no signs of subchondral bone degeneration. Notably, collagen II immunohistochemical staining showed that the regenerated tissue was primarily hyaline cartilage. These cellular observations were supported with a set of biomechanical tests that showed equivalency for Young's modulus – a measurement of rigidity and resistance to resistance – in the

Biopen and prefabricated groups, only measuring slightly below the host cartilage. These findings suggest promising evidence for meaningful chondral restoration and provide insights into the scaffold's durability as well [75].

They were also able to elucidate the ideal photoinitiator as well as overcome challenges with cytotoxicity and compromised cell viability resulting from the photocrosslinking process. Of the photoinitiators studied, lithium acylphosphinate (LAP) produced the highest modulus values. While sub-analyses revealed marginal cell toxicity from just LAP exposure, when coupled with UV light exposure, cell viability was greatly hampered. To resolve this issue, the authors developed a co-axial extrusion approach, designed to deliver the cells in a core–shell structure. The inner core inhabited non-crosslinked cell-laden hydrogel, while the outer shell was solely crosslinked GelMA/HAMa. This compartmentalization strategy successfully protected the cells from toxicity while simultaneously providing structural stability. Moreover, co-axial printing seems to be more efficacious than mono-axial printing both immediately after printing and over the course of the culture period, which may inform future studies that require strategies to preserve delicate cell-biomaterial viability [76].

They also demonstrated how discussions on cell sourcing need to be more conclusive. In a subsequent study, they prepared a human infrapatellar adipose-derived stem cell (hADSCs)-ladened GelMA/HAMa bioscaffold from OA patients undergoing knee arthroplasty. Profile analysis of the hADSCs revealed that at least 90% of these cells expressed CD90, CD73, CD105 CD29, CD49, and CD44, which all uniquely play a role in chondrogenesis, especially CD44 as it is a crucial hyaluronan receptor involved early chondrogenesis [104, 105]. Additionally, the expression of hematopoietic markers was less than 3%, which is significant as it confirms the homogeneity of the cell population. Gene expression analysis revealed increased expression of collagen type I and II, aggrecan, and Sox-9 markers. These markers are critical because the corresponding proteins are largely present in the ECM of chondrocytes and play various supportive roles. While the ideal cell source for cartilage regeneration needs further exploration, these findings suggest that hADSCs sourced from the infrapatellar fat pad bolster chondrogenesis in patients with varying degrees of OA [105].

Bone regeneration differs from cartilage regeneration in several aspects: cellular composition, ECM, vascularization, and mechanical properties; expectedly, in situ strategies differ accordingly. In bone regeneration, the focus is on fabricating bioscaffolds that promote osteogenesis, but also facilitate mineralization and vascularization, unlike cartilage, which has minimal need for hematopoietic capabilities. The literature on tissue engineering approaches to osteogenic regeneration advocate for a bottom-up processing strategy – a brick-by-brick regenerative approach – as compared to a top-down strategy that relies on pre-engineering biomaterials in laboratories followed by surgical implantation. The evidence is consistent with this hypothesis; in the context of bone regeneration, the bottom-up approach optimizes scaffold-tissue morphological alignment and integration and also better replicates native tissue and cellular distribution [95].

Cohen et al. in 2010 designed a protocol for chondral and osteochondral regeneration using geometric feedback-based strategies [106]. They developed alginate hydrogels ladened with demineralized bone matrix (DBM) for the treatment of bone defects. Their substrate of choice was a bovine femur, which was imaged by CT to understand its native geometric qualities and how it changed after structural defect production. In the osteochondral case, the alginate hydrogel was responsible for treating the chondral defect, while a DBM cap was employed to facilitate osteogenic repair. DBM is osteoconductive because it is composed of bone formation-promoting materials like collagen and growth factors; moreover, it facilitates osteoblast migration and proliferation. In addition, it is biocompatible and osteoinductive by virtue of its ability to facilitate stem cell differentiation into osteoblasts as well.

After each treated case with the bioprinted alginate hydrogel-DBM construct, geometric fidelity was assessed by measuring mean error (ME) and spread error (SE). ME is a measurement of how closely the dimensions captured by the CT scan and the fabricated construct are aligned among the treated cases. SE is a measurement of how dispersed the printed construct was around the defect area among the treated cases. Their observations revealed that in each case, the fabricated construct nearly matched the geometry of the defect. The chondral defect exhibited a ME of 0.0 ± 0.2 mm, whereas the osteochondral defect was 0.1 ± 0.1 mm, both falling within the accepted clinical range for graft acceptance. There were varying levels of agreement with respect to SE: the former group exhibited 42.8% of the surface points falling within the error envelope, while the latter had 83.6% of the surface points within the error envelope. Importantly, the upward layering of the construct likely contributed to this error; potentially, with further optimization of printing mechanics and materials, improved results may be observed [106].

However, another challenge with typical imaging techniques is the lack of precise visualization of the intricacies of defect areas, which can also be a contributor to error. Consequently, the morphological features are largely ambiguous and lead to suboptimal pre-surgical planning – possibly affecting perioperative and postoperative outcomes if executed in clinical applications. 3D scanning may offer a solution to this problem by using 3D digital models of the injured site to extract information previously undetected by current imaging techniques. One group used three models to replicate various orthopedic pathologies: large segmental defects of long bones, chondral lesions, and a freeform fracture of the femoral condyle. They developed a 3D scanned model of the defect area that was processed using Boolean operations before being imported into a 3D bioprinter. The results of the study provided promising results on its feasibility for bone and cartilage restoration. Their digital models provided an accurate understanding of the morphological features of the defect area. When compared to the corresponding uninjured parts, their biofabricated scaffolds had very similar measurements with minimal variation [107].

In the case of severe traumatic injuries that require swift clinical responses, currently proposed in situ fails to consider other factors, like bacterial infections at the injured site. One group tested a scaffold containing poly(caprolactone) (PCL), zinc oxide (ZnO), and hydroxyapatite (HAp) microparticles. PCL is a biodegradable polymer that is known for its durability and slow degradation rate. ZnO is a known antimicrobial; the release of zinc ions disrupts the cell membrane of bacteria. HAp is a ceramic bioactive compound and has similarities to the mineral properties of native bone tissue, which makes it conducive for osteoblast proliferation. The authors utilized a partially automated handled printer called "PenBone" on a porcine jawbone. Their findings revealed that when exposed to methicillin-resistant *Staphylococcus aureus*, the composites' antimicrobial performance was meaningful, especially at 1% ZnO. Other supplemental findings showed that HAp was conducive to improved biocompatibility and increased protein absorption. Moreover, at higher concentrations of HAp, MSCs were more proliferative. Interestingly, this combination of materials in the context of in situ printing for bone regeneration has not been investigated. Future studies need to focus on composite formulation and optimization to improve bioscaffold performance and ensure it can perform under a variety of contexts [40].

12.5.2 Skeletal Muscle Regeneration

Human beings have three main muscle types: smooth muscle, cardiac muscle, and SM. Each of these muscle groups plays an integral role in maintaining normal functioning. The former groups regulate involuntary movements of physiological processes like digestion and proper rhythmic contraction of our hearts, respectively. On the other hand, SM plays a number of physiological roles, namely, generating voluntary movement, maintaining glucose homeostasis, and providing structural support and protection to our bones and internal organs [108]. While SM has a remarkable ability to regenerate, its ability to do so is greatly hampered in cases of VML [109, 110]. VML is characterized by *en bloc* – in other words, complete or total – SM degeneration and typically occurs in response to physical trauma to the body: severe accidents, disease, or surgical interventions [41, 77]. VML is also associated with corollary functional impairments, significantly decreasing one's ability to perform day-to-day activities and, by extension, reducing their quality of life [110]. Spontaneous healing from VML is challenging because the condition often produces extensive fibrotic scarring and incomplete tissue regeneration, which hinders muscle contractile ability and the fusion of newly formed muscle fibers [77]. While VML is common, current treatment options are lacking.

Functional muscle transfer or muscle flap transplantation is the current standard of care. It involves the use of microsurgical techniques to extract a graft from one area of the body and then insert it onto the injured site [111]. While the literature reports some efficacy of this approach, it has several limitations: limited donor tissue availability, high risk for complications at the donor site, incomplete restoration of native muscle functionality, and high failure rates [77]. Moreover, due to the complexity of the surgery, only highly specialized surgical teams are qualified to perform them; naturally, this reduces its accessibility to patients as only tertiary care centers would have this capability. Prosthetic bracing is another treatment option for VML, which involves the utilization of a medical device to provide external support and stability to the affected limb. While not the most efficacious solution, it is a non-invasive, personalizable option that can facilitate improved mobility and function. The major drawback is that it is prohibitively expensive and is unlikely to restore native functionality – or at least to the same degree as a successful muscle transfer. VML is debilitating and often leads to hard-to-treat pain and permanent disability [77, 110]. For these reasons, regenerative medicine may provide insights into new treatments for patients with VML.

Currently, tissue engineering approaches utilize cell delivery, growth factor therapy, or scaffolding techniques to promote tissue regeneration. Cell delivery involves culturing of different cell types with various origins to promote differentiation and proliferation at the site of injury. Some examples include myogenic cells, like satellite cells and myoblasts – useful for their ability to form contractile muscle fibers – and non-myogenic cells, like MSCs, which are known for their regenerative and immunomodulatory potential. However, cell delivery is unreliable because of an insufficient understanding related to cell sourcing heterogeneity, optimal dosing, and scalable extraction and proliferation protocols. Growth factors, on the other hand, are cell products – like proteins and peptides – that can facilitate natural healing of SMs. Common growth factors include hepatocyte growth factor, insulin-like growth-1 (IGF-1) factor, and vascular endothelial growth factor (VEGF). There is evidence in the literature that shows the use of these growth factors to promote angiogenesis and myogenic differentiation and even reduce the progression of apoptosis. However, the primary obstacle of isolated growth factor therapy is their short half-lives. Therefore, scaffolding techniques in combination with the abovementioned strategies offer the benefit of programmability allowing these pro-regenerative cells and cell-products to be released in a precise and controlled manner [112].

VML injuries are prone to causing profound inflammatory reactions and initiate a cascade of myogenic loss, reducing the SMs' ability for spontaneous regeneration. When treated in an untimely manner, the inflammation could produce fibrotic scarring and profoundly reduce SM functionally. Therefore, Russell and colleagues in 2020 sought to understand the feasibility of a cell-laden gelatin methacrylate (GelMA) hydrogel in an in situ application for SM regeneration. The long-term aim is to produce a rapid response system that will limit and possibly regenerate permanent damage and disability [77].

The first stage of their experimentation elucidated the structural, chemical, and rheological feasibility of acellular GelMA hydrogels for in situ treatment of SM injury. The rationale for choosing GelMA was its swift crosslinking capability, biocompatibility, and ability to be printable on non-flat surfaces. Moreover, several factors influence the hydrogel quality. The authors determined that GelMA could be optimally modified by changing its concentration and physical and chemical properties, which can ultimately influence the final hydrogel's mechanical properties, rate of degradation, and even its cellular response. Their findings suggest that a medium-level methacrylate GelMA is the optimal formulation for muscle tissue applications. The second stage of their experimentation was to facilitate myofibril and myotube formation via the delivery of muscle cells ladened in the GelMA hydrogel and assess tissue adhesion and inflammation at the site of injury when compared to the control – a collagen glycosaminoglycan (CGC) scaffold, a widely employed implanted scaffold for various soft tissue injuries. They hypothesized that because there is limited evidence on new muscle regeneration following scaffold implantation, there may be scope to modulate the status of the remnant SM at the target site. With this in mind, they evaluated fiber size and hypertrophy in the remnant muscle following the treatment of a C2C12 myoblast-laden GelMA hydrogel [77].

Four murine groups were prepared: (i) a healthy uninjured group, (ii) a VML-injured group without any treatment, (iii) a VML-injured group treated with an in situ printed C2C12-laden GelMA scaffold, and (iv) a VML-injured group treated with an implanted C2C12-laden CGC scaffold. VML injury was induced by an SM resection. Interestingly, they found a statistically significant difference in the central surface area of the muscle fibers between the murine groups. Expectedly, the remnant muscle exhibited hypertrophy as a physiological response to the trauma. However, interestingly, the VML-injured group that received the in situ treatment produced muscle tissue that was 25% larger than the injured group without treatment. Their findings suggest that muscle hypertrophy can be triggered when a suitable scaffold is applied in addition to the physiological trauma response [77].

Building on their work, Quint and colleagues in 2021 developed a handheld bioprinter that can deliver a VEGF-bound nanoclay within a hydrogel to improve muscle functionality in a similar murine VML injury model. First, they bound VEGF to a cell-delivery vehicle called Laponite®: a 2D, artificially manufactured nanoclay disk. Its unique geometry produces a dual charge distribution, whereby the clay surface has a permanent negative charge and the edges a permanent positive charge. This quality is advantageous for its conduciveness to electrostatically bind to proteins and retain protein functionality upon binding. Many studies have used this formation in other models for its biocompatibility as well; upon degradation, its remnant ionic byproducts are nontoxic [41, 113]. Secondly, they incorporated the VEGF-bound Laponite® to a GelMA hydrogel – the final nanocomposite termed "Muscle Ink." The authors prepared four treatment groups of mice: (i) uninjured with no treatment, (ii) VML-injured mice with no treatment, (iii) VML-injured mice with VEGF-integrated Muscle Ink (VEGF+), and (iv) VML-injured mice without VEGF-integrated Muscle Ink (VEGF−). The induced VML injuries were in the form of a bilateral resection of the hind-limb quadriceps [41].

First, they sought to elucidate the functional in vivo impact of Muscle Ink (VEGF+); to do so, they prepared an exercise test on a treadmill eight weeks post-injury treatment. They found that all VML-injured mice had lower maximum speeds than the uninjured groups, but the VEGF+ group had the highest maximum speed of the injured cohorts. Interestingly, the VEGF+ and uninjured groups reported a statistically insignificant difference in maximum speeds. This is a notable finding as a characteristic quality of VML is severe loss of native muscle functionality, but the VEGF+ group's near return to baseline is indicative of a pro-regenerative effect of the VEGF treatment. Likewise, maximum distance traveled was also used as a proxy for functionality. Expectedly, the uninjured group had the greatest maximum distance out of all the murine cohorts; however, the VEGF+ group ran twice the maximum distance than the VEGF− and untreated VML-injured groups, with no significant difference in results between the latter groups. These findings suggest that the sustained release of VEGF improved functional performance in the VEGF-treated groups, lending to its hypertrophic and anti-fibrotic capabilities [41].

Further analysis was conducted to understand the impact of VEGF-loaded Muscle Ink on a histological level. Cross-sections of the VML-induced quadriceps were collected. The authors found that the Muscle Ink was still bound to the wound site demonstrating its long-term adhesiveness as well as its durability to withstand rigorous physical exercise. Moreover, immunohistochemical staining revealed that VEGF+ had a pro-vascular effect. While the mean number of blood vessels was higher in the injured groups as compared to the uninjured group, the VEGF+ mice had a statistically significant increase in CD31-stained capillaries relative to the VML-injured mice. These analyses were coupled with trichrome staining to illustrate the effect of VEGF on fibrosis, as indicated by a blue stain on collagen – the typical marker for fibrosis. The VEGF+ group had lower fibrotic scarring than the untreated VML-injured mice. This observation is significant because fibrosis is an indication for ongoing inflammation, which inevitably leads to further muscle breakdown; however,

VEGF was able to control fibrotic development without the need for an anti-fibrotic agent [41].

Overall, in situ bioprinting approaches to SM regeneration are an incredibly promising area of inquiry. While the emerging literature is limited, the abovementioned studies evidence the possibility of this approach in clinical applications. Integration of this form of treatment into standard of care would have profound benefits: improve treatment precision and allow for more opportunities to provide patient-specific care; reduce adverse surgical outcomes like fibrotic scarring and loss of functionality; improve the healing rate and recovery; develop an avenue to print clinically relevant scaffolds at scale; offer a minimally invasive procedure and would reduce the need to perform time-consuming and expensive diagnostic testing, reducing the current burden on healthcare utilization.

12.5.3 Skin and Wound Regeneration

The skin is the largest organ in our body and exists as a layered structure; the three primary divisions are the epidermis, dermis, and hypodermis. It plays a plethora of structural and functional roles, including protection against opportunistic infection and dehydration. While the skin is incredibly resistant and self-reparative, full-thickness wounds are problematic because they destroy the abovementioned layers of the skin, in some instances, permanently. In less severe cases, the layers of the skin are still damaged, but not completely, which leaves scope for healing; however, it remains a challenge due to lengthy healing times and leaving the skin vulnerable during this period. The natural healing process is gradual dermal reconstitution and then re-epithelialization. Treatments are aimed to expedite these natural healing processes and ensure complete protection and closure of the wound [80].

The current gold standard is autologous split-thickness grafting, which involves the selection of a secondary donor site on the patient's body, extracting a split-thickness graft from the donor site, and then reapplying it to the site of injury. While this approach is helpful in regenerating tissue in some injuries, it is inadequate for more serious cases; moreover, it is also reliant on a limited availability of donor skin. Allografts are like autologous grafting, but it involves extracting donor tissue from a different recipient. The main drawbacks are the potential for immune rejection, the need for patients to be placed on immunosuppressive medication, and success not being guaranteed, potentially leading to subsequent procedures. Acellular dermal substitutes are a third domain of treatment, which utilize polymeric scaffolds – typically crosslinked, freeze-dried collagen dressing – but they are expensive and usually produce poor cosmetic outcomes [114].

Recent emergence of bioprinting techniques may alleviate these limitations by direct deposition of biomaterials to the damaged area, improved precision of treatment, and high replicability of native tissue [114]. Several studies have demonstrated the efficacy of prefabricated skin tissues in vitro [115–117]. Previous investigations have suggested that protein-based biopolymers are most advantageous for biocompatibility, structural similarity to dermal ECM, and optimizing cell interaction and signaling. The most popular proteins of interest are collagen I, the most abundant protein in the dermis, and fibrin, a crucial protein in wound healing. However, these biomaterials have poor printability because of their low viscosity and lengthy gelation times [80]. With this in mind, newer investigations sought to explore optimal cell sources and scaffold materials for use in in situ applications.

One group investigated the benefit of MSC-ladened fibrin-HA bioscaffolds for the treatment of burn wounds on porcine models. They found improved wound healing and enhanced proliferative capacity and printability with the addition of HA to the hydrogel. In addition, treatment with this composite demonstrated evidence for the reduction in inflammation, fibrotic scarring, and contraction – which can cause contracture lesions and hamper tissue regeneration. Another group studied amniotic fluid-derived stem cells (AFS) for their immunomodulatory properties, proliferative capacity, multipotency, and low-level immunogenicity [114, 118]. While MSCs have demonstrated therapeutic potential, these authors opted concurrently to study AFS because it is simpler to isolate, has a short doubling period, and does not require supportive feeder layers. The AFS group exhibited near-complete wound closure, as well as high contraction re-epithelialization. When compared to the MSC group, the AFS group demonstrated greater vascular diameter and density. Further histological analysis also showed improved blood vessel maturation and reduced extravasation of red blood cells. The authors also found that the AFS contained more growth factors at higher concentrations. While the etiology is still unclear, they hypothesized that wound closure and neovascularization may be due to the secretion of these growth factors in lieu of direct cell-tissue integration [114].

Hakimi and colleagues demonstrated an in situ printing approach to skin wounds using planar biomaterials and skin tissue sheets. Their approach was efficacious in both murine and porcine models; however, the profundity of their investigation is the insights into how the physical and mechanical properties of the bioscaffold can be modified to optimize outcomes. Achieving uniform thickness is imperative to ensure even distribution of the therapeutic products and a consistent environment for tissue growth and regeneration. However, certain factors must be considered to avoid suboptimal outcomes. Herein is a summary of those considerations [80]:

- Flow rate is a determinant of sheet thickness, but the viscosity of the crosslinking solution influences their relationship

- Bioink identity also plays a role by virtue of their gelation kinetics. The general rule of thumb: faster gelation quickens the gel-solidification process
  - Alginate-based sheets have high crosslinking rates when exposed to ions, whereas fibrinogen-based sheets do so when exposed to enzymes that convert it into fibrin, which produces the gel
  - Presence of HA increases viscosity which reduces the gelation rate
- Young's modulus – a measure of rigidity or stiffness – and elongation at break – a measure of flexibility – depend on bioink material and composition
- Topography of site of injury needs to be assessed, which will inform the type of crosslinker and sheet patterning (meshed, striped, or spotted) required to achieve optimal functional outcomes

For instance, Ying and team developed an ergonomic design for a handheld bioprinter that delivered pore-forming bioscaffolds characteristic of a two-phase emulsion bioink composed of gelatin methacryloyl (GelMA) and gelatin. They discovered that emulsion-based bioinks allow the capability to adjust pore size depending on the clinical application. There were three parameters that were found to influence pore size: polymer concentrations, phase ratios, and mixing time. Their porous scaffold demonstrated the ability to better chaperone oxygen and nutrients and was conducive for cell proliferation and migration to promote wound healing [43].

Finally, preliminary studies on robotics-assisted in situ printing offer promising results that may be translatable in clinical settings. One group developed a single-nozzle, multi-degree-of-freedom bioprinter with 3D scanning capabilities. Previous literature demonstrated the combination of epidermal stem cells (Epi-SCs) and skin-derived precursors (SKPs) can promote hair follicle regeneration [119]. Epi-SCs and SKPs perform the best when cultured in special environments, like collagen; however, as previously discussed, collagen and similar large proteins lack printability. To this end, they sought to understand to what extent GelMA, ladened with Epi-SCs and SKPs, could promote de novo follicular regeneration. Histological analysis coupled with the examination of key marker expression confirmed the successful growth of hair follicles. Not only did this study provide support for the use of LAB, but it also provided insights into the use of dermal papillae stem cell-embedded GelMA hydrogels for complex soft tissue restoration [39].

12.5.4 Dental Pulp Regeneration

A vital component in dental physiology is the dental pulp: loose unmineralized connective tissue found within a long chamber composed of dentin, enamel, and cementum in the center of the tooth and arranged in a mineralized matrix. Its functionality ranges from aiding in tooth development to providing nutritional, sensory, and defensive support to our teeth. These functions are performed by virtue of its rich vascularization, giving it access to oxygen and nutrients; high innervation, allowing perception of changes in temperature, pressure, or pain; and production of specialized cells, like odontoblasts, that help form protective structural proteins like dentin and help defend against microbial intrusion [82, 120, 121]. Despite these protective mechanisms, there are many pathologies that threaten our dental health. The most common oral diseases and reasons for tooth loss are dental caries and periodontitis – left untreated, can severely degrade the dental pulp [122].

Dental caries is a multifactorial disease primarily mediated by the biofilm – the pH-modulating oral microbiota – and the nature of one's diet – typically high in sugars and acidic contents. Its pathophysiology involves the formation of carious lesions on the dental hard tissues, which may or may not be cavitated, and the demineralization of the tooth surface. On the other hand, periodontitis is an advanced inflammatory condition that has some genetic predisposition but is also mediated by lifestyle factors, like smoking, and underlying comorbidities, like diabetes. Clinically, periodontitis may manifest as gingival tissue inflammation, presence of calculi, and abnormal teeth mobility [122]. Pulpal damage is an adverse clinical manifestation of these diseases; particularly, with deep carious lesions, inflammation can become irreversible, the infected teeth are more susceptible to bacterial infections, and the dental pulp can become necrotic. Preserving pulp vitality is especially important for immature permanent teeth because the root is not yet fully developed [123].

Standard treatment for dental caries involves excision of the infected tissue, decontamination of the root canal, and sealing the emptied chamber with a biocompatible inert filler material [124]. Despite a high success rate, there are several limitations with this approach, namely, the inability to restore the aforementioned features of the dental pulp. For these reasons, regenerative endodontic procedures (REPs) have become an emerging area of inquiry with the primary objective of restoring pulp vitality. For instance, treatments can involve injection of stem cells between the periapical region and root canal to promote vascularization. However, results only showed production of pulp-like products – as noted by the absence of odontoblasts – and not true recovery of pulpal vitality. Other REP approaches have also shown evidence for faster healing rates, thickened root walls, and reduced apical foramen, but subsequent studies found critical limitations. Histological assessments showed that there are primarily formations of ectopic bone, cementum, and periodontal ligaments instead of dental pulp. Moreover, immunohistochemical data revealed a greater expression of osteogenic markers in treated teeth than in mineralized tissue, indicating the REP's propensity for reparation or healing rather than a truly regenerative function [125]. To this end, tissue engineering approaches are being explored to ameliorate the aforementioned challenges.

Research in additive manufacturing for dental pulp regeneration requires many considerations, but a significant contributor to success is the appropriate selection of biomaterials that are biocompatible and can optimize cell-biomaterial interactions. Athirasala et al. in 2017 proposed a novel in vitro engineering approach using prevascularized, cell-laden GelMA hydrogel pulp-like tissue. The hydrogels were ladened with odontoblast-like cells (OD21). Upon injecting these prevascularized tissue constructs in root canals, the authors observed successful monolayer formation and angiogenesis. Higher hydrogel polymer concentrations also enhanced the viability, spreading, and proliferation of the OD21 cells, as well as endothelial cell spreading and monolayer formation. However, a significant drawback to their proposed method was the lack of vasculogenesis and presence of other peripheral components like nerves and fibroblasts [120]. Re-establishing vasculogenesis is integral to bolster pulp vitality; recently, a research group proposed an in situ method with promising results on this front.

Duarte et al. in 2020 developed a novel portable bioprinting approach focused on maintaining pulp vitality and promoting vasculogenesis [123]. Their strategy was built on milestone findings in emerging literature. To summarize:

- Human dental pulp stem cells (hDPSCs) are proliferative and self-renewing. They also have a diverse differentiation profile as evidenced by their ability to form pulp- and bone-like tissues as well as reparative dentin [126].
- hDPSCs are especially attractive because they are non-invasively extracted and can differentiate into odontoblast cells – a characteristic feature of dental pulp [127].
- Moreover, they also play a role in pulp tissue formation and vascularization [128].
- Interestingly, co-culturing human umbilical vein endothelial cells (HUVECs) with hDPSCs enhances angiogenesis and odontogenesis while stabilizing capillary-like structures [129].

Their methodology involved two experimental setups: hydrogel preparation and cell extraction, and then, bovine tooth preparation. The study was designed to assess the performance of vascular tube formation in bioprintable and non-bioprintable (control) bioinks. hDPSCs and HUVECs were provided at the RWTH Aachen University Hospital from the Department of Oral and Maxillofacial Surgery and Clinic for Gynecology and Obstetrics, respectively. They isolated hDPSCs from extracted third molars of patients aged 18–26; concurrently, HUVECs were extracted from umbilical cords provided by the clinic. Bovine teeth were the model of choice and were used to perform the contraction analysis of the cell-loaded hydrogels. Depending on the tooth size, the printing process took one to two seconds; the contraction analysis showed a complete homogeneous distribution of the bioink with no signs of contraction with the printed bioinks [123].

Campos and colleagues undertook a previous investigation seeking to understand the factors influencing printing resolution and stem cell integrity. Their findings suggested that cell viability could be optimally maintained using an on-demand handheld approach without negatively affecting the cell functionality. This finding informed their current hypothesis that vascular tube formation could be accomplished using an in situ approach. Unlike the abovementioned studies that showed poor evidence for vasculogenesis, Campos and colleagues' results were incredibly promising in this regard: all tested samples – using both bioprintable and non-bioprintable inks – demonstrated successful vascular tube formation. Qualitatively this was shown through the elongation of the endothelial cells, which was not the case previously where those cells remained rounded – an indicator for failed vasculogenesis [123].

Their findings also support the need to investigate how rheological and structural properties of hydrogel bioinks influence treatment efficacy. Optimal treatment outcomes rely on the long-term stability of the cell-loaded bioink, marked by minimal contraction of the filling material. Contraction is an indicator for greater risk of reinfection and consequently, delayed treatment. The extent to which contractions occur depends on the materials' properties and their response to external factors like mechanical stress; hence, understanding cell-biomaterial interactions is crucial to optimize outcomes. In this study, contraction was observed in both tested control hydrogels – 0.5% fibrin and 0.3% type I collagen – after 14 days in in vitro culture. Rheologically, this observation was supported by two changes: an increased loss modulus (LM) and storage modulus (SM) with the fibrin and type I collagen bioink combination [123].

The LM is a reflection to what extent the material dissipates energy as it deforms under stress, like chewing and biting. If a material demonstrates an increased LM it means that it is more susceptible to deformation and energy dissipation which can cause a reduction in structural integrity, alter functional properties, and impair marginal integrity, which can lead to the seal breaking and expose the site to bacteria or cause microleakage [130, 131]. Similarly, the SM indicates to what extent a material can store elastic energy and provides insights into the materials' rigidity. An increased SM would entail improved structural integrity – by virtue of more rigid material; better load-bearing capabilities, reducing the risk of potential fractures; and overall, improved long-term stability. Interestingly, Campos and colleagues observed a constant LM with an increased SM with the bioprintable bioinks: 0.2% type I collagen and 0.5% agarose. The higher SM is characteristic of more solid-like gels with better mechanical properties as evidenced by Weng et al. [132]. These findings were supported by Campos and team by the shape fidelity of the printed filling and the absence of noticeable shrinkage [123].

Finally, the proposed in situ method of bioink injection into the root canal offers several other benefits in relation to traditional techniques like casting or pipetting.

The in situ approach uses a nanoliter ejection tip that offers precise filling volumes, minimizing bioink loss. This is particularly advantageous in cases involving viscous fluids as pipetting often results in inaccuracies during aspiration or discharge. In addition, in situ dispensing allows excellent penetration depth at a rate of 20 m/s, which maximizes the therapeutic potential of the treatment. Pipetting and casting often trap air bubbles within the cavity; while this may be ameliorated with gentle shaking or waiting for the ink to settle, the proposed method is significantly more efficient and reliable [123].

In closing, Campos and colleagues' work was the first time an in situ approach has been proposed in the treatment of dental root canals, with promising results in vasculogenesis and optimizing biological functionality. Future investigations in this arena will need to address several unanswered questions:

- With rheological, structural, and functional considerations in mind, how can the current bioink combinations be further optimized?
- To what extent can the proposed approach produce revascularization and regeneration in an in vivo model?
- Can this treatment modality be validated in different in vivo models?
- Furthermore, will this approach prove to be a valuable investment to undergo the rigors of clinical randomized trials?

While these questions are incredibly important considerations, research to advance this field of in situ bioprinting in a dental context can improve treatment outcomes.

12.6 Future Directions

12.6.1 4D Bioprinting

Four-dimensional bioprinting builds on 3D in situ scaffolds – where the fourth dimension is the dynamic change of these structures over time. Such structures are responsive to various stimuli, such as temperature, pH, electric field, acoustic signals, and humidity [133]. The advantages that these scaffolds provide are numerous. One example could be a time-based controlled drug delivery. For instance, it is possible to print droplets that carry antibiotics, nanoparticles with antimicrobial particles, or growth factors. When desired, these droplets can then be subject to a pre-specified stimulus that induces a structural deformation. This structural change causes the droplets to release their contents in the desired time to provide antibiotics to the defect site, nanoparticles to prevent microbial attachment to the printed scaffold, or growth factors to stimulate tissue integration [134] (Figure 12.5).

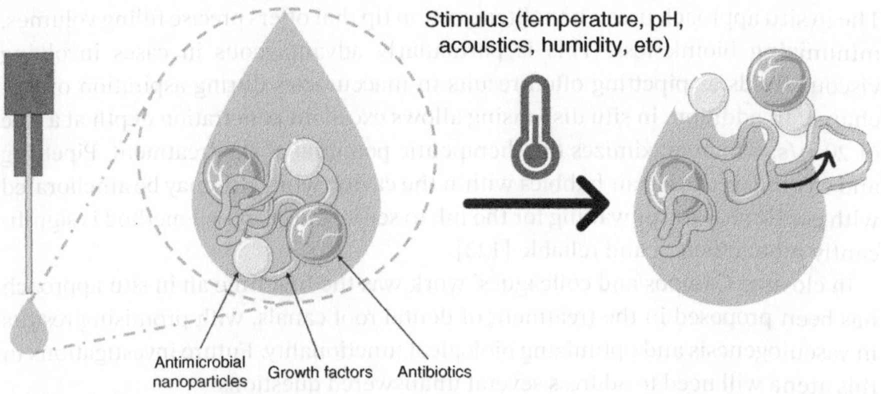

Figure 12.5 *4D bioprinting approaches.* Bioprint droplets that contain antibiotics, antimicrobial nanoparticles, and growth factors are subject to a pre-specified stimulus, in this case, a change in temperature. The stimulus causes a conformational change in the structure of the droplet causing it to release its internal contents that are vital in the scaffold function. *Source:* Parts of the figure were drawn by using pictures from Servier Medical Art. Servier Medical Art by Servier is licensed under a Creative Commons Attribution 3.0 Unported License (https://creativecommons.org/licenses/by/3.0/).

12.6.2 Open-Source Collaboration

Some of the main challenges faced by in situ bioprinting can be addressed by creating open-source libraries and encouraging idea-sharing within 3D printing communities [135]. The high costs of commercially available in situ bioprinters, the lack of crosslinking guidance databases, and the overall deficiency in tissue-based 3D models are all issues that can be addressed through open-source initiatives. Moreover, low-cost imaging modalities should be implemented to reduce the high cost associated with in situ bioprinting and thereby increase the translational potential of in situ 3D bioprinting to healthcare systems.

12.6.3 Integration with Machine Learning

As mentioned in previous sections, an essential aspect of integrating a biofabricated scaffold is to have a similar structure to the native tissue. This requirement becomes more critical in complex defects where the injury may cross multiple layers of tissue. One such example is complex skin and soft tissue injuries. In such cases, machine learning algorithms need to be implemented after scans of the defect to segment the different layers of tissues. Then, advanced multi-tissue layering paths are generated, directing automated or semi-automated printers to print parallel to the native tissues. If this can be achieved, cells and growth factors of corresponding tissues can act within the scaffold and integrate into the native tissue. While such a machine learning algorithm has been designed for skin and soft tissues [136], it would be invaluable to integrate such a model for all the other different tissues. For example, bioprinting of cortical and cancellous bone, or bioprinting of the colon in its different layers.

12.7 Conclusion

A thorough examination of in situ 3D bioprinting reveals this innovative technology's enormous potential to reshape biomedical research and healthcare. Although it currently confronts challenges, such as controlling functional vascularization development and the expense of commercially available in situ bioprinters, the benefits of in situ bioprinting are indisputable.

In situ bioprinting enables precision in cellular deposition, facilitating the exact placement of different cell types in complex patterns – an essential requirement for replicating intricate organs. It encourages integration with host tissue by stimulating host vascularization to produce capillaries around the tissue construct, eliminating the need for transplantation. By depositing cells directly onto their natural environment, it accurately replicates the in vivo environment, ensuring high-fidelity replication of the extracellular milieu. Moreover, in situ bioprinting may reduce the risk of graft rejection and failure rates associated with HLA mismatches and ABO disparities, thereby diminishing the immune response. From a regulatory standpoint, in situ bioprinting may encounter fewer barriers to clinical translation than traditional tissue engineering methods, due to the lack of a transplantation requirement. This technology also presents opportunities for personalized medicine, given its ability to precisely reproduce individual organ defects.

The potential for topical or localized drug delivery using in situ bioprinting is evident, thanks to its high-resolution printing system and micron-level precision deposition capability. The technology has already achieved success in regenerating tissues such as skin, bones, and cartilage.

In recent years, in situ bioprinters have been engineered to accommodate the challenges associated with printing on living tissue, such as the physiological movements of the heart and lungs. These advancements have paved a path toward commercialization and integration of in situ bioprinting into healthcare systems.

In conclusion, in situ 3D bioprinting is a promising technology, poised to transform the future of biomedical research and healthcare. As this field continues to evolve, we anticipate more breakthroughs that will bring us closer to the reality of personalized medicine and in situ tissue and organ regeneration. The future of in situ 3D bioprinting is bright, with vast and thrilling potential applications.

Acknowledgments

Varun Srikanth – contributing to Section 12.5; Hui Xin Zhang – contributing to Section 6; Gerard K. Kim – referencing and formatting.

References

1 Ebefors, K., Bergwall, L., and Nyström, J. (2022). The glomerulus according to the mesangium. *Front. Med.* 26 (8): 740527.

2 Truter, D., Chellan, N., Strijdom, H. et al. (2018). Histomorphological changes in the pancreas and kidney and histopathological changes in the liver in male Wistar rats on antiretroviral therapy and melatonin treatment. *Acta Histochem.* 120 (4): 347–355.

3 Arackal, A. and Alsayouri, K. (2023). Histology, heart. In: *StatPearls*. Treasure Island (FL): StatPearls Publishing http://www.ncbi.nlm.nih.gov/books/NBK545143/.

4 Zuev, D.M., Nguyen, A.K., Putlyaev, V.I., and Narayan, R.J. (2020). 3D printing and bioprinting using multiphoton lithography. *Bioprinting* 20: e00090.

5 Rouwkema, J., Rivron, N.C., and Van Blitterswijk, C.A. (2008). Vascularization in tissue engineering. *Trends Biotechnol.* 26 (8): 434–441.

6 Carmeliet, P. and Jain, R.K. (2000). Angiogenesis in cancer and other diseases. *Nature* 407 (6801): 249–257.

7 Nerem, R.M. and Seliktar, D. (2001). Vascular tissue engineering. *Annu. Rev. Biomed. Eng.* 3 (1): 225–243.

8 Black, A.F., Berthod, F., L'Heureux, N. et al. (1998). In vitro reconstruction of a human capillary-like network in a tissue-engineered skin equivalent. *FASEB J.* 12 (13): 1331–1340.

9 Kelm, J.M., Djonov, V., Hoerstrup, S.P. et al. (2006). Tissue-transplant fusion and vascularization of myocardial microtissues and macrotissues implanted into chicken embryos and rats. *Tissue Eng.* 12 (9): 2541–2553.

10 Kolesky, D.B., Homan, K.A., Skylar-Scott, M.A., and Lewis, J.A. (2016). Three-dimensional bioprinting of thick vascularized tissues. *Proc. Natl. Acad. Sci. USA* 113 (12): 3179–3184.

11 Levenberg, S., Rouwkema, J., Macdonald, M. et al. (2005). Engineering vascularized skeletal muscle tissue. *Nat. Biotechnol.* 23 (7): 879–884.

12 Nichol, J.W., Koshy, S.T., Bae, H. et al. (2010). Cell-laden microengineered gelatin methacrylate hydrogels. *Biomaterials* 31 (21): 5536–5544.

13 Benning, L., Gutzweiler, L., Tröndle, K. et al. (2018). Assessment of hydrogels for bioprinting of endothelial cells: bioprinting of endothelial cells. *J. Biomed. Mater. Res.* 106 (4): 935–947.

14 Gafni, Y., Zilberman, Y., Ophir, Z. et al. (2006). Design of a filamentous polymeric scaffold for in vivo guided angiogenesis. *Tissue Eng.* 12 (11): 3021–3034.

15 Morgan, J.P., Delnero, P.F., Zheng, Y. et al. (2013). Formation of microvascular networks in vitro. *Nat. Protoc.* 8 (9): 1820–1836.

16 Woodfield, T.B.F., Malda, J., De Wijn, J. et al. (2004). Design of porous scaffolds for cartilage tissue engineering using a three-dimensional fiber-deposition technique. *Biomaterials* 25 (18): 4149–4161.

17 Lee, A., Hudson, A.R., Shiwarski, D.J. et al. (2019). 3D bioprinting of collagen to rebuild components of the human heart. *Science* 365 (6452): 482–487.

18 Bouillard, M., Boucher, G., Ortas, J.F. et al. (2019). High production rate of single-photon and two-photon Fock states for quantum state engineering. *Opt. Express* 27 (3): 3113.

19 Ingber, D.E., Mow, V.C., Butler, D. et al. (2006). Tissue engineering and developmental biology: going biomimetic. *Tissue Eng.* 12 (12): 3265–3283.

20 Koos, C., Vorreau, P., Vallaitis, T. et al. (2009). All-optical high-speed signal processing with silicon–organic hybrid slot waveguides. *Nature Photon.* 3 (4): 216–219.

21 Loewner, S., Heene, S., Baroth, T. et al. (2022). Recent advances in melt electro writing for tissue engineering for 3D printing of microporous scaffolds for tissue engineering. *Front. Bioeng. Biotechnol.* 17 (10): 896719.

22 Ehrbar, M., Metters, A., Zammaretti, P. et al. (2005). Endothelial cell proliferation and progenitor maturation by fibrin-bound VEGF variants with differential susceptibilities to local cellular activity. *J. Control. Release* 101 (1–3): 93–109.

23 Kneser, U., Polykandriotis, E., Ohnolz, J. et al. (2006). Engineering of vascularized transplantable bone tissues: induction of axial vascularization in an osteoconductive matrix using an arteriovenous loop. *Tissue Eng.* 12 (7): 1721–1731.

24 Shen, M., Wang, L., Gao, Y. et al. (2022). 3D bioprinting of in situ vascularized tissue engineered bone for repairing large segmental bone defects. *Materials Today Bio.* 16: 100382.

25 Nishida-Aoki, N. and Gujral, T.S. (2019). Emerging approaches to study cell-cell interactions in tumor microenvironment. *Oncotarget* 10 (7): 785–797.

26 Strand, D.W. and Hayward, S.W. (2010). Modeling stromal-epithelial interactions in disease progression. *Discov. Med.* 9 (49): 504–511.

27 Weng, T., Zhang, W., Xia, Y. et al. (2021). 3D bioprinting for skin tissue engineering: current status and perspectives. *J. Tissue Eng.* 12: 204173142110285.

28 Zhang, M., Zhang, C., Li, Z. et al. (2023). Advances in 3D skin bioprinting for wound healing and disease modeling. *Regener. Biomater.* 17 (10): rbac105.

29 Reddy, M.S.B., Ponnamma, D., Choudhary, R., and Sadasivuni, K.K. (2021). A comparative review of natural and synthetic biopolymer composite scaffolds. *Polymers* 13 (7): 1105.

30 Oberweis, C.V., Marchal, J.A., López-Ruiz, E., and Gálvez-Martín, P. (2020). A worldwide overview of regulatory frameworks for tissue-based products. *Tissue Eng. B Rev.* 26 (2): 181–196.

31 Zheng, Y.L. (2016). Some ethical concerns about human induced pluripotent stem cells. *Sci. Eng. Ethics* 22 (5): 1277–1284.

32 Beatty, P.G., Clift, R.A., Mickelson, E.M. et al. (1985). Marrow transplantation from related donors other than HLA-identical siblings. *N. Engl. J. Med.* 313 (13): 765–771.

33 Hutt, D. (2018). Engraftment, graft failure, and rejection. In: *The European Blood and Marrow Transplantation Textbook for Nurses* (ed. M. Kenyon and A. Babic), 259–270. Cham: Springer International Publishing http://link.springer.com/10.1007/978-3-319-50026-3_13.

34 Mattsson, J., Ringdén, O., and Storb, R. (2008). Graft failure after allogeneic hematopoietic cell transplantation. *Biol. Blood Marrow Transplant.* 14 (1): 165–170.

35 Government of Canada (2022). Biologic and radiopharmaceutical drugs directorate. https://www.canada.ca/en/health-canada/corporate/about-health-canada/branches-agencies/health-products-food-branch/biologic-radiopharmaceutical-drugs-directorate.html (accessed 28 June 2023).

36 Halioua-Haubold, C.L., Peyer, J.G., Smith, J.A. et al. (2017). Regulatory considerations for gene therapy products in the US, EU, and Japan. *Yale J. Biol. Med.* 90 (4): 683–693.

37 Schleidgen, S., Klingler, C., Bertram, T. et al. (2013). What is personalized medicine: sharpening a vague term based on a systematic literature review. *BMC Med. Ethics.* 14 (1): 55.

38 Fortunato, G.M., Sigismondi, S., Nicoletta, M. et al. (2023). Analysis of the robotic-based in situ bioprinting workflow for the regeneration of damaged tissues through a case study. *Bioengineering* 10 (5): 560.

39 Chen, H., Ma, X., Gao, T. et al. (2023). Robot-assisted in situ bioprinting of gelatin methacrylate hydrogels with stem cells induces hair follicle-inclusive skin regeneration. *Biomed. Pharmacother.* 158: 114140.

40 Mostafavi, A., Abdullah, T., Russell, C.S. et al. (2021). In situ printing of scaffolds for reconstruction of bone defects. *Acta Biomater.* 127: 313–326.

41 Quint, J.P., Mostafavi, A., Endo, Y. et al. (2021). In vivo printing of nanoenabled scaffolds for the treatment of skeletal muscle injuries. *Adv. Healthcare Mater.* 10 (10): 2002152.

42 Xu, J., Feng, Q., Lin, S. et al. (2019). Injectable stem cell-laden supramolecular hydrogels enhance in situ osteochondral regeneration via the sustained co-delivery of hydrophilic and hydrophobic chondrogenic molecules. *Biomaterials* 210: 51–61.

43 Ying, G., Manríquez, J., Wu, D. et al. (2020). An open-source handheld extruder loaded with pore-forming bioink for in situ wound dressing. *Mater. Today Bio* 8: 100074.

44 Benson, H.A.E., Grice, J.E., Mohammed, Y. et al. (2019). Topical and transdermal drug delivery: from simple potions to smart technologies. *CDD* 16 (5): 444–460.

45 Pandey, M., Choudhury, H., Fern, J.L.C. et al. (2020). 3D printing for oral drug delivery: a new tool to customize drug delivery. *Drug Deliv. Transl. Res.* 10 (4): 986–1001.

46 Bom, S., Santos, C., Barros, R. et al. (2020). Effects of starch incorporation on the physicochemical properties and release kinetics of alginate-based 3D hydrogel patches for topical delivery. *Pharmaceutics* 12 (8): 719.

47 Goyanes, A., Det-Amornrat, U., Wang, J. et al. (2016). 3D scanning and 3D printing as innovative technologies for fabricating personalized topical drug delivery systems. *J. Control. Release* 234: 41–48.

48 Domínguez-Robles, J., Martin, N., Fong, M. et al. (2019). Antioxidant PLA composites containing lignin for 3D printing applications: a potential material for healthcare applications. *Pharmaceutics* 11 (4): 165.

49 Muwaffak, Z., Goyanes, A., Clark, V. et al. (2017). Patient-specific 3D scanned and 3D printed antimicrobial polycaprolactone wound dressings. *Int. J. Pharm.* 527 (1–2): 161–170.

50 Economidou, S., Pissinato Pere, C., Okereke, M., and Douroumis, D. (2021). Optimisation of design and manufacturing parameters of 3D printed solid microneedles for improved strength, sharpness, and drug delivery. *Micromachines* 12 (2): 117.

51 Luzuriaga, M.A., Berry, D.R., Reagan, J.C. et al. (2018). Biodegradable 3D printed polymer microneedles for transdermal drug delivery. *Lab Chip* 18 (8): 1223–1230.

52 Xenikakis, I., Tzimtzimis, M., Tsongas, K. et al. (2019). Fabrication and finite element analysis of stereolithographic 3D printed microneedles for transdermal delivery of model dyes across human skin in vitro. *Eur. J. Pharm. Sci.* 137: 104976.

53 Yeung, C., Chen, S., King, B. et al. (2019). A 3D-printed microfluidic-enabled hollow microneedle architecture for transdermal drug delivery. *Biomicrofluidics* 13 (6): 064125.

54 Mao, S., Man, J., Wang, J. et al. (2022). Research progress and challenges of bioprinting in wound dressing and healing: bibliometrics-based analysis and perspectives. *Int. J. Bioprinting* 9 (2): https://doi.org/10.18063/ijb.v9i2.653 https://ijb.sg/index.php/int-j-bioprinting/article/view/653.

55 Singh, S., Choudhury, D., Yu, F. et al. (2020). In situ bioprinting – Bioprinting from benchside to bedside? *Acta Biomater.* 101: 14–25.

56 Samandari, M., Mostafavi, A., Quint, J. et al. (2022). In situ bioprinting: intraoperative implementation of regenerative medicine. *Trends Biotechnol.* 40(10): 1229–1247.

57 Zandrini, T., Florczak, S., Levato, R., and Ovsianikov, A. (2023). Breaking the resolution limits of 3D bioprinting: future opportunities and present challenges. *Trends Biotechnol.* 41 (5): 604–614.

58 Zhao, W., Chen, H., Zhang, Y. et al. (2022). Adaptive multi-degree-of-freedom in situ bioprinting robot for hair-follicle-inclusive skin repair: a preliminary study conducted in mice. *Bioeng. Transl. Med.* 7 (3): e10303. https://onlinelibrary.wiley.com/doi/10.1002/btm2.10303.

59 Ma, K., Zhao, T., Yang, L. et al. (2020). Application of robotic-assisted in situ 3D printing in cartilage regeneration with HAMA hydrogel: an in vivo study. *J. Adv. Res.* 23: 123–132.

60 Keriquel, V., Guillemot, F., Arnault, I. et al. (2010). In vivo bioprinting for computer- and robotic-assisted medical intervention: preliminary study in mice. *Biofabrication* 2 (1): 014101.

61 Akilbekova, D. and Mektepbayeva, D. (2017). Patient specific in situ 3D printing. In: *3D Printing in Medicine* (ed. D.M. Kalaskar), 91–113. Elsevier https://linkinghub.elsevier.com/retrieve/pii/B9780081007174000041.

62 Zhao, W., Hu, C., and Xu, T. (2023). In vivo bioprinting: broadening the therapeutic horizon for tissue injuries. *Bioactive Materials.* 25: 201–222.

63 Fortunato, G.M., Batoni, E., Bonatti, A.F. et al. (2022). Surface reconstruction and tissue recognition for robotic-based in situ bioprinting. *Bioprinting* 26: e00195.

64 Zhao, W., Hu, C., Xu, T. et al. (2022). Subaqueous bioprinting: a novel strategy for fetal membrane repair with 7-axis robot-assisted minimally invasive surgery. *Adv. Funct. Mater.* 32 (51): 2207496.

65 Fortunato, G.M., Rossi, G., Bonatti, A.F. et al. (2021). Robotic platform and path planning algorithm for in situ bioprinting. *Bioprinting* 22: e00139.

66 Zhou, C., Yang, Y., Wang, J. et al. (2021). Ferromagnetic soft catheter robots for minimally invasive bioprinting. *Nat. Commun.* 12 (1): 5072.

67 Yang, Y., Yu, Z., Lu, X. et al. (2023). Minimally invasive bioprinting for in situ liver regeneration. *Bioact. Mater.* 26: 465–477.

68 Thai, M.T., Phan, P.T., Tran, H.A. et al. (2023). Advanced soft robotic system for in situ 3D bioprinting and endoscopic surgery. *Adv. Sci.* 10 (12): 2205656.

69 Zhao, W. and Xu, T. (2020). Preliminary engineering for in situ in vivo bioprinting: a novel micro bioprinting platform for in situ in vivo bioprinting at a gastric wound site. *Biofabrication* 12 (4): 045020.

70 Simeunovic, A., Wolf, K., Tierling, K., and Hoelzle, D.J. (2022). A surgical robot for intracorporeal additive manufacturing of tissue engineering constructs. *IEEE Rob. Autom. Lett.* 7 (3): 7495–7502.

71 Urciuolo, A., Poli, I., Brandolino, L. et al. (2020). Intravital three-dimensional bioprinting. *Nat. Biomed. Eng.* 4 (9): 901–915.

72 Li, H., Cheng, F., Orgill, D.P. et al. (2021). Handheld bioprinting strategies for in situ wound dressing. Jang J, editor. *Essays Biochem.* 65 (3): 533–543.

73 Pazhouhnia, Z., Beheshtizadeh, N., Namini, M.S., and Lotfibakhshaiesh, N. (2022). Portable hand-held bioprinters promote in situ tissue regeneration. *Bioeng. Transl. Med.* 7 (3): https://onlinelibrary.wiley.com/doi/10.1002/btm2.10307.

74 O'Connell, C.D., Di Bella, C., Thompson, F. et al. (2016). Development of the biopen: a handheld device for surgical printing of adipose stem cells at a chondral wound site. *Biofabrication* 8 (1): 015019.

75 Di Bella, C., Duchi, S., O'Connell, C.D. et al. (2018). In situ handheld three-dimensional bioprinting for cartilage regeneration. *J. Tissue Eng. Regen. Med.* 12 (3): 611–621.

76 Duchi, S., Onofrillo, C., O'Connell, C.D. et al. (2017). Handheld co-axial bioprinting: application to in situ surgical cartilage repair. *Sci. Rep.* 7 (1): 5837.

77 Russell, C.S., Mostafavi, A., Quint, J.P. et al. (2020). In situ printing of adhesive hydrogel scaffolds for the treatment of skeletal muscle injuries. *ACS Appl. Bio. Mater.* 3 (3): 1568–1579.

78 Zheng, Z., Eglin, D., Alini, M. et al. (2021). Visible light-induced 3D bioprinting technologies and corresponding bioink materials for tissue engineering: a review. *Engineering* 7 (7): 966–978.

79 Al-Abboodi, A., Zhang, S., Al-Saady, M. et al. (2019). Printing in situ tissue sealant with visible-light-crosslinked porous hydrogel. *Biomed. Mater.* 14 (4): 045010.

80 Hakimi, N., Cheng, R., Leng, L. et al. (2018). Handheld skin printer: in situ formation of planar biomaterials and tissues. *Lab Chip* 18 (10): 1440–1451.

81 Xie, M., Shi, Y., Zhang, C. et al. (2022). In situ 3D bioprinting with bioconcrete bioink. *Nat. Commun.* 13 (1): 3597.

82 Li, R., Zhao, Y., Zheng, Z. et al. (2023). Bioinks adapted for in situ bioprinting scenarios of defect sites: a review. *RSC Adv.* 13 (11): 7153–7167.

83 Cao, P., Tao, L., Gong, J. et al. (2021). 4D printing of a sodium alginate hydrogel with step-wise shape deformation based on variation of crosslinking density. *ACS Appl. Polym. Mater.* 3 (12): 6167–6175.

84 Yang, X., Lu, Z., Wu, H. et al. (2018). Collagen-alginate as bioink for three-dimensional (3D) cell printing based cartilage tissue engineering. *Mater. Sci. Eng. C* 83: 195–201.

85 Liu, Y., Luo, X., Wu, W. et al. (2021). Dual cure (thermal/photo) composite hydrogel derived from chitosan/collagen for in situ 3D bioprinting. *Int. J. Biol. Macromol.* 182: 689–700.

86 Thomas, L.V., Rahul, V.G., and Nair, P.D. (2017). Effect of stiffness of chitosan-hyaluronic acid dialdehyde hydrogels on the viability and growth of encapsulated chondrocytes. *Int. J. Biol. Macromol.* 104: 1925–1935.

87 Yu, S., Ji, Y., Guo, C. et al. (2021). A dual-cross-linked hydrogel based on hyaluronic acid/gelatin tethered via tannic acid: mechanical properties' enhancement and stability control. *Iran. Polym. J.* 30 (3): 307–317.

88 Kim, S.D., Jin, S., Kim, S. et al. (2022). Tyramine-functionalized alginate-collagen hybrid hydrogel inks for 3D-bioprinting. *Polymers* 14 (15): 3173.

89 Liu, S., Huang, D., Hu, Y. et al. (2020). Sodium alginate/collagen composite multiscale porous scaffolds containing poly(ε-caprolactone) microspheres fabricated based on additive manufacturing technology. *RSC Adv.* 10 (64): 39241–39250.

90 Freeman, S., Calabro, S., Williams, R. et al. (2022). Bioink formulation and machine learning-empowered bioprinting optimization. *Front Bioeng Biotechnol.* 13 (10): 913579.

91 GhavamiNejad, A., Ashammakhi, N., Wu, X.Y., and Khademhosseini, A. (2020). Crosslinking strategies for 3D bioprinting of polymeric hydrogels. *Small* 16 (35): 2002931.

92 Hajikhani, A., Scocozza, F., Conti, M. et al. (2019). Experimental characterization and computational modeling of hydrogel cross-linking for bioprinting applications. *Int. J. Artif. Organs.* 42 (10): 548–557.

93 Amukarimi, S. and Mozafari, M. (2021). 4D bioprinting of tissues and organs. *Bioprinting* 23: e00161.

94 Park, Y.L., Park, K., and Cha, J.M. (2021). 3D-bioprinting strategies based on in situ bone-healing mechanism for vascularized bone tissue engineering. *Micromachines* 12 (3): 287.

95 Keriquel, V., Oliveira, H., Rémy, M. et al. (2017). In situ printing of mesenchymal stromal cells, by laser-assisted bioprinting, for in vivo bone regeneration applications. *Sci. Rep.* 7 (1): 1778.

96 Barceló, X., Eichholz, K.F., Garcia, O., and Kelly, D.J. (2022). Tuning the degradation rate of alginate-based bioinks for bioprinting functional cartilage tissue. *Biomedicines* 10 (7): 1621.

97 Raees, S., Ullah, F., Javed, F. et al. (2023). Classification, processing, and applications of bioink and 3D bioprinting: a detailed review. *Int. J. Biol. Macromol.* 232: 123476.

98 Highley, C.B., Rodell, C.B., and Burdick, J.A. (2015). Direct 3D printing of shear-thinning hydrogels into self-healing hydrogels. *Adv. Mater.* 27 (34): 5075–5079.

99 Solanki, K., Shanmugasundaram, S., Shetty, N., and Kim, S.J. (2021). Articular cartilage repair & joint preservation: a review of the current status of biological approach. *J. Clin. Orthop. Trauma* 22: 101602.

100 Poole, A.R., Kojima, T., Yasuda, T. et al. (2001). Composition and structure of articular cartilage: a template for tissue repair. *Clin. Orthop. Relat. Res.* 391: S26–S33.

101 Chung, C. and Burdick, J.A. (2008). Engineering cartilage tissue. *Adv. Drug Deliv. Rev.* 60 (2): 243–262.

102 Sharma, B., Fermanian, S., Gibson, M. et al. (2013). Human cartilage repair with a photoreactive adhesive-hydrogel composite. *Sci. Transl. Med.* 5 (167): 167ra6. https://www.science.org/doi/10.1126/scitranslmed.3004838.

103 Ye, K., Felimban, R., Moulton, S.E. et al. (2013). Bioengineering of articular cartilage: past, present and future. *Regen. Med.* 8 (3): 333–349.

104 Knudson, C.B. (2003). Hyaluronan and CD44: strategic players for cell-matrix interactions during chondrogenesis and matrix assembly. *Birth Defects Res. C Embryo Today* 69 (2): 174–196.

105 Onofrillo, C., Duchi, S., O'Connell, C.D. et al. (2018). Biofabrication of human articular cartilage: a path towards the development of a clinical treatment. *Biofabrication* 10 (4): 045006.

106 Cohen, D.L., Lipton, J.I., Bonassar, L.J., and Lipson, H. (2010). Additive manufacturing for in situ repair of osteochondral defects. *Biofabrication* 2 (3): 035004.

107 Li, L., Yu, F., Shi, J. et al. (2017). In situ repair of bone and cartilage defects using 3D scanning and 3D printing. *Sci. Rep.* 7 (1): 9416.

108 Cezar, C.A., Roche, E.T., Vandenburgh, H.H. et al. (2016). Biologic-free mechanically induced muscle regeneration. *Proc. Natl. Acad. Sci. USA* 113 (6): 1534–1539.

109 Almada, A.E. and Wagers, A.J. (2016). Molecular circuitry of stem cell fate in skeletal muscle regeneration, ageing and disease. *Nat. Rev. Mol. Cell Biol.* 17 (5): 267–279.

110 Nuutila, K., Sakthivel, D., Kruse, C. et al. (2017). Gene expression profiling of skeletal muscle after volumetric muscle loss: Gene expression profile of a nonhealing skeletal muscle. *Wound Repair Regen.* 25 (3): 408–413.

111 Garg, K., Ward, C.L., Rathbone, C.R., and Corona, B.T. (2014). Transplantation of devitalized muscle scaffolds is insufficient for appreciable de novo muscle fiber regeneration after volumetric muscle loss injury. *Cell Tissue Res.* 358 (3): 857–873.

112 Qazi, T.H., Duda, G.N., Ort, M.J. et al. (2019). Cell therapy to improve regeneration of skeletal muscle injuries. *J. Cachexia. Sarcopenia Muscle* 10 (3): 501–516.

113 Gaharwar, A.K., Cross, L.M., Peak, C.W. et al. (2019). 2D nanoclay for biomedical applications: regenerative medicine, therapeutic delivery, and additive manufacturing. *Adv. Mater.* 31 (23): 1900332.

114 Skardal, A., Mack, D., Kapetanovic, E. et al. (2012). Bioprinted amniotic fluid-derived stem cells accelerate healing of large skin wounds. *Stem Cells Transl. Med.* 1 (11): 792–802.

115 Cubo, N., Garcia, M., Del Cañizo, J.F. et al. (2016). 3D bioprinting of functional human skin: production and in vivo analysis. *Biofabrication* 9 (1): 015006.

116 Lee, V., Singh, G., Trasatti, J.P. et al. (2014). Design and fabrication of human skin by three-dimensional bioprinting. *Tissue Eng. Part C Methods* 20 (6): 473–484.

117 Pourchet, L.J., Thepot, A., Albouy, M. et al. (2017). Human skin 3D bioprinting using scaffold-free approach. *Adv Healthcare Mater.* 6 (4): 1601101.

118 Maxson, S., Lopez, E.A., Yoo, D. et al. (2012). Concise review: role of mesenchymal stem cells in wound repair. *Stem Cells Transl. Med.* 1 (2): 142–149.

119 Wang, X., Wang, X., Liu, J. et al. (2016). Hair follicle and sebaceous gland de novo regeneration with cultured epidermal stem cells and skin-derived precursors. *Stem Cells Transl. Med.* 5 (12): 1695–1706.

120 Athirasala, A., Lins, F., Tahayeri, A. et al. (2017). A novel strategy to engineer pre-vascularized full-length dental pulp-like tissue constructs. *Sci. Rep.* 7 (1): 3323.

121 Yu, C. and Abbott, P. (2007). An overview of the dental pulp: its functions and responses to injury. *Aust. Dent. J.* 52: S4–S6.

122 Laudenbach, J.M. and Kumar, S.S. (2020). Common dental and periodontal diseases. *Dermatol. Clin.* 38 (4): 413–420.

123 Duarte Campos, D.F., Zhang, S., Kreimendahl, F. et al. (2020). Hand-held bioprinting for *de novo* vascular formation applicable to dental pulp regeneration. *Connect. Tissue Res.* 61 (2): 205–215.

124 Lee, D.K., Kim, S.V., Limansubroto, A.N. et al. (2015). Nanodiamond–gutta percha composite biomaterials for root canal therapy. *ACS Nano* 9 (11): 11490–11501.

125 Wei, X., Yang, M., Yue, L. et al. (2022). Expert consensus on regenerative endodontic procedures. *Int. J. Oral. Sci.* 14 (1): 55.

126 Gronthos, S., Mankani, M., Brahim, J. et al. (2000). Postnatal human dental pulp stem cells (DPSCs) in vitro and in vivo. *Proc. Natl. Acad. Sci. USA* 97 (25): 13625–13630.

127 Couble, M.L., Farges, J.C., Bleicher, F. et al. (2000). Odontoblast differentiation of human dental pulp cells in explant cultures. *Calcif. Tissue Int.* 66 (2): 129–138.

128 Nait Lechguer, A., Kuchler-Bopp, S., Hu, B. et al. (2008). Vascularization of engineered teeth. *J. Dent. Res.* 87 (12): 1138–1143.

129 Dissanayaka, W.L., Zhu, L., Hargreaves, K.M. et al. (2014). Scaffold-free prevascularized microtissue spheroids for pulp regeneration. *J. Dent. Res.* 93 (12): 1296–1303.

130 Soares, P.A., de Seixas, J.R., Albuquerque, P.B. et al. (2015). Development and characterization of a new hydrogel based on galactomannan and κ-carrageenan. *Carbohydr. Polym.* 134: 673–679.

131 Xing, Q., Yates, K., Vogt, C. et al. (2014). Increasing mechanical strength of gelatin hydrogels by divalent metal ion removal. *Sci. Rep.* 4 (1): 4706.

132 Weng, L., Chen, X., and Chen, W. (2007). Rheological characterization of in situ crosslinkable hydrogels formulated from oxidized dextran and *N*-carboxyethyl chitosan. *Biomacromolecules* 8 (4): 1109–1115.

133 Ashammakhi, N., Ahadian, S., Zengjie, F. et al. (2018). Advances and future perspectives in 4D bioprinting. *Biotechnol. J.* 13 (12): 1800148.

134 Gao, B., Yang, Q., Zhao, X. et al. (2016). 4D bioprinting for biomedical applications. *Trends Biotechnol.* 34 (9): 746–756.

135 Bharadwaj, T. and Verma, D. (2021). Open source bioprinters: revolutionizing the accessibility of biofabrication. *Bioprinting* 23: e00155.

136 Wang, H., Lian, Q., Li, D. et al. (2021). Multi-tissue layering and path planning of in situ bioprinting for complex skin and soft tissue defects. *RPJ* 27 (2): 321–332.

13

Importance of Machine Learning in 3D Bioprinting

Shohreh Vanaei[1], *Saeedeh Vanaei*[2], *Michèle Kanhonou*[3], *Sofiane Khelladi*[4], *Abbas Tcharkhtchi*[5], *and Hamid Reza Vanaei*[3,4]

[1] *Department of Bioengineering, Northeastern University, Boston, MA, USA*
[2] *Department of Mechanical Industrial and Manufacturing Engineering, University of Toledo, Toledo, OH, USA*
[3] *Léonard de Vinci Pôle Universitaire, Research Center, Paris La Défense, France*
[4] *Arts et Métiers Institute of Technology, CNAM, LIFSE, HESAM University, Paris, France*
[5] *Arts et Métiers Institute of Technology, CNRS, CNAM, PIMM, HESAM University, Paris, France*

13.1 Introduction

The concept and creation of three-dimensional (3D) objects from computer-aided design (CAD) data have been increasingly developed during the past decades. The so-called procedure refers to a group of techniques covered under the term additive manufacturing (AM) [1]. Up to now, several techniques such as stereolithography (SLA), selective laser sintering (SLS), and fused deposition modeling (FDM) have been developed to construct metals, ceramics, polymers, and alloys [2, 3]. The invention of SLA in 1983 spurred researchers to advance, culminating in the patenting of FDM in 1989 [4]. Nowadays, FDM is mostly known as 3D printing and is defined as a process of layer-by-layer deposition of material in successive X–Y planes through Z direction [5]. Despite the several applications of 3D printing, such as aerospace, automotive, and biomedical, anatomical models have been inevitably constructed during the past four decades. The process of printing 3D constructs implementing biological materials was first developed two decades ago and is known as 3D bioprinting [6].

3D bioprinting is a pioneering technique for the fabrication of 3D objects with biocompatible materials, particularly living cells such as bioinks [7]. The difference between 3D printing and 3D bioprinting relies on the type of materials used in the process, as mentioned. In 3D printing of polymers or polymer-based

3D Bioprinting from Lab to Industry, First Edition. Edited by Prosenjit Saha, Sabu Thomas, Jinku Kim, and Manojit Ghosh.

composites, a thermoplastic extrudes in a defined manner and then solidifies to fabricate the desired object, while in 3D bioprinting, a biological material (e.g., bioink) is printed. In this case, the selection of the material plays a crucial role and thus biocompatibility is an important criterion in 3D bioprinting. Nowadays, one of the most important challenges in this process is the defect detection of the 3D-printed objects in real-time, and to provide a suitable environment for cell growth. In this way, it is required to find a solution to optimize the process. Determining the best process variables related to 3D bioprinting machines further with the applied materials and their characteristics is still challenging and thus finding the optimized values is inevitable.

ML is a tool that takes data as input and identifies patterns and makes decisions without human involvement, as opposed to artificial intelligence (AI) which imitates human thinking [8]. There are several studies for optimizing the AM techniques, notably the 3D printing process, by focusing on the process variables and implementing ML algorithms to obtain the optimized values for fabricating high-quality parts [9–11]. As a solution, ML could be applied to 3D bioprinting process to be able to combine recent computing techniques with cutting-edge medical technologies and to achieve better quality using the 3D bioprinting machine [12].

This chapter is focused on the incorporation of ML and 3D bioprinting and the importance of using ML for the optimization of the properties of the constructed objects. We concentrate on literature that presents the role of ML in 3D printing of biomedical objects for the replacement/repairing of damaged organs and tissues.

13.2 3D Bioprinting

13.2.1 Fundamentals of 3D Bioprinting

3D bioprinting is the use of biomaterials along with other necessary materials, including cells and growth factors, to create 3D structures to be used as organs in human body and is applied in tissue engineering, repairing, studying diseases, and drug testing. In the past decades, it has evolved from tissue generation and technologies, such as laser direct write (LDW) to inkjet techniques, in situ skin printing, and large-scale tissue printing. At its early stage, 3D bioprinting was limited to scaffolds, but its use expanded to tissue and organ fabrication and cancer treatment. 3D bioprinting techniques are generally categorized into laser, extrusion, and droplet techniques to deposit bioinks in a layer-by-layer process. Each technology has its advantages and challenges. Laser-based bioprinters use laser as energy source for the deposition of bioinks. This method has been mostly used in tissue engineering. Parameters involved in this technique, including biomaterial viscosity and surface tension, can be optimized and tuned to change the resolution. Nevertheless, laser-based bioprinting is a high-cost technique. Extrusion-based technique incorporates force to extrude the bioink from the nozzle. The continuous deposition inherent in this method brings about better control over the process. However, the use of this method is limited to materials with lower density. In the droplet technique, a defined volume of the bioink in the shape of droplets is ejected from the nozzle and deposited on the substrate. The source of ejection can be piezoelectric, thermal, etc. Droplet techniques hold the benefits of high resolution, low fabrication cost, and a relatively higher rate. However, these methods are limited to low viscosity materials and there is the disadvantage of Nozzle clogging [13, 14].

13.2.2 Materials and Technologies in 3D Bioprinting

13.2.2.1 Different Materials in 3D Bioprinting

Biocompatibility and printability are the most important criteria for the materials (more precisely, biomaterials) being used in 3D bioprinting process. As they are mostly implemented for replacement or repair of damaged body organs, their chemical properties, along with other related characteristics, are important to be taken into account for the desired parts. Biomaterials are mainly categorized into four main groups: polymers, metals, ceramics, and composites. The most applicable material in 3D bioprinting is polymer, either synthetic polymers or natural polymers [15]. Among the synthetic polymers, there is a wide range of materials such as polyether ether ketone (PEEK), polyglycolic acid (PGA), acrylonitrile butadiene styrene (ABS), and polylactic acid (PLA) that are briefly summarized in Table 13.1.

The mentioned materials have various applications in 3D bioprinting such as scaffold printing, muscle tissue printing, and organ printing. Scaffolds are the basic unit in tissue engineering and the fabrication procedure is the prerequisite in tissue engineering. The strength and mechanical behavior of polymers in 3D bioprinting make them one of the most applicable materials for muscle tissue printing. Rapid construction of organs for fast replacement of a damaged one is the most important advantage of organ printing that is applicable in engineered regeneration and is broadly being used in the field.

Bioink(s), which refers to biocompatible materials comprising living cells, is the most applicable material in 3D bioprinting. They are a combination of biomaterial, hydrogel, and living cells that are applicable in engineered regeneration as they create a hydrogel environment. The key fact of 3D bioprinting is the cell proliferation that is accessible using bioinks, which are classified as follows: structural bioinks, fugitive bioinks, functional bioinks, and support bioinks. Both natural and synthetic polymers are applicable for the mentioned bioinks. In fact,

Table 13.1 Summary of common materials in 3D bioprinting.

Nanomaterial	Application
PCL/HA	Enhancement of bioactivity
PCL/Alginate	Maintenance in scaffold structure
PLA/HA	High mechanical strength High elastic modulus
ABS, PLA	Biocompatible Suitable melting point Application in tissue engineering
PEG/PCL	Increase in cell growth Restraint of tissue mineralization
PEG hydrogel	Capable of hydrogel formation Reduce in viscosity of hydrogel
PEG	Used as bioink Acceptable elasticity modulus Photo-crosslinkable Water soluble

Source: Vanaei et al. [16].

the most important characteristic of polymers is their formability that well explains the requirement of the utilized materials for 3D bioprinting. Accordingly, it is important to use hydrophilic polymers to cover the cell-friendly environment. As an example, hydrogels are good candidates due to their capacity to save sufficient water in the desired structure [17]. The formation of bioinks and their most common applications are schematically presented in Figure 13.1.

Given the mentioned explanations, there are various types of bioinks. Hyaluronic acid (HA) is a bioink that was first produced from bovine eyes. Karl Meyer and John Palmer made a significant discovery when they extracted a glycosaminoglycan from bovine eyes, which they named HA. HA is a polysaccharide with a high molecular weight and serves as the main component of the extracellular matrix (ECM). This material has gained great importance in the field of tissue engineering due to its remarkable properties, including biodegradability, bioresorbability, biocompatibility, nonadhesive nature, nonthrombogenic nature, and nonimmunogenicity. Additionally, HA has the potential to form flexible hydrogels, making it even more versatile for various applications. One of the notable applications of HA is in 3D bioprinting, specifically in the form of hydrogels. Hydrogels are 3D networks of crosslinked polymers that can absorb and retain large amounts of water. They mimic the natural environment of tissues and

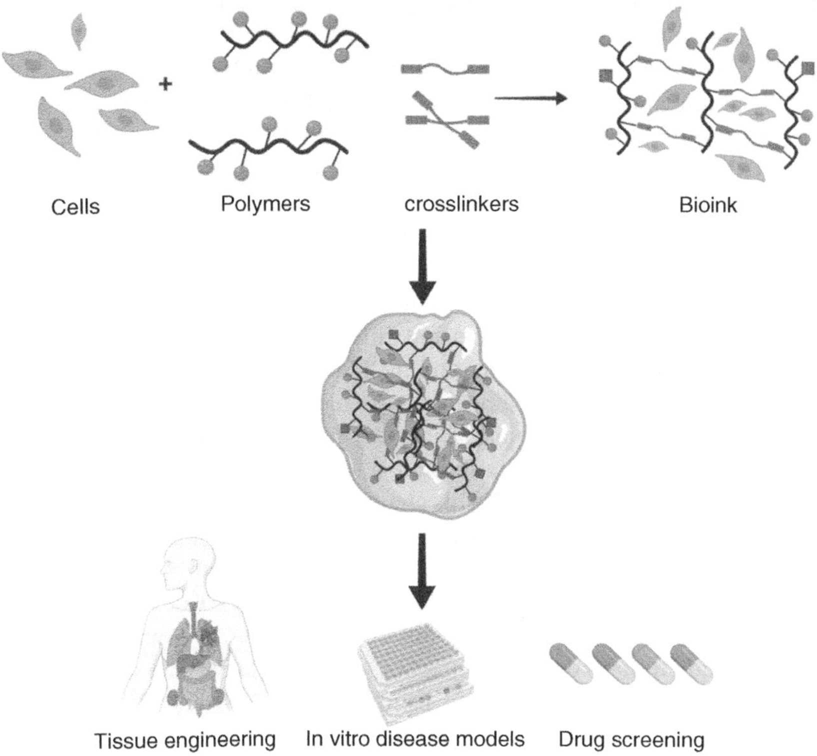

Figure 13.1 Formation of bioinks and their medical applications. *Source:* Vanaei et al. [16].

provide a suitable substrate for cells to grow and differentiate. HA-based hydrogels have been used in 3D bioprinting due to their unique properties. In a study by Highley et al., an innovative approach for fabricating a supramolecular hydrogel using HA was proposed. This was achieved through the utilization of extrusion-based 3D bioprinting techniques. The supramolecular hydrogel formed by this method exhibits low viscosity under mechanical deformation. This is attributed to the presence of noncovalent bonds within the hydrogel structure, which provide reversible properties. These reversible bonds allow the hydrogel to maintain its structural integrity during the printing process and enable it to be reshaped or reformed when necessary. So, the use of HA in 3D bioprinting holds tremendous potential for tissue engineering applications. Its biocompatibility, biodegradability, and ability to form flexible hydrogels make it a suitable material for creating complex tissue structures with high precision. The development of HA-based hydrogels through 3D bioprinting techniques opens up new possibilities for creating functional tissues and organs in the field of regenerative medicine [18].

Collagen is another bioink that is broadly used in 3D bioprinting. It is a widely studied biocompatible polymer in the field of bioprinting, particularly in tissue engineering. It is the main component of the ECM found in most tissues, and plays a crucial role in providing structural support. Collagen is a natural protein that exhibits a triple-helical structure, making it highly compatible with biological systems. Consequently, collagen scaffolds are known to have minimal immunological reactions, reducing the risk of adverse responses when used in biomedical applications. One of the significant advantages of collagen is its ability to enhance cell growth, adhesion, and attachment. These properties are crucial for tissue engineering, as they promote the integration of cells into the scaffold and support their proliferation and differentiation. Collagen's biological activity and its ability to mimic the natural ECM environment make it an attractive choice for creating tissue-engineered constructs. However, collagen type I, which is commonly used in 3D bioprinting, has certain limitations. At low temperatures, collagen remains in a liquid state and only forms a fibrous structure when the temperature is increased. This characteristic poses a challenge for 3D bioprinting, as the collagen bioink needs to maintain its shape stability during the printing process. The complete gelation of collagen can also be a time-consuming process, often taking up to 30 minutes at 37 °C. To overcome these limitations, researchers have explored different strategies. Diamantides et al. conducted a study to investigate the effect of riboflavin photo-crosslinking and pH on the rheological properties and printability of collagen bioinks. They found that the pH of the bioink played a crucial role in the shape stability of the printed structures during gelation. However, the rate of gelation did not significantly affect the printability of collagen bioinks [19]. By optimizing the pH conditions and exploring crosslinking techniques, researchers aim to improve the printability and structural stability of collagen-based bioinks. These advancements are crucial for successful 3D bioprinting, as they enable the precise fabrication of complex tissue structures that closely resemble native tissues. So, collagen is a highly biocompatible polymer, widely used in bioprinting and tissue engineering. While it offers numerous advantages, such as minimal immunological reactions and enhanced cell interactions, the low-temperature gelation and shape stability of collagen bioinks present challenges in 3D bioprinting. Ongoing research focuses on optimizing the gelation process and exploring crosslinking methods to improve the printability and structural integrity of collagen-based constructs.

Gelatin is also categorized as a bioink, which is a natural protein derived from collagen hydrolysis. It is a noncytotoxic and water-soluble material used as a biomaterial for regenerative medicine. Despite the various types of bioinks with different characteristics, multimaterial bioinks are used nowadays with the objective of increasing the printability of the applied materials. Several efforts have been taken into account to produce multimaterial bioinks such as

Table 13.2 Summary of the most recent studies on bioinks.

Type of bioink	Application
Gelatin-methacrylamide	Tissue engineering
Alginate-PCL	Tissue engineering
PEG-gelatin	Tissue engineering
Collagen	Tissue engineering
Collagen-chitosan	Tissue engineering
Hyaluronic acid	Tissue engineering
Alginate	Tissue bioprinting
Collagen-gelatin	Tissue engineering
Alginate-PCL fibers	Bone organ engineering
Alginate-chitosan	Neural tissue
PLA-gelatin	Living tissue constructs
Hyaluronic acid-gelatin	High viable liver construct

Source: S. Vanaei et al. [16].

considering collagen-alginate in tissue engineering, alginate-methylcellulose as a hydrogel, or gelatin methacrylate with the objective of cell adhesion improvement [20]. A summary of the most recent studies on bioinks is presented in Table 13.2.

13.2.2.2 Different Technologies/Techniques in 3D Bioprinting

There are several bioprinting techniques based on the type of material, jetting, or extruding system, and also the desired application. Researchers and industries have proposed different categories. Through the studies from different sectors, they are all based on the flowchart indicated in Figure 13.2.

Seemingly, bioprinting techniques are considered as extrusion-based, droplet-based, and photo-curing. In extrusion-based technique, the material (more precisely bioink) is extruded from a nozzle to be deposited layer-by-layer. Accordingly, the bioink is compressed by a pneumatic force to be extruded through a nozzle. Then, the bioink is compressed using a micro pump to push the bioink. Finally, the bioink is screwed into the system [21].

Considering the mentioned explanation, there are still several parameters that have an impact on the 3D printed bioinks and thus it is crucial to take them into account for the construction of high-quality parts. Notwithstanding the fact that extrusion-based 3D bioprinting is the most common technique that is broadly used worldwide, photo-curing was first developed as a 3D printing technique and has started to be implanted in biomedical applications. As an example,

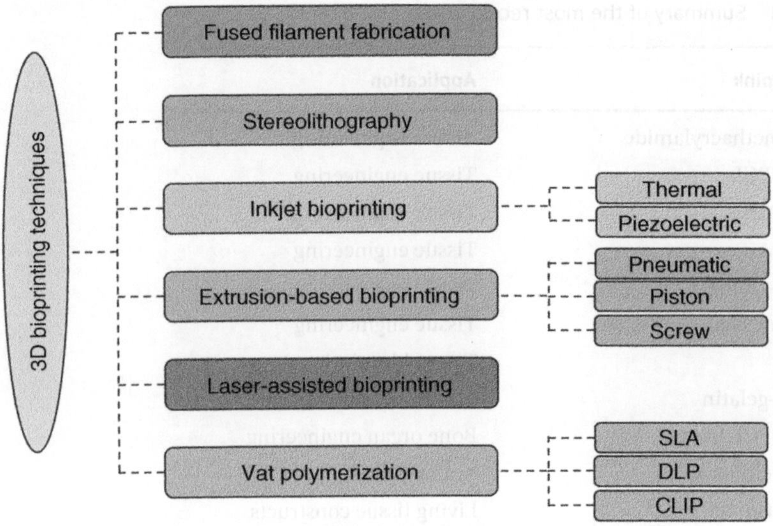

Figure 13.2 Flowchart of the 3D bioprinting techniques. *Source:* Vanaei et al. [16].

stereolithography (SLA), SLS, digital light processing (DLP), etc. are the most common methods that have been previously developed and are being utilized nowadays as a 3D bioprinting technique [22].

Aside from extrusion-based bioprinters, jetting systems are broadly utilized in 3D bioprinting process, particularly for microfluidic chips. In fact, the characteristics of the bioink (e.g., flexibility, biocompatibility, etc.) determine the type of jetting system to be implanted and thus there are various syringe pumps in this printing system.

13.2.3 Application of 3D Bioprinting

As briefly discussed, there are various applications into which 3D bioprinting could be integrated, and thus the 3D bioprinters are capable of playing their role. There are different categories according to the literature; however, they could be mainly divided into tissue engineering (regenerative medicine) and biomedical application. The point is that 3D bioprinting could be integrated into the fabrication of different organs as well as tissue structures for the aforementioned applications.

In vivo and ex vivo models are being constructed with the integration of human cells in order to consider the microfluidic organ-on-chip models that are mostly constructed using 3D bioprinting. Laser-assisted bioprinting is frequently implanted in the production of 3D bioprinted cancer models and the objective is to trace the movement of biomaterial in the cancer cell environment [23]. Blood vessel printing has been progressively integrated into the fabrication of artificial vessels, and there are different methodologies in this regard. Xu et al. have used silicon ink to 3D print a scaffold to produce blood vessels [24]. Peng et al. have implemented 3D bioprinting to fabricate scaffolds to be applied in drug delivery [25]. Further applications have been implemented, such as organ transplantation, pharmaceutical screening, etc.; however, tissue engineering has been widely developed in recent years. The most common tissues that could be cited are cartilage tissue engineering, bone tissue engineering, neural tissue engineering, etc. [26, 27].

13.3 Machine Learning in 3D Bioprinting

ML, one of the subsets of AI, is progressively developing nowadays. The main objective of this field is to predict the efficiency of a system, the characteristics of a fabricated object, the performance and strength of a structure, etc., which is usually based on previous experiences and data. The point is that the data set should be trained first by the existing algorithm of ML to generate the desired model, and then use the generated model to develop it with new dataset to be trained for better prediction of the desired objective [28].

The most common ML techniques are supervised ML, unsupervised ML, and reinforcement learning. The most applicable ML technique is supervised ML which is based on labeled dataset including the inputs and outputs, and the objective is to create a link by the ML algorithm between the input and output data. An example related to the first stage of implementing supervised ML was email spam and image processing. To be more precise, the applied dataset is divided into two training and test datasets. The first group is for training the developed model and the second group is for testing the model and also verifying the utilized algorithms. The main challenge is the data collection for model preparation, which is why the unsupervised ML technique is applicable as it does not require labeled data set or a specific data category. So, the ML algorithms are required to find the pattern for the implemented dataset. Furthermore, there is also another technique called semi-supervised ML, which is a mix of supervised and unsupervised MLs. The objective of this technique is the possibility of improving the performance of the supervised model using unlabeled data. In reinforcement ML, the objective is to learn by rewarding or punishing based on the outcome of the action instead of training data like in supervised and unsupervised ML, and thus the training of the model would be repeated for better reward achievement. It is worth mentioning that reinforcement ML covers both supervised and unsupervised ML as it is required to collect data and define the awards/punishment rules. In what follows, the importance of ML and its application in 3D bioprinting will be thoroughly discussed.

13.3.1 3D Bioprinting Process Using ML

In 3D bioprinting, there are three main steps: pre-bioprinting, bioprinting, and post-bioprinting. In the first step, it is required to choose the applicable biomaterial as well as the utilized cell. Then, the CAD file needs to be defined and bioimaging should be done in parallel to start the printing procedure. In the second step, appropriate technique should be taken into account based on the type of material that has been chosen. In the third step, real monitoring of the final product should be done to verify the maturation of the printed structure and, consequently, the state of the cells. Besides the material steps, the selection of suitable process variables is inevitable, and thus constructing a functional structure requires the interaction of all engaged variables.

To do this, optimization procedures should be implemented to achieve high-quality printed objects [29], and ML techniques are promising techniques for optimization and printing complex geometries. Moreover, researchers have applied different methods that are the basis for the development of bioinks. Lee et al. have developed a bioink using multiple regression to predict its printability [30]. With reference to the work of Lee et al., Shi et al. optimized the process variables using ML algorithms by applying fully connected neural network (NN) to perform an interaction of parameters. They have found that the applied technique gives precise cell arrangement [31].

Accordingly, ML algorithms can be applied to a set of parameters of the 3D bioprinting machine to further predict the part quality with the characteristics of the bioinks. To be more precise, these parameters could be related to nozzle geometry and the rheology of the applied bioink. An example is shown in Figure 13.3 regarding the application of ML in 3D bioprinting. As can be seen, the input of the

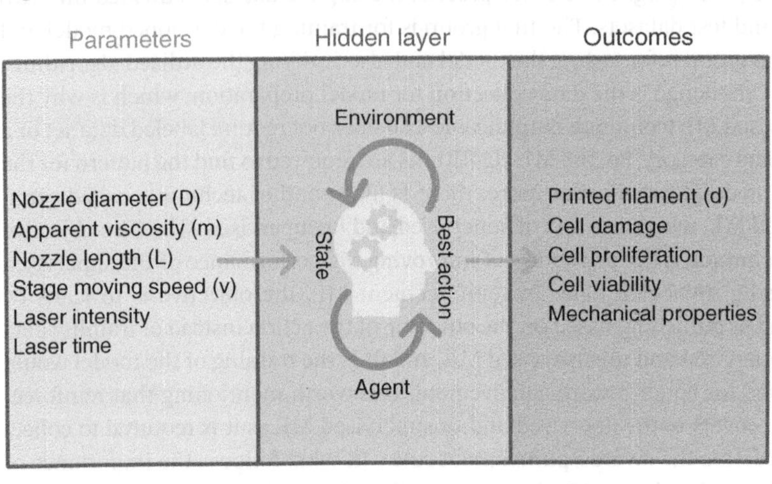

Figure 13.3 Representing the way to apply NN for optimization purposes in 3D bioprinting. *Source:* Shin et al. [32]. MDPI / Licensed Under CC BY 4.0.

NN comprises the process variables and the hidden layers cover the ML algorithms and also the structure of the NN. By training the NN using the input data, the output would be the characteristics of the printed structure for optimization purposes.

13.3.2 Application of ML in 3D Bioprinting

The application of ML in 3D bioprinting has been progressively developed in different ways, such as the prediction of the material's characteristics, the detection of manufacturing defects, and the dimensional accuracy of the printed structure. As discussed, ML is a great tool for predicting the material characteristics and thus optimizing the process variables. On the other hand, ML could be useful for defect detection in the case of microstructure errors or the position of the cells by monitoring the process. Tissue-generated scaffolds are one of the most common examples of how ML could be integrated, as they have complex geometries and cell growth should take place in a well-printed structure [33].

Seemingly, in situ process monitoring based on image analysis could be a solution for optimization of the 3D printed part quality as well as the process variables. So, the integration of the process variables to ML could be taken into account as one of the inputs, and the captured image can also be entered for training the NN. This would result in a trained NN based on real-time monitoring, and consequently be implemented for optimization objectives.

Both traditional ML and deep learning (DL) methods can be utilized for the analysis of the collected images. In the ML approach, 53 features are extracted from the images, and these features are subsequently used as inputs for model construction. On the other hand, DL directly employs the images themselves to identify underlying patterns for accomplishing the same objective. In the context of DL methods, one of the most prominent approaches is convolutional neural network (CNN). CNN has been effectively employed in various applications, including EHD bioprinting, to distinguish between different modes of cone shapes. In particular, it has the capability to recognize the standard cone mode for stable fabrication or identify deformed cones, indicating the necessity for adaptive adjustments to printing parameters. To assess and compare the performance of various CNN algorithms, two specific architectures were put to use: a custom-designed four-layer CNN network and a pre-trained ResNeXt-52 network. These networks were employed to evaluate the quality of fibers, the printing pattern, and the precise positional information of the deposition layers in the EBB (electrohydrodynamic bioprinting) process. Additionally, for benchmarking purposes, a linear support vector machine (SVM) model was developed. In this model, the features used were histograms of oriented gradients extracted from the deposition images. The performances of these three methods were subsequently compared, and the results demonstrated that the pre-trained ResNeXt-52 network achieved the highest accuracy in detecting anomalies related to fiber continuity, regularity, and surface uniformity across all anomaly cases. This outcome highlights the network's effectiveness in identifying printing defects, distinguishing various printout patterns, and accurately determining the location information of the printed layers. The successful detection of printing defects, analysis of printout patterns, and precise location determination through DL methods like CNN can significantly streamline the process of dynamic parameter tuning. By leveraging these insights, researchers and practitioners can make informed adjustments to printing parameters in real-time, ensuring improved bioprinting outcomes and enhancing the overall efficiency and effectiveness of the process [34, 35].

In the context of real-time process monitoring, it is frequently possible to assign quantitative or qualitative labels to the captured images. This enables the evaluation of parameters like fiber quality, diameter, and interfiber spacing. However, labeling more complex aspects, such as flow patterns and droplet evolution, presents challenges. These phenomena involve intricate spatial and temporal information that is difficult to label manually. To tackle this challenge, unsupervised ML techniques have been introduced, including dynamic recurrent neural networks (DRNN). These methods can forecast the spatial and temporal attributes of flow patterns and droplet evolution in droplet-based inkjet printing. Unlike supervised learning, which relies on labeled data, unsupervised ML methods derive insights from unlabeled data, enabling them to uncover patterns and make predictions without the need for explicit labeling. It is worth noting that the prediction task associated with understanding droplet formation, motion, and jetting behaviors is computationally intensive. It requires the ML model to learn from vast amounts of unlabeled data, extracting meaningful representations and patterns autonomously. This self-learning process using unlabeled data is necessary to train the model to make accurate predictions about the complex dynamics of droplets. Through the utilization of unsupervised ML methods such as DRNN, researchers endeavor to surmount the constraints of manual labeling and attain a more profound comprehension of the flow patterns and droplet evolution in droplet-based inkjet printing. These predictions can provide valuable insights into the behavior of the printing process, aiding in process optimization and control. However, due to the computational complexity and the need for extensive unlabeled data, implementing these self-learning approaches can be challenging and resource-intensive. Nevertheless, the potential advantages of precisely forecasting the spatial and temporal details of flow patterns and droplet evolution warrant investments in both efforts and computational resources. This knowledge can enhance in situ process monitoring, enabling improved control, optimization, and quality assurance in droplet-based inkjet printing and related applications [36].

Despite the wide application of ML in process optimization, it plays an important role in cell performance. In this regard, prediction of cell behavior and its viability is considered a key point at this stage. Regarding the mechanism of part fabrication and the presence of extruders, the extrusion force could be the point of failure of the cells. Therefore, choosing suitable parameters and also high-performance bioinks can help in constructing high-quality parts, as mentioned. Therefore, the output of the NN would be cell viability by using the prediction from ML algorithms that are integrated into the NN. Another application of ML in cell performance is the interaction of the cells including several bioinks. Biocompatibility, their physicochemical behavior, etc. play an important role in this regard [37].

The utilization of ML to recognize cell types, phenotypes, and shapes has received substantial attention. ML has demonstrated superior performance compared to human experts in tasks such as segmenting and classifying cells and nuclei in biological images, as observed in multiple studies. Inspired by these advancements, researchers have started exploring the use of ML for analyzing cell-microenvironment interactions. This includes investigating interactions between cells and scaffolds, cell-to-cell interactions, and interactions between cells and materials.

Confocal laser scanning microscopy (CLSM) is a widely used tool in biological research for visualizing cell behavior and exploring interactions between cells and their microenvironment. Yet, extracting cell shape and morphological features from these images through conventional statistical techniques can pose difficulties. In response to this challenge, ML algorithms have been introduced as a means to quantitatively analyze and furnish researchers with this valuable information. SVM) is one such ML algorithm that has been utilized to quantify the diverse morphologies of human bone marrow-derived mesenchymal stem cells (hBMSCs) populations using CLSM images. These morphologies have been effectively linked to specific microenvironments such as poly(ε-caprolactone) (PCL) fibrous substrates and PCL spin-coated films. The SVM has also displayed the capability to discern the influence of bioprinted construct morphology on cell shape phenotypes. By extracting metrics pertaining to cellular and subcellular morphometry, the SVM model developed in this study accurately classified substrates as either woven PCL mesh with precision-stacked microscale fibers or nonwoven mesh with randomly oriented fibers. It is worth noting that the quality and reliability of the SVM models rely heavily on the composition and size of the training dataset. Regrettably, previous studies failed to specify the size and diversity of the datasets they employed. This lack of information gives rise to concerns regarding the robustness and generalizability of the proposed SVM models when dealing with images obtained from a range of substrate morphologies. In summary, ML algorithms, particularly the SVM, have proven to be valuable tools for extracting and analyzing cell shape and morphological features from CLSM images. These approaches have shown promise for associating cell phenotypes with specific microenvironments and substrate topologies. However, the adequacy and generalizability of the SVM models may be influenced by the size and diversity of the training datasets, which should be considered for reliable identification and accurate predictions.

13.4 Challenges in 3D Bioprinting Process Using ML

ML methods offer valuable tools for developing classification, regression, and segmentation models in the context of bioprinting processes and bioprinted constructs. These ML models provide a systematic approach to diagnosing and understanding uncontrollable factors that may influence the bioprinting process. By analyzing and interpreting large datasets, ML algorithms can identify patterns and correlations that may not be evident through traditional methods. The integration of ML in bioprinting empowers researchers to uphold the dependability of the bioprinting process, ensuring precision in forecasting and control of critical variables. ML models can be utilized to optimize biomaterials and bioinks, as well as process parameters, leading to improved print quality and consistency. These models contribute to process optimization, reducing the likelihood of errors and ensuring reliable and reproducible bioprinted constructs. Furthermore, ML methods have shown promise in customizing bioprinted constructs and manipulating their cell culture responses. By analyzing the relationships between biomaterial properties, printing parameters, and desired outcomes, ML algorithms can guide the design of tailored constructs that meet specific requirements. This capability holds great potential for applications such as tissue engineering and regenerative medicine, where precise control over the cellular microenvironment is crucial for successful outcomes. Nonetheless, it is crucial to recognize that there are multiple challenges and issues that must be resolved before delving deeper into and widely adopting ML in bioprinting. These challenges include the need for robust and reliable datasets, the interpretability and transparency of ML models, ethical considerations, and regulatory aspects. Additionally, the complexity and variability of biological systems pose unique challenges for ML algorithms, requiring careful validation and optimization. Addressing these challenges and concerns is crucial for the responsible and effective integration of ML in bioprinting. By doing so, researchers can unlock the full potential of ML to advance the field, enabling precise and personalized bioprinted constructs that meet the needs of various applications in healthcare and beyond.

The effectiveness of ML models in the bioprinting field is heavily dependent on the quality of the available datasets. Comprehensive and consistent datasets play a pivotal role in the development of ML models that can be scaled to encompass a broader spectrum of material properties or adapted for diverse bioprinting technologies. Regrettably, the current quality of datasets falls short of meeting this requirement. One major challenge with the existing datasets is the presence of significant noise and bias. These datasets are frequently gathered from a variety of bioprinters and operational protocols, resulting in inconsistencies and data variations. Even when datasets are collected under the same experimental setup, using identical material composition and process parameters, they can still be influenced by uncontrollable factors inherent in micro/nanoscale bioprinting. As a result, directly utilizing such datasets can undermine the effectiveness and reliability of the ML models developed using them. Surprisingly, despite the importance of dataset quality, there has been limited exploration of dataset cleaning and data normalization techniques specifically tailored for bioprinting. The process of cleaning datasets involves identifying and removing outliers, correcting errors, and mitigating biases that can distort the data. Additionally, data normalization techniques aim to standardize and transform the data to ensure consistency and comparability across different datasets. By addressing dataset cleaning and data normalization challenges in bioprinting, researchers can enhance the reliability and effectiveness of ML models. This includes developing robust methodologies to identify and eliminate noise, bias, and other sources of data inconsistencies. Furthermore, establishing standardized procedures for data collection and normalization can contribute to the creation of high-quality datasets that enable the development of more accurate and reliable ML models. Addressing these concerns related to dataset quality is essential for advancing the field of bioprinting and leveraging the full potential of ML. By improving dataset cleaning and normalization techniques, researchers can ensure that ML models in bioprinting perform optimally and provide meaningful insights for applications such as tissue engineering and regenerative medicine.

13.5 Future Outlook

Expanding the utilization of ML in 3D printing has significant potential to enhance its performance and applications. However, several factors have impeded the widespread adoption of ML in the specific context of 3D bioprinting. One crucial factor is the limited availability of data for bioprinting. ML algorithms rely on abundant data to ensure precise predictions and process optimization. In contrast, traditional 3D printing benefits from a more extensive collection of data and information for analysis. Furthermore, 3D bioprinting is a relatively recent area of study when compared to traditional 3D printing. The bioprinting process itself encounters diverse obstacles, such as concerns surrounding material composition, ensuring cell viability, facilitating tissue integration, and adhering to regulatory requirements. These factors have somewhat constrained the utilization of ML in the field of bioprinting up until now.

However, there is an optimistic expectation that ML will play a crucial role in the advancement of bioprinting. The author presents a viewpoint on how ML can be effectively implemented in this field. ML algorithms offer opportunities to enhance and analyze multiple facets of the bioprinting process, including the improvement of surface quality, detection of flaws, assurance of dimensional precision, and the design of materials with specific properties. By leveraging ML in these ways, the existing challenges can be addressed, and the bioprinting process can be optimized, resulting in outcomes that are more dependable and efficient.

The utilization of ML techniques in the field of bioprinting offers valuable advantages to researchers and practitioners. By harnessing the power of ML algorithms, they can gain data-driven insights and make predictions that may not be readily attainable through traditional approaches. ML enables the identification of intricate patterns, correlations, and optimizations within bioprinting data, leading to a deeper understanding of the underlying processes. As the field of bioprinting progresses and accumulates more data, the integration of ML is anticipated to play a pivotal role in enhancing the capabilities and impact of bioprinting technologies. ML can assist in improving the accuracy and precision of bioprinting processes by uncovering hidden relationships and identifying subtle factors that contribute to successful outcomes. ML algorithms can analyze large datasets, detect complex patterns, and uncover intricate relationships between various parameters in bioprinting. This enables researchers to optimize critical factors such as material selection, printing parameters, cell behavior, and scaffold design. ML also has the potential to facilitate real-time monitoring and feedback control systems, allowing for dynamic adjustments during the printing process to ensure desired outcomes. Furthermore, ML can aid in the prediction of cell behavior, tissue development, and the overall performance of printed constructs. By leveraging ML models, researchers can explore virtual simulations, predict the impact of different variables on the printed structures, and optimize the bioprinting process accordingly. This data-driven approach provides valuable guidance for decision-making and improves the overall efficiency and reliability of bioprinting. As the availability of data continues to increase, ML algorithms will become more sophisticated, accurate, and capable of handling complex bioprinting challenges. The integration of ML techniques with bioprinting methodologies holds great promise for advancing tissue engineering, regenerative medicine, and personalized healthcare, ultimately leading to the development of functional and biocompatible 3D-printed tissues and organs.

13.6 Summary and Conclusion

Over the past few decades, 3D bioprinting has opened up new avenues in the field of tissue regeneration and organ fabrication. Despite significant efforts, the development of suitable bioinks and tissue constructs remains a complex task.

Enhancing the resolution of the structures, preserving cell viability, and reducing fabrication time are the primary challenges in formulating novel bioinks and advancing 3D bioprinting techniques.

In this chapter, the fundamental concepts of 3D bioprinting, including materials and techniques, are discussed. Furthermore, the potential role of ML in optimizing 3D bioprinting techniques is explored. Emphasis is placed on identifying the challenges associated with 3D bioprinting and the need for optimization. With the rapid advancements in ML, there is an opportunity to leverage its capabilities to improve the accuracy and precision of the 3D bioprinting process. Therefore, combining 3D bioprinting with ML techniques presents a promising research area that can accelerate future advancements in the field.

References

1 Wohlers, T. (2014). Wohlers report 2017: 3D printing and additive manufacturing state of the industry: annual worldwide progress report.

2 Yao, S.-S., Jin, F.L., Rhee, K.Y. et al. (2018). Recent advances in carbon-fiber-reinforced thermoplastic composites: a review. *Compos. Part B* 142: 241–250.

3 Rodrı´guez, J.F., Thomas, J.P., and Renaud, J.E. (2003). Design of fused-deposition ABS components for stiffness and strength. *J. Mech. Des.* 125 (3): 545–551.

4 Crump, S.S. (1991). Fast, precise, safe prototypes with FDM. *ASME, PED* 50: 53–60.

5 Vanaei, H.R., Shirinbayan, M., Costa, S.F. et al. (2021). Experimental study of PLA thermal behavior during fused filament fabrication. *J. Appl. Polym. Sci.* 138 (4): 49747.

6 Thayer, P., Martinez, H., and Gatenholm, E. (2020). History and trends of 3D bioprinting. 3D bioprinting. In: *Principles and Protocols* (ed. Jeremy M. Crook), 3–18. Springer.

7 Jovic, T.H., Combellack, E.J., Jessop, Z.M. et al. (2020). 3D bioprinting and the future of surgery. *Front. Surg.* 7: 609836.

8 Dabbagh, S.R., Rabbi, F., Doğan, Z. et al. (2020). Machine learning-enabled multiplexed microfluidic sensors. *Biomicrofluidics* 14 (6): 61506.

9 Ruberu, K., Senadeera, M., Rana, S. et al. (2021). Coupling machine learning with 3D bioprinting to fast track optimisation of extrusion printing. *Appl. Mater. Today* 22: 100914.

10 Bone, J.M., Childs, C.M., Menon, A. et al. (2020). Hierarchical machine learning for high-fidelity 3D printed biopolymers. *ACS Biomater. Sci. Eng.* 6 (12): 7021–7031.

11 Fu, Z., Angeline, V., and Sun, W. (2021). Evaluation of printing parameters on 3D extrusion printing of pluronic hydrogels and machine learning guided parameter recommendation. *Int. J. Bioprinting* 7 (4): 343.

12 Ji, C., Mandania, R., Liu, J. et al. (2018). Incorporating risk in field services operational planning process. *Artificial Intelligence XXXV: 38th SGAI International Conference on Artificial Intelligence, AI 2018*, Cambridge, UK (11–13 December 2018), Proceedings 38. Springer.

13 Vijayavenkataraman, S., Yan, W.C., Lu, W.F. et al. (2018). 3D bioprinting of tissues and organs for regenerative medicine. *Adv. Drug Deliv. Rev.* 132: 296–332.

14 Murphy, S.V. and Atala, A. (2014). 3D bioprinting of tissues and organs. *Nat. Biotechnol.* 32 (8): 773–785.

15 Atala, A. and Yoo, J.J. (2015). *Essentials of 3D Biofabrication and Translation*. Academic Press.

16 Vanaei, S., Parizi, M.S., Salemizadehparizi, F. et al. (2021). An overview on materials and techniques in 3D bioprinting toward biomedical application. *Eng. Regener.* 2: 1–18.

17 Raza, F., Zafar, H., Zhu, Y. et al. (2018). A review on recent advances in stabilizing peptides/proteins upon fabrication in hydrogels from biodegradable polymers. *Pharmaceutics* 10 (1): 16.

18 Camci-Unal, G., Cuttica, D., Annabi, N. et al. (2013). Synthesis and characterization of hybrid hyaluronic acid-gelatin hydrogels. *Biomacromolecules* 14 (4): 1085–1092.

19 Guvendiren, M., Molde, J., Soares, R.M. et al. (2016). Designing biomaterials for 3D printing. *ACS Biomater. Sci. Eng.* 2 (10): 1679–1693.

20 Rowley, J.A., Madlambayan, G., and Mooney, D.J. (1999). Alginate hydrogels as synthetic extracellular matrix materials. *Biomaterials* 20 (1): 45–53.

21 Gu, Z., Fu, J., Lin, H. et al. (2020). Development of 3D bioprinting: From printing methods to biomedical applications. *Asian J. Pharm. Sci.* 15 (5): 529–557.

22 Quan, H., Zhang, T., Xu, H. et al. (2020). Photo-curing 3D printing technique and its challenges. *Biomed. Mater.* 5 (1): 110–115.

23 Datta, P., Dey, M., Ataie, Z. et al. (2020). 3D bioprinting for reconstituting the cancer microenvironment. *npj Precis. Oncol.* 4 (1): 18.

24 Xu, Y., Hu, Y., Liu, C. et al. (2018). A novel strategy for creating tissue-engineered biomimetic blood vessels using 3D bioprinting technology. *Materials* 11 (9): 1581.

25 Peng, W., Datta, P., Ayan, B. et al. (2017). 3D bioprinting for drug discovery and development in pharmaceutics. *Acta Biomater.* 57: 26–46.

26 Gu, B.K., Choi, D.J., Park, S.J. et al. (2018). 3D bioprinting technologies for tissue engineering applications. In: *Cutting-Edge Enabling Technologies for Regenerative Medicine* (ed. H.J. Chun, C.H. Park, I.K. Kwon, and G. Khang), 15–28. Springer.

27 Kundu, J., Shim, J.H., Jang, J. et al. (2015). An additive manufacturing-based PCL–alginate–chondrocyte bioprinted scaffold for cartilage tissue engineering. *J. Tissue Eng. Regen. Med.* 9 (11): 1286–1297.

28 Francis, L. (2014). Unsupervised learning. In: *Predictive Modeling Applications in Actuarial Science*, vol. 1, 280–312. Cambridge University Press.

29 Passamai, V., Dernowsek, J.A., Nogueira, J. et al. (2016). From 3D bioprinters to a fully integrated organ biofabrication line. *J. Phys. Conf. Ser.* 705: 012010. IOP Publishing.

30 Lee, J., Oh, S.J., An, S.H. et al. (2020). Machine learning-based design strategy for 3D printable bioink: elastic modulus and yield stress determine printability. *Biofabrication* 12 (3): 035018.

31 Shi, J., Song, J., Song, B. et al. (2019). Multi-objective optimization design through machine learning for drop-on-demand bioprinting. *Engineering* 5 (3): 586–593.

32 Shin, J., Lee, Y., Li, Z. et al. (2022). Optimized 3D bioprinting technology based on machine learning: a review of recent trends and advances. *Micromachines* 13 (3): 363.

33 Xie, M., Gao, Q., Zhao, H. et al. (2019). Electro-assisted bioprinting of low-concentration GelMA microdroplets. *Small* 15 (4): 1804216.

34 Jin, Z., Zhang, Z., Shao, X. et al. (2021). *Monitoring Anomalies in 3D Bioprinting with Deep Neural Networks.* ACS Biomaterials Science & Engineering.

35 Sun, J., Jing, L., Fan, X. et al. (2019). Electrohydrodynamic printing process monitoring by microscopic image identification. *Int. J. Bioprinting* 5 (1): 164.

36 Huang, J., Segura, L.J., Wang, T. et al. (2020). Unsupervised learning for the droplet evolution prediction and process dynamics understanding in inkjet printing. *Addit. Manuf.* 35: 101197.

37 McQuin, C., Goodman, A., Chernyshev, V. et al. (2018). *CellProfiler 3.0: next-generation image processing for biology. PLoS Biol.* 16 (7): e2005970.

14

Advanced Bioprinting for the Future

D.A. Gouripriya[1], Soumyadeep Bera[1], Jaideep Adhikari[2], Poonam Debnath[1], Prosenjit Saha[1], and Sabu Thomas[3,4,5]

[1] *Centre for Interdisciplinary Sciences, JIS Institute of Advance Studies and Research, Kolkata, West Bengal, India*
[2] *School of Advanced Materials, Green Energy and Sensor Systems, Indian Institute of Engineering Science and Technology, Howrah, West Bengal, India*
[3] *School of Energy Materials, School of Nanoscience and Nanotechnology, School of Polymer Science and Technology, School of Chemical Science and International and Inter University Centre for Nanoscience and Technology (IIUCNN), Mahatma Gandhi University, Kottayam, Kerala, India*
[4] *Department of Chemical Sciences, University of Johannesburg, Doornfontein, Johannesburg, South Africa*
[5] *TrEST Research Park, Sreekariyam, Trivandrum, Kerala, India*

14.1 Introduction

Bioprinting is the key to unlock, revolutionary possibilities in biomedical sciences. It is a cutting-edge technology, which is used to print living biological cells. Bioprinting has evolved from a standard 2D cell deposition process to an advanced 3D fabrication technique in tissue engineering and organ printing. It has evolved significantly and continues to hold immense potential for the future due to the scope of research in this field. As the complexity of human anatomy and organs presents challenges in artificial replication in the laboratory, it leads to continuous research and exploration of advanced technologies such as 4D, 5D, and 6D printing and later possibly to the cell-laden version, multimaterial, electrospinning, gravity-assisted bioprinting, and many more. Today, bioprinting has reached a remarkable stage where researchers can print and grow human tissues and organs for potential implantation in medical treatments, which aims to address the critical shortage of organs for transplantation. According to "ORGAN India," the organ donation rate in India is 0.86 per million population, and approximately 1 lakh organs are needed annually. Barely 2–3% of the demand is met, and many die due to organ failure [1]. However, in coming years, it appears that human lung

3D Bioprinting from Lab to Industry, First Edition. Edited by Prosenjit Saha, Sabu Thomas, Jinku Kim, and Manojit Ghosh.

transplants will become a reality because, in animal models, a human lung scaffold with 200 million alveoli and 4000 km of functional capillaries with oxygen exchange capacity has been 3D printed by the United Therapeutics Corporation [2]. Much research is ongoing for the development of fully functional human organs, and some researchers are successful in developing stages. Precisely, people may have the option to acquire 3D-bioprinted organs customized to their specific needs, revolutionizing health care and reducing organ-related mortality [3]. Although the initial cost of 3D-bioprinted organs might be high due to advanced technology, patient specificity, and on-time availability, mass production is expected to make them more affordable. Alongside organ printing, organ-on-a-chip (OOC) technology plays a significant role in the biomedical field of pharmaceutical companies, vaccine developers, and scientists to test drugs and understand their performance and side effects, reducing the need for animal testing. In 2021, around 4.8 million animals were used for research testing in South Korea. Dr Alexander Brill developed a vein-on-a-chip model to understand the mechanisms of blood clot formation [4], which would accelerate wound-healing research.

Bioprinting has even reached space, where it is expected to serve astronauts and explore the effects of gravity on bioprinting. Scientists are intrigued by the potential of gravity-assisted bioprinting, which introduces new dimensions to the technique. 4D, 5D, and 6D printings are advanced versions that incorporate extra dimensions for more precise and sophisticated printing of various shape-changeable implants, drug-eluting stents, and complex-curved implant structures. Some other advanced techniques are electrospinning-combined bioprinting, comingled bioprinting, and multimaterial printing. This chapter briefly talks about all the advanced, ongoing, and upcoming techniques to bring revolutionary developments in the field of medical science.

14.2 Electrospinning and Bioprinting

Bioprinting has many advantages and offers a controllable spatial resolution. Nevertheless, it still faces some weaknesses in mechanical properties and the printing process-induced impact on cell viability. So, researchers thought of combining more than one fabrication technique to improve the quality of the printed scaffold. Thus, experiments were carried out by combining electrospinning and bioprinting to compensate for shortcomings of individual processes [5]. In this regard, Qianheng Jin et al. tried to fabricate blood vessels with small diameters by combining electrospinning and a rotary bioprinter. Here, a monolayer of endothelial cells (ECs) is formed on the surface of electrospun polycaprolactone (PCL), and a layer of gelatin methacrylate (GelMA) with smooth muscle cells in the horizontal axis of rotation is printed through bioprinting. Both cell types showed good viability and superior properties to natural blood vessels in strength and burst pressure [6]. Teresa Carranza et al. tried to develop multilayered structures by combining 3D bioprinting and electrospinning (ES) using gelatin and polyvinyl alcohol (PVA) as the ES ink and gelatin and chitin as the bioprintable ink. Bioprinting provides scaffolds with an environment similar to native tissues, and ES can produce nano- and porous fibrous structures with a high surface area-to-volume ratio, facilitating cell adhesion and growth [7]. Faraz Fazal et al. developed a modified hybrid setup with one bioprinting head and two electrospinning heads. Helical tubular constructs were printed around a rotating needle by the bioprinting head and two electrospinning needles were used for co-electrospinning two different materials [8]. Xingliang Dai et al. combined 3D bioprinting and electrospinning to develop tissue-like constructs with neurosecretory functions. Hydrogel with sodium alginate/gelatin/fibrinogen was used for bioprinting, which was seeded with neurosecretory cells, and an electrospun layer of polylactic acid (PLA)/gelatin was used to cover it. The resultant structure showed excellent neurosecretory function, high activity, and mechanical strength, which is more significant than the individual hydrogel system [9]. Young Soo Yu et al. constructed a new trachea implant similar to native tissue by 3D printing C-shaped cartilage grooves and a 250 μm-thick layer composed of electrospun PCL nanofibers. The incorporation of electrospun fibers and the cartilage helped in the adhesion of trachea mucosal cells and the proliferation of chondrocytes, respectively. The implant showed excellent rotational and bending response with the face [10]. Sang Jin Lee et al. printed artificial blood vessels by blending chitosan and electrospun PCL nanofibers and covering them with PCL strands. The resulting blood vessels exhibited excellent mechanical properties [11]. Isabel Mayoral et al. have initiated the advanced stage of combining 3D printing with electrospinning to create patient-specific vascular patches by tying together the concepts of mechanical engineering, tissue engineering, and computational fluid dynamics. The fused deposition modeling (FDM) will help create the geometry of the patch, and the electrospinning technique will give the 3D support for the vascular smooth cells. Overall, the process highlights the potential of developing patient-specific patches with excellent remodeling and repair ability [12].

14.3 4D Printing

4D printing is a unique and emerging technology for fabricating intricate structures for specialized applications. 4D printing has emerged from the novel 3D printing technology, which utilizes the printing head and platform to move in three dimensions (*X*, *Y*, and *Z*), similar to the 3D printing, along with an extra dimension as time. The significance of time as an extra dimension is to affect the change in shape of the models after the printing process. This time-dependent property will show its effect when subjected to external stimuli like temperature, moisture, magnetic and electric field, light, solution, etc [13]. In 2013, MIT researchers combined smart materials with 3D printing technology to give birth to 4D printing.

4D printing assures to address the shortcomings of 3D printing [14] with potential application in medical implants and devices, tissue engineering, drug delivery and even in food industry. The process dictates printing of smart materials using 3D printing technologies. Material selection, model design, printing conditions, and successful activation are major factors influencing the transformation of 4D printable models [15]. SMPs (shape memory polymers) and LCNs (liquid crystal networks) are widely used to make complex shape-changing structures using single or multimaterial printing. These smart materials can show properties like biodegradability, flexibility, heat processability, and cross-linking ability.

SMPs are a class of polymers that can be given a temporary shape and transmitted back to their original form while exposed to external stimuli. SMPs possess transition (T_{trans}) and permanent (T_{perm}) temperature points allowing them to revert to their original form or undergo permanent deformation. SMPs are known to be able to store multiple different shapes in memory [16]. Some examples of SMPs include PLA, thermoplastic styrol block-copolymer (TPS), and acrylonitrile butadiene rubber (ABS).

Liquid crystalline networks (LCNs) are a class of polymers under smart materials with liquid-like flexibility and crystal-like orientation due to the presence of mesogens. Mesogens are liquid crystalline material, and mesophase is a state which occurs between the crystalline solid and isotropic liquid states. Alteration of mesogen groups, orientation of mesogens, and distribution of cross-linking can lead to many deformation possibilities, including extension, contraction, bending, twisting, or combination. Their nature is similar to SMPs, where they can be temporarily shaped and successfully regain their original shapes. LCN can reversibly deform between ordered (LC phase) and disordered (disordered state) shapes. There are different stimuli to trigger the transition [17], with stimulus concentration affecting the outcome. Table 14.1 includes different categories of stimuli based on their interaction with the matter (physical, chemical, and biological) along with their potential applications and respective mechanisms.

AI is a revolutionary tool, which will transform the human lifestyle shortly. It is used in computation, data analysis, continuous monitoring, and many more tasks surpassing human capabilities. AI integrated with 4D printing is going to be a revolutionary development in the field of biomedicals. For cardiovascular instruments, AI-driven approaches typically involve material inherent property (extracted from backlog data), shape optimization of 4D printing model with favorable selection of material resources, ideal manufacturing methods, and achievable optimized rheology through varying degrees of extrusion [29].

Table 14.1 Different categories of stimulus with application and respective mechanism.

SL no.	Stimuli type	Stimuli	Polymer	Application	Mechanism
A.	Physical	1) Temperature	An hydrogel made from methacrylate poly (ethylene oxide), poly (propylene oxide) and poly (ethylene oxide) triblock copolymers [18]	Smart valves and soft actuators	Thermoresponsive polymers possess transition (T_{trans}) and permanent (T_{perm}) temperature points allowing them to revert to their original form or undergo permanent deformation. The T_{trans} is lower than the T_{perm}.
			Double-layer structure made of poly(*N*-isopropyl-lacrylamide) and poly(caprolactone) [19]	Encapsulation and release application of cells	
		2) Liquid/moisture	Poly (ethylene glycol)-based bilayered hydrogel [20]	Cell encapsulation for insulin production	Absorption and desorption of liquid by hydrogels lead to swelling and deswelling, which are responsible for making reversible deformations.
		3) Light	Tricolor component construct containing veroyellow (yellow) and verocyan (blue) fibers, which are digital SMPs held in tango (sky blue) matrix [21]	Actuators and biomedical application	Different colors absorb different energies at different wavelengths of electromagnetic radiation. When red color illuminated, verocyan (blue) absorbed that light and the temperature increased above *Tg*, which softens the stretched verocyan (blue) fibers and shortened it to the original length; whereas yellow fibers remained unaffected by the red color light (i.e., below *Tg*); resulting in no change in length. The multilayer composite bent as a result of the length difference between the blue and yellow fibers.
		4) Magnetic field	PLA/Fe_3O_4 composite [22]	Potential application in biological and medical field	An alternating magnetic field stimulates the magnetic nanoparticles to vibrate and produce heat for shape-changing application.

(*Continued*)

Table 14.1 (Continued)

Sl. no.	Stimuli type	Stimuli	Polymer	Application	Mechanism
		5) Electric field	Hing flap has VeroClear at both the ends and a flexible part in the middle (mixture of VeroClear and Agilus30) connected with resistive wires (Constantan, copper–nickel alloy with a resistance of 3.90 Ω/m and diameter of 0.4 mm) [23]	Actuators	This also works on increasing the temperature on the surface by applying a certain voltage to the resistive wires for few seconds, which will make the structure deformed from the temporary shape to the original shape (SMPs are a class of polymers, which deform on application of heat).
B.	Chemical	1) pH	(2-vinyl pyridine) (P2VP) [24]	Controlled-drug release, self-folding devices, providing drug encapsulation and directional release, tissue engineering, etc.	The described group of polymers features ionizable functional groups and falls under the classification of polybasics and polyacids. These polymers undergo reversible transitions marked by pH-dependent protonation and deprotonation. The polymer-charging process induces electrostatic repulsion, resulting in the elongation of polymer chains and the disintegration of their globular structure upon neutralization of functional groups.
		2) Ionic concentration	Alginate, mannuronic and glucuronic acid residues are physically cross-linked with divalent cations [(Ca^{2+}), (Ba^{2+}), (Mg^{2+})] [25]	Wound healing, drug delivery, and tissue engineering, etc.	The concentration of ions plays a crucial role in influencing hydrogel swelling and, consequently, altering the shape during swelling. Ions play a vital role in establishing physical interactions, particularly ionic bonds, within charged hydrogels. In particular, divalent cations show a strong degree of coordination with these divalent ions by binding to guluronate blocks inside the alginate chains with an exclusive affinity. One polymer's guluronate blocks form junctions with neighboring guluronate blocks, resulting in the formation of a gel structure.

C.	Biological	1) Glucose	CNT/GOx/Pt-NP/polyaniline (Graphene oxide and platinum nanoparticles placed between a CNT nanotube; and polyaniline network is printed over CNT/GOx/Pt-NP to form resistive bridge between CNT electrodes) [26, 27]	Glucose monitoring and insulin delivery	The sensor undergoes catalysis by GOx in the presence of glucose, generating hydrogen peroxide (H_2O_2) as a by-product. PtNP further catalyzes H_2O_2, inducing local acidification. With a detection limit of two millimoles (mM), the pH-responsive conductivity of polyaniline functions as a chemiresistive sensor, allowing for the quantitative evaluation of glucose concentration for diabetes management. In addition, rise in pH triggers hydrogel swelling, leading to the release of insulin.
		2) Enzymes	Gelatin methacryloyl (GelMA) surface decorated with magnetite nanoparticles for magnetic actuation and collagenase and MMP2 enzymes. [28]	Targeted release of anti-inflammatory drug	The drug-loaded hydrogel (microrobots) undergoes enzymatic degradation, facilitating on-demand drug delivery to specific regions in the body. These magnetically driven microrobotic devices are completely eliminated and leave no measurable toxic residues when collagenase and MMP2 enzymes are present.

Optimized printing processes include a three-step assessment procedure [30], which are as follows:

- Establish an efficient process control,
- Assessment of efficacy in prefabrication stages, and
- Automatic defect detection with closed-loop control.

Moreover, due to technical advancements and interdisciplinary approaches, it can be predicted that AI can soon autonomously create and simulate structures on its own with mere requirements about the structure, which will be a great step toward developments in many fields.

Controlled drug release is another potential application in biomedical engineering, where 4D printing holds a grip due to its unique shape-changing properties. Scientists had used α-zein, a cereal protein from maize, which depicted considerable potential for use in drug-eluting biodegradable ureteral stents. [31] To prepare zein gel loaded with ciprofloxacin and doxorubicin, zein (20 and 30% w/v) was dissolved in 80% v/v ethanol in water and printed on a supporting bath of different water concentrations with suspended Carbopol 980 in ureteral stent shape. The hydrophobic nature of zein allows it to self-assemble in water and self-tune itself with functions for printed constructs. After the implantation of the drug-eluting biodegradable ureteral stent in a diseased kidney, it showed good secretion and excretion functions after a few days, owning to its porous zein fibrous structure, which also enhances biodegradability.

The translation of 4D printing to 4D bioprinting has also been achieved, and since the morphing needs to be carried out at the physiological temperature of the human body and neutral pH for cell survival, the pool of suitable stimuli gets restricted. Primarily, swelling/deswelling is the popular method for shape morphing. Additionally, the replicability of shape morphing becomes further challenging with natural biopolymers and very few combinations are reported in the literature.

14.4 5D and 6D Printing

5D printing is misunderstood to have five dimensions; however, it has only three dimensions, similar to 3D printing, along with two additional printing angles, making it unique and creating a difference between 3D and 5D with added benefits. A 5D printer is a hybrid tool, which can be used for both additive and subtractive manufacturing. It prints very complex structures using movements of the printing head and bed in all the three Cartesians (like 3D printing), along with two extra rotational axis, which allow the printer bed to rotate at two different angles on its axis or enable the printer head and bed to rotate independently on

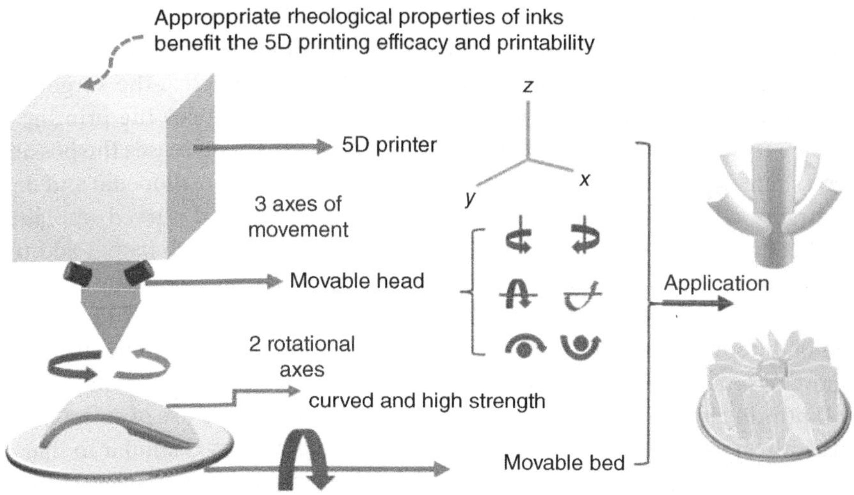

Figure 14.1 Pictorial representation of 5D printing. *Source:* Cheng et al. [35] / Reprinted from Elsevier.

their axis in two different directions (collectively two printing angles) as depicted in Figure 14.1 [32, 33]. This opens great possibilities for printing potential; complex structures having curved layers create definite concave shapes with high precision, which allow printing of a larger object than the print bed size in a single time, all while staying within the confines of the design constraints. It is found that a 5D-printed model is five times stronger than a 3D-printed model and does not need any additional postprocessing. It also does not need supporting materials due to its multidimensional printing, which consumes 25% less material than 3D printing [33]; therefore, complex-shaped models with additional support material can be printed using 5D printing technology. The input data will be in the form of a CAD file, which will be further sliced by various softwares to divide the model into thousands of layers, and the generated G-codes for every point exist in the 3D matrix converting it into a stereolithography (.stl) file. Finally, this file will be read by the printer to print [34]. This technique is very useful in the biomedical applications to print implants and medicines, food industry, automobile, and other fields. 5D printers or 5-axis printing requires relatively fewer materials, and formulation of cell-laden bioink is needed for bioprinting.

Despite its promising advantages, 5D printing does come with some limitations. First, the inclusion of two extra axes adds to the cost of the technology. Second, operating a 5D printer requires precise software and hardware, making it necessary to have highly skilled human resources for its operation and maintenance [32].

Some operations are way too serious and complex for surgeons to handle due to the distorted anatomy and invasion of surrounding structures, making the surgery more complex. Therefore, medical practitioners practice the surgery on 5D-printed models before performing the actual one. It enables the printing of complex structures with extra details and perfection, which increases the possible success rate of very complex surgeries [36]. On the contrary, orthopedic and dental surgery demands an essential requirement of strong and curved implants/prosthetics, which cannot be printed through 3D printing due to its flat additive manufacturing technique, which is inferior to 5D in terms of strength [36]. It is also an emerging technique in the food industry, wherein it is used to decorate food in different structures to create attraction. Rheology of the edible inks is an important criterion that can affect the final quality and structure of food [35].

6D printing is a result of the combination of 5D printing and use of smart materials with shape-changing ability. The entire printing process is similar to that of 5D printing, involving three dimensions and two extra printing angles, with an added feature of printables to change their shape on the application of different stimulus. It is regarded as the progeny of 5D printing technology using fused deposition modeling technology and the idea was first picturized by Georgantzinos et al. in 2021 [37]. In other words, we can say that the 6D printing is the advanced version of the 4D printing. Hence, translation to bioprinting is also possible.

6D printing eliminates shortcomings of 4D printing, including need for supporting material, postprocess requirements, absence of extra printing degrees of freedom, need for advanced processing routes, lack of structural complexity, and coherence and capabilities of the final smart structure. The main advantage of this method is the opportunity to print in multidimensions with smart materials, which will provide a better complex design with structural integrity due to the multidimensional access resulting in more flexible products with less material consumption. It is anticipated to possess the high capability of producing stronger and better-quality products, while minimizing processing times but the only limitation faced is the cost of manufacturing, which may vary depending on the size of the production.

The superiority of 6D printing can be explained with examples shown in Figure 14.2 and 14.3; Figure 14.2 depicts the dimensions constituting the 6D and in Figure 14.3 the left image depicts the 4D-printed model, which is horizontally flat-printed with two types of polymers, whereas the right image is printed with inclined V-type layers through 6D printing. The V-type layer will favor easy deformation (bending) with the highest flexibility during external stimulation compared with the former [37]. Technological advancement and superior quality make this process a potential technique to be adopted by many high-end industries where accuracy is essential, like in the electronics, medical, and aerospace sectors, as well as in components for energy applications. As discussed in 4D

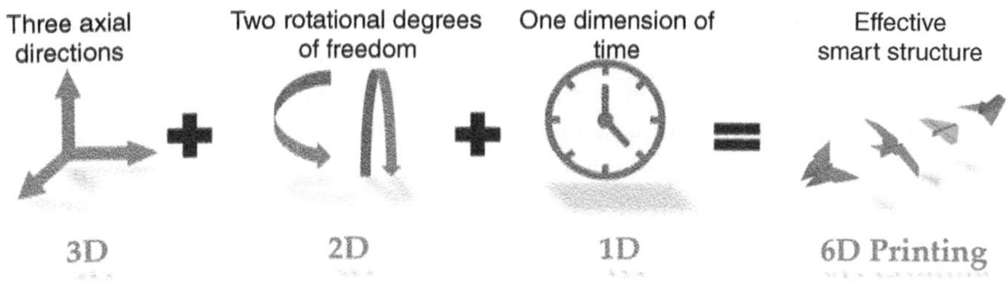

Figure 14.2 Contribution of different dimensions in 6D printing. *Source:* Georgantzinos et al. [37] / MDPI / Licensed Under CC BY 4.0.

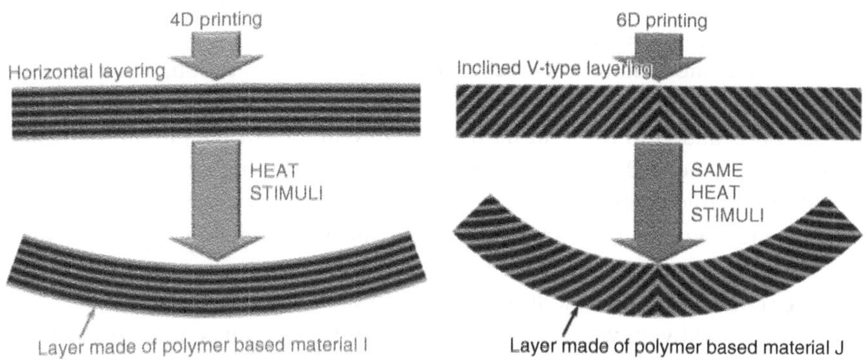

Figure 14.3 Flexibility of 4D- and 6D-printed models. *Source:* Georgantzinos et al. [37] / MDPI / Licensed Under CC BY 4.0.

bioprinting, shape morphing with cell-laden inks is complex, which becomes further challenging with natural polymers. The same could be predicted for 6D bioprinting.

14.5 Organ Printing

Organ printing is a future possibility in the medical field and a ray of hope for all the patients waiting to receive an organ. The need for organ printing artificially in a lab comes from the lack of organ donors [38], and the lack of donors comes from the sudden and unique organ requirements of the patient undergoing transplantation. The donor organs must entirely fulfill all criteria for the successful transplantation of the organ. And this happens with the help of 3D bioprinting, which is the an effective method used for fabricating organs effectively in laboratory, due to its additive manufacture approach, multimaterial printing along with the tailoring possibility. For printing an organ, cell samples from the patient are collected, and the cells are grown in an in-vitro condition with a suitable environment. These cells are then fed with nutrients, proteins, growth factors, ECM (extracellular matrix), and other essential materials and mixed with a gel that acts as a glue [3] to print in a layer-by-layer approach in the form of an organ-shaped scaffold where growth and proliferation of cells turn into tissues and gradually to a matured organ. A scaffold is a three-dimensional model with interconnections among all units, printed through 3D printing technique with or without cells. Researchers have succeeded in developing tissues of many organs, including the heart, kidney, pancreas, brain, multilayered skin, bones, muscle structures, blood vessels, eye cornea, and other organs [2]. Both donated and 3D-printed organs have pros and cons; and Table 14.2 describes the advantages and disadvantages of the donated and 3D-printed organs [39].

According to the Health Resources and Services Administration, on an average, 17 people per day die due to unavailability of organ donors, and every 10 minutes one person is added to the waiting list. More than 90% of people on the transplantation list needed a kidney, and dialysis was a high-cost option [38]. A California-based company named Organovo became successful in engineering 3D-bioprinted human liver and kidney and made it available for commercial use [3]. However, these organs are not available for human transplantation and are under clinical trials to test drugs and vaccines.

Sometimes a fully grown organ is not required if the organ is not entirely damaged; instead, a regenerative medicine-based therapy is enough to save the patient. This therapy involves using a stem cell patch that can repair the damaged tissue by myocardial infarction without invasive surgery, and this patch can be printed through 3D printing technology [40]. Professor Alexander Gaggl developed a 3D-printed cranial prosthesis made of PEEK polymer, and surgeons successfully implanted the prosthesis to treat craniostenosis, one of the first in Europe. In craniostenosis, the skull gets deformed at the infant stage of life. As the brain grows with the infant, there is an increased intracranial pressure, which can lead to blindness, mental disability, and psychological stress to patients due to deformed skull and its appearance [41]. In San Antonio, Texas, Dr. Arturo Bonilla successfully implanted an outer ear in a 20-year-old woman, who was born without her right ear [2].

The requirement of human skin is also urgent as many people have to undergo plastic surgeries due to accidents or burnt skin. In conventional plastic surgeries, the skin is taken from other parts of the body, yet sometimes the coverage of the damaged area is so extensive that managing skin from other parts is not possible. A French company, "Poeitis," developed an artificial "bio-printed, human full-thickness skin model" named "Poieskin," mimicking the dermal and epidermal layers of the skin along with fibroblasts, collagen, and keratinocytes present in the

Table 14.2 Advantages and disadvantages of donated and 3D-printed organs.

Type of organ	Advantages	Disadvantages
1) Donated organ	1) It is natural and original, and most importantly, it was functioning earlier in a human body, which favors high acceptance possibility. 2) Testing of organs before transplanting is not required. 3) The life of the organ can be assured once it is accepted by the acceptor's body. 4) No risk is involved regarding the quality of the organ.	1) Donor organs have to be perfectly matched with all attributes for successful transplantation. 2) Difficult to find a donor. 3) Testing and medications are required regularly until the receiver adapts to the new externalities. 4) Patients have to wait for a long time for organs.
2) 3D-printed organ	1) Organs can be customized and fabricated according to the needs of the patient. 2) Easily available compared with donated organs. 3) It has the potential to provide organs at a huge scale to a large population in the near future.	1) Even after matching all attributes, the successiveness of the transplantation cannot be assured because it is a prototype of the real organ, and is an artificially mimicked structure in which functionalities are incorporated using technology. 2) The life of an organ cannot be assured in real time. 3) A huge risk is involved regarding the successiveness of transplantation. 4) High cost 5) Medical practitioners face challenges in convincing their patients that printed organs will yield identical results to conventional methods due to trust issues.

skin, thus making it fully functional. Currently, this skin is used by various cosmetic industries to test their products and is not available for clinical treatments [42]. Skylar-Scott mentioned that modern bioprinting is a slow process and will take too long to print a single heart due to the presence of billions of cells in it. Therefore, he and his team developed a method to speed up the process by printing clusters of thousands of cells in a condensed form called organoids. They have also developed a bio-printed, tube-like structure similar to the human vein, which can pump fluid on its own [35].

14.6 Vascularized Organ on a Chip

In the advanced era of medical treatment and biomedicine, research is carried out to enhance the quality of diagnosis and target treatment at the molecular and organ levels. Current disease models lack to identify and rectify the cause of disease at the cellular and molecular levels. So, an advanced modeling technique is to be followed for next-generation treatment of vascular diseases [43]. OOC technology is a 3D cell culture setup, which can be developed into a tissue or organ for clinical drug testing and disease studies. Yet, there are some challenges as it cannot mimic the exact tissue environment, and there is difficulty in producing blood vessels and necessary vasculature outside the living organism. Blood vessels and the circulatory pathway are needed to transport oxygen and nutrients for tissues, so a vascularized OOC is necessary to replicate tissues and their physiological functions. The heart-on-a-chip model is an example of creating vasculature under such conditions. [44]. A vascularized glioblastoma multiforme (GBM)-on-a-chip model was created by Giulia Silvani et al. using 3D bioprinting and microfluidics technologies. The system can recreate the brain tumor microenvironment by mimicking the physiological stress and cellular interactions [45]. Jiasheng Yan et al. claimed that OOC has three essential technologies. First, using 3D bioprinting to create microscaffolds; second, the use of microfluidic chip technology to regulate parameters, like blood perfusion and shear stress; and third, conditions, like the oxygen gradient and the metabolism of nutrients and waste products, can be controlled, thanks to cultivation technology. These three factors work together to make effective results out of OOC technology [46]. Fabrication methods of OOC depend upon the target characteristics of the model. The process starts with designing the OOC model in 3D and then manufacturing the structure using suitable materials, followed by tissue construction using a microfluidic or bioprinting approach [47]. Several models have been suggested for the vascularization of OOC, including the angiogenesis or endothelial barrier and vasculogenesis models. The angiogenesis model is based on forming new capillaries from existing blood vessels and is formulated in 3D by carefully positioning the ECs. The vasculogenesis model is based on differentiating ECs from new vascular networks.

3D bioprinting is becoming relevant in the scenario as it can produce patterned structures with microchannels to place cells carefully, and it can create heterogeneous complex tissues and organs. Converging 3D bioprinting and OOC technology can be used for creating micro-architecture of human tissues and organs as shown in Reference [48]To quote an example, Figure 14.4 represents a lung on a chip model is fabricated by mimiciking a structure similar to actual human respiratory pathway. The crossection model was designed through 3D printer and an inlet and outlet was given for the flow of media and a portal for seeding the cells. In the final step,the behavior of cells in the chip model is analyse. Jesus Shrestha et al. created a straightforward lung-on-a-chip model to mimic the in vivo

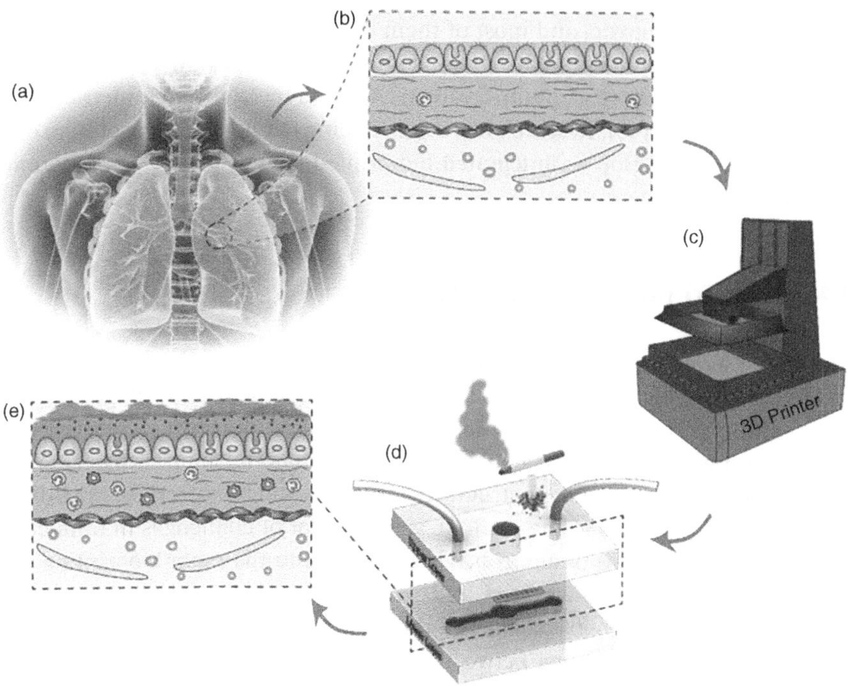

Figure 14.4 Lung-on-a-chip model design and fabrication. (a) A schematic representation of human respiratory system (b) Crossection of air pathway (c) fabrication of chip model using 3D printer (d) The chip model containing outer layer with an inlet and outlet for media transport and an open well for seeding cells (e) After the cells starts flowing the effect of the model on the cells are analysed. *Source:* Shrestha et al. [49] / with permission of Elsevier.

conditions of the airway at an air–liquid interface. In the chip, Calu-3 cells were cultured and the results showed promising properties for an in-vitro model that is suitable for toxicology studies and drug delivery examinations [49]. Currently, there are four main OOC structures, namely microchannels, culture chambers, membrane structures, or multistructured organ chips, based on their target function and cell culture process.

Moreover, the microfluidic OOC facilitates the integration of sensors for the real-time monitoring and responses of the tissues to the drugs. OOC technology has been used in simulating cardiovascular, brain, lung and respiratory systems, liver, gut, breast, bone and cartilage systems, and skin studies. Yet, the complete integration has not been fully developed. The perfusion of cells in OOC is done by embedded micropumps, which increase its fabrication cost and complexity. Paper-based cell culture platforms and gravity-driven or auto-perfusion techniques have been relied on to mitigate existing disadvantages [50]. Another drawback faced by OOC is the selection of material. Most common resin materials used for precise printing can be cytotoxic, and most of them cannot be autoclave sterilized. OOC technology faces challenges with organ compatibility of multiorgan chips, chips with multiple systems (e.g., pumps, valves, etc.), and the control of microenvironmental parameters [46]. Research has been ongoing to find a proper setup to clear all barriers and develop an integrated bio-printed OOC platform for functional tissues [44].

14.7 Multimaterial Bioprinting

Printing of biomaterials with a single component requires multiple optimization processes to produce a stable scaffold. Still, it faces some disadvantages in functional and mechanical aspects, and to mitigate these demerits, scientists have tried printing with multiple components. Multimaterial bioprinting (MMBP) enables to print heterogeneous structures with cells and biomaterials. MMBP can bring essential qualities to the scaffold, yet face a lot of challenges in mimicking the properties of native tissue. It has been applied to the main bioprinting processes such as inkjet, light-assisted, and extrusion-based bioprinting (EBB). It can be found that MMBP in extrusion and inkjet techniques are easy to carry out by facilitating smooth swapping between different biomaterials; however, the process lacks resolution. Meanwhile, MMBP in light-assisted technologies offers high resolution, yet still, some drawbacks exist, such as the necessity of photocross-linkable material for bio inks and complex switching mechanisms [51]. MMBP offers the benefit of replicating complex tissue constructs as it has properties of multiple components, including native components like hydrogels that provide ECM-like environment, which can be combined with synthetic components to provide mechanical robustness [52]. A completely vascularized skin patch model and a three-layer tubular vessel model were printed using multinozzle MMBP. Three different bio inks were used to make three different layers, and the resultant model was found to fit closely with the inner and outer layers showing ideal morphology [51]. MMBP with GelMA and PCL was tried for nasal reconstruction and examined for biological activity and mechanical integrity. It was found that the bio-printed construct showed excellent neocartilage formation and similar mechanical properties as nasal cartilage [53].

MMBP is classified based on the deposition mechanism as side-by-side printing, core-shell printing, and advanced techniques, such as embedded bioprinting and pre-set nozzle bioprinting. The initial attempt of MMBP was side-by-side printing, in which materials are deposited alternately in the same layer or separate layers while printing [54]. Side-by-side printing can be used for developing scaffolds for cartilage, bone, musculoskeletal tissue engineering, tracheal tubes, and aortic valves by carefully tuning mechanical and biological characteristics demanded by respective tissues. However, its applicability is reduced by the slow fabrication speed due to the switching of nozzle heads containing different materials. In core-shell printing, also known as coaxial printing, the nozzle is arranged concentrically [55] to deposit materials simultaneously to form hollow vascular fiber constructs for efficient transfer of nutrients and gases. It is possible to print hollow fibers with inbuilt microchannels through core-shell printing; nevertheless, this hollow structure can result in weak scaffolds. The next version of MMBP could change the extrusion material without changing the nozzle; it enables changing the material so quickly that different materials can be incorporated into the same fiber. In 2017, Liu et al. demonstrated that multiple materials can be printed continuously without changing the printer head. Here, the main nozzle comprises seven minute capillaries, which can be activated digitally by controlled pneumatic pressure [56]. Another technique in the MMBP process is embedded bioprinting, where the bio ink material is embedded in a supporting/sacrificial material, which is removed after printing. The supporting material is used for increasing the printing fidelity and decreasing the printing time as given in Figure 14.5 [55] and will retain its 3D shape in vivo. Pre-set nozzle printing is the next advanced technique, where a cartridge filled with the material is connected to the end of a syringe, which forms a core-shell fiber. It can print tissues, like the spinal cord, capillaries, blood vessels, and hepatic lobules.

The scaffolds developed by MMBP should undergo proper testing and characterization to understand the influence of materials, printing parameters, and processes on the resulting scaffolds. One of the factors that could not find a way out is the lack of spatial control of materials in the longitudinal and circumferential directions. Having proper control in the spatial deposition and continuous printing of multiple materials can give us the advantage of superior scaffolds for

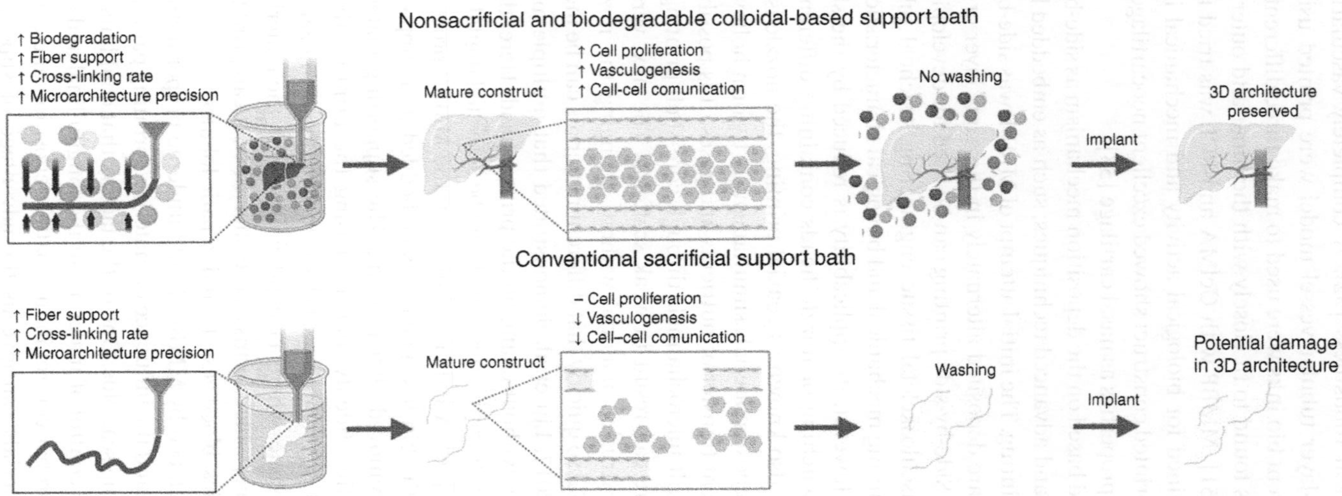

Figure 14.5 Combination of microfluidic multimaterial printing. *Source:* Hassan et al. [55] / ACS.

advanced applications [54]. Self-healing biodegradable colloidal gels can be used as a support material to improve the spatial organization and shape fidelity of the printed construct and develop interconnected porous structures. The interconnected porous structure facilitates gas and nutrient transport and promotes rapid cell infiltration and proliferation [55].

14.8 Printing in Microgravity

EBB is the most popular method, which allows spatial control in depositing materials to bring out customized structures. However, it lacks mechanical stability while fabricating, especially, high-aspect ratio structures. Printing hollow tubes with EBB for high length-to-width ratio structures without using a sacrificial material is difficult. Bhavya Khilnani et al. tried to print a high-aspect ratio structure using gravity as the fifth dimension along with *X*, *Y*, *Z*, and time [57]. The GBM-on-a-chip model was evaluated for the influence of microgravity in cellular response, showing a significant cell morphological and mechanotransduction response [45]. Printing in microgravity is gaining importance in space exploration and studies, especially in space health care. Microgravity and high-end radiation significantly affect organs and tissues, such as skeletal, renal, and cardiac systems. It can lead to the deprivation of health of an astronaut, like losing bone and muscle mass [58]. Scaffolds for bone tissue engineering have been tried to print in microgravity to treat bone-associated disorders caused by altered gravity. Gravity significantly affects the growth and differentiation of cells through physiological and mechanical forces [59]. Numerous techniques, including magnetic levitation, rotating wall vessel (RWV), clinostat, random positioning machine (RPM), and others, can simulate microgravity conditions. To negate the gravitational effect in clinostat setups, the system is rotated perpendicular to the gravity vector. First, RPMs involve rotating the system around two rotational axes within a mount, while continuously varying the rotational direction and velocity. Second, RWVs are fluid-filled and function similar to clinostats by causing their constituent parts to fall through circular pathways. Last, magnetic field forces are employed in magnetic levitation setups to lift the study system. Among all methods used to generate microgravity settings, magnetic levitation is widely accepted as it does not include rotation of the system and application of shear stress. To create the weightless environment needed to investigate the impact of gravity on biological systems, magnetic levitation makes use of negative magnetophoresis [60]. With the aid of gadolinium-based solutions, a 3D cell culture capable of in situ self-assembly was created in microgravity under magnetic levitation conditions [61, 62]. In another research, a custom device was designed for fabricating 3D cartilage tissue constructs in the International Space Station (ISS) [63].

Printing in microgravity becomes relevant in a condition where it can introduce efficient tissue engineering methods for improving the quality of treatment for astronauts in long-term, future space missions [64], which is very useful in space health care. Yet, it still takes high expense to fabricate the setup. Also, limitations of the bioprinting setup to find the influence of microgravity on biological systems should be addressed. Moreover, growing organs and tissues in space should clear all ethical permissions and regulations [57].

As microgravity-assisted bioprinting is a very recently realized technology, printing in space or microgravity chamber is a costlier process because of which research is limited. In July 2019, the bio fabrication facility (BFF) could successfully print tissue constructs consisting of human heart cells as well as human menisci in the ISS Lab. Findings showed that microgravity facilitates printing by preventing the collapse of structures with holes, caverns, and tunnels and by allowing organs to develop without the need for scaffolding. Because there is no convection, buoyancy, or segmentation, it is simple to create 3D structures in microgravity. This allows for the use of a bio ink with low viscosity, and the cells and nutrients stay where they are placed [65].

The European Space Agency's Exploration Program recently granted Redwire Space NV a €14-million contract to design and develop the 3D-biosystem facility. It can be a novel way to learn about interactions between cells; develop vascularized tissues with the goal of developing transplantable, fully functional organs for tissue engineering; and create organ models for drug efficacy and toxicity testing [66]. In this program, NASA, Redwire Space NV, and the Uniformed Services University of the Health Sciences Center for Biotechnology (4DBio3) sent a modified BFF to the ISS in November 2022 with the goal of printing a human knee meniscus while in orbit [67].

14.9 In Vivo Bioprinting

All the strategies discussed belong to bioprinting which is done outside the living system. Here, printing and growth of cells are done by simulating the condition of living tissue, and the effectiveness is studied. The process still lacks a successful representation of the living tissue environment, and the proliferation trend cannot be relied upon. Also,the culture conditions are susceptible and highly prone to contamination. Perfect synchronization between thelaboratorycultured tissues and the implantation site should exist. To improve the quality of bioprinting in implantation applications, a new attempt is made to print directly into the defect, termed in vivo bioprinting or in situ bioprinting. It has been applied in skin, bone, and cartilage tissue engineering fields. The scaffolds are attached to existing tissues in the target site by in situ cross-linking to ensure proper fitting and mechanical stability. The main aim of in vivo bioprinting is to allow the proliferation of the cells in the host native environment so that it will not suffer rejection. It must be ensured that the in vivo-printed construct possesses proper shape fidelity and oxygen and nutrient transport to support proliferation. Contrary to in vitro bioprinting, where external stimuli can be applied for cross-linking and structural stability, in vivo bioprinting does not prefer external factors due to concern for the living tissues and cells in the host environment [68]. The selection of biomaterials for in vivo bioprinting is also important because the compatibility and cytotoxicity of the material determine the success of bio-printed tissues. Another factor to consider is the precision and deposition quality of the material at the target point. Often artificial intelligence and design softwares are used to take high-resolution images of the defect sites, and the printing pattern is decided with the help of CAD-CAM [68]. It has been reported that in situ bioprinting with microalga (*Chlorella pyrenoidosa*) undergoes oxygenic photosynthesis and provides oxygen under exposure to light, helping cellular functionalities [69]. Recently, multiple-axis bioprinting, up to six axes, has been introduced to achieve excellent printing resolution and geometrical and topographical accuracies of printed constructs as given in Figure 14.6.

Our body has excellent immune sensitivity and response. So, introducing an external body into the host environment and enabling the conditions for the growth of cells to match with native tissues is still a challenge. It provides only limited control over the stability and growth rate as safety concerns exist to not affect other living tissues. Therefore, the cost of implementation of in vivo bio-printed organ constructs is still out of the hands of economically weaker sections.

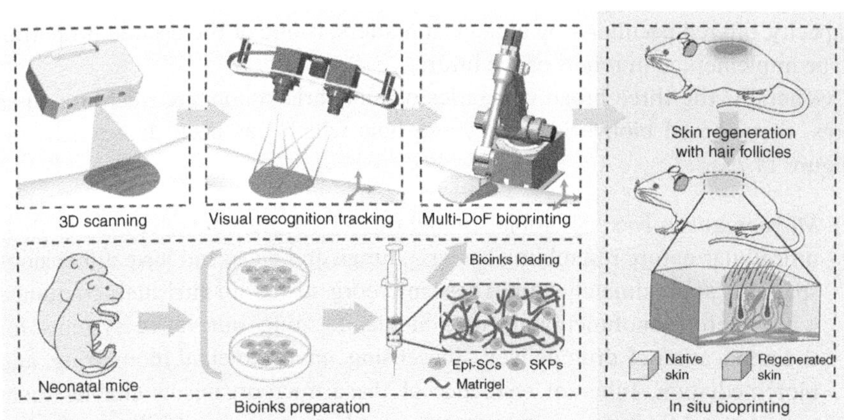

Figure 14.6 Multiple axis bioprinting process. *Source:* Zhao et al. [70] / John Wiley & Sons.

14.10 Biohybrid Robots

The biohybrid robot provides a platform to synergize the characteristics of biological entities (cells and tissues that can possess self-healing, adaptability, and sensing) and artificial materials (smart and soft) for obtaining more efficient and complex systems [71, 72]. Creating this class of robot primarily aims to utilize muscle cells for movement, replacing motors, gears, and cables. The robot employs proteins and nutrients to generate its internal energy, enabling it to function autonomously because synthetic materials do not have the ability to fully mimic the complexity of human muscle structures along with the force generated from the hierarchical arrangement of the tissue through self-generated energy. Also, the living materials hold promise for future and sustainable robotics due to intrinsic softness, higher environmental safety, compatibility, and more efficient energy conversion processes than traditional robots. Their greater flexibility allows them to move like living organisms to "interact with the human body" safely than machines. Inspired by biological systems, engineers created incredibly effective mobile robots that can operate in actual environments [73]. These robots are powered by artificial muscles that are capable of mimicking the performance of native muscle tissues, and these mobile robots are considered one of the most investigated applications in soft robotics [72]. In case of soft robots, advantages are substantial. Devices that use compounds found in their surroundings can produce electricity, such as Rossiter's ROWBOT. NASA is currently looking into the creation of soft robots to explore other solar systems [73]. As nature is the best designer, using natural tools in robotics can solve a lot of major issues, including damage tolerance, power consumption, and adaptability in complex, unstructured environments for real-world applications [74]. Research is ongoing in the field of biohybrid robots to make them fully autonomous, intelligent, and self-assembledRegenerative capacity, offered peculiarly by biological matters, is one of the desired properties to be implemented in robots of the future.

Generally, the three broad categories of biohybrid robots are microorganism bots, tissue-based biohybrid and cyborg bots which has been depicted in the Figure 14.7:

a) *Microorganism-bots*: Microbial biohybrid robots are characterized by their unicellular nature resembling bacteria, fungi, and algae and have dimensions spanning approximately 1 μm. These microorganisms are intricately combined with micro-/nanofunctional components, forming innovative systems for the application of drug delivery, biosensing, environmental monitoring, and bioremediation. Inherent resilience of these microorganisms enables them to withstand fluctuations in temperature, pH, nutrient availability, and other environmental conditions [75, 76].

b) *Tissue-based biohybrid bots*: Tissue-based biohybrid bots are the only hybrid bots that can be fabricated through 3D printing techniques using live cells and proliferate it in a laboratorycondition for growing it into a tissue.
   i) Human-induced pluripotent stem cells can be used to generate cardiomyocytes [77], which showcase outstanding contractile properties and have the capacity to undergo millions of actuation cycles without requiring rest to prevent fatigue [78].
   ii) Invertebrate tissues derived from mammals and birds, such as insect antennae [79] or sea slug muscle [80], can survive in natural settings and continue to function when cultured for an extended period of time.
   iii) Skeletal muscle tissues are of particular interest in robotics due to their effective contractile behavior, controllable beating, and self-regenerative ability [81].

c) *Cyborg*: This final class of biohybrid robots uses an animal's body as a scaffold to manipulate it robotically through artificial enhancement to accomplish a desired task, while maintaining control over its movement. Then, using outside stimuli, three modes of locomotion – walking/running, flying, and swimming – can be controlled.

Among all three bots, tissue-based biohybrid bots are our concern because of their 3D printability creating a room where these bots can be tailored to accommodate the situation and not the situation according to the bot. Hence, using this arrangement of muscle tissues, the robot can mimic the actions of a human organ or can be controlled externally. Naturally, the living tissues get contracted or stimulated due to the electrical signal generated by the brain; similarly, external stimulations, like electrical pulse, light, heat, and magnetic induction can be introduced to contract the biohybrid bots.

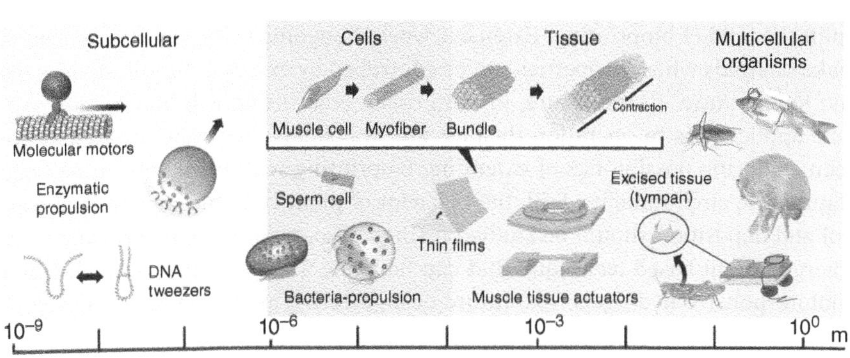

Figure 14.7 Multiscale biohybridization. *Source:* Filippi et al. [82] / PNAS.

The stimulations can be tethered or nontethered. To enable pulsed electrical stimulation in robotics, electrodes must be present in the medium or a tether must be utilized, restricting potential applications. Robot actuation can be manipulated by varying the stimulation's frequency and intensity. The electrodes can be integrated into the same robot skeleton or placed close to the robot [83, 84]. But as mentioned earlier, pulsed stimulation limits the control, therefore, more complex control can be achieved by harnessing the biological neurons' natural computation capabilities, which is extremely desirable to use the body's natural electrical stimulation from motor neurons to contract the muscles; this nontethered stimulation provides high degree of freedom to the bot. Recent research has shown that neuromuscular tissue circuits can be grown [85] or harvested [80] to create the fundamental motor systems needed for biohybrid robots. To fully realize these neuron–muscle systems' potential, precise fabrication along with modeling and control approaches is essential; therefore, 3D printing comes into the picture. Light pulses can stimulate genetically altered neurons, which then use those signals to detect external stimuli and send signals to motor neurons, which in turn activate the muscles. In other words, these are sentient machines that can sense their surroundings and make choices [73]. Sarkar et.al showed that the control over contractions of muscles can be done locally through optogenetically modified skeletal muscle cells by applying concentrated light pulses and potentially applied in drug screening platforms [86]. Recently, a team of researchers have developed a remote-controlled walking robot that gets power from lab-grown muscle tissue and light. It is composed of mouse muscle cells, soft 3D-printed scaffolds, and wireless LED control chips. The walking mode of the robot gets activated using LED lights, which stimulates the contraction of muscles [86].

14.11 Conclusion and Future Perspectives

In recent years, many innovative techniques and materials have been tried to uplift the field of bioprinting. Extensive work is ongoing using smart polymers to make scaffolds whose properties can be controlled by external stimuli, such as pH and temperature. In the future, in vitro organ bioprinting will be used for real-time applications in operation theaters and war fields. Recently, scientists have been exploring possibilities of extending bioprinting with plant cells to recreate plant tissue environments. Still, the behavior of plant cells and their spatial control and deposition should be studied further. Also, volumetric bioprinting is an emerging light-based technique that can fabricate 3D structures with cell-laden photoresponsive hydrogels. The future of bioprinting awaits successful development of highly complex organs and tissue constructs, which will be available on demand and replace defective ones.

References

1 ORGAN India Situation in India & Data. https://www.organindia.org/overviews/know-organ-donation/.

2 Barber, C. (2023). 3D-printed organs may soon be a reality. Looking ahead, we'll not need donor hearts, FORTUNE WELL. https://fortune.com/well/2023/02/15/3d-printed-organs-may-soon-be-a-reality/ (accessed 16 February 2023).

3 Global Data (2022). 3D-printed organs and their affordability. Medical Device Network. https://www.medicaldevice-network.com/comment/3d-printed-organs-affordability/ (accessed 15 June 2022).

4 Wood, I. (2023). Organ-on-a-chip device aids the study of thrombosis without animal models. Drug Target Review. https://www.drugtargetreview.com/news/109464/organ-on-a-chip-device-aids-the-study-of-thrombosis-without-animal-models/ (accessed 27 April 2023).

5 Wang, D., Xu, Y., Li, Q., and Turng, L.S. (2020). Artificial small-diameter blood vessels: materials, fabrication, surface modification, mechanical properties, and bioactive functionalities. *J. Mater. Chem. B* 8 (9): 1801–1822.

6 Jin, Q., Fu, Y., Zhang, G. et al. (2022). Nanofiber electrospinning combined with rotary bioprinting for fabricating small-diameter vessels with endothelium and smooth muscle. *Compos. Part B* 1 (234): 109691.

7 Carranza Fernandino, T., Uranga Gama, J., Irastorza, A. et al. Combination of 3D printing and electrospinning to develop chitin/gelatin/PVA scaffolds.

8 Fazal, F., Sanchez, F.J., Waqas, M. et al. (2021). A modified 3D printer as a hybrid bioprinting-electrospinning system for use in vascular tissue engineering applications. *Med. Eng. Phys.* 1 (94): 52–60.

9 Dai, X., Tian, X., Gu, S. et al. (2023). Hybrid biofabrication of neurosecretory structures as a model for neurosecretion. *Int. J. Bioprinting* 9 (2): 659.

10 Yu, Y.S., Ahn, C.B., Son, K.H., and Lee, J.W. (2021). Motility improvement of biomimetic trachea scaffold via hybrid 3D-bioprinting technology. *Polymers* 13 (6): 971.

11 Lee, S.J., Heo, D.N., Park, J.S. et al. (2015). Characterization and preparation of bio-tubular scaffolds for fabricating artificial vascular grafts by combining electrospinning and a 3D printing system. *Phys. Chem. Chem. Phys.* 17 (5): 2996–2999.

12 Mayoral, I., Bevilacqua, E., Gómez, G. et al. (2022). Tissue engineered in-vitro vascular patch fabrication using hybrid 3D printing and electrospinning. *Mater. Today Bio* 1 (14): 100252.

13 Meng, Q. and Hu, J. (2009). A review of shape memory polymer composites and blends. *Compos. A: Appl. Sci. Manuf.* 40 (11): 1661–1672.

14 Gao, B., Yang, Q., Zhao, X. et al. (2016). 4D bioprinting for biomedical applications. *Trends Biotechnol.* 34 (9): 746–756.

15 Cerbe, F., Sinapius, M., and Böl, M. (2022). Methodology for FDM 4D printing with thermo-responsive SMPs. *Mater. Today Proc.* https://doi.org/10.1016/j.matpr.2022.11.440.

16 Lendlein, A. and Kelch, S. (2002). Shape-memory polymers. *Angew. Chem. Int. Ed.* 41 (12): 2034–2057.

17 Xiao, Y.Y., Jiang, Z.C., Hou, J.B., and Zhao, Y. (2021). Desynchronized liquid crystalline network actuators with deformation reversal capability. *Nat. Commun.* 12 (1): 624.

18 Dutta, S. and Cohn, D. (2017). Temperature and pH responsive 3D printed scaffolds. *J. Mater. Chem. B* 5 (48): 9514–9521.

19 Stoychev, G., Puretskiy, N., and Ionov, L. (2011). Self-folding all-polymer thermoresponsive microcapsules. *Soft Matter* 7 (7): 3277–3279.

20 Jamal, M., Kadam, S.S., Xiao, R. et al. (2013). Bio-origami hydrogel scaffolds composed of photocrosslinked PEG bilayers. *Adv. Healthcare Mater.* 2 (8): 1142–1150.

21 Jeong, H.Y., Woo, B.H., Kim, N., and Jun, Y.C. (2020). Multicolor 4D printing of shape-memory polymers for light-induced selective heating and remote actuation. *Sci. Rep.* 10 (1): 6258.

22 Zhang, F., Wang, L., Zheng, Z. et al. (2019). Magnetic programming of 4D printed shape memory composite structures. *Compos. A Appl.Sci. Manuf.* 105571.

23 Akbari, S., Sakhaei, A.H., Kowsari, K. et al. (2018). Enhanced multimaterial 4D printing with active hinges. *Smart Mater. Struct.* 27 (6): 065027.

24 Symes, M.D., Kitson, P.J., Yan, J. et al. (2012). Integrated 3D-printed reactionware for chemical synthesis and analysis. *Nat. Chem.* 4 (5): 349–354.

25 Hu, W., Wang, Z., Xiao, Y. et al. (2019). Advances in crosslinking strategies of biomedical hydrogels. *Biomater. Sci.* 7 (3): 843–855.

26 Song, E., da Costa, T.H., and Choi, J.W. (2017). A chemiresistive glucose sensor fabricated by inkjet printing. *Microsyst. Technol.* 23: 3505–3511.

27 Podual, K., Doyle, F.J. III, and Peppas, N.A. (2000). Glucose-sensitivity of glucose oxidase-containing cationic copolymer hydrogels having poly (ethylene glycol) grafts. *J. Control. Release* 67 (1): 9–17.

28 Wang, X., Qin, X.H., Hu, C. et al. (2018). 3D printed enzymatically biodegradable soft helical microswimmers. *Adv. Funct. Mater.* 28 (45): 1804107.

29 Hanaphy, P. AI in 3D printing: accelerating new material discovery and qualification. 3D Printing Industry. https://3dprintingindustry.com/news/ai-in-3d-printing-accelerating-new-material-discovery-and-qualification-207924/ (accessed 21 July 2022).

30 Vasilev, C. (2022). How can AI overcome 3D printing defects? AZoM.Com. https://www.azom.com/article.aspx?ArticleID=21252 (accessed 24 January 2022).

31 Zhang, Y., Raza, A., Xue, Y.Q. et al. (2023). Water-responsive 4D printing based on self-assembly of hydrophobic protein "Zein" for the control of degradation rate and drug release. *Bioact. Mater.* 1 (23): 343–352.

32 Haleem, A. and Javaid, M. (2019). Expected applications of five-dimensional (5D) printing in the medical field. *Curr. Med. Res. Pract.* 9: 208.

33 Pinninti, R. and Devi, P. (2018). Review on the advancements to additive manufacturing-4D and 5D Printing. *Int. J. Mech. Prod. Eng. Res. Dev.* 8: 397–402.

34 Zeijderveld, J.V. (2018). 5D printing: a new branch of additive manufacturing. Sculpteoreport.

35 Cheng, Y., Fu, Y., Ma, L. et al. (2022). Rheology of edible food inks from 2D/3D/4D printing, and its role in future 5D/6D printing. *Food Hydrocoll.* 1 (132): 107855.

36 Gillaspie, E.A., Matsumoto, J.S., Morris, N.E. et al. (2016). From 3-dimensional printing to 5-dimensional printing: enhancing thoracic surgical planning and resection of complex tumors. *Ann. Thorac. Surg.* 101 (5): 1958–1962.

37 Georgantzinos, S.K., Giannopoulos, G.I., and Bakalis, P.A. (2021). Additive manufacturing for effective smart structures: the idea of 6D printing. *J. Comp. Sci.* 5 (5): 119.

38 Rogers, K. (2023). When we'll be able to 3D-print organs and who will be able to afford them, CNN Health. https://edition.cnn.com/2022/06/10/health/3d-printed-organs-bioprinting-life-itself-wellness-scn/index.html (accessed 10 March 2023).

39 Sreekala, P., Suresh, M., and Priyadarsini, S.L. (2020). 3D organ printing: review on operational challenges and constraints. *Mater. Today Proc.* 33 (7): 4703–4707.

40 Avery, S. (2023). Cardiopatch, a project that relies on 3D printing to treat heart problems, 3D natives. https://www.3dnatives.com/en/cardiopatch-project-relies-on-3d-printing-treat-heart-problems-020520236/ (accessed 2 May 2023).

41 Avery, S. (2023). A 3D printed cranial prosthesis was implanted in a patient. 3D natives. https://www.3dnatives.com/en/a-3d-printed-cranial-prosthesis-was-implanted-in-a-patient-070420236/ (accessed 7 April 2023).

42 Alksnė, M. (2023). 3D printed organs: the future of organ transplantation? Health news. https://healthnews.com/health-conditions/genetic-disorders/3d-printed-organs-the-future-of-organ-transplantation/ (accessed 23 March 2023).

43 Gold, K., Gaharwar, A.K., and Jain, A. (2019). Emerging trends in multiscale modeling of vascular pathophysiology: organ-on-a-chip and 3D printing. *Biomaterials* 1 (196): 2–17.

44 Yu, F. and Choudhury, D. (2019). Microfluidic bioprinting for organ-on-a-chip models. *Drug Discov. Today* 24 (6): 1248–1257.

45 Silvani, G., Basirun, C., Wu, H. et al. (2021). A 3D-bioprinted vascularized glioblastoma-on-a-chip for studying the impact of simulated microgravity as a novel pre-clinical approach in brain tumor therapy. *Adv. Ther.* 4 (11): 2100106.

46 Yan, J., Li, Z., Guo, J. et al. (2022). Organ-on-a-chip: a new tool for in vitro research. *Biosens. Bioelectron.* 15 (216): 114626.

47 Aazmi, A., Zhou, H., Li, Y. et al. (2022). Engineered vasculature for organ-on-a-chip systems. *Engineering* 1 (9): 131–147.

48 Yi, H.G., Lee, H., and Cho, D.W. (2017). 3D printing of organs-on-chips. *Bioengineering* 4 (1): 10.

49 Shrestha, J., Ghadiri, M., Shanmugavel, M. et al. (2019). A rapidly prototyped lung-on-a-chip model using 3D-printed molds. *Organs-on-a-Chip.* 1 (1): 100001.

50 Rahmani Dabbagh, S., Rezapour Sarabi, M., Birtek, M.T. et al. (2023). 3D bioprinted organ-on-chips. *Aggregate* 4 (1): e197.

51 Zhou, H., Liu, P., Gao, Z. et al. (2022). Simultaneous multimaterial multimethod bioprinting. *Bio-Des. Manuf.* 5 (3): 433–436.

52 Ashammakhi, N., Ahadian, S., Xu, C. et al. (2019). Bioinks and bioprinting technologies to make heterogeneous and biomimetic tissue constructs. *Mater. Today Bio* 1 (1): 100008.

53 Ruiz-Cantu, L., Gleadall, A., Faris, C. et al. (2020). Multi-material 3D bioprinting of porous constructs for cartilage regeneration. *Mater. Sci. Eng. C* 1 (109): 110578.

54 Betancourt, N. and Chen, X. (2022). Review of extrusion-based multi-material bioprinting processes. *Bioprinting* 1 (25): e00189.

55 Hassan, S., Gomez-Reyes, E., Enciso-Martinez, E. et al. (2022). Tunable and compartmentalized multimaterial bioprinting for complex living tissue constructs. *ACS Appl. Mater. Interfaces* 14 (46): 51602–51618.

56 Liu, W., Zhang, Y.S., Heinrich, M.A. et al. (2017). Rapid continuous multimaterial extrusion bioprinting. *Adv. Mat.* 29 (3): 1604630.

57 Khilnani, B., Leon, K., Pino, C. et al. (2023). Gravity-assisted 3D bioprinting: using gravity as a design input to build high aspect-ratio structures. *Bioprinting* 1 (32): e00277.

58 Rezapour Sarabi, M., Yetisen, A.K., and Tasoglu, S. (2023). Bioprinting in microgravity. *ACS Biomater. Sci. Eng.* 9: 3074.

59 Mochi, F., Scatena, E., Rodriguez, D. et al. (2022). Scaffold-based bone tissue engineering in microgravity: potential, concerns and implications. *npj Microgravity* 8 (1): 45.

60 Anil-Inevi, M., Sarigil, O., Kizilkaya, M. et al. (2020). Stem cell culture under simulated microgravity. *Cell Bio. Trans. Med.* 10: Stem Cells in Tissue Regeneration.: 105–132.

61 Sarabi, M.R., Yetisen, A.K., and Tasoglu, S. (2022). Magnetic levitation for space exploration. *Trends Biotechnol.* 40: 915.

62 Anil-Inevi, M., Yaman, S., Yildiz, A.A. et al. (2018). Biofabrication of in situ self-assembled 3D cell cultures in a weightlessness environment generated using magnetic levitation. *Sci. Rep.* 8 (1): 7239.

63 Parfenov, V.A., Khesuani, Y.D., Petrov, S.V. et al. (2020). Magnetic levitational bioassembly of 3D tissue construct in space. *Sci. Adv.* 6 (29): eaba4174.

64 Van Ombergen, A., Chalupa-Gantner, F., Chansoria, P. et al. (2023). 3D bioprinting in microgravity: opportunities, challenges, and possible applications in space. *Adv. Healthcare Mater.* 23: 2300443.

65 Listek, V. (2020). Bioprinting in microgravity: where do we stand? 3D print.com. https://3dprint.com/274114/bioprinting-in-microgravity-where-do-we-stand/ (accessed 22 October 2020).

66 Wakefield, E. (2023). Redwire space awarded €14 million to develop 3D-biosystem facility, Voxel Matters. https://www.voxelmatters.com/redwire-space-awarded-e14-million-to-develop-3d-biosystem-facility/ (accessed 11 May 2023).

67 Cowing, K. (2023). A 3D bioprinter will print a human meniscus on the ISS. SPACEREF*. https://spaceref.com/space-stations/a-3d-bioprinter-will-print-a-human-meniscus-on-the-iss/ (accessed 14 March 2023).

68 Singh, S., Choudhury, D., Yu, F. et al. (2020). In situ bioprinting–bioprinting from benchside to bedside? *Acta Biomater.* 1 (101): 14–25.

69 Wang, X., Yang, C., Yu, Y., and Zhao, Y. (2022). In situ 3D bioprinting living photosynthetic scaffolds for autotrophic wound healing. *Research* 2022: 9794745.

70 Zhao, W., Chen, H., Zhang, Y. et al. (2022). Adaptive multi-degree-of-freedom in situ bioprinting robot for hair-follicle-inclusive skin repair: a preliminary study conducted in mice. *Bioeng. Transl. Med.* 7 (3): e10303.

71 Appiah, C., Arndt, C., Siemsen, K. et al. (2019). Living materials herald a new era in soft robotics. *Adv. Mater.* 31 (36): 1807747.

72 Sun, L., Yu, Y., Chen, Z. et al. (2020). Biohybrid robotics with living cell actuation. *Chem. Soc. Rev.* 49 (12): 4043–4069.

73 Yanes, J. (2019). Biohybrid robots, the next step in the robotic revolution. OpenMind BBVA. https://www.bbvaopenmind.com/en/technology/robotics/biohybrid-robots-the-robotic-revolution/ (accessed 21 October 2019).

74 Yang, G.Z., Bellingham, J., Dupont, P.E. et al. (2018). The grand challenges of sicience robotics. *Sci. Robot.* 3 (14): eaar7650.

75 Martel, S., Tremblay, C.C., Ngakeng, S., and Langlois, G. (2006). Controlled manipulation and actuation of micro-objects with magnetotactic bacteria. *Appl. Phys. Lett.* 89: 233904.

76 Steager, E., Kim, C.-B., Patel, J. et al. (2007). Control of microfabricated structures powered by flagellated bacteria using phototaxis. *Appl. Phys. Lett.* 90: 263901.

77 Lee, K.Y., Park, S.J., Matthews, D.G. et al. (2022). An autonomously swimming biohybrid fish designed with human cardiac biophysics. *Science* 375 (6581): 639–647.

78 Kim, J., Park, J., Yang, S. et al. (2007). Establishment of a fabrication method for a long-term actuated hybrid cell robot. *Lab Chip* 7 (11): 1504–1508.

79 Terutsuki, D., Uchida, T., Fukui, C. et al. (2021). Electroantennography-based bio-hybrid odor-detecting drone using silkmoth antennae for odor source localization. *JoVE* 174: e62895.

80 Webster, V.A., Young, F.R., Patel, J.M. et al. (2017). 3D-printed biohybrid robots powered by neuromuscular tissue circuits from Aplysia californica. *Biomimetic and Biohybrid Systems: 6th International Conference, Living Machines 2017*, Stanford, CA, USA, (26–28 July 2017), Proceedings 6 2017 (pp. 475–486). Springer International Publishing.

81 Cvetkovic, C., Raman, R., Chan, V. et al. (2014). Three-dimensionally printed biological machines powered by skeletal muscle. *Proc. Natl. Acad. Sci.* 111 (28): 10125–10130.

82 Filippi, M., Yasa, O., Kamm, R.D. et al. (2022). Will microfluidics enable functionally integrated biohybrid robots? *Proc. Natl. Acad. Sci.* 119 (35): e2200741119.

83 Mestre, R., Patiño, T., Barceló, X. et al. (2019). Force modulation and adaptability of 3D-bioprinted biological actuators based on skeletal muscle tissue. *Adv. Mater. Technol.* 4 (2): 1800631.

84 Kim, Y., Pagan-Diaz, G., Gapinske, L. et al. (2020). Integration of graphene electrodes with 3D skeletal muscle tissue models. *Adv. Healthcare Mater.* 9 (4): 1901137.

85 Aydin, O., Zhang, X., Nuethong, S. et al. (2019). Neuromuscular actuation of biohybrid motile bots. *Proc. Natl. Acad. Sci.* 116 (40): 19841–19847.

86 Sakar, M.S., Neal, D., Boudou, T. et al. (2012). Formation and optogenetic control of engineered 3D skeletal muscle bioactuators. *Lab Chip* 12 (23): 4976–4985.

15

Nanomaterials for Designing Functional Properties of Bioinks

Laila Hussein[1], Mostafa Mabrouk[1], Mohamed G. Farahat[2], and Hanan H. Beherei[1]

[1] *Refractories, Ceramics and Building Materials Department, Advanced Materials, Technology and Mineral Resources Research Institute, National Research Centre, Cairo, Egypt*
[2] *Botany and Microbiology Department, Faculty of Science, Cairo University, Giza, Egypt*

15.1 3D-Bioprinting

The field of bioprinting is expanding and having an extraordinary influence on medical and pharmaceutical sciences [1]. In order to create bioengineered constructions, a procedure known as "bioprinting" involves printing both viable cells and biomaterials with a specified layer-by-layer stacking structure [2]. It provides excellent spatial precision for the implantation of cells, proteins, DNA, medication particles, growth factors, and biologically active particles to better direct the development of new tissue. With the development of physiologically accurate tissue constructs, models, and organs, as well as organs-on-a-chip models for medicine and pharmaceutics, this powerful technology seems to hold more potential [3]. A subset of traditional three-dimensional (3D) printing called 3D bioprinting has the potential to outperform current 3D-printed tissue engineering techniques. The difference between 3D bioprinting and the simple 3D printing of biomaterials entails the placement of cell-containing bioinks in predetermined patterns, where they combine and mature to form sizable tissues [4–6].

15.1.1 Bioinks for 3D-Bioprinting

3D bioprinting is a significant subfield of 3D printing that combines biology and 3D printing together [7]. In this field, biological organs or tissues are created using bioinks with the intention of closely resembling their natural counterparts in

3D Bioprinting from Lab to Industry, First Edition. Edited by Prosenjit Saha, Sabu Thomas, Jinku Kim, and Manojit Ghosh.

terms of form and functionality [8]. Bioink, a natural or synthetic biomaterial containing living cells, serves as the starting material of 3D bioprinting procedures [5]. As a result, choosing the right bioinks for 3D bioprinting is both a fundamental consideration and one of the largest obstacles [9].

15.1.2 Designing of Functional Bioinks

15.1.2.1 Desirable Properties of Bioinks

It is necessary for the appropriate bioink formulation to meet all the patient's needs. Printability, mechanical strength, degradation, and functionalizing ability are the major measures of material qualities. Bioactivity, cytocompatibility, and biocompatibility are the three main biological needs for tissue replacement. These biological needs along with material properties are the main requirements for an ideal bioink (Figure 15.1). Printability is the most crucial factor to consider when analyzing the qualities of a material. The two components of printability are (i) bioink formulation's processability and [10] print fidelity, which are related to the printed construct's mechanical strength and ability to retain a 3D structure after printing. Printability is influenced by crosslinking type, surface tension, and solution viscosity, which directly decide the quality of the produced structures. Viscosity is a key factor in the formulation of bioink because it has an impact on both print accuracy and cell encapsulation capability. The polymer solution's high viscosity makes it less likely to flow readily, and the printed structure shape will be preserved after printing. However, because they need greater pressures to flow, the gauge size and

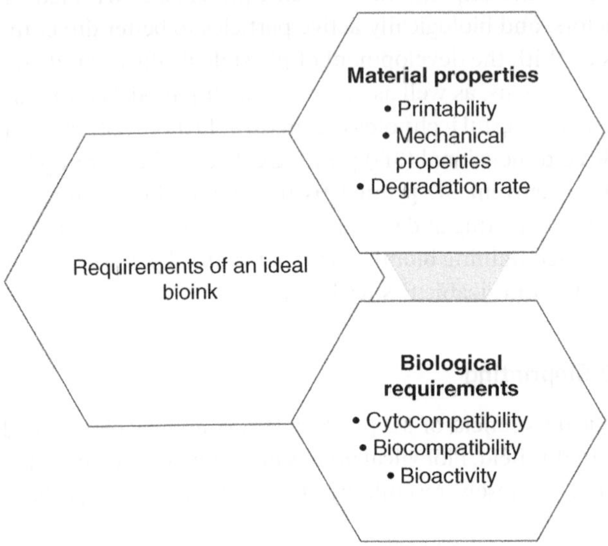

Figure 15.1 Desirable Properties of Bioinks.

smallest practicable print size are constrained for direct ink writing (DIW). In this context, Tirella et al. [11] used pressure-assisted microfabrication to evaluate the processing window for alginate hydrogels (DIW technique). They were successful in creating a 3D phase diagram that illustrates how bioink viscosity, print speed, and applied pressure interact to provide great print fidelity [11]. To work with various bioprinters, it is preferred that the bioink formulation has a configurable viscosity. For instance, the viscosity of bioinks for inkjet or droplet-based bioprinters is close to 10 mPa [12]. The range of bioink viscosities for extrusion-based DIW bioprinting is 30–6×107 mPas [13–15]. The bioink viscosity for laser-assisted bioprinting is between 1 and 300 mPas [13, 16]. The shear-thinning property is sought for high-viscosity bioinks used in extrusion- and droplet-based printing to make up for the significant shear stress brought on by the high viscosity [17].

The overall mechanics, or the feasible stiffness, is crucial for controlling and directing cellular function in addition to helping to build self-supporting structures. Degradation allows cells to gradually replace the printed construct with their own extracellular matrix (ECM), which is crucial for the effective integration of the implanted scaffolds. The bioink and the breakdown products should not contain substances that cause an inflammatory host interaction. Incorporating biochemical cues, or bioactivity, to control cellular behavior such as adhesion, migration, and differentiation requires functionalizability [17].

15.1.2.2 Limitations of Bioinks

Several obstacles still exist in this developing sector, although the achieved advancements in the fabrication of new bioink materials, the creation of new bioinks is a crucial breakthrough in materials technology. The major goals of 3D bioprinting are to increase cell–cell connections and decrease cell loss, vascularize large-scale tissue constructs, improve mechanical features, and develop a thorough understanding of new substances for sustaining 3D bioprinted constructs. The main problem of 3D bioprinting is the simulation of biological characteristics, such as mimicking the ECM, cell proliferation, and growth, alongside gel swelling, printability, shape retention over an extended period of time or until tissue regeneration, adhesion of consecutive layers, and their maximum number. Combinations of polymers and other crosslinking processes are being used to address these issues [18].

15.2 Designing Functional Bioinks Using Nanoscale Biomaterials

The incorporation of nano-biomaterials into bioink results in colloidal mediums, which enhance the characteristics of hydrogels including their structural stability, printability, and biological features [19, 20]. The network formed by the covalently or physically crosslinked nanomaterials or nanostructures provides the nanocomposite bioink with its properties. For example, the network can provide the bioink with mechanical strength, shape stability, and the ability to hold cells in place. Customization of the bioinks' characteristics is made possible by the variation of nano-biomaterials [6, 21]. Functional nanocomposite bioinks are employed in different fields of tissue engineering to give strength, stability, and bioactivity to the printed matrix. Nanocomposite bioinks also enhance the qualities of electrical, thermal, optical, and magnetic fields [22]. The compatibility and bioactivity of biomaterials toward the system is one of the vital criteria upon which nano-biomaterials are selected as a component of the bioink. The suspension ability in the ink system is just as important as biocompatibility with the cells in the bioink and the implanted biological environment. Although the primary goal of using nano-biomaterials is to increase functional applications, structural stability like facilitating cell-to-cell communication, promoting ECM formation, and improving cell proliferation are also important [23]. Researchers have highlighted that stem cells can differentiate into different cell types when they are bioprinted with nanocomposite bioinks.

15.2.1 Types of Nano-Biomaterials

Both organic and inorganic sources of nano-biomaterials are employed to create nanocomposite bioinks (Figure 15.2). Because they are bioactive, ceramics like silicates and calcium phosphates were commonly used as nanofillers for 3D scaffolds. Carbon-based nanomaterials (graphene and carbon nanotubes (CNTs)) are

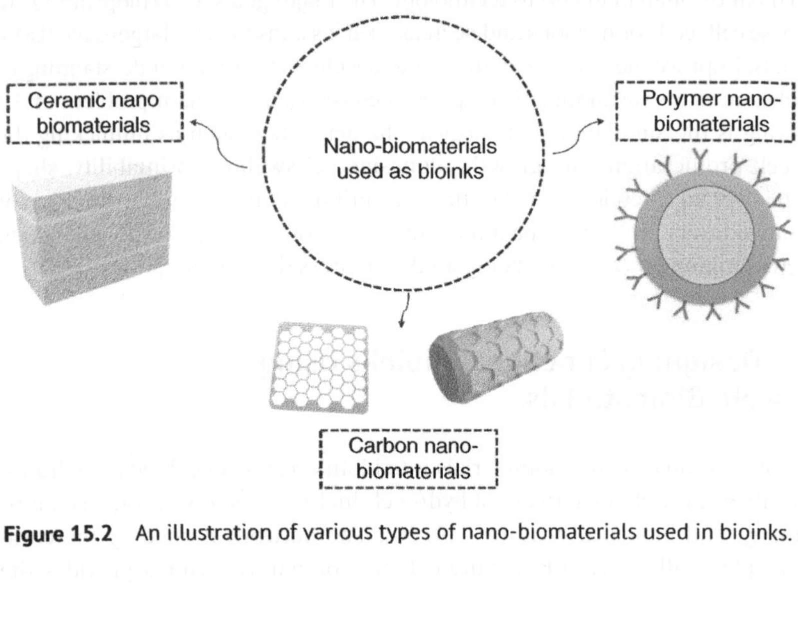

Figure 15.2 An illustration of various types of nano-biomaterials used in bioinks.

being studied because of their excellent mechanical properties and ability to conduct electricity, which can improve communication between the cells of the printed structure. The most common nanomaterials employed in the area of polymer nano-biomaterials are nanocellulose and its derivatives. Recently, bioink was reinforced with collagen nanofibers made from decellularized tissue for controlled delivery of the biomolecules to the cells. Polymeric nanostructures, such as liposomes, micelles, nanogels, and core–shell, were frequently utilized [24]. Researchers have also studied the combination of nanomaterials to achieve the desired multifunctional properties [25, 26].

15.2.1.1 Ceramic NPs (CNs)

Ceramics are excellent options for complex tissue repair and regeneration due to their remarkable biocompatibility and osteoconductivity, notably for the restoration of bone gaps and defects. Bioactive glasses (BGs) and calcium phosphates are among the ceramic NPs that have been utilized for 3D bioprinting [27–33]. During bone tissue engineering (BTE), CNs are introduced into cell-laden hydrogels for sustained structural support and guided osteogenesis because they resemble natural bone matrix and degrade more slowly than hydrogels.

Bioactive Glass Nanoparticles

Biodegradable BGNPs are promising materials for use in BTE. These nanoparticles have several advantages, including their customized composition, controlled degradability in the body, and ions that can induce cellular responses in bone and other tissues [34]. Alginate dialdehyde-gelatin (ADA/GEL) was impregnated with BGNPs and BGNPsSr to improve cell adhesion and proliferation. This enhances the formation of apatite-like precipitates on the surfaces of these scaffolds after incubation in simulated body fluid (SBF) that resembles bone formation. The strontium ions released from the BGNPsSr were known to accelerate bone regeneration [35]. In addition, the gelation period of the bioink was shortened significantly by adding more BGNPs.

Moreover, to promote the growth of bone tissue, poly(ethylene glycol) dimethacrylate (PEGDMA) and human mesenchymal stem cells (hMSCs) generated from bone marrow were used to 3D print BG and HAp NPs [36]. The PEG/HAp group had the highest viability after 21 days of cultivation and the scaffolds had improved mechanical features. The PEG/BG group had the lowest percentage of hMSCs that were viable, suggesting that BG may be harmful to cells. Additionally, RUNX2 was the highest showing that hMSC differentiation in PEG/BG was slower than in PEG/HAp. The ALP activity and ECM in hMSCs printed with PEG/HAp were all increased in the PEG/HAp group. Therefore, in bioprinted bone constructions, HAp encouraged hMSCs osteogenesis more so than BG [27].

Calcium Phosphate Nanoparticles

A class of substances known as calcium phosphate (CaP) [37] contains Ca^{2+} and PO_4-^3. CaP has been widely employed for tissue regeneration applications because it has excellent osteoconductivity and an inorganic phase that is similar to genuine bone. This section introduces the uses of various CaP types as bioinks for 3D-bioprinting, including HAp [30–33] and beta-tricalcium phosphate (β-TCP) [29]. Another reinforcement agent is β-TCP, which causes hASC differentiation without the need for an osteogenic media. A porous 3D scaffold for bone tissue regeneration was developed using collagen along with β-TCP composite bioink loaded with hASCs. After adding β-TCP to the bioinks composite, the collagen fibrils were disturbed, which decreased the gelation point. The 20% β-TCP displayed equivalent printability and cell survivability to pure collagen cell-laden structure. Bioinks with more than 20% of β-TCP have a higher viscosity and cause high tissue damage in addition to causing the nozzle to have high shear stress. Even though there was an increase in the compressive modulus of the structure resulting from the printing of the 20% weight β-TCP, it was still insufficient to be used in the load-bearing zone. The 20% β-TCP scaffold enhanced cell mineralization more than pure collagen without being cultured in an osteogenic media. Additionally, compared to controls, it displayed a greater level of osteogenic markers. An induced microenvironment for cell differentiation was produced as a result of β-TCP being released into the medium, leading to the increased osteogenic differentiation of hASCs. As a result, collagen/β-TCP-based composites allowed the possibility of bone regeneration in areas of minimal load up and also as a temporary biological replacement [29].

Cell–Material Interaction

BGs and CaPs are appropriate as bone substitute materials because they share chemical similarities with inorganic matrix of bones. Because of their biodegradability, osteoconductivity, and bioactivity, they have undergone substantial research [38]. Nanotechnology has led to the synthesis of nano-ceramics with a greater surface area (SA): volume and multi-functionality [36]. Bioinks containing cells were supplemented with ceramic-based nanoparticles to encourage osteogenesis and rebuild bone and cartilage. During bone and cartilage renewal, there is a complicated interplay between cells, substances, and the microenvironment. The characteristics of scaffold materials, such as their morphology, size, surface chemistry, and chemical composition, can affect cell growth, differentiation, and blood vessel development. BGNPs are a type of nanoparticle that can be absorbed by cells and might contain Ca, P, Na, Si, Mg, K, and O. The Ca and Si that are dissolved from the BGNPs have the ability to enhance osteoblast growth without affecting the cellular viability. The number of cells that reached the G2/M phase in the cell cycle as well as the cells that expressed higher levels of cyclin B1 and E proteins were higher when exposed to higher calcium levels of BNGPs [39]. The expression of one of the osteoblastic-specific miRNAs, miR-30c, was enhanced by the BGNPs, which then decreased the expression of TGIF2 and HDAC4 and increased the expression of RUNX2 [40]. The dissolution products were discovered to increase angiogenesis and production of VEGF both in vitro and in vivo [41, 42].

N-CaP are naturally absorbing ceramics, and their ionic exchange kinetics and protein absorption are controlled by their content, surface chemistry, and topographical features. This has an impact on cellular functioning [43]. When nano-CaP is exposed to body fluids, it releases calcium and phosphate ions, which help in the formation of apatite crystals on the scaffolds' surfaces. These crystals provide a suitable surface for cells to attach, and the cells spread out in a healthy, polygonal shape. The nano-apatite structure encourages the growth of stem cells [44]. Nano-CaP enhanced the development of osteogenic cells by increasing the expression of osteogenic genes, such as BMP2 and RUNX2. These genes are responsible for the production of ECM components and biomineralization, which are essential for bone formation. [36, 44].

The nanoscale features of BGNPs and nano-CaP, such as their structure, crystallinity, and roughness, influence how cells interact with them. First, the large SA of nanostructured BGNPs and nano-CaP provides additional binding sites for cells to adhere to. Additionally, these nanoparticles can absorb and release osteogenic growth factors (OGFs), which can stimulate osteogenesis [44]. Second, there are nanometric interactions between BGNPs and N-CaP, biological molecules, and cells. CaP's nano-crystallinity, BGNPs CaP's, and nano topography, which attach cell membrane receptors, activate pathways involved in osteogenesis, and ultimately control cell proliferation and differentiation, have a direct impact on how cells behave [36, 44].

15.2.1.2 Carbon Nano-Biomaterials

Carbon-based nanomaterials that possess low dimensions are composed of carbon atoms organized in a specified pattern. Fullerenes, CNTs, graphene, nanodiamonds, and carbon dots are among these materials. Nanodiamonds have sp3-hybridized carbon, while carbon dots have a mix of sp2- and sp3-hybridized carbon atoms. Fullerene, CNTs, and graphene are mostly made up of sp2 carbon atoms [45]. Because of their large SA, specialized visual characteristics, thermoelectric conductivity, and high mechanical strength, carbonaceous nanoparticles have been widely used in biomedical applications such as drug administration, reinforcing additives, and cellular sensors. Carbon nanoparticles are commonly used in biomedical applications [46].

Carbon Nanotubes

CNTs are categorized according to the number of their layers into single-walled CNTs (SWCNTs) or multi-walled CNTs (MWCNTs). SWCNTs have a high aspect ratio and are turned over on themselves to form a cylinder. They are composed of thick single-atom sheets of carbon atoms linked by sp2 bonds and organized in hexadic structural units. MWCNTs are roll-up structures made up of numerous concentric hexagonal structural elements. SWCNTs have lengths of up to 1 m and diameters ranging between 0.4 and 2 nm. MWCNTs have lengths of several micrometers with diameters ranging from 2 to 100 nm [47]. They have been employed in 3D bioprinting of flexible electronics, specifically designing encapsulated cells, vasculature, and cardiac tissue constructs, because of their unique structure and characteristics [48].

Single-walled CNTs Electronic devices that are used in biomedical applications need to be stretchable, foldable, bendable, and biocompatible. Therefore, they should possess an intimate and conformal contact with tissues and adapt to the tissue. Inks derived from CNTs are attractive for the creation of electronics with some degree of flexibility because they offer excellent electromechanical conductivity. DNA was used to create a suspension of SWCNTs in HA to reduce toxicity and overcome the lack of cell-binding sites of CNT-based inks. The large SA of SWCNTs allowed them to interact with DNA and HA chains simultaneously, resulting in the development of long-lasting fibers. The fibers had exceptional mechanical strength. The electrical conductivity of the fibers was greater than that of chitosan/SWCNTs fibers containing 0.3% SWCNTs and DNA/SWCNTs fibers with 0.4% SWCNTs. Cardiomyocyte-infused GelMA hydrogel was created by stacking the fibers in three dimensions. Within the conductive hydrogels, the heart cells matured to a significant degree. During the 10 days of cell culturing, no cytotoxic effect was observed after using the HA- and DNA-coated SWCNTs [49]. The obtained CNT hybrid materials are suitable for functionalized 3D-printed scaffolds as well as foldable and flexible biosensors [48].

Multi-walled CNTs Coaxial bio-printing technique can be used to generate vascular conduits, this could be done b using MWSCNTs/alginate bioink [50, 51]. A successful perfusion was possible when a vascular conduit was directly printed with smooth, well-defined walls. The presence of MWCNTs raised the tensile stress as well as the elastic modulus while lowering the strain without a significant difference between MWCNT concentrations. The strong adhesion between the interfaces of MWCNTs and the alginate matrix may be the reason for the enhancement of mechanical characteristics. For creating blood arteries or adding extra vasculatures, the reinforced vascular conduits were helpful. It was noticed that there was no discernible difference between the control and MWCNT-reinforced scaffolds in terms of cell viability. Cells were damaged during the printing process, which reduced their vitality after printing. Even though cells were recovered after three days of in vitro cell culture, only a small percentage of cells have the ability to endure and operate properly in long-term culture. When human coronary artery smooth muscle cells (HCASMCs) were used in MWCNTs-coated tubes, they suffered damage and eventually died after six weeks [50, 51].

A promising method for treating myocardial infarction was the bioprinting of cardiac patches that encourage cell growth [52]. Cardiac patches were created with intricate characteristics to closely resemble the myocardium's ECM for quick tissue integration. In the meanwhile, they ought to be conductive, elastic, stable, bioactive, and vascular. For this, a hybrid cardiac patch was 3D printed that contained HCAECs. COOH was used for the functionalization of MWCNTs, and then it was added to methacrylated collagen (MeCol) and an alginate matrix in order to lessen their cytotoxicity. The viscoelastic behavior and electrical conductivity of MeCol were enhanced by a highly interconnected nanofibrous network of the functionalized MWCNTs. The storage modulus and compressive elastic modulus of MeCol and alginate were improved by the addition of MWCNTs. Additionally, within seven to 10 days in vitro, considerable cell proliferation, migration, and differentiation with lumen-like development were seen [52]. Uncertainty remained regarding the signal transduction process at the cardiomyocyte/MWCNT contact. The electrophysiological situation in vivo could not be mimicked by the patch's electrical activity [48].

Graphene and Graphene Oxide

G is a mono-carbon-atom-thick sheet with a large SA, remarkable mechanical flexibility, and the potential for chemical functionalization, which has attracted a great deal of interest in the biological field [46]. Graphene is a single layer of carbon atoms that are arranged in a hexagonal lattice. This two-dimensional sp2 hybridization gives graphene its unique properties, such as high electrical conductivity and mechanical strength. GO is a derivative of graphene that is created by the oxidation of graphite. GO has a number of chemical groups, including hydroxyl, carboxylic, and epoxy, which allow it to interact with a variety of molecules [53]. The size spans from a few nanometers to one nanometer. G and GO have been used as bioink additives for the tissue engineering of neurons, bones, and cartilage [48].

Graphene Murine neural stem cells (NSCs) were added to polyurethane along with graphene stabilized with pluronic (G-P) to create a bioink for the 3D printing of neural tissue constructions [10]. Higher rate of cell survival was assigned to polyurethane/G-P scaffolds. After seven days of culture, a minor addition of graphene (25 ppm) increased cell survival and multiplication, for the abovementioned specimen and a 150% rise for pure polyurethane. From day 7 until day 14, cell viability was preserved in polyurethane/G-P, while it deteriorated in pure polyurethane. In comparison to polyurethane, the oxygen consumption rate (OCR) in PU/G-P was double. About 3.5 times more ATP was produced, and non-mitochondrial-related oxygen was consumed in polyurethane/G-P than in pure polyurethane. The sp2 structure of graphene, which offers a stable structure for free electrons, might be related to the modulation effect. This in turn indicates that graphene had a stimulating effect on the neural differentiation of NSCs. In terms of protein expression, the expression of tubulin and GFAP was also noted [54]. The development of polyurethane/G-P composite bioink that contains neural cells offered an appropriate method for bioprinting neural tissue structures.

In the presence or absence of graphene, there was no discernible cytotoxicity when NSCs were cultivated on a 2D surface. The addition of graphene to hydrogels that contained cells resulted in a minor slowing of cell proliferation. NSCs enclosed in hydrogels showed neuron development and neurite elongation after two weeks of culture, showing promise for nerve repair. However, there was no difference in the expression of the neuronal markers GFAP and beta-tubulin III with or without graphene [55].

Graphene Oxide To create nerve tissue, GO was combined with PU and NSCs. When compared to the unchanged amount of G-P, GO was more hazardous to NSCs, resulting in reduced cell viability and survival rates. Although there were no appreciable variations in the amounts of beta-tubulin and GFAP gene expression for PU/GO and PU, it was still possible to clearly see their expression in the proteins. NSCs in PU, on the other hand, showed no expression. It may be said that the addition of GO to PU hydrogel had a minimally positive influence on NSCs' ability to differentiate into neural tissue [54].

On the other hand, it was reported that alginate/hMSCs bioink was enhanced with the presence of GO that promoted bone renewal. It was found to act as a protection for hMSCs from oxidative stress due to the presence of alginate/GO coating. GO was able to counterbalance harmful ROS and improve cell survival signals, thanks to its ability to bind proteins and its antioxidant properties. The presence of GO in the scaffolds enhanced the differentiation of osteocytes after seven days, which resulted in high deposition of calcium, greater ALP activity, and also higher expression of markers involved in osteogenesis (ALP, RUNX2, and OPN). Alizarin staining was used extensively to show that the hMSC-filled alginate/GO scaffolds had mineralized. The addition of 0.5 mg to each 1 ml of GO in 3% (w : v) alginate improved the hMSCs' printability, stability, and ability to induce osteogenesis for BTE applications [53].

Given that, these alginate concentrations impede cell spreading and reduce osteogenic differentiation, it was proposed that a low-density Alg/(Gel) hydrogel with an optimal GO content would increase scaffold integrity and boost hMSC osteogenic differentiation. hMSCs and GO (0, 0.5, 1, 2 mg/ml) were mixed with an Alg/(Gel) hydrogel (0.8/4.1% (w:v)). The addition of GO in a concentration-dependent manner improved the viscosities and shear-thinning properties. After the initial shear sequence, 1GO (1 mg/ml) showed a shear recovery rate of 97.09%, compared to 0GO's 37.39%, 0.5GO's 79.55%, and 2GO's 51.08%. After one day of culture, the compressive moduli of 1GO and 2GO scaffolds were much higher than those of 0GO. In all groups, the presence of GO increased cell spread. GO has a considerable adsorption capacity that aided in the promotion of early cell proliferation due to stacking and electrostatic interactions. With 1GO being the greatest, GO induced the differentiation of osteocytes of hMSCs with elevated osteogenic gene expression. Due to the harmful ROS produced by oxidative stress, 2GO prevented osteogenic cell development. Mineral deposition was improved by the inclusion of GO, with 1GO reaching its greatest level after 42 days of culture. The potential uses in BTE were proven by 3D bioprinted 1GO scaffolds with Alg/Gel [55]

The procedure used HRP enzyme-mediated crosslinking to create hydroxyphenyl propionic acid conjugated with gelatin (GHPA) hydrogels containing GO (GO/GHPA) to provide a favorable environment for the growth and differentiation of myoblasts [56]. When GO was added, it boosted molecular interaction and enhanced hydrogel heat stability. Constructs made from 3D-printed C2C12 myoblasts maintained good fidelity and enabled cell spreading and proliferation. Within a 3D-printed construct, multinucleated cells displayed mature myotubes. C2C12 cells spontaneously differentiated into myoblasts as a result of GO's cytoskeleton reorganization, actin filament aggregation, and alignment, alongside upregulation of muscle-specific proteins. Additionally, the shear stress experienced during 3D printing helped to align polymer fibers and cells, which was beneficial for myogenesis. Cell-filled GO/GHPA constructions created using 3D printing technology may be used as therapeutic scaffolds for the creation of skeletal muscle tissue [56].

GO has a high drug-loading capacity due to its large specific SA and robust ion-exchange ability. After being loaded into GO, BMP7 was suspended in CS and Col solution. In comparison to a hydrogel, BMP7 was released from GO/Col/CS quicker and was thereafter sustained for at least 30 days. The mechanical strength and rate of degradation of 3D-printed scaffolds were unaffected by the addition of GO. After 96 hours, GO/Col/CS had higher cell viability than the other groups. With sp2 structure, GO could interact with cell membranes by binding with peptides' helical domains and affecting cell metabolism and signaling pathways. GO could block the Rank/Rankl/OPG pathway that suppresses an inflammatory response with reduced inflammatory factor NF-B [57].

15.2.1.3 Polymer Nano-Biomaterials

Cellulose and Its Derivatives

A wide variety of polysaccharides, such as pectins, cellulose, and pectins, have drawn significant scientific attention for tissue engineering because they mimic the glycan component of natural ECM [58–61]. Among these, cellulose stands out due to its value as a hydrogel matrix and its inclusion of nano reinforcements [62]. Cellulose acetate and methyl cellulose are utilized to make printable ink, whereas nanocrystals are employed to improve the functional properties of various hydrogels and extrudable polymers [63]. The FDA has approved methylcellulose as a sedentary ingredient [64]. When combined with alginate hydrogel, carboxymethyl cellulose demonstrated good printability and form retention along with great cell survival [65]. In another study, a combination of methyl cellulose and alginate at 3% and 9% ($^{w}/_{v}$) each showed good extrusion, shape retention, and thixotropic properties [66]. With the right solidification process and drying circumstances, the cellulose nanofibril hydrogel may produce a very stable 3D structure even in dried conditions [67]. Commercially available nano-fibrillated cellulose (NFC) (2% $^{w}/_{w}$) and sodium alginate (0.5% $^{w}/_{w}$) are used to create the biocompatible 3D bioink product CELLINK® [68]. Pure alginate hydrogel's printability and shape fidelity can be greatly enhanced by NFC when printing intricate structures like the ear or meniscus [69]. Surface modification of NFC was done to alleviate the loading restriction brought on by its extensively crosslinked structure [70]. Bioink made of dialdehyde cellulose nanocrystals (DAC) and gelatin demonstrated low swelling ratio, good shape consistency, print accuracy, and cell viability while having a compression breaking strength that was more than 41 times higher (1.695 MPa) than control gelatin (0.041 MPa) [71].

Moreover, glycerin was used by Leppiniemi et al. to preserve the structure of the alginate/nanocellulose mixture after printing [10]. Du et al. define cellulose nanocrystals and cellulose nanofibril-coated hydrogels as the two key application sectors, with drug delivery and 3D bioprinting being the two main application applications. Because cellulose cannot be entirely removed by the biological system of the animals, the data suggest long-term inflammatory reactions and dose-dependent cytotoxicity. The majority of experiments utilizing cellulose nanoparticles indicate neither biotoxicity nor genotoxicity. Du et al. define cellulose nanocrystals and cellulose nanofibril-coated hydrogels as the two key application sectors, with drug delivery and 3D bioprinting being the two main application applications. Because cellulose cannot be entirely removed by the biological system of the animals, the data suggest long-term inflammatory reactions and dose-dependent cytotoxicity. The majority of experiments utilizing cellulose nanoparticles indicate neither biotoxicity nor genotoxicity [72]. The production of nanocellulose may be done in a variety of ways. However,

TEMPO-oxidized nanocellulose, on the other hand, has reduced endotoxin contamination (lipopolysaccharides), which reduces inflammatory reactions [73].

Collagen Nanofibers

In decellularized ECM, the main protein of the fibrous matrix is collagen [74]. Self-assembled collagen nanofibers are used to make filament lattice structures for 3D printing [75]. These structures have good mechanical properties and excellent biocompatibility, even with 95% porosity with adequate mechanical features [76]. Collagen-based 3D structures guarantee prolonged biomolecule delivery for stable cell development and closely resemble the extracellular architecture [77]. Bioink made of collagen nanofibers was used to bioprint parenchymal vascularized heart tissue. Parenchymal vascularized heart tissue is tissue that has both muscle cells and blood vessels. The printed blood artery network in the human heart is still in its early stages of development and requires more accuracy in imaging and printing [78].

15.2.2 Smart Multifunctional Nanobioinks

Therapeutic compounds have been delivered to tissues using a variety of forms, such as nanoparticles, nanofibers, microspheres, fillers, and films. These forms have the ability to stimulate cell growth, affect cell differentiation and proliferation, and regulate the secretion of ECM [79]. Examples of drugs that can promote bone growth include statins [80], osteoprotegerin [81], v3 integrin antagonists [82], cathepsin K inhibitors [83], parathyroid hormone [84], transforming growth factor, and bone morphogenetic protein (BMP) [85]. The development of blood vessels is regulated by a number of growth factors, including vascular endothelial growth factors (VEGFs), fibroblast growth factors (FGFs), hepatocyte growth factors (HGFs), and platelet-derived growth factors (PDGFs) [79]. When 3D bioprinting functional tissues, it is important to assess and improve the biophysical and biochemical properties of the hydrogel. These properties have a significant impact on the behavior and functionality of the tissue [86]. The overall quality of the printed constructions is significantly influenced by the bioink selection. Hydrogel-based bioinks, for example, can function as both an architectural substrate for manufactured tissue and an environment for encapsulated cells, empowering them to regulate how they function [87]. Drug-releasing particles identified in composite bioinks are referred to as "smart bioinks" with an emphasis on bioinks with controlled drug release. Numerous studies have proven that biomolecules such as medications and growth hormones may be administered locally [88].

Research activities focused on atomistic and molecular-level concepts and characteristics are referred to as nanotechnology [89]. Constructing items or materials smaller than 100 nm is the goal of nanotechnology [90]. As shown in (Figure 15.3),

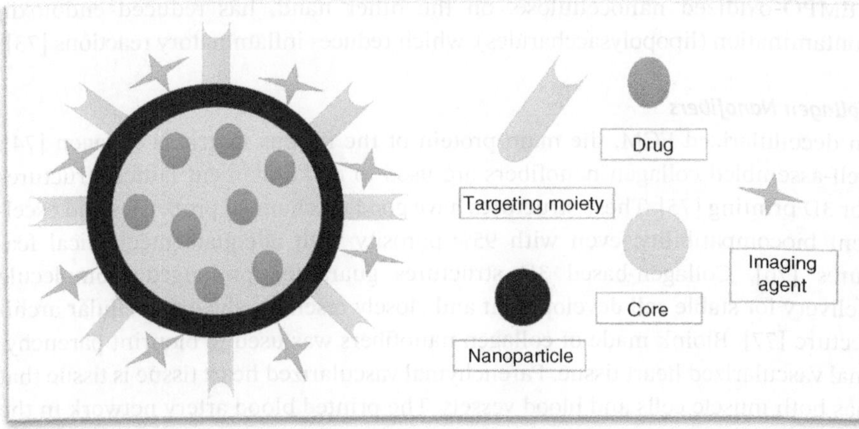

Figure 15.3 A smart multifunctional drug-loaded nanoparticle is a representation of a nanoparticle that is designed to deliver drugs to specific cells or tissues in the body and can also be used to image tissues or to deliver genes to cells. *Source:* Adapted from Maan et al. [88].

these nanomaterials can be made up of nanoparticles, which, in contrast to very thin particles of natural origin, are manufactured on purpose. At the nanoscale, individual molecules and their interactions become more important than the bulk macroscopic properties of a material or device. This is because the macroscopic physical and chemical features are controlled by their basic molecular structure [91]. Heid and Boccaccini investigated composite hydrogel bioinks for 3D bioprinting that contain bioreactive inorganic fillers (BIFs) [92]. Isotropic and anisotropic silicates, such as BGs and nanoclays, as well as calcium phosphates, such as HA, can induce in situ crosslinking effects and promote or direct cell line differentiation, which are BIFs. HA gives the matrix extra functionality and in situ crosslinking effects. Most BIFs can promote cell adhesion and proliferation due to their diverse chemical compositions, and some of them can also affect the differentiation of stem cells, progenitor cells, or cell lines. For instance, bioceramics, chondrogenic cells, angiogenic cells, adipogenic cells, silica nanoparticles, and CNTs all have an impact on osteogenic cells and neurogenic responses [92].

Gungor-Ozkerim et al. provide a comprehensive overview of many bioinks that are currently used for bioprinting, such as tri-composite bioinks that contain bioactive compounds [93]. To create printable semi-synthetic extracellular matrix (sECM) hydrogels, they used gelatin, gold nanoparticles (AuNPs), and thiol-modified biomacromonomers derived from hyaluronic acid (HA) as dynamically crosslinkable components to form AuNP-sECMs, a hydrogel with covalent intra- and inter-gel connections. NIH 3T3 fibroblasts were encapsulated in AuNP-sECMs to 3D bioprint tubular tissue constructs. Over the course of four weeks, the cells multiplied and secreted ECM, which made the cellular rings opaque. Positive control tissues showed a similar staining pattern. However, the collagen content of the tissue structures suggests that the bioprinted cells changed their environment [94]. Similarly, Zeng et al. [95] developed a thermosensitive injectable hydrogel composed of glycerophosphate-bound chitosan (CGP) and dopamine-modified alginate (Alg-DA). Gold nanorods (AuNRs) coated with polydopamine (AuNR-PDA) were integrated into the hydrogel through mussel-mimetic adhesion. Because of the strong contacts between PDA and polymers, AuNR-PDA could be fixed uniformly and securely in the composite hydrogel, preventing overheating or leaking. In vitro cytotoxicity tests showed that the composite hydrogel was biocompatible with normal cells and mouse fibroblasts, but it suppressed human hepatocellular carcinoma (HepG2) cells. In vivo antitumor tests (PTT) showed that the CGP/Alg-DA/AuNR composite hydrogel significantly prevented tumor formation under a variety of photothermal therapies [95].

Gao et al. co-printed bone marrow-derived hMSCs alongside nanoparticles of BG, hydroxyapatite (HAp), or the two suspended in a PEGDMA scaffold that was synthesized concurrently to stimulate osteogenesis [27]. After 21 days in culture, hMSCs that interacted with HAp showed the highest cell survival and compressive modulus compared to the other groups. Biochemical analyses revealed that the PEG-HAp scaffold exhibited the highest levels of collagen deposition, ALP activity, and collagen production. Therefore, HAp nanoparticles promoted hMSC osteogenesis in bioprinted tissues better than BG [27]. In order to assemble nano-hydroxyapatite (nHAp) and human osteoprogenitors (HOPs) in two and three dimensions in a culture medium, Catros et al. employed LaBP [96]. The physicochemical properties of nHAp were maintained during LaBP, and after 15 days, cell proliferation was confirmed. Castro et al. adopted the use of two biologically inspired nanomaterials, core–shell poly(lactic-co-glycolic acid) (PLGA) nanospheres and PLGA nanospheres [97] and nanocrystalline hydroxyapatite (nHAp). After combining these two substances with chondrogenic transforming growth factor (TGF-1) for prolonged release, they bioprinted them into a porous and highly interconnected osteochondral scaffold with distinct nano-to-microstructure and temporally bioactive factor gradients. Human bone marrow-derived MSC adhesion, proliferation, and osteochondral differentiation were enhanced in vitro by the biomimetic graded 3D bioprinted osteochondral construct. After combining these two substances with chondrogenic transforming growth factor (TGF-1) for prolonged release, they bioprinted them into a porous and highly interconnected osteochondral scaffold with distinct nano-to-microstructure and temporally bioactive factor gradients. Human bone marrow-derived MSC adhesion, proliferation, and osteochondral differentiation were enhanced in vitro by the biomimetic-graded 3D bioprinted osteochondral construct [98].

15.3 Synthesis and Tailoring the Properties of Nanobioinks

There is no question that nano-biomaterials can enhance a bioink's physical characteristics. Although their biological activities are limited beyond ideal levels, nanofillers also directly interfere with crosslinking and processability, which limits their use in printed bioink [99]. Throughout the printing process, the ink's consistent viscoelastic properties and printability should be maintained. Additional research on chemicals, and cells that have been encapsulated is important to investigate the steadiness of the nano-biomaterials. Researchers have used a variety of ways to create nanocomposite bioinks, including direct mixing, melt mixing, in situ nanocomposite creation, and biomimetic mineralization. The majority of works employ mixing prior to crosslinking or gel-forming processes. Each method's ability to increase cell loading and decrease (or optimize) the dosage of nano-biomaterials is still constrained. Researchers most frequently use direct mixing at the solution stage prior to crosslinking; however, because mixing rheology has received less attention, direct mixing frequently lacks homogeneity, which reduces functional activities. In situ mixing can solve direct mixing issues; however, it can occasionally compromise porosity and homogeneous crosslinking [100].

Although biomimetic (in vivo) mineralization takes longer, the outcome is superior [101]. In this process, mineralization or the production of nanoparticles takes place under physiological circumstances and is influenced by ions found in physiological fluids. The procedure might not be beneficial if the strength is insufficient after printing. Once more, in situ and biomineralization are only applicable to areas where nanoparticles can develop after printing. The most appropriate method when taking these into consideration is normal direct mixing with targeted attention to address the associated issues. As a single polymer cannot provide all the needed qualities, a mixture of polymers is frequently employed for bioprinting. As a result, the hydrogel matrices must be dual crosslinked for this to be performed. As the viscosity and crosslinked network are gradually increased to confine the cells in the polymer network, this is also useful in preventing cell sedimentation [102]. The in situ primary crosslinking can be started by the nanoparticles, which helps to establish the necessary physical properties for printing the hydrogel. Later on, the printed structure may add secondary crosslinking. Another method involves printing and mixing two streams of droplets that intersect each other at the same time [103]. Additionally, as cells are the most crucial element of bioink, a number of approaches are used to reduce the mechanical, thermal, or chemical stresses on the cells, enhancing their survival as well as their capacity to form tissues after printing. Cells are often combined in the decisive step right before printing, so they will not be exposed for too long. In many circumstances, the pressures are greater if cells are added directly to the bioink, and before the cells are mixed with the ink, cell aggregations may occur [104].

ECM naturally surrounds cells within the body and serves as a scaffold for cell growth and proliferation [105]. There are two main subclasses that make up the ECM. Hyaluronic acid is a polysaccharide glycosaminoglycan that functions as a mediator for nutrition transfer and cell growth. Second, fibrous proteins primarily serve as adhesives and structural supports. The interactions between cells and their ECM have a significant impact on how cells behave and frequently determine where they end up. An in vitro model can show indicators of the earliest cell replies to implantation or chemotherapy, even if it is challenging to simulate the complicated body response to biomaterials or medications. This emphasizes how crucial the surface characteristics of 3D bioprinted objects are at the nanoscale. Microgravity bioreactors, one of the 3D cell-based model [106], scaffolds made of synthetic and natural polymers, as well as micro- and nano-porous substrate [37], create a system where different stimuli can be applied to live cells in vitro to measure their effects. These models' structures can be created at the nanoscale to closely resemble the biological settings seen in different human parts, including tissues, cancers, and blood arteries [107].

15.3.1 Choice of Bioink

The selection of a bioink material is one of the most difficult aspects of developing a 3D bioprinted structure. A mixture of cells and materials called "bioinks" can be processed using automated bio-fabrication technologies. The biomaterials that can act as a vehicle for the delivery of cells during formulation are called bioink components. The functional chemical and mechanical qualities needed to create tissue constructs should be provided by these materials. These substances specify the biocompatibility, rate of deterioration, and functional chemical groups that enable potential surface modification. In the realm of regenerative medicine, a wide range of materials are now being researched for repair and regeneration. They can be divided into several major groups, such as naturally occurring polymers that are frequently extracted from animal or human tissues [108–110], artificial compounds [111, 112], or a combination of synthetic and natural materials [113, 114]. Natural polymers' resemblance to the human ECM and innate bioactivity make them perfect for 3D bioprinting [108–110]. The ability to customize synthetic polymers with specific physical properties for a given application is one of their key benefits. However, frequently their applications in the field of regenerative medicine are hampered by their reduced biocompatibility, toxic products, and loss of mechanical capabilities during neo-tissue replacement. Due to their shear-thinning capabilities, mixtures of natural and synthetic materials are frequently used [115]. Additionally, the sort of bioprinting technology that can be employed depends on the bioinks selected [116].

15.3.2 Porosity

The 3D structure of the scaffold, in addition to the materials used, is essential in determining how cells respond. Porous networks with high porosity and connectivity are typically preferred [117, 118]. With conventional scaffold manufacturing procedures, it might be difficult to create a well-ordered porous structure with the ideal internal design. Research is currently moving quickly away from using synthetic implants and tissue transplants and toward a technique called tissue engineering that uses porous, degradable scaffolds combined with living cells or chemicals to repair tissues [119, 120]. In order to permit the cells to act as planned and produce tissues and organs of the correct shape and size, the scaffold bioink must be carefully chosen [121].

15.3.3 Surface Properties

Cellular responses are significantly modulated by the surface's physical and chemical properties as well as the underlying artificial environment. This consists of all the characteristics of the cell and surface interfaces, including geometric shape, toughness, and the presence of sticky ligands, which greatly influence the behavior of the cell [122]. Combining bioink with ECM proteins such as fibronectin or collagen is a simple and effective technique to replicate the structure of the ECM. However, for extremely basic research that has to carefully regulate the content and order of the surface modification [123], the reliability of the drawn conclusions may be compromised by the ambiguity with which the real cell adhesive ligands are displayed in protein and polymeric adhesive surfaces [124]. To clear up this uncertainty and reveal how cells perceive various physical cues, surfaces with strict molecularly designed chemistry are required [122, 125]. Controlled production of sticky ligands has been suggested as a remedy by immobilizing the ligands on surfaces in the form of a monolayer [125, 126]. To mimic the role of the ECM in cell attachment, the bioink assembly can be adjusted by sticky ligands such as arginylglycylaspartic acid (RGD) peptides. Using two-dimensional models, it has been established that even modest changes to the types, spacing, surrounding microenvironment, and organization of RGD adhesive ligands have a major influence on the cellular phenotype [125]. For instance, Kato and Mrksich's study [127] showed how the type of RGD adhesive ligands affected the nucleation rate and growth of focal adhesion in fibroblasts. In this study, a linear or cyclic Arg-Gly-Asp peptide was exposed to 3T3 Swiss fibroblasts at equal densities using self-assembled monolayers of alkanethiolates connected to tri-(ethylene glycol)-terminated alkanethiol molecules. The significance of managing sticky ligand structure was supported by this investigation. This is consistent with the findings of Xiao and Truskey [128], who showed that immobilized cyclic RGD peptides had a higher level of cell attachment than linear peptides.

Numerous research has looked into the significance of the RGD–RGD spacing on cell activities in addition to the type [129–131]. The effect of RGD's density on silicon surface and how it affects cellular performance was investigated [129]. In this study, undecylic acid and various ratios of 1-amino-hexa-ethylene-oxide-mono-methyl-ether (EO6) to 1-amino-hexa-ethylene-oxide were applied to silicon surfaces. Surfaces with different ligand spacing were created by attaching RGD ligands to hydroxyl groups. The findings of this study showed that a variety of RGD concentrations controlled the number of cells, extending, union, and migratory capacity of endothelial cells, as well as the effectiveness of signal transmission. The surface of the glass was changed by AuNPs in both well-ordered and disorderly patterns in the initial study by Spatz and coworkers [132]. To prevent nonspecific cell attachment, PEG-silane 2-methoxy-polyethyleneoxy-propyl-trimethoxysilane molecules have been added to the gaps between nanoparticles on the surface [131, 133]. Thiol groups were used to encapsulate AuNPs with c-RGDfK-thiol ligands. Despite the fact that the typical receptor spacing is greater than 70 nm, the arrangement of RGD ligands on the substrate affects integrin clustering and adhesion. Different cellular processes depend on the expression of RGD ligands as well as the groups that surround these peptides. The effect of the length of the oligo(ethylene glycol) groups related to the self-assembled monolayers enclosing the RGD peptides on cell adhesion and spreading was investigated by Houseman and Mrksich [134]. In this study, thiols with tri, tetra, penta, or hexa(ethylene glycol) units were used to vary the average distance between the glycol groups and the peptide ligand while leaving the structure of the background monolayer untouched. The research showed that as oligo(ethylene glycol) molecule length rose, cell adhesion to RGD ligands decreased. These experiments also demonstrated how precisely several aspects of cellular function are regulated depending on the degree of adhesion motif presentation. These investigations further emphasized how crucial it is to use carefully monitored chemistry when treating surfaces for cell assays. It was demonstrated how the nanoscale arrangement of sticky ligands affects cell responses to medicines in a thorough investigation employing electrochemistry and fluorescence microscopy [135]. This is significant because it illustrates how crucial surface characteristics are for drug testing. However, research into the impact of the bioinks' surface characteristics on 3D bioprinted systems is still in its preliminary stages. The material itself caused the most variations, for example, chitosan scaffolds with bigger pores led to more proinflammatory cytokine secretion [121].

15.4 Nanobioinks and Tissue Engineering

One of the most efficient tactics to address the global organ scarcity situation is tissue engineering [136, 137]. It attempts to create functional tissue for a range of uses, such as drug testing, biosensing, and regenerative medicine [136, 138]. Tissue engineering utilizing 3D bioprinting techniques improves tissue regeneration by carefully balancing the materials (scaffolds), suitable cells, and the right biochemical components. The literature has mostly focused on two methods for biologically producing tissues and organs, bioprinting with and without scaffolds [139–141]. Figure 15.4 provides an illustration of how the usage of nanocomposite bioinks improves different functional, physical, and biological qualities. It is difficult to satisfy the gel matrix's rheological property criteria by using polymer gels because the selection of acceptable polymer matrices is constrained when cell survival is considered. A few nanomaterials have been utilized for this purpose with success. Other advantages include printability, mechanical reinforcements, and the stability of the post-printing structure. Conductive nanomaterials stimulate excitable cells including heart, nerve, and muscle cells to aid in the development of new tissue. Increased growth, proliferation, and ECM secretion are the results of the unique differentiation pathways and cellular activities that are triggered by the release of ions and biomolecules from nano-biomaterials [104].

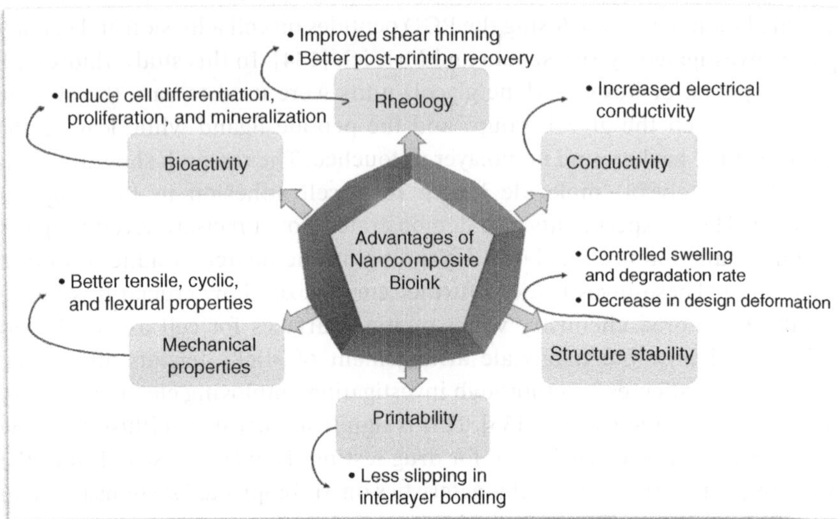

Figure 15.4 Designing functional features for nanocomposite bioinks with nanobiomaterials. *Source:* Adapted from Bhattacharyya et al. [104], Gaharwar et al. [24].

15.4.1 Tissue Engineering of Bone and Cartilage

With specified placement of cells, ECM, and biomolecules together with the vascular system, 3D bioprinting has the ability to provide answers to challenging issues in orthopedics. Nanocomposite bioinks have been used most frequently in the tissue engineering of bone and cartilage. Nanocomposite inks perform significantly better in terms of strength than pure hydrogels. Because these components are present in the ECM of bone, CaP-based NPs are frequently used in bone printing [142]. The osteogenesis of hMSC in nHAp/PEGDA bioinks encompassing RGDS peptide is significantly enhanced by using low-intensity ultrasonic radiation (1.5 MHz) [143]. In addition, demineralized insoluble dentin powder [73] and polymeric nanofibers [18] are employed to enhance biological, stiffness, and rheological qualities. GO was employed to make bone marrow-derived stem cells for chondrogenic differentiation [144]. For craniofacial bone abnormalities, in one study, a dual channel 3D bioprinting was used to create bone graft with a supra-structure called PCL/HAp. Vascularization was encouraged utilizing hydrogel bioinks containing endothelial lineage cells during seven days of brief hypoxia circumstances. The development of a vascular system within the bone construct in less than 21 days was demonstrated by subcutaneous implantation in dorsal nude mice [145]. Stem cells and chondrocytes are the most typical cell sources for cartilage bioprinting. Although hydrogels constructed from several natural and synthetic polymers have been investigated for cartilage tissue creation, employing pure hydrogels makes it challenging to attain the necessary mechanical characteristics and 3D printability [146]. As a result, several nanofillers have been employed to enhance shape consistency while mimicking the characteristics of genuine cartilage. Intervertebral disc and meniscus fibrocartilage tissues can bear an excessive amount of pressure; as a result, cellulose nanofibrils or HAp nanoparticles are utilized to increase structural stability [147]. For the creation of meniscus scaffolds, carbon nanoparticles are also utilized [148].

15.4.2 Tissue Engineering of Nerves

The utilization of various cells, biomaterials, and biomolecules in 3D bioprinting benefits both the central and peripheral nervous systems. 3D bioprinted nerve guiding conduits consisting of alginate-HAp, gelMA, and PEGDA has been discovered to be capable of motor and sensory nerve tissue engineering and wider gap repair [149]. To create complicated brain architectures, a great deal of research has been done on metal nanoparticles, polymers that transmit electricity like polypyrrole, and nanofillers like G and CNF [150]. The 4D bioprinting method advanced with the addition of smart polymers to the bioink hydrogel, successfully transmitting action to the efferent limbs [151]. The challenging 3-layered tubular nerve tube construction can be made with multiple bioprinter nozzles with ease.

The researchers first tried inkjet bioprinting with low-viscosity ink, but extrusion-based bioplotting finally took over to create such large 3D structures [152].

15.4.3 Tissue Engineering in Cardiovascular Disease

The most viable alternative to heart implantation given its donor, immunological, and dualism restrictions is 3D bioprinting. Only a cell-laden scaffold structure allows for rhythmic tightening and expanding movements of cardiomyocytes in the cardiovascular tissue. Additionally, this method is best suited to precisely build the intricate, multiscale vascular network needed for cardiovascular regeneration, including the myocardium and heart valves [153]. To advance a 3D bioprinted construct to the clinical trial stage, however, significant improvements must be made to its electrophysical, hydrodynamic (including hemodynamic and thrombogenesis), and mechanical capabilities [154]. For increasing the conductivity and sensing capabilities of cardiac tissue engineering, some nanomaterials have been reported including AuNRs and carbon NPs [155]. Due to the electrical bridging, AuNRs also improve cardiac cell adhesion, self-organization, and synchronized contraction in bioprinting. After the bioprinted heart tissue has been implanted, infections may be avoided by using iron oxide-like antibacterial nanoparticles [156]. Small-diameter vascular transplantation using polymer-based vessels is not effective due to thrombosis and intimal hyperplasia. The inner lining is made up of endothelial cells, whereas the primary wall body and outer layers are made up of smooth muscle cells and fibroblasts, which support vessel structure. Given the current printing resolutions and accuracy, it is challenging to mimic their arrangements in a 3D-printed vessel with a small diameter [157]. Since the natural vessel has better biomechanical qualities than successfully printed ones, developing suitable formulations with enhanced biomechanical features is the main focus of the current research [158].

15.5 Future Outlook

The multipurpose features of nanocomposite hydrogels offer a probable future direction. Smart polymers mixed with conducting nanofillers like CNT, carbon nanofiber, or graphene are examples of possible bioinks for 4D bioprinting, which can alter printed constructions in a time-dependent way [159, 160]. Along with improving hydrogel materials' physical and chemical characteristics, nanoparticles also increase cellular activity [161]. Nanoparticle-induced patterning enables rapid and high-throughput cell assembly. Techniques such as magnetic self-assembly or sonic integration can be used to construct scaffolds in a certain direction [162]. For 3D bioprinting, a number of interesting nano-biomaterials have not yet been thoroughly fully investigated, including collagen nanofiber, bioglass, and sugar-glass. Collagen nanofiber generated from decellularized tissue is anticipated to grow into patient-specific 3D-printed scaffolds [73]. With the right biomolecules and nanoparticles, the 3D-printed cells may perform the desired biological activities. In situ 3D bioprinting is now possible because of the development of portable multichannel bioprinters. Although big organ printing and rapid organ maturation make it difficult to design a bioink that is appropriate for such in situ processing, work in this area is advancing as the entire process becomes more automated [163].

References

1 Ozbolat, I.T. (2015). Bioprinting scale-up tissue and organ constructs for transplantation. *Trends Biotechnol.* 33 (7): 395–400.

2 Ozbolat, I.T. (2015). Scaffold-based or scaffold-free bioprinting: competing or complementing approaches? *J. Nanotechnol. Eng. Med.* 6 (2): 024701.

3 Ozbolat, I.T., Peng, W., and Ozbolat, V. (2016). Application areas of 3D bioprinting. *Drug Discov. Today* 21 (8): 1257–1271.

4 Ramadan, Q. and Zourob, M. (2021). 3D bioprinting at the frontier of regenerative medicine, pharmaceutical, and food industries. *Front. Med. Technol.* 2: 607648.

5 Hospodiuk, M., Dey, M., Sosnoski, D., and Ozbolat, I.T. (2017). The bioink: a comprehensive review on bioprintable materials. *Biotechnol. Adv.* 35 (2): 217–239.

6 Zhang, X. and Zhang, Y. (2015). Tissue engineering applications of three-dimensional bioprinting. *Cell Biochem. Biophys.* 72: 777–782.

7 Munaz, A., Vadivelu, R.K., John, J.S. et al. (2016). Three-dimensional printing of biological matters. *J. Sci.: Adv. Mater. Devices* 1 (1): 1–17.

8 Matai, I., Kaur, G., Seyedsalehi, A. et al. (2020). Progress in 3D bioprinting technology for tissue/organ regenerative engineering. *Biomaterials* 226: 119536.

9 Mao, Q., Wang, Y., Li, Y. et al. (2020). Fabrication of liver microtissue with liver decellularized extracellular matrix (dECM) bioink by digital light processing (DLP) bioprinting. *Mater. Sci. Eng. C* 109: 110625.

10 Leppiniemi, J., Lahtinen, P., Paajanen, A. et al. (2017). 3D-printable bioactivated nanocellulose–alginate hydrogels. *ACS Appl. Mater. Interfaces* 9 (26): 21959–21970.

11 Tirella, A., Orsini, A., Vozzi, G., and Ahluwalia, A. (2009). A phase diagram for microfabrication of geometrically controlled hydrogel scaffolds. *Biofabrication* 1 (4): 045002.

12 Gudapati, H., Dey, M., and Ozbolat, I. (2016). A comprehensive review on droplet-based bioprinting: past, present and future. *Biomaterials* 102: 20–42.

13 Hölzl, K., Lin, S., Tytgat, L. et al. (2016). Bioink properties before, during and after 3D bioprinting. *Biofabrication* 8 (3): 032002.

14 Ozbolat, I.T. and Hospodiuk, M. (2016). Current advances and future perspectives in extrusion-based bioprinting. *Biomaterials* 76: 321–343.

15 Ozbolat, I.T., Moncal, K.K., and Gudapati, H. (2017). Evaluation of bioprinter technologies. *Addit. Manuf.* 13: 179–200.

16 Guillotin, B., Souquet, A., Catros, S. et al. (2010). Laser assisted bioprinting of engineered tissue with high cell density and microscale organization. *Biomaterials* 31 (28): 7250–7256.

17 Wang, H., Yu, H., Zhou, X. et al. (2022). An overview of extracellular matrix-based bioinks for 3D bioprinting. *Front. Bioeng. Biotechnol.* 10: 905438.

18 Ashammakhi, N., Ahadian, S., Xu, C. et al. (2019). Bioinks and bioprinting technologies to make heterogeneous and biomimetic tissue constructs. *Mater. Today Bio* 1: 100008.

19 Carrow, J.K. and Gaharwar, A.K. (2015). Bioinspired polymeric nanocomposites for regenerative medicine. *Macromol. Chem. Phys.* 216 (3): 248–264.

20 Rastogi, P. and Kandasubramanian, B. (2019). Review of alginate-based hydrogel bioprinting for application in tissue engineering. *Biofabrication* 11 (4): 042001.

21 Yang, T., Hu, Y., Wang, C., and Binks, B.P. (2017). Fabrication of hierarchical macroporous biocompatible scaffolds by combining pickering high internal phase emulsion templates with three-dimensional printing. *ACS Appl. Mater. Interfaces* 9 (27): 22950–22958.

22 Leberfinger, A.N., Dinda, S., Wu, Y. et al. (2019). Bioprinting functional tissues. *Acta Biomater.* 95: 32–49.

23 Mora-Boza, A. and Lopez-Donaire, M.L. (2018). Preparation of polymeric and composite scaffolds by 3D bioprinting. *Adv. Exp. Med. Biol.* 1058: 221–245.

24 Gaharwar, A.K., Peppas, N.A., and Khademhosseini, A. (2014). Nanocomposite hydrogels for biomedical applications. *Biotechnol. Bioeng.* 111 (3): 441–453.

25 Chimene, D., Lennox, K.K., Kaunas, R.R., and Gaharwar, A.K. (2016). Advanced bioinks for 3D printing: a materials science perspective. *Ann. Biomed. Eng.* 44: 2090–2102.

26 Farahani, R.D., Dubé, M., and Therriault, D. (2016). Three-dimensional printing of multifunctional nanocomposites: manufacturing techniques and applications. *Adv. Mater.* 28 (28): 5794–5821.

27 Gao, G., Schilling, A.F., Yonezawa, T. et al. (2014). Bioactive nanoparticles stimulate bone tissue formation in bioprinted three-dimensional scaffold and human mesenchymal stem cells. *Biotechnol. J.* 9 (10): 1304–1311.

28 Leite, Á.J., Sarker, B., Zehnder, T. et al. (2016). Bioplotting of a bioactive alginate dialdehyde-gelatin composite hydrogel containing bioactive glass nanoparticles. *Biofabrication* 8 (3): 035005.

29 Kim, W. and Kim, G. (2019). Collagen/bioceramic-based composite bioink to fabricate a porous 3D hASCs-laden structure for bone tissue regeneration. *Biofabrication* 12 (1): 015007.

30 Fan, Y., Shi, T., Yue, X. et al. (2019). 3D composite cell printing gelatin/sodium alginate/n-HAP bioscaffold. Journal of Physics: Conference Series. IOP Publishing.

31 Wenz, A., Borchers, K., Tovar, G.E., and Kluger, P.J. (2017). Bone matrix production in hydroxyapatite-modified hydrogels suitable for bone bioprinting. *Biofabrication* 9 (4): 044103.

32 You, F., Chen, X., Cooper, D. et al. (2018). Homogeneous hydroxyapatite/alginate composite hydrogel promotes calcified cartilage matrix deposition with potential for three-dimensional bioprinting. *Biofabrication* 11 (1): 015015.

33 Wüst, S., Godla, M.E., Müller, R., and Hofmann, S. (2014). Tunable hydrogel composite with two-step processing in combination with innovative hardware upgrade for cell-based three-dimensional bioprinting. *Acta Biomater.* 10 (2): 630–640.

34 Kolan, K.C., Semon, J.A., Bindbeutel, A.T. et al. (2020). Bioprinting with bioactive glass loaded polylactic acid composite and human adipose stem cells. *Bioprinting* 18: e00075.

35 Hoppe, A., Sarker, B., Detsch, R. et al. (2014). In vitro reactivity of Sr-containing bioactive glass (type 1393) nanoparticles. *J. Non-Cryst. Solids* 387: 41–46.

36 Sethu, S.N., Namashivayam, S., Devendran, S. et al. (2017). Nanoceramics on osteoblast proliferation and differentiation in bone tissue engineering. *Int. J. Biol. Macromol.* 98: 67–74.

37 Powers, M.J., Domansky, K., Kaazempur-Mofrad, M.R. et al. (2002). A microfabricated array bioreactor for perfused 3D liver culture. *Biotechnol. Bioeng.* 78 (3): 257–269.

38 Vallet-Regí, M. and Ruiz-Hernández, E. (2011). Bioceramics: from bone regeneration to cancer nanomedicine. *Adv. Mater.* 23 (44): 5177–5218.

39 Moorthi, A., Parihar, P., Saravanan, S. et al. (2014). Effects of silica and calcium levels in nanobioglass ceramic particles on osteoblast proliferation. *Mater. Sci. Eng. C* 43: 458–464.

40 Moorthi, A., Vimalraj, S., Avani, C. et al. (2013). Expression of microRNA-30c and its target genes in human osteoblastic cells by nano-bioglass ceramic-treatment. *Int. J. Biol. Macromol.* 56: 181–185.

41 Day, R.M. (2005). Bioactive glass stimulates the secretion of angiogenic growth factors and angiogenesis in vitro. *Tissue Eng.* 11 (5–6): 768–777.

42 Gerhardt, L.-C. and Boccaccini, A.R. (2010). Bioactive glass and glass-ceramic scaffolds for bone tissue engineering. *Materials* 3 (7): 3867–3910.

43 Barrère, F., van Blitterswijk, C.A., and de Groot, K. (2006). Bone regeneration: molecular and cellular interactions with calcium phosphate ceramics. *Int. J. Nanomedicine* 1 (3): 317.

44 Wang, P., Zhao, L., Liu, J. et al. (2014). Bone tissue engineering via nanostructured calcium phosphate biomaterials and stem cells. *Bone Res.* 2 (1): 1–13.
45 Hong, G., Diao, S., Antaris, A.L., and Dai, H. (2015). Carbon nanomaterials for biological imaging and nanomedicinal therapy. *Chem. Rev.* 115 (19): 10816–10906.
46 Cha, C., Shin, S.R., Annabi, N. et al. (2013). Carbon-based nanomaterials: multifunctional materials for biomedical engineering. *ACS Nano* 7 (4): 2891–2897.
47 Amenta, V. and Aschberger, K. (2015). Carbon nanotubes: potential medical applications and safety concerns. *Wiley Interdiscip. Rev. Nanomed. Nanobiotechnol.* 7 (3): 371–386.
48 Cai, Y., Chang, S.Y., Gan, S.W. et al. (2022). Nanocomposite bioinks for 3D bioprinting. *Acta Biomater.* 151: 45.
49 Shin, S.R., Farzad, R., Tamayol, A. et al. (2016). A bioactive carbon nanotube-based ink for printing 2D and 3D flexible electronics. *Adv. Mater.* 28 (17): 3280–3289.
50 Zhang, Y., Yu, Y., Dolati, F., and Ozbolat, I.T. (2014). Effect of multiwall carbon nanotube reinforcement on coaxially extruded cellular vascular conduits. *Mater. Sci. Eng. C* 39: 126–133.
51 Dolati, F., Yu, Y., Zhang, Y. et al. (2014). In vitro evaluation of carbon-nanotube-reinforced bioprintable vascular conduits. *Nanotechnology* 25 (14): 145101.
52 Izadifar, M., Chapman, D., Babyn, P. et al. (2018). UV-assisted 3D bioprinting of nanoreinforced hybrid cardiac patch for myocardial tissue engineering. *Tissue Eng. Part C Methods* 24 (2): 74–88.
53 Choe, G., Oh, S., Seok, J.M. et al. (2019). Graphene oxide/alginate composites as novel bioinks for three-dimensional mesenchymal stem cell printing and bone regeneration applications. *Nanoscale* 11 (48): 23275–23285.
54 Huang, C.-T., Shrestha, L.K., Ariga, K., and Hsu, S.-h. (2017). A graphene–polyurethane composite hydrogel as a potential bioink for 3D bioprinting and differentiation of neural stem cells. *J. Mater. Chem. B* 5 (44): 8854–8864.
55 Zhang, J., Eyisoylu, H., Qin, X.-H. et al. (2021). 3D bioprinting of graphene oxide-incorporated cell-laden bone mimicking scaffolds for promoting scaffold fidelity, osteogenic differentiation and mineralization. *Acta Biomater.* 121: 637–652.
56 Kang, M.S., Kang, J.I., Le Thi, P. et al. (2021). Three-dimensional printable gelatin hydrogels incorporating graphene oxide to enable spontaneous myogenic differentiation. *ACS Macro Lett.* 10 (4): 426–432.
57 Cheng, Z., Landish, B., Chi, Z. et al. (2018). 3D printing hydrogel with graphene oxide is functional in cartilage protection by influencing the signal pathway of Rank/Rankl/OPG. *Mater. Sci. Eng. C* 82: 244–252.
58 Piras, C.C., Fernández-Prieto, S., and De Borggraeve, W.M. (2017). Nanocellulosic materials as bioinks for 3D bioprinting. *Biomater. Sci.* 5 (10): 1988–1992.

59 Janarthanan, G., Tran, H.N., Cha, E. et al. (2020). 3D printable and injectable lactoferrin-loaded carboxymethyl cellulose-glycol chitosan hydrogels for tissue engineering applications. *Mater. Sci. Eng. C* 113: 111008.

60 Janarthanan, G., Shin, H.S., Kim, I.-G. et al. (2020). Self-crosslinking hyaluronic acid–carboxymethylcellulose hydrogel enhances multilayered 3D-printed construct shape integrity and mechanical stability for soft tissue engineering. *Biofabrication* 12 (4): 045026.

61 Aljohani, W., Ullah, M.W., Zhang, X., and Yang, G. (2018). Bioprinting and its applications in tissue engineering and regenerative medicine. *Int. J. Biol. Macromol.* 107: 261–275.

62 Dai, L., Cheng, T., Duan, C. et al. (2019). 3D printing using plant-derived cellulose and its derivatives: a review. *Carbohydr. Polym.* 203: 71–86.

63 Wang, Q., Sun, J., Yao, Q. et al. (2018). 3D printing with cellulose materials. *Cellulose* 25: 4275–4301.

64 Ahlfeld, T., Köhler, T., Czichy, C. et al. (2018). A methylcellulose hydrogel as support for 3D plotting of complex shaped calcium phosphate scaffolds. *Gels* 4 (3): 68.

65 Habib, A., Sathish, V., Mallik, S., and Khoda, B. (2018). 3D printability of alginate-carboxymethyl cellulose hydrogel. *Materials* 11 (3): 454.

66 Li, H., Tan, Y.J., Leong, K.F., and Li, L. (2017). 3D bioprinting of highly thixotropic alginate/methylcellulose hydrogel with strong interface bonding. *ACS Appl. Mater. Interfaces* 9 (23): 20086–20097.

67 Håkansson, K.M., Henriksson, I.C., de la Peña, V.C. et al. (2016). Solidification of 3D printed nanofibril hydrogels into functional 3D cellulose structures. *Adv. Mater. Technol.* 1 (7): 1600096.

68 Sultan, S., Siqueira, G., Zimmermann, T., and Mathew, A.P. (2017). 3D printing of nano-cellulosic biomaterials for medical applications. *Curr. Opin. Biomed. Eng.* 2: 29–34.

69 Markstedt, K., Mantas, A., Tournier, I. et al. (2015). 3D bioprinting human chondrocytes with nanocellulose–alginate bioink for cartilage tissue engineering applications. *Biomacromolecules* 16 (5): 1489–1496.

70 Siqueira, G., Kokkinis, D., Libanori, R. et al. (2017). Cellulose nanocrystal inks for 3D printing of textured cellular architectures. *Adv. Funct. Mater.* 27 (12): 1604619.

71 Jiang, Y., Zhou, J., Yang, Z. et al. (2018). Dialdehyde cellulose nanocrystal/gelatin hydrogel optimized for 3D printing applications. *J. Mater. Sci.* 53: 11883–11900.

72 Du, H., Liu, W., Zhang, M. et al. (2019). Cellulose nanocrystals and cellulose nanofibrils based hydrogels for biomedical applications. *Carbohydr. Polym.* 209: 130–144.

73 Chinga-Carrasco, G. (2018). Potential and limitations of nanocelluloses as components in biocomposite inks for three-dimensional bioprinting and for biomedical devices. *Biomacromolecules* 19 (3): 701–711.

74 Kim, B.S., Kim, H., Gao, G. et al. (2017). Decellularized extracellular matrix: a step towards the next generation source for bioink manufacturing. *Biofabrication* 9 (3): 034104.

75 Marques, C., Diogo, G., Pina, S. et al. (2019). Collagen-based bioinks for hard tissue engineering applications: a comprehensive review. *J. Mater. Sci. Mater. Med.* 30: 1–12.

76 Shin, K.-H., Kim, J.-W., Koh, Y.-H., and Kim, H.-E. (2015). Novel self-assembly-induced 3D plotting for macro/nano-porous collagen scaffolds comprised of nanofibrous collagen filaments. *Mater. Lett.* 143: 265–268.

77 Sorushanova, A., Delgado, L.M., Wu, Z. et al. (2019). The collagen suprafamily: from biosynthesis to advanced biomaterial development. *Adv. Mater.* 31 (1): 1801651.

78 Noor, N., Shapira, A., Edri, R. et al. (2019). 3D printing of personalized thick and perfusable cardiac patches and hearts. *Adv. Sci.* 6 (11): 1900344.

79 Mei, Q., Rao, J., Bei, H.P. et al. (2021). 3D bioprinting photo-crosslinkable hydrogels for bone and cartilage repair. *Int. J. Bioprinting* 7 (3): 367.

80 Gao, G., Schilling, A.F., Hubbell, K. et al. (2015). Improved properties of bone and cartilage tissue from 3D inkjet-bioprinted human mesenchymal stem cells by simultaneous deposition and photocrosslinking in PEG-GelMA. *Biotechnol. Lett.* 37: 2349–2355.

81 Doschak, M.R., Kucharski, C.M., Wright, J.E. et al. (2009). Improved bone delivery of osteoprotegerin by bisphosphonate conjugation in a rat model of osteoarthritis. *Mol. Pharm.* 6 (2): 634–640.

82 Park, D., Park, C.-W., Choi, Y. et al. (2016). A novel small-molecule PPI inhibitor targeting integrin αvβ3-osteopontin interface blocks bone resorption in vitro and prevents bone loss in mice. *Biomaterials* 98: 131–142.

83 Teno, N., Masuya, K., Ehara, T. et al. (2008). Effect of cathepsin K inhibitors on bone resorption. *J. Med. Chem.* 51 (17): 5459–5462.

84 Arrighi, I., Mark, S., Alvisi, M. et al. (2009). Bone healing induced by local delivery of an engineered parathyroid hormone prodrug. *Biomaterials* 30 (9): 1763–1771.

85 Rodan, G.A. and Martin, T.J. (2000). Therapeutic approaches to bone diseases. *Science* 289 (5484): 1508–1514.

86 Pei, M., Mao, J., Xu, W. et al. (2019). Photocrosslinkable chitosan hydrogels and their biomedical applications. *J. Polym. Sci. A Polym. Chem.* 57 (18): 1862–1871.

87 Yi, H.-G., Kim, H., Kwon, J. et al. (2021). Application of 3D bioprinting in the prevention and the therapy for human diseases. *Signal Transduction Targeted Ther.* 6 (1): 177.

88 Maan, Z., Masri, N.Z., and Willerth, S.M. (2022). Smart bioinks for the printing of human tissue models. *Biomolecules* 12 (1): 141.

89 Narducci, D. (2007). An introduction to nanotechnologies: What's in it for us? *Vet. Res. Commun.* 31: 131–137.

90 Weiss, J., Takhistov, P., and McClements, D.J. (2006). Functional materials in food nanotechnology. *J. Food Sci.* 71 (9): R107–R116.

91 Saini, R., Saini, S., and Sharma, S. (2010). Nanotechnology: the future medicine. *J. Cutan. Aesthet. Surg.* 3 (1): 32.

92 Heid, S. and Boccaccini, A.R. (2020). Advancing bioinks for 3D bioprinting using reactive fillers: a review. *Acta Biomater.* 113: 1–22.

93 Gungor-Ozkerim, P.S., Inci, I., Zhang, Y.S. et al. (2018). Bioinks for 3D bioprinting: an overview. *Biomater. Sci.* 6 (5): 915–946.

94 Skardal, A., Zhang, J., McCoard, L. et al. (2010). Dynamically crosslinked gold nanoparticle–hyaluronan hydrogels. *Adv. Mater.* 22 (42): 4736–4740.

95 Zeng, J., Shi, D., Gu, Y. et al. (2019). Injectable and near-infrared-responsive hydrogels encapsulating dopamine-stabilized gold nanorods with long photothermal activity controlled for tumor therapy. *Biomacromolecules* 20 (9): 3375–3384.

96 Catros, S., Fricain, J.-C., Guillotin, B. et al. (2011). Laser-assisted bioprinting for creating on-demand patterns of human osteoprogenitor cells and nano-hydroxyapatite. *Biofabrication* 3 (2): 025001.

97 de la Vega, L., Lee, C., Sharma, R. et al. (2019). 3D bioprinting models of neural tissues: the current state of the field and future directions. *Brain Res. Bull.* 150: 240–249.

98 Castro, N.J. and O'brien J, Zhang LG. (2015). Integrating biologically inspired nanomaterials and table-top stereolithography for 3D printed biomimetic osteochondral scaffolds. *Nanoscale* 7 (33): 14010–14022.

99 Nadgorny, M. and Ameli, A. (2018). Functional polymers and nanocomposites for 3D printing of smart structures and devices. *ACS Appl. Mater. Interfaces* 10 (21): 17489–17507.

100 Luo, Y., Lode, A., Wu, C. et al. (2015). Alginate/nanohydroxyapatite scaffolds with designed core/shell structures fabricated by 3D plotting and in situ mineralization for bone tissue engineering. *ACS Appl. Mater. Interfaces* 7 (12): 6541–6549.

101 Abouzeid, R.E., Khiari, R., Beneventi, D., and Dufresne, A. (2018). Biomimetic mineralization of three-dimensional printed alginate/TEMPO-oxidized cellulose nanofibril scaffolds for bone tissue engineering. *Biomacromolecules* 19 (11): 4442–4452.

102 Dubbin, K., Hori, Y., Lewis, K.K., and Heilshorn, S.C. (2016). Dual-stage crosslinking of a gel-phase bioink improves cell viability and homogeneity for 3D bioprinting. *Adv. Healthcare Mater.* 5 (19): 2488–2492.

103 Christensen, K., Compaan, A., Chai, W. et al. (2017). In situ printing-then-mixing for biological structure fabrication using intersecting jets. *ACS Biomater. Sci. Eng.* 3 (12): 3687–3694.

104 Bhattacharyya, A., Janarthanan, G., and Noh, I. (2021). Nano-biomaterials for designing functional bioinks towards complex tissue and organ regeneration in 3D bioprinting. *Addit. Manuf.* 37: 101639.

105 Lukashev, M.E. and Werb, Z. (1998). ECM signalling: orchestrating cell behaviour and misbehaviour. *Trends Cell Biol.* 8 (11): 437–441.

106 Jessup, J., Goodwin, T., and Spaulding, G. (1993). Prospects for use of microgravity-based bioreactors to study three-dimensional host – tumor interactions in human neoplasia. *J. Cell. Biochem.* 51 (3): 290–300.

107 Edmondson, R., Broglie, J.J., Adcock, A.F., and Yang, L. (2014). Three-dimensional cell culture systems and their applications in drug discovery and cell-based biosensors. *Assay Drug Dev. Technol.* 12 (4): 207–218.

108 Malafaya, P.B., Silva, G.A., and Reis, R.L. (2007). Natural–origin polymers as carriers and scaffolds for biomolecules and cell delivery in tissue engineering applications. *Adv. Drug Deliv. Rev.* 59 (4–5): 207–233.

109 Sell, S.A., Wolfe, P.S., Garg, K. et al. (2010). The use of natural polymers in tissue engineering: a focus on electrospun extracellular matrix analogues. *Polymers* 2 (4): 522–553.

110 Dang, J.M. and Leong, K.W. (2006). Natural polymers for gene delivery and tissue engineering. *Adv. Drug Deliv. Rev.* 58 (4): 487–499.

111 Pereira, T.F., Oliveira, M.F., Maia, I.A. et al. (2012). 3D printing of poly (3-hydroxybutyrate) porous structures using selective laser sintering. *Macromol. Symp.* Wiley Online Library. https://doi.org/10.1002/masy.201100237.

112 Daly, S.R., Fathi, A., Bahramian, B. et al. (2018). A green process for the purification of biodegradable poly (β-hydroxybutyrate). *J. Supercrit. Fluids* 135: 84–90.

113 Manavitehrani, I., Fathi, A., Wang, Y. et al. (2017). Fabrication of a biodegradable implant with tunable characteristics for bone implant applications. *Biomacromolecules* 18 (6): 1736–1746.

114 Almeida, C.R., Serra, T., Oliveira, M.I. et al. (2014). Impact of 3-D printed PLA-and chitosan-based scaffolds on human monocyte/macrophage responses: unraveling the effect of 3-D structures on inflammation. *Acta Biomater.* 10 (2): 613–622.

115 Zhang, Y.S., Yue, K., Aleman, J. et al. (2017). 3D bioprinting for tissue and organ fabrication. *Ann. Biomed. Eng.* 45: 148–163.

116 Jose, R.R., Rodriguez, M.J., Dixon, T.A. et al. (2016). Evolution of bioinks and additive manufacturing technologies for 3D bioprinting. *ACS Biomater. Sci. Eng.* 2 (10): 1662–1678.

117 Hollister, S.J. (2005). Porous scaffold design for tissue engineering. *Nat. Mater.* 4 (7): 518–524.

118 Zadpoor, A.A. (2015). Bone tissue regeneration: the role of scaffold geometry. *Biomater. Sci.* 3 (2): 231–245.

119 Rezwan, K., Chen, Q., Blaker, J.J., and Boccaccini, A.R. (2006). Biodegradable and bioactive porous polymer/inorganic composite scaffolds for bone tissue engineering. *Biomaterials* 27 (18): 3413–3431.

120 Messner, K. and Gillquist, J. (1993). Synthetic implants for the repair of osteochondral defects of the medial femoral condyle: a biomechanical and histological evaluation in the rabbit knee. *Biomaterials* 14 (7): 513–521.

121 Manavitehrani, I., Nasiri, N., and Parviz, M. (2019). Nanomaterials in 3D bioprinting: synthesis, modification, and application. In: *Harnessing Nanoscale Surface Interactions* (ed. D. Henry), 149–172. Elsevier.

122 Mrksich, M. (2002). What can surface chemistry do for cell biology? *Curr. Opin. Chem. Biol.* 6 (6): 794–797.

123 Mrksich, M. (1998). Tailored substrates for studies of attached cell culture. *Cell. Mol. Life Sci.* 54: 653–662.

124 Keselowsky, B.G., Collard, D.M., and García, A.J. (2005). Integrin binding specificity regulates biomaterial surface chemistry effects on cell differentiation. *Proc. Natl. Acad. Sci.* 102 (17): 5953–5957.

125 Gooding, J.J., Parker, S.G., Lu, Y., and Gaus, K. (2014). Molecularly engineered surfaces for cell biology: from static to dynamic surfaces. *Langmuir* 30 (12): 3290–3302.

126 Hersel, U., Dahmen, C., and Kessler, H. (2003). RGD modified polymers: biomaterials for stimulated cell adhesion and beyond. *Biomaterials* 24 (24): 4385–4415.

127 Kato, M. and Mrksich, M. (2004). Using model substrates to study the dependence of focal adhesion formation on the affinity of integrin – ligand complexes. *Biochemistry* 43 (10): 2699–2707.

128 Xiao, Y. and Truskey, G.A. (1996). Effect of receptor-ligand affinity on the strength of endothelial cell adhesion. *Biophys. J.* 71 (5): 2869–2884.

129 Le Saux, G., Magenau, A., Gunaratnam, K. et al. (2011). Spacing of integrin ligands influences signal transduction in endothelial cells. *Biophys. J.* 101 (4): 764–773.

130 Cavalcanti-Adam, E.A., Volberg, T., Micoulet, A. et al. (2007). Cell spreading and focal adhesion dynamics are regulated by spacing of integrin ligands. *Biophys. J.* 92 (8): 2964–2974.

131 Arnold, M., Cavalcanti-Adam, E.A., Glass, R. et al. (2004). Activation of integrin function by nanopatterned adhesive interfaces. *ChemPhysChem* 5 (3): 383–388.

132 Huang, J., Grater, S.V., Corbellini, F. et al. (2009). Impact of order and disorder in RGD nanopatterns on cell adhesion. *Nano Lett.* 9 (3): 1111–1116.

133 Bluemmel, J., Perschmann, N., Aydin, D. et al. (2007). Protein repellent properties of covalently attached PEG coatings on nanostructured SiO2-based interfaces. *Biomaterials* 28 (32): 4739–4747.

134 Houseman, B. and Mrksich, M. (2001). The microenvironment of immobilized Arg-Gly-Asp peptides is an important determinant of cell adhesion. *Biomaterials* 22 (9): 943–955.

135 Parviz, M., Gaus, K., and Gooding, J.J. (2017). Simultaneous impedance spectroscopy and fluorescence microscopy for the real-time monitoring of the response of cells to drugs. *Chem. Sci.* 8 (3): 1831–1840.

136 Nasiri, N., Ceramidas, A., Mukherjee, S. et al. (2016). Ultra-porous nanoparticle networks: a biomimetic coating morphology for enhanced cellular response and infiltration. *Sci. Rep.* 6 (1): 24305.

137 Nerem, R. (2000). Tissue engineering: confronting the transplantation crisis. *Proc. Inst. Mech. Eng. H J. Eng. Med.* 214 (1): 95–99.

138 Bhadriraju, K. and Chen, C.S. (2002). Engineering cellular microenvironments to improve cell-based drug testing. *Drug Discov. Today* 7 (11): 612–620.

139 Tan, Y., Richards, D.J., Trusk, T.C. et al. (2014). 3D printing facilitated scaffold-free tissue unit fabrication. *Biofabrication* 6 (2): 024111.

140 Norotte, C., Marga, F.S., Niklason, L.E., and Forgacs, G. (2009). Scaffold-free vascular tissue engineering using bioprinting. *Biomaterials* 30 (30): 5910–5917.

141 Yu, Y., Moncal, K.K., Li, J. et al. (2016). Three-dimensional bioprinting using self-assembling scalable scaffold-free "tissue strands" as a new bioink. *Sci. Rep.* 6 (1): 28714.

142 Cross, L.M., Thakur, A., Jalili, N.A. et al. (2016). Nanoengineered biomaterials for repair and regeneration of orthopedic tissue interfaces. *Acta Biomater.* 42: 2–17.

143 Zhou, X., Castro, N.J., Zhu, W. et al. (2016). Improved human bone marrow mesenchymal stem cell osteogenesis in 3D bioprinted tissue scaffolds with low intensity pulsed ultrasound stimulation. *Sci. Rep.* 6 (1): 32876.

144 Zhou, X., Nowicki, M., Cui, H. et al. (2017). 3D bioprinted graphene oxide-incorporated matrix for promoting chondrogenic differentiation of human bone marrow mesenchymal stem cells. *Carbon* 116: 615–624.

145 Kuss, M.A., Harms, R., Wu, S. et al. (2017). Short-term hypoxic preconditioning promotes prevascularization in 3D bioprinted bone constructs with stromal vascular fraction derived cells. *RSC Adv.* 7 (47): 29312–29320.

146 De Mori, A., Peña Fernández, M., Blunn, G. et al. (2018). 3D printing and electrospinning of composite hydrogels for cartilage and bone tissue engineering. *Polymers* 10 (3): 285.

147 Costa, J.B., Silva-Correia, J., Reis, R.L., and Oliveira, J.M. (2018). Recent advances on 3D printing of patient-specific implants for fibrocartilage tissue regeneration. *J. 3D Print. Med.* 2 (3): 129–140.

148 Pillai, M.M., Gopinathan, J., Selvakumar, R., and Bhattacharyya, A. (2018). Human knee meniscus regeneration strategies: a review on recent advances. *Curr. Osteoporos. Rep.* 16: 224–235.

149 Houshyar, S., Bhattacharyya, A., and Shanks, R. (2019). Peripheral nerve conduit: materials and structures. *ACS Chem. Neurosci.* 10 (8): 3349–3365.

150 Vijayavenkataraman, S., Vialli, N., Fuh, J.Y., and Lu, W.F. (2019). Conductive collagen/polypyrrole-b-polycaprolactone hydrogel for bioprinting of neural tissue constructs. *Int. J. Bioprinting* 5 (2.1): 229.

151 Lee, S.J., Esworthy, T., Stake, S. et al. (2018). Advances in 3D bioprinting for neural tissue engineering. *Adv. Biosyst.* 2 (4): 1700213.

152 Zhuang, P., An, J., Tan, L.P., and Chua, C.K. (2018). The current status of 3D bioprinting for neural tissue models.

153 Cui, H., Miao, S., Esworthy, T. et al. (2018). 3D bioprinting for cardiovascular regeneration and pharmacology. *Adv. Drug Deliv. Rev.* 132: 252–269.

154 Duan, B. (2017). State-of-the-art review of 3D bioprinting for cardiovascular tissue engineering. *Ann. Biomed. Eng.* 45: 195–209.

155 Roy, A., Saxena, V., and Pandey, L.M. (2018). 3D printing for cardiovascular tissue engineering: a review. *Mater. Technol.* 33 (6): 433–442.

156 Hu, J.B., Tomov, M.L., Buikema, J.W. et al. (2018). Cardiovascular tissue bioprinting: physical and chemical processes. *Appl. Phys. Rev.* 5 (4): 041106.

157 Zhang, Z., Wang, B., Hui, D. et al. (2017). 3D bioprinting of soft materials-based regenerative vascular structures and tissues. *Compos. Part B* 123: 279–291.

158 Li, X., Liu, L., Zhang, X., and Xu, T. (2018). Research and development of 3D printed vasculature constructs. *Biofabrication* 10 (3): 032002.

159 Mitchell, A., Lafont, U., Hołyńska, M., and Semprimoschnig, C. (2018). Additive manufacturing – a review of 4D printing and future applications. *Addit. Manuf.* 24: 606–626.

160 Li, Y.-C., Zhang, Y.S., Akpek, A. et al. (2016). 4D bioprinting: the next-generation technology for biofabrication enabled by stimuli-responsive materials. *Biofabrication* 9 (1): 012001.

161 Gaharwar, A.K., Arpanaei, A., Andresen, T.L., and Dolatshahi-Pirouz, A. (2016). 3D biomaterial microarrays for regenerative medicine: current state-of-the-art, emerging directions and future trends. *Adv. Mater.* 28 (4): 771–781.

162 Arslan-Yildiz, A., El Assal, R., Chen, P. et al. (2016). Towards artificial tissue models: past, present, and future of 3D bioprinting. *Biofabrication* 8 (1): 014103.

163 Ashammakhi, N., Ahadian, S., Pountos, I. et al. (2019). In situ three-dimensional printing for reparative and regenerative therapy. *Biomed. Microdevices* 21: 1–6.

16

3D Bioprinting from Lab to Industry

Saeedeh Vanaei[1], *Shohreh Vanaei*[2], *Michèle Kanhonou*[3], *Abbas Tcharkhtchi*[4], *and Hamid Reza Vanaei*[3,5]

[1] *Department of Mechanical Industrial and Manufacturing Engineering, University of Toledo, Toledo, OH, USA*
[2] *Department of Bioengineering, Northeastern University, Boston, MA, USA*
[3] *Léonard de Vinci Pôle Universitaire, Research Center, Paris La Défense, France*
[4] *Arts et Métiers Institute of Technology, CNRS, CNAM, PIMM, HESAM University, Paris, France*
[5] *Arts et Métiers Institute of Technology, CNAM, LIFSE, HESAM University, Paris, France*

16.1 Introduction

The 3D printing process, also known as additive manufacturing, revolutionizes the traditional methods of creating three-dimensional (3D) objects. It operates by building objects layer by layer, directly from digital designs or files. This technology has garnered significant attention, particularly in the era of Industry 4.0, which focuses on the integration of advanced automation, data exchange, and smart technologies into manufacturing processes. The adoption of 3D printing is steadily growing among various companies seeking to optimize their business processes and gain a competitive edge [1, 2]. This technology offers several advantages that traditional manufacturing methods cannot match. Some of the key benefits include:

- *Design flexibility*: 3D printing enables intricate and complex designs that would be challenging or impossible to achieve using traditional manufacturing techniques. This level of design freedom allows companies to create innovative products and optimize their functionality.
- *Cost-effectiveness*: For certain types of products, 3D printing can be more cost-effective, especially for low-volume production or customized items. It eliminates the need for expensive tooling, molds, and complex assembly processes.

3D Bioprinting from Lab to Industry, First Edition. Edited by Prosenjit Saha, Sabu Thomas, Jinku Kim, and Manojit Ghosh.

- *Rapid prototyping*: 3D printing enables the quick and cost-efficient creation of prototypes. Companies can iterate designs rapidly, reducing the time to market for new products.
- *On-demand manufacturing*: With 3D printing, companies can produce products on demand, reducing inventory costs and waste. This on-demand manufacturing approach is particularly useful for spare parts, which can be printed when needed.
- *Sustainability*: In some cases, 3D printing can be more environmentally friendly than traditional manufacturing methods. It often generates less waste and uses fewer materials, making it a greener option for specific applications.

Despite the numerous advantages, transitioning 3D printing technology from the laboratory to widespread industry adoption and successful commercialization comes with challenges. Some of these challenges include material limitations, production speed, quality control, intellectual property, and the skill gap. Despite these challenges, 3D printing has already made significant strides in various fields. Industries such as aerospace, automotive, biomedical, and consumer goods have embraced 3D printing for prototyping, custom manufacturing, and the production of specialized parts. As the technology continues to mature and address its limitations, it is likely to become even more pervasive, transforming the way products are designed, manufactured, and distributed across industries. The successful commercialization of 3D printing in these fields serves as a testament to its potential and paves the way for further advancements and integration in Industry 4.0 and beyond [3].

3D bioprinting is a cutting-edge technology that takes the principles of traditional 3D printing and applies them to the fields of tissue engineering (TE) and regenerative medicine. Instead of using conventional thermoplastic materials, 3D bioprinting utilizes biocompatible materials, especially living cells, to create 3D structures, often referred to as "bioinks." The fundamental distinction between 3D printing and 3D bioprinting lies in the types of materials employed during the fabrication process. In traditional 3D printing, polymers or polymer-based composites, which are typically non-living materials, are used. These materials are usually in a solid state, and a thermoplastic extrusion process is employed. The thermoplastic material is heated and then precisely deposited layer by layer to build the final object, which subsequently solidifies to form the desired 3D structure.

The potential applications of 3D bioprinting are groundbreaking. The technology holds immense promise in the field of regenerative medicine, as it allows for the creation of complex tissues and organs that could be used for transplantation or to study disease mechanisms in a controlled environment. It has the potential to revolutionize healthcare by providing patient-specific solutions for tissue and organ replacement, reducing the need for organ transplantation waiting lists and the risk of rejection. In addition to its medical applications, 3D bioprinting also finds utility in pharmaceutical research and drug testing. By creating 3D tissue models that mimic human organs or specific disease conditions, researchers can conduct more accurate and ethical drug testing, leading to the development of safer and more effective medications.

In the realm of industrial manufacturing, 3D bioprinting has emerged as a standout technology due to its unique advantages, including quickness, innovation, and ease of use. Unlike traditional manufacturing methods, 3D bioprinting allows for the rapid creation of 3D objects directly from digital designs, eliminating the need for complex tooling and lengthy setup processes. The potential of 3D bioprinting and how it can be applied in biomedical and healthcare applications are discussed in the following sections.

16.2 3D Bioprinting and Its Historical Point of View

When dealing with patients experiencing cardiac failure, researchers have a wide range of medical devices available to stabilize the patient and, in the best-case scenario, slow down the advancement of cardiac dysfunction. Although these interventions can offer temporary relief, the ultimate objective in treating cardiac failure is to replace the failing heart with a viable one through allograft transplantation. This biological remedy holds the promise of being a more enduring and efficient solution. Lately, the field of regenerative medicine has witnessed remarkable advancements. Cutting-edge techniques are under development, presenting a groundbreaking approach: using 3D bioprinting and assembling a fully functional heart from biological and cellular precursors. The concept involves combining biological elements with artificial components to engineer the heart. This extraordinary feat has given rise to the term "biofabricated," describing these assembled biologic replacement parts that blend living tissues with engineered materials [4, 5].

3D bioprinting technology has made remarkable progress and has moved beyond its initial stages. The realization of an implantable total biofabricated heart is becoming increasingly feasible, thanks to the dedicated efforts of researchers and scientists. This innovative concept is inching closer to reality as they work diligently to achieve it. An implantable biofabricated heart holds the potential to address the scarcity of suitable donor organs while also offering a fully personalized and customized solution for patients dealing with cardiac failure. The path to creating a complete biofabricated heart is a challenging and interdisciplinary effort, requiring the expertise of professionals in 3D printing, TE, regenerative medicine, and cardiology. There are still hurdles to overcome, including ensuring

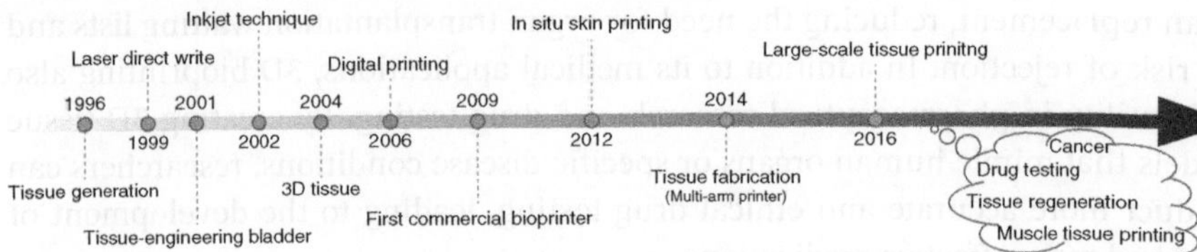

Figure 16.1 Progress and potential of 3D bioprinting from lab to industry.

seamless integration of the biofabricated heart with the recipient's circulatory system, ensuring its long-term viability, and achieving immune compatibility. Despite these challenges, the advancements made so far are incredibly encouraging and stand as a testament to the extraordinary capabilities of 3D bioprinting technology in the field of regenerative medicine, offering great promise for the future [6, 7].

As we move forward, ongoing research and development will undoubtedly play a pivotal role in bringing this revolutionary technology to fruition. The possibility of biofabricated organs, particularly a complete biofabricated heart, holds the power to revolutionize cardiac care and introduce a new era of personalized medicine, providing hope for numerous patients grappling with cardiac failure. Accordingly, the most important progress in this field for the last three decades is presented in Figure 16.1.

16.3 Potential of 3D Bioprinting from Lab to Industry

When developing new materials, companies often need to create prototypes to test and refine their designs before moving to full-scale production. 3D printing excels in this aspect because it enables the rapid production of prototypes, allowing companies to iterate and make design changes quickly. Compared to traditional manufacturing methods, where tooling and molds might need to be adjusted or recreated for design changes, 3D printing offers a more cost-effective and time-efficient approach to prototyping. The versatility and flexibility of 3D printing also contribute to its prominence in industrial manufacturing. It can handle a wide variety of materials, ranging from plastics and metals to ceramics and even biological materials for certain applications like 3D bioprinting. This adaptability allows manufacturers to choose the most suitable material for their specific product requirements. The technology is finding its way into the production of an extensive array of objects, showcasing its diverse capabilities. In the dental industry, 3D bioprinting is used to create dental prostheses, such as crowns and dentures, with high precision and customized for each patient's unique dental structure.

While 3D bioprinting continues to revolutionize industrial manufacturing, it is essential to acknowledge that it may not replace all traditional manufacturing methods outright. Certain high-volume, mass-production scenarios might still be better served by conventional techniques. However, as the technology advances and becomes more cost-effective, we can expect to see even more applications and integration of 3D bioprinting in the manufacturing landscape. 3D bioprinting has established itself as a game-changer in industrial manufacturing. Its speed, innovation, and user-friendly nature make it an attractive choice for prototyping and producing a wide range of objects. From dental prostheses to even aerospace components, 3D bioprinting is reshaping various industries and opening up new possibilities for customization, efficiency, and sustainable manufacturing practices. As the technology evolves, its impact on the manufacturing sector is likely to grow, paving the way for further advancements in design and production processes.

16.4 The Diversity of 3D Bioprinting

The remarkable potential of 3D bioprinting in regenerative medicine and TE is evident through its diverse applications. This state-of-the-art technology offers a wide range of methodologies, each tailored to address specific medical challenges. Researchers can utilize the versatility of 3D bioprinting to create intricate tissue structures by precisely arranging cells and biomaterials. The availability of diverse biomaterials and cell types further expands the opportunities, enabling the fabrication of tissue models and organ-like structures that closely emulate the complexity of natural biological tissues. From regenerating damaged tissues to developing personalized medicine for individual patients, 3D bioprinting brings new possibilities to healthcare. The ability to engineer tissues with varying levels of complexity and functionalities highlights the immense diversity within this field, instilling optimism for its potential to advance regenerative therapies and revolutionize modern medicine. In what follows, some of the most important applications of 3D bioprinting have been presented.

16.4.1 Medical Devices

In recent years, 3D bioprinting technology has made significant strides and has begun to revolutionize various industries, including medical equipment manufacturing. One area where 3D printing has shown immense potential is in the production of surgical guides for medical procedures. These 3D-printed surgical guides have become a game-changer in the medical device industry, offering several advantages over traditional manufacturing methods. One of the most crucial applications of 3D-printed surgical guides is in dentistry, where their use has been established for over a decade. In this field, dental surgeons have been utilizing 3D-printed guides to enhance the accuracy and precision of dental implant placement. By incorporating 3D printing technology, dentists can create custom-made surgical guides that perfectly fit a patient's unique anatomy. These guides help ensure the optimal positioning and alignment of dental implants, resulting in improved outcomes and reduced risks during the procedure.

Beyond dentistry, 3D-printed surgical guides have also found applications in various other surgical disciplines, such as spinal surgery, neurosurgery, maxillofacial surgery, and orthopedic surgery. The ability to tailor-make surgical guides for each patient has opened up new possibilities for complex and delicate procedures in these areas. In spinal surgery, for example, 3D-printed surgical guides can be used to assist surgeons in accurately placing screws and other spinal implants. These guides are designed based on a patient's specific spinal anatomy, enabling precise positioning and reducing the potential for complications. Similarly, in neurosurgery, 3D-printed surgical guides aid in planning and conducting intricate brain surgeries. They provide detailed maps of the patient's brain, helping surgeons navigate critical structures with enhanced precision and safety.

In maxillofacial surgery, the use of 3D-printed surgical guides has improved the outcomes of procedures such as jaw reconstruction and facial implants. The guides assist in achieving optimal facial symmetry and alignment while minimizing surgical invasiveness. Furthermore, in orthopedic surgery, 3D-printed surgical guides have shown great promise for procedures like joint replacements. By utilizing patient-specific guides, orthopedic surgeons can accurately position and align implants, leading to improved joint function and reduced wear and tear over time. The benefits of 3D-printed surgical guides extend beyond precision alone. The ability to pre-plan surgeries using 3D models based on a patient's medical imaging data allows surgeons to identify potential challenges and develop effective strategies before the actual operation. This level of preparation can lead to shorter surgery times, reduced anesthesia exposure for patients, and potentially faster recovery periods. Moreover, 3D bioprinting offers a relatively quick and cost-effective way to produce these custom surgical guides. This approach eliminates the need for costly and time-consuming manual fabrication processes, making personalized medicine more accessible to a broader range of patients.

Given the abovementioned explanations, the integration of 3D printing technology into medical equipment manufacturing has opened up exciting possibilities in the field of surgical guides. With applications ranging from dentistry to complex neurosurgery, 3D-printed surgical guides offer enhanced accuracy, improved patient outcomes, and greater efficiency in the operating room. As technology continues to advance, it is likely that 3D printing will continue to play an increasingly vital role in medical procedures, further shaping the landscape of modern healthcare.

The clinical study conducted by Putzier et al. [8] represents a significant advancement in the field of medical technology, particularly in the use of 3D printing for surgical guides. The primary objective of the study was to leverage 3D printing technology to create custom-made surgical guides that would optimize functionality for the thoracic and lumbar segments of the spine, taking into account the unique anatomy of each individual patient. The ultimate goal was to achieve maximum stability, user-friendliness, and an optimal fit with the surfaces of the spines during surgical procedures. Severe scoliosis patients face complex challenges during surgery, and the placement of pedicle screws is a critical aspect of their treatment. Precision is crucial to avoid potential complications and ensure successful outcomes. Traditional methods of surgical guidance may not always offer the level of accuracy required in such complex cases. This study aimed to address these challenges by developing and testing 3D-printed surgical guides tailored specifically for patients with severe scoliosis. By utilizing 3D bioprinting technology, the researchers were able to create personalized guides based on the precise anatomical data of each patient's spine, allowing for a patient-specific approach to surgery.

Their findings revealed that the newly developed positioning guide proved to be an invaluable navigational tool for the implantation of pedicle screws in severe scoliosis patients. The level of customization provided by the 3D-printed guides led to enhanced precision during surgery, resulting in better implant placement and overall outcomes for the patients. Some key aspects of the study's findings are listed as follows:

- *Enhanced precision*: The use of 3D-printed surgical guides allowed for a high level of precision in the placement of pedicle screws. The guides were designed to fit the unique curvature and shape of each patient's spine, reducing the risk of misplacement and associated complications.
- *Improved stability*: The customized surgical guides provided enhanced stability during the implantation process. This stability is crucial in severe scoliosis cases, where the spine's alignment is significantly altered, making traditional surgical approaches more challenging.
- *User-friendly approach*: The design and utilization of the 3D-printed surgical guides offered a user-friendly and intuitive experience for surgeons. The guides facilitated accurate navigation, making the surgery more efficient and less time-consuming.
- Optimal fit: The study demonstrated that the 3D-printed guides achieved an optimal fit with the surfaces of the patients' spines. This close fit is essential for

the successful integration of the pedicle screws and reduces the risk of postoperative complications.

Overall, this clinical study highlighted the immense potential of 3D bioprinting technology in improving surgical outcomes for severe scoliosis patients. The patient-specific approach provided by the customized surgical guides has the potential to revolutionize the treatment of complex spinal conditions, making surgeries safer, more efficient and ultimately leading to better patient outcomes. As research in this area continues to progress, it is likely that 3D bioprinting will play an increasingly important role in personalized medicine and surgical interventions for various medical conditions.

16.4.2 Drug Development

3D bioprinting has been making waves in various industries, and one of its exciting applications is drug printing, which has the potential to revolutionize the pharmaceutical field. This innovative method allows for the production of patient-specific medicine by customizing the composition and dosage required for each individual. The advantages of drug printing go beyond its cost-effectiveness, as it can utilize a wide range of pharmaceutically accepted polymers like polylactic acid, cellulose, and methacrylate to create personalized oral dosage forms. The core benefit of drug printing lies in its ability to tailor medications to suit the unique needs of each patient. By precisely adjusting the composition and dosage of the drug, it becomes possible to optimize its effectiveness while minimizing the risk of side effects. This personalized approach enhances the overall therapeutic outcome, ensuring that patients receive medications that are specifically designed to address their individual medical conditions [9].

Moreover, drug printing offers the potential to create safer medications. Since the dosage and formulation are customized for each patient, the likelihood of adverse reactions due to improper dosing or unsuitable ingredients is significantly reduced. This aspect of drug printing holds great promise for patients with complex medical conditions or those who have shown sensitivity to certain pharmaceutical ingredients in the past. One of the remarkable advantages of drug printing is its versatility with pharmaceutically accepted polymers. These materials have been extensively studied and approved for use in medical applications, ensuring the safety and compatibility of the 3D-printed oral dosage forms. This compatibility allows pharmaceutical researchers to explore a wide range of drug formulations and delivery methods, further expanding the possibilities for personalized medicine.

For pediatric patients, drug printing presents a particularly exciting prospect. Children often struggle with taking medications due to factors such as pill size, taste, and fear of swallowing. With drug printing, medications can be produced in various colors, shapes, and designs, offering a child-friendly approach. This customization makes medicine more appealing and less intimidating to young patients, ultimately leading to improved medication adherence and better treatment outcomes. Despite its immense potential, drug printing is still an evolving technology that requires careful research, development, and regulatory oversight. Ensuring the quality, safety, and reliability of 3D-printed medications are essential considerations in their implementation. Rigorous testing and adherence to pharmaceutical standards are necessary to validate the effectiveness and safety of drug printing in clinical settings [10].

Presumably, drug printing using 3D bioprinting technology represents a groundbreaking advancement in the pharmaceutical field. Its capacity to create patient-specific medications based on individual needs, along with the low cost and utilization of pharmaceutically accepted polymers, makes it a promising avenue for personalized medicine. Drug printing has the potential to provide safer and more effective drugs tailored to each patient, ultimately improving treatment outcomes and patient satisfaction. For children, the customization of medication color, shape, and design offers a child-friendly approach, making it easier for young patients to comply with their treatment regimens. While further research and regulatory efforts are required, the future of drug printing holds immense potential for transforming the landscape of pharmaceuticals and delivering better healthcare solutions to patients of all ages.

The innovative drug delivery approach described here represents a significant advancement in medical technology, but it is currently in the midst of a rigorous research and development process. Before this method can be made available to patients, it must undergo a thorough evaluation and gain approval from regulatory bodies, which can be a time-consuming and meticulous procedure. This stringent regulatory process is crucial to ensuring the safety, efficacy, and quality of the new drug delivery method before it can be implemented in clinical settings. Once the necessary clinical trials have been successfully completed, the process of obtaining regulatory approvals can take a considerable amount of time, potentially up to five years. This extended timeline reflects the meticulous scrutiny and careful evaluation that new medical technologies undergo to meet the stringent standards set by regulatory authorities. Looking ahead to the future, if this innovative drug delivery approach receives the necessary approvals, it holds the potential to transform the way patients manage their medications, particularly those with multiple health conditions. The concept of a single 3D-printed pill that encompasses multiple medications tailored to an individual's specific health needs is both promising and intriguing. This personalized approach to drug administration could revolutionize patient care, simplifying complex medication regimens and enhancing treatment adherence. The idea of replacing a multitude of pills, each with its own specific dosage and schedule, with a single daily 3D-printed pill has the potential to improve patient compliance significantly. Managing multiple medications can be challenging for individuals, particularly those with chronic health conditions or complex treatment plans. By consolidating these medications into one convenient and personalized polypill, patients could experience improved convenience, reduced chances of missed doses, and better overall medication adherence.

In this potential future scenario, 3D bioprinters at local pharmacies might become a common sight. Pharmacists could receive prescriptions for customized polypills based on a patient's unique medical requirements and print them on-site. This on-demand production of personalized medications could lead to better patient outcomes and greater patient satisfaction, as individuals receive treatments tailored specifically to their needs. However, it is important to acknowledge that while the future of 3D-printed polypills is promising, there are still challenges to overcome. Ensuring the safety, accuracy, and consistency of 3D-printed medications will be a critical aspect of their implementation. Additionally, addressing regulatory concerns and establishing robust quality control measures will be essential to building trust in this new drug delivery method.

The innovative drug delivery approach currently under research has the potential to revolutionize patient care by providing personalized, 3D-printed polypills. While the approval process may be time-consuming, the potential benefits of this technology are immense, offering simplified medication regimens and improved treatment adherence for patients with multiple health conditions. As technology and regulatory processes progress, the day when patients can receive custom-made polypills from their local pharmacies may not be too far off, ushering in a new era of personalized medicine.

16.4.3 Tissue Engineering

TE is a cutting-edge field that aims to promote tissue regeneration through the use of specialized constructs called scaffolds. These scaffolds serve as biocompatible and biodegradable frameworks that are infused with cells and growth factors. Once implanted into the body, they facilitate tissue rejuvenation and healing. The scaffold's crucial role lies in providing an appropriate environment for cell growth, proliferation, and signaling, as well as allowing space for the secretion and remodeling of the extracellular matrix (ECM) to occur, all of which are essential for tissue repair. Designing a 3D scaffold requires careful consideration of its macro-, micro-, and nano-level features. The macrostructure pertains to the overall shape and geometry of the scaffold, which can be tailored to suit individual patients' specific needs, such as anatomical features or injury sites. The microstructure, on the other hand, is determined by factors like porosity, pore interconnectivity, and pore size, which influence cell infiltration and nutrient transport within the scaffold. Meanwhile, the nanostructure encompasses surface features that influence cell attachment, proliferation, and differentiation, thus directing the tissue regeneration process.

3D bioprinting technology has emerged as a valuable tool for fabricating scaffolds with precise control over their macro-, micro-, and nano-level characteristics. Through 3D bioprinting, it becomes possible to customize the scaffold's overall shape and structure, adapting it to the requirements of individual patients. Furthermore, the porosity and interconnectivity of the scaffold's microstructure can be finely tuned, promoting optimal cell infiltration and nutrient diffusion. For post-processing, 3D bioprinting enables easy modifications of the scaffold's nanostructure, allowing researchers to tailor surface features to suit specific cellular responses.

One noteworthy application of 3D bioprinting in TE is the creation of synthetic skin for transplantation in burn patients. 3D-printed skin offers a promising alternative for patients with extensive burns, as it can closely mimic the structure and function of native skin. Additionally, 3D-printed skin can be employed to test chemical, pharmaceutical, and cosmetic products, providing a valuable tool for assessing their safety and efficacy. In addition to skin, researchers have explored the use of 3D bioprinting to mimic other complex tissues, such as heart valves, bones, teeth, and cartilage. By precisely engineering the macrostructure, microstructure, and nanostructure of the scaffolds, it becomes feasible to develop biomimetic constructs that closely resemble native tissues, opening up new possibilities for regenerative medicine and patient-specific treatments.

3D bioprinting represents a groundbreaking technology with the potential to revolutionize TE. This cutting-edge method enables the simultaneous positioning of biomaterials and living cells in a layer-by-layer fashion, allowing the construction of complex 3D structures that closely mimic the morphology and structure of natural biological tissues. In a noteworthy study conducted by Maiullari et al. [11], they demonstrated the power of 3D bioprinting by designing a multicellular structure for cardiac tissue modeling. This structure was composed of two types of cells: human umbilical vein endothelial cells (HUVECs) and induced pluripotent cell-derived cardiomyocytes (iPSC-CMs). The combination of these cells aimed to replicate the intricate composition of the heart tissue.

To fabricate this cardiac tissue model, the researchers employed hydrogel strands containing a mixture of alginate and PEG-fibrinogen. These hydrogel strands served as the biomaterial matrix that encapsulated the HUVECs and iPSC-CMs. The team then utilized a microfluidic printhead, a key component of the 3D bioprinting process, to precisely extrude the hydrogel strands along with the living cells. This enabled the precise 3D spatial deposition of the cells within the hydrogel matrix while maintaining a high level of printing fidelity and resolution. The in vivo implantation of the 3D-printed cardiac tissue model yielded remarkable results. The 3D-printed blood vessel-like structures, containing the integrated HUVECs and iPSC-CMs, demonstrated superior integration with the host's own blood vessels. This finding indicates that the 3D-printed tissue successfully assimilated with the surrounding native tissue, a critical factor for the long-term success of TE applications.

The implications of this research are profound. By successfully creating complex multicellular structures with high printing precision and cellular viability, 3D bioprinting opens up new possibilities for regenerative medicine and tissue replacement therapies. The ability to generate personalized tissues and organs tailored to individual patients holds great promise for addressing a wide range of medical conditions and injuries. However, despite these exciting developments, there are still challenges to overcome before 3D bioprinting becomes a routine clinical practice. Scaling up the technology to create larger and more complex tissues or organs is a significant hurdle, as is ensuring the functionality and long-term stability of the bioprinted constructs. Additionally, regulatory considerations and ethical implications must be carefully addressed as this technology advances.

As researchers continue to explore and refine 3D bioprinting techniques, we can anticipate significant advancements in regenerative medicine, offering new hope for patients in need of tissue and organ replacements in the future.

16.5 3D Bioprinting and Human Hearts

Microvascular 3D bioprinted constructs hold great promise for regenerative medicine and TE, as they aim to create functional tissues with a network of blood vessels to support cell viability and tissue function. However, one of the major limitations of this technology lies in the dimensions of the blood vessels that can be directly assembled during the bioprinting process. The most common method of 3D bioprinting involves using pen tips with a time-pressure microfluidic delivery system to deposit materials onto a substrate. These pen tips typically have a blunt needle orifice through which cell-laden bioinks and binding or crosslinkable polymers are delivered. As a result, the structures formed using this method are cylindrical in shape and have diameters not smaller than 100 μm. The problem arises when considering the scale of the components of the microcirculation – the smallest blood vessels in the body. Arterioles, which carry oxygenated blood away from the heart, have diameters ranging from 20 to 80 μm. Venules, which carry deoxygenated blood toward the heart, typically range from 30 to 100 μm in diameter. Capillaries, which are the tiniest blood vessels where oxygen and nutrient exchange with tissues occur, have diameters between 4 and 12 μm. Due to the current limitations of the 3D bioprinting technique, the blood vessels created in the bioprinted constructs are significantly larger than the native microvasculature, which can impede proper tissue function. To overcome this limitation, the bioprinted constructs undergo vasculogenesis and angiogenesis after the bioprinting step. Vasculogenesis refers to the formation of new blood vessels from precursor cells present in the bioink. Angiogenesis, on the other hand, involves the sprouting and branching of blood vessels from pre-existing ones. Together, these processes help the bioprinted cells organize and form a competent and functional microcirculatory network within the construct. One approach to improve the formation of blood vessels in bioprinted constructs is by incorporating microvascular fragments derived from adipose tissue into 3D collagen gels. These microvascular fragments contain a collection of cells and structures that can support vasculogenesis and angiogenesis in vitro. By providing a more natural microenvironment, these 3D collagen gels filled with microvascular fragments can enhance the formation of a functional vascular network within the bioprinted construct [12].

In this regard, 3D bioprinting of human hearts includes the mentioned features due to their specific structures. The process described outlines a highly sophisticated and personalized approach to creating patient-specific human hearts using a combination of advanced medical technologies, including medical imaging, stem cell research, 3D bioprinting, and TE. The ultimate goal is to produce fully functional human hearts that can be transplanted into patients without the need for immunosuppression therapy, as the bioprinted hearts are composed of the patient's own cells, ensuring immune tolerance. It starts by obtaining detailed MRI images of the patient's heart, allowing for the generation of a comprehensive 3D map that is specific to that individual. This personalized heart map serves as a blueprint for the subsequent steps in the process. Next, a skin biopsy is taken from the patient, from which dermal fibroblasts are isolated. These dermal fibroblasts are then transformed into induced pluripotent stem (iPS) cells. iPS cells are a type of stem cells that possess the remarkable ability to differentiate into virtually any cell type found in the human body. In this case, the iPS cells are reprogrammed to differentiate into contracting cardiomyocytes, the specialized cells responsible for generating the rhythmic contractions of the heart [13].

In an ideal scenario, the reprogramming process would extend to include other specific cell types necessary for the proper functioning of the heart. This includes conducting pacemakers and Purkinje cells, which coordinate the heart's contractions, as well as cells of the vascular system, such as smooth muscle cells, endothelial cells, and cardiac fibroblasts, which are essential components of the blood vessels and heart structure. After successful reprogramming, the cells are coupled with custom-formulated bioinks, which are unique for different cell types. These bioinks are carefully designed formulations of biomaterials, additives, growth factors, and hormones that provide the optimal environment for cell survival, growth, and maturation. The bioinks are used during the bioprinting process to create patient-specific human hearts layer by layer, precisely replicating the complex architecture of the heart. Once the bioprinting is complete, the hearts are subjected to a specific culturing process. Initially, they are cultured under static conditions for several days to allow the cells to grow and interact. Subsequently, the hearts undergo bioreactor culture and conditioning, which provide a controlled environment to support heart muscle development and maturation. This conditioning process is critical for ensuring that the bioprinted hearts reach a functional and mature state, ready for transplantation. Throughout the entire process, custom sensors are employed to monitor and measure various parameters in real time. These sensors help to assess the viability of the cells and tissues and evaluate the functional performance of the bioprinted hearts. Metrics such as left ventricle pressure and electrocardiogram properties are recorded to ensure that the hearts meet the required functional standards for transplantation [14].

One of the main benefits of this approach is that the bioprinted hearts are derived from the patient's own cells. This results in immune tolerance, where the patient's immune system does not recognize the transplanted heart as foreign, thereby preventing rejection. As a result, the need for immunosuppression therapy, typically required in organ transplantation to avoid rejection and complications, is eliminated. This groundbreaking process marks a significant advancement in regenerative medicine and transplantation. The integration of patient-specific medical imaging, stem cell technology, 3D bioprinting, and TE enables researchers to strive for functional and immunocompatible human hearts for clinical transplantation. Although the technology is still in the research and development phase, it holds tremendous potential for the future of organ transplantation, offering renewed hope to patients requiring life-saving heart transplants.

16.6 3D Bioprinting and Microfluidic Organ-on-a-Chip Models

An organ-on-a-chip is an innovative microfluidic cell culture device that mimics the physiological conditions of tissues and organs at a microscale level. It is created using microchip manufacturing methods and comprises continuously perfused chambers containing living cells arranged to simulate the complex architecture and functionality of tissues and organs. Unlike conventional 2D or 3D culture systems, organ-on-a-chip devices can replicate the multicellular structures, tissue-tissue interfaces, physicochemical microenvironments, and vascular perfusion found in the human body, resulting in higher levels of tissue and organ functionality. These devices offer several key advantages over traditional cultural methods. By replicating the native tissue and organ microenvironments, organ-on-a-chip models can provide a more accurate representation of how cells interact and respond to stimuli in the body. They also enable researchers to perform high-resolution, real-time imaging and in vitro analysis of the biochemical, genetic, and metabolic activities of living cells in a functional tissue and organ context. The technology's potential extends to a wide range of applications in scientific research. Organ-on-a-chip platforms hold great promise in advancing the study of tissue development, organ physiology, and disease etiology. Researchers can use these models to investigate complex cellular interactions, disease progression, and the effects of drugs on specific tissues or organs [15, 16].

In the field of drug discovery and development, organ-on-a-chip technology is particularly valuable. It allows researchers to gain insights into the molecular mechanisms of drug action, prioritize potential lead candidates for further investigation, and evaluate drug toxicity in a more physiologically relevant context. This can lead to more efficient drug development processes and improved safety profiles for potential therapeutic agents. Additionally, organ-on-a-chip devices play a crucial role in identifying potential biomarkers. These biomarkers can serve as indicators of disease progression or treatment response, aiding in the diagnosis and monitoring of various medical conditions. Organ-on-a-chip technology represents a groundbreaking approach that bridges the gap between traditional cell culture systems and the complexity of the human body. Its ability to recreate tissue and organ functionality at a microscale level opens up new avenues for understanding human physiology, disease mechanisms, and drug responses. As the technology continues to evolve and gain widespread adoption, it holds immense promise for revolutionizing medical research and drug development, ultimately leading to improved healthcare outcomes for patients worldwide [17–20].

Bioprinting represents a groundbreaking technology for constructing supportive frameworks used in tissue growth. Meanwhile, microfluidic organ-on-a-chip has emerged as a valuable platform with widespread applications, especially in drug screening and pathological studies. These organ-on-a-chip models are specifically designed to mimic the structural, microenvironmental, and physiological functions of human organs. Recently, the field of bioprinting has been harnessed to create organ-on-a-chip models, benefiting from its capability to simultaneously print multiple materials and cell types with high spatial precision and consistency. This unique capability allows for the development of biomimetic microenvironments with complex 3D structures. The integration of microfluidic and bioprinting technologies has enabled the direct printing of functional vascularized tissue structures, facilitating fluid flow to transport nutrients, facilitate gaseous exchange, and remove waste [18, 21]. A brief timeline regarding the development of the organ-on-a-cheap models and bioprinting is indicated in Figure 16.2.

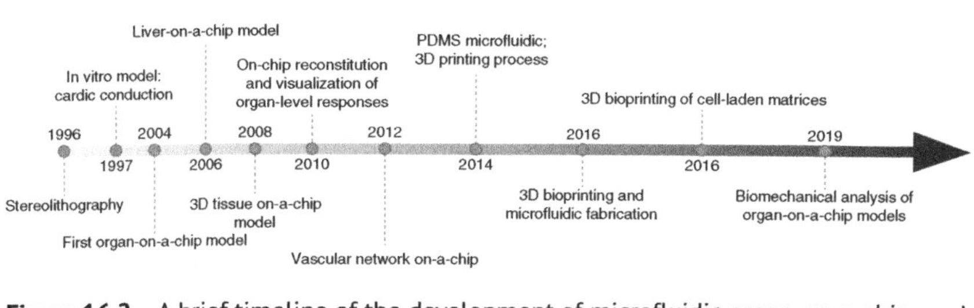

Figure 16.2 A brief timeline of the development of microfluidic organ-on-a-chip models and 3D bioprinting.

16.7 Future Developments

The ongoing progress in 3D printing technology is leading to marginal improvements as techniques become more refined and competition intensifies in the market. Merely creating smaller printers will no longer be enough to gain a competitive advantage. Instead, factors like reliability, a diverse range of scaffolds, and various cell sources will emerge as the primary economic differentiators. Market analysts are already discussing the next phase of bioprinting, suggesting that incorporating an additional dimension will bring about a fundamental transformation in how bioprinting is utilized. This indicates that advancements in bioprinting have the potential to revolutionize the healthcare industry, offering unprecedented opportunities for medical applications. However, fully realizing the potential of 3D bioprinting necessitates significant investment and a steadfast commitment to drive research forward and eventually apply these breakthroughs in clinical settings. While the journey toward clinical translation may present challenges, it holds immense promise for reshaping the approach and practice of healthcare.

References

1 Vanaei, H.R., Khelladi, S., and Tcharkhtchi, A. (2022). Roadmap: numerical-experimental investigation and optimization of 3D-printed parts using response surface methodology. *Materials* 15 (20): 7193.

2 Vanaei, H. (2021). *Etude et Modélisation de la rhéologie des polymères au cours du procédé FDM (Fabrication additive)*. Paris: HESAM.

3 Mousavi Nejad, Z., Zamanian, A., Saeidifar, M. et al. (2021). 3D bioprinting of polycaprolactone-based scaffolds for pulp-dentin regeneration: investigation of physicochemical and biological behavior. *Polymers* 13 (24): 4442.

4 Smith, C.M., Stone, A.L., Parkhill, R.L. et al. (2004). Three-dimensional bioassembly tool for generating viable tissue-engineered constructs. *Tissue Eng.* 10 (9–10): 1566–1576.

5 Jakab, K., Neagu, A., Mironov, V. et al. (2004). Organ printing: fiction or science. *Biorheology* 41 (3–4): 371–375.

6 Mandrycky, C., Wang, Z., Kim, K. et al. (2016). 3D bioprinting for engineering complex tissues. *Biotechnol. Adv.* 34 (4): 422–434.

7 Smith, C.M., Christian, J.J., Warren, W.L. et al. (2007). Characterizing environmental factors that impact the viability of tissue-engineered constructs fabricated by a direct-write bioassembly tool. *Tissue Eng.* 13 (2): 373–383.

8 Putzier, M., Strube, P., Cecchinato, R. et al. (2017). A new navigational tool for pedicle screw placement in patients with severe scoliosis. *Clin. Spine Surg.* 30 (4): E430–E439.

9 Tabriz, A.G., Nandi, U., Hurt, A.P. et al. (2021). 3D printed bilayer tablet with dual controlled drug release for tuberculosis treatment. *Int. J. Pharm.* 593: 120147.

10 Aimar, A., Palermo, A., and Innocenti, B. (2019). The role of 3D printing in medical applications: a state of the art. *J. Healthcare Eng.* 2019: 1.

11 Maiullari, F., Costantini, M., Milan, M. et al. (2018). A multi-cellular 3D bioprinting approach for vascularized heart tissue engineering based on HUVECs and iPSC-derived cardiomyocytes. *Sci. Rep.* 8 (1): 13532.

12 Hoying, J.B., Boswell, C.A., and Williams, S.K. (1996). Angiogenic potential of microvessel fragments established in three-dimensional collagen gels. *In Vitro Cell. Dev. Biol. Anim.* 32: 409–419.

13 Duncker, D.J., Koller, A., Merkus, D. et al. (2015). Regulation of coronary blood flow in health and ischemic heart disease. *Prog. Cardiovasc. Dis.* 57 (5): 409–422.

14 Thai, H.M., Juneman, E., Lancaster, J. et al. (2009). Implantation of a three-dimensional fibroblast matrix improves left ventricular function and blood flow after acute myocardial infarction. *Cell Transplant.* 18 (3): 283–296.

15 Bein, A., Shin, W., Jalili-Firoozinezhad, S. et al. (2018). Microfluidic organ-on-a-chip models of human intestine. *Cell. Mol. Gastroenterol. Hepatol.* 5 (4): 659–668.

16 Van Duinen, V., Trietsch, S.J., Joore, J. et al. (2015). Microfluidic 3D cell culture: from tools to tissue models. *Curr. Opin. Biotechnol.* 35: 118–126.

17 Sontheimer-Phelps, A., Hassell, B.A., and Ingber, D.E. (2019). Modelling cancer in microfluidic human organs-on-chips. *Nat. Rev. Cancer* 19 (2): 65–81.

18 Zhang, B., Korolj, A., Lai, B.F.L. et al. (2018). Advances in organ-on-a-chip engineering. *Nat. Rev. Mater.* 3 (8): 257–278.

19 Mi, S., Du, Z., Xu, Y. et al. (2018). The crossing and integration between microfluidic technology and 3D printing for organ-on-chips. *J. Mater. Chem. B* 6 (39): 6191–6206.

20 Lee, A., Hudson, A.R., Shiwarski, D.J. et al. (2019). 3D bioprinting of collagen to rebuild components of the human heart. *Science* 365 (6452): 482–487.

21 Jose, R.R., Rodriguez, M.J., Dixon, T.A. et al. (2016). Evolution of bioinks and additive manufacturing technologies for 3D bioprinting. *ACS Biomater. Sci. Eng.* 2 (10): 1662–1678.

Index

3D Bioprinting from Lab to Industry, First Edition. Edited by Prosenjit Saha, Sabu Thomas, Jinku Kim, and Manojit Ghosh.

d

t

Printed and bound by CPI Group (UK) Ltd, Croydon, CR0 4YY
17/09/2026
14964255-0005